T0192731

Andreas M. Hinz • Sandi Klavžar • Ciril Petr

The Tower of Hanoi – Myths and Maths

Second Edition

Foreword by Ian Stewart

 Birkhäuser

Andreas M. Hinz
Faculty of Mathematics,
Computer Science and Statistics,
LMU München
Munich, Germany

Sandi Klavžar
Faculty of Mathematics and Physics
University of Ljubljana
Ljubljana, Slovenia

Ciril Petr
Faculty of Natural Sciences and Mathematics
University of Maribor
Maribor, Slovenia

ISBN 978-3-030-08856-9   ISBN 978-3-319-73779-9   (eBook)
https://doi.org/10.1007/978-3-319-73779-9

Printed on acid-free paper

This book is published under the imprint Birkhäuser, www.birkhauser-science.com by the registered company Springer International Publishing AG part of Springer Nature.
The registered company address is: Gewerbestrasse 11, 6330 Cham, Switzerland

Foreword

by Ian Stewart

I know when I first came across the Tower of Hanoi because I still have a copy of the book that I found it in: *Riddles in Mathematics* by Eugene P. Northrop, first published in 1944. My copy, bought in 1960 when I was fourteen years old, was a Penguin reprint. I devoured the book, and copied the ideas that especially intrigued me into a notebook, alongside other mathematical oddities. About a hundred pages further into Northrop's book I found another mathematical oddity: Wacław Sierpiński's example of a curve that crosses itself at every point. That, too, went into the notebook.

It took nearly thirty years for me to become aware that these two curious structures are intimately related, and another year to discover that several others had already spotted the connection. At the time, I was writing the monthly column on mathematical recreations for *Scientific American*, following in the footsteps of the inimitable Martin Gardner. In fact, I was the fourth person to write the column. Gardner had featured the Tower of Hanoi, of course; for instance, it appears in his book *Mathematical Puzzles and Diversions*.

Seeking a topic for the column, I decided to revisit an old favourite, and started rethinking what I knew about the Tower of Hanoi. By then I was aware that the mathematical essence of many puzzles of that general kind—rearranging objects according to fixed rules—can often be understood using the state diagram. This is a network whose nodes represent possible states of the puzzle and whose edges correspond to permissible moves. I wondered what the state diagram of the Tower of Hanoi looked like. I probably should have thought about the structure of the puzzle, which is recursive. To solve it, forget the bottom disc, move the remaining ones to an empty peg (the same puzzle with one disc fewer), move the bottom disc, and put the rest back on top. So the solution for, say, five discs reduces to that for four, which in turn reduces to that for three, then two, then one, then zero. But with no discs at all, the puzzle is trivial.

Instead of thinking, I wrote down all possible states for the Tower of Hanoi with three discs, listed the legal moves, and drew the diagram. It was a bit messy, but after some rearrangement it suddenly took on an elegant shape. In fact, it looked remarkably like one of the stages in the construction of Sierpiński's curve. This couldn't possibly be coincidence, and once I'd noticed this remarkable resemblance, it was then straightforward to work out where it came from: the recursive structure of the puzzle.

Several other people had already noticed this fact independently. But shortly after my rediscovery I was in Kyoto at the International Congress of Mathematicians. Andreas Hinz introduced himself and told me that he had used the connection with the Tower of Hanoi to calculate the average distance between any two points of Sierpiński's curve. It is precisely $466/885$ of the diameter. This is an extraordinary result—a rational number, but a fairly complicated one, and far from obvious.

This wonderful calculation is just one of the innumerable treasures in this fascinating book. It starts with the best account I have ever read of the history of the puzzle and its intriguing relatives. It investigates the mathematics of the puzzle and discusses a number of variations on the Tower of Hanoi theme. This new edition has been updated with the latest discoveries, including Thierry Bousch's impressive proof that the conjectured minimum number of moves to solve the four-tower version is correct. And to drive home how even the simplest of mathematical concepts can propel us into deep waters, it ends with a list of currently unsolved problems. The authors have done an amazing job, and the world of recreational mathematics has a brilliant new jewel in its crown.

Preface

The British mathematician Ian Stewart pointed out in [395, p. 89] that "Mathematics intrigues people for at least three different reasons: because it is fun, because it is beautiful, or because it is useful." Careful as mathematicians are, he wrote "at least", and we would like to add (at least) one other feature, namely "surprising". The Tower of Hanoi (TH) puzzle is a microcosmos of mathematics. It appears in different forms as a recreational game, thus fulfilling the fun aspect; it shows relations to Indian verses and Italian mosaics via its beautiful pictorial representation as an esthetic graph, it has found practical applications in psychological tests and its theory is linked with technical codes and phenomena in physics.

The authors are in particular amazed by numerous popular and professional (mathematical) books that display the puzzle on their covers. However, most of these books discuss only well-established basic results on the TH with incomplete arguments. On the other hand, in the last decades the TH became an object of numerous—some of them quite deep—investigations in mathematics, computer science, and neuropsychology, to mention just central scientific fields of interest. The authors have acted frequently as reviewers for submitted manuscripts on topics related to the TH and noted a lack of awareness of existing literature and a jumble of notation—we are tempted to talk about a Tower of Babel! We hope that this book can serve as a base for future research using a somewhat unified language.

More serious were the errors or mathematical myths appearing in manuscripts and even published papers (which did *not* go through our hands). Some "obvious assumptions" turned out to be questionable or simply wrong. Here is where many mathematical surprises will show up. Also astonishing are examples of how the mathematical model of a difficult puzzle, like the *Chinese rings*, can turn its solution into a triviality. A central theme of our book, however, is the meanwhile notorious *Frame-Stewart conjecture*, a claim of optimality of a certain solution strategy for what has been called *The Reve's puzzle*. Despite many attempts and even allegations of proofs, this had been[1] an open problem for more than 70 years.

Apart from describing the state of the art of its mathematical theory and applications, we will also present the historical development of the TH from its

---

[1] "has been" in the original preface

invention in the 19th century by the French number theorist Édouard Lucas. Although we are not professional historians of science, we nevertheless take historical remarks and comments seriously. During our research we encountered many errors or historical myths in literature, mainly stemming from the authors copying statements from other authors. We therefore looked into original sources whenever we could get hold of them.

Our guideline for citing other authors' papers was to include "the first and the best" (if these were two). The first, of course, means the first to our current state of knowledge, and the best means the best to our (current) taste.

This book is also intended to render homage to Édouard Lucas and one of his favorite themes, namely recreational mathematics in their role in mathematical education. The historical fact that games and puzzles in general and the TH in particular have demonstrated their utility is universally recognized (see, e.g., [383, 173]) more than 100 years after Lucas's highly praised book series started with [283].

Myths

Along the way we deal with numerous myths that have been created since the puzzle appeared on the market in 1883. These myths include mathematical misconceptions which turned out to be quite persistent, despite the fact that with a mathematically adequate approach it is not hard to clarify them entirely. A particular goal of this book is henceforth to act as a myth buster.

Prerequisites

A book of this size can not be fully self-contained. Therefore we assume some basic mathematical skills and do not explain fundamental concepts such as sets, sequences or functions, for which we refer the reader to standard textbooks like [156, 122, 370, 38, 398]. Special technical knowledge of any mathematical field is not necessary, however. Central topics of discrete mathematics, namely combinatorics, graph theory, and algorithmics are covered, for instance, in [270, 54], [432, 60, 104], and [247, 306], respectively. However, we will not follow notational conventions of any of these strictly, but provide some definitions in a glossary at the end of the book. Each term appearing in the glossary is put in **bold face** when it occurs for the first time in the text. This is mostly done in Chapter 0, which serves as a gentle introduction to ideas, concepts and notation of the central themes of the book. This chapter is written rather informally, but the reader should not be discouraged when encountering difficult passages in later chapters, because they will be followed by easier parts throughout the book.

The reader must also not be afraid of mathematical formulas. They shape the language of science, and some statements can only be expressed unambiguously when expressed in symbols. In a book of this size the finiteness of the number of symbols like letters and signs is a real limitation. Even if capitals and lower case, Greek and Roman characters are employed, we eventually run out of them. Therefore, in order to keep the resort to indices moderate, we re-use letters for sometimes quite different objects. Although a number of these are kept rather stable globally, like n for the number of discs in the TH or names of special sequences like Gros's g, many will only denote the same thing locally, e.g., in a section. We hope that this will not cause too much confusion. In case of doubt we refer to the indexes at the end of the book.

Algorithms

The TH has attracted the interest of computer scientists in recent decades, albeit with a widespread lack of rigor. This poses another challenge to the mathematician who was told by Donald Knuth in [245, p. 709] that "It has often been said that a person doesn't really understand something until he teaches it to someone else. Actually a person doesn't really understand something until he can teach it to a computer, i.e. express it as an algorithm." We will therefore provide provably correct algorithms throughout the chapters. Algorithms are also crucial for human problem solvers, differing from those directed to machines by the general human deficiency of a limited memory.

Exercises

Édouard Lucas begins his masterpiece *Théorie des nombres* [288, iii] with a (slightly corrected) citation from a letter of Carl Friedrich Gauss to Sophie Germain dated 30 April 1807 ("jour de ma naissance"): "Le goût pour les sciences abstraites, en général, et surtout pour les mystères des nombres, est fort rare; on ne s'en étonne pas. Les charmes enchanteurs de cette sublime science ne se décèlent dans toute leur beauté qu'à ceux qui ont le courage de l'approfondir."[2] Sad as it is that the first sentence is still true after more than 200 years, the second sentence, as applied to all of mathematics, will always be true. Just as it is impossible to get an authentic impression of what it means to stand on top of a sizeable mountain from reading a book on mountaineering without taking the effort to climb up oneself, a mathematics book has always to be read with paper and pencil in reach. The readers of our book are advised to solve the excercises posed throughout the chapters. They give additional insights into the topic, fill missing details, and challenge our skills. All exercises are addressed in the body of the text. They are of different grades of difficulty, but should be treatable at the place where they are cited. At least, they should then be *read*, because they may also contain new definitions and statements needed in the sequel. We collect hints and solutions to the problems at the end of the book, because we think that the reader has the right to know that the writers were able to solve them.

---

[2]"The taste for abstract sciences, in general, and in particular for the mysteries of numbers, is very rare; this doesn't come as a surprise. The enchanting charms of that sublime science do not disclose themselves in all their beauty but to those who have the courage to delve into it."

Contents

The book is organized into ten chapters. As already mentioned, Chapter 0 introduces the central themes of the book and describes related historical developments. Chapter 1 is concerned with the Chinese rings puzzle. It is interesting in its own right and leads to a mathematical model that is a prototype for an approach to analyzing the TH. The subsequent chapter studies the classical TH with three pegs. The most general problem solved in this chapter is how to find an optimal sequence of moves to reach an arbitrary regular state from another regular state. An important subproblem solved is whether the largest disc moves once or twice (or not at all). Then, in Chapter 3, we further generalize the task to reach a given regular state from an irregular one. The basic tool for our investigations is a class of graphs that we call *Hanoi graphs*. A variant of these, the so-called *Sierpiński graphs*, is introduced in Chapter 4 as a new and useful approach to Hanoi problems.

The second part of the book, starting from Chapter 5, can be understood as a study of variants of the TH. We begin with the famous The Reve's puzzle and, more generally, the TH with more than three pegs. The central role is played by the notorious Frame-Stewart conjecture which has been open since 1941. Computer experiments are also described that further indicate the inherent difficulty of the problem. We continue with a chapter in which we formally discuss the meaning of the notion of a variant of the TH. Among the variants treated we point out the *Tower of Antwerpen* and the *Bottleneck TH*. A special chapter is devoted to the *Tower of London*, invented in 1982 by T. Shallice, which has received an astonishing amount of attention in the psychology of problem solving and in neuropsychology, but which also gives rise to some deep mathematical statements about the corresponding *London graphs*. Chapter 8 treats TH type puzzles with oriented disc moves, variants which, together with the more-pegs versions, have received the broadest attention in mathematics literature among all TH variants studied.

In the final chapter we recapitulate open problems and conjectures encountered in the book in order to provide stimulation for those who want to pass their time expediently waiting for some Brahmins to finish a divine task.

Educational aims

With an appropriate selection from the material, the book is suitable as a text for courses at the undergraduate or graduate level. We believe that it is also a convenient accompaniment to mathematical circles. The numerous exercises should be useful for these purposes. Themes from the book have been employed by the authors as a leitmotif for courses in discrete mathematics, specifically by A. M. H. at the LMU Munich and in block courses at the University of Maribor and by S. K. at the University of Ljubljana. The playful nature of the subject lends itself to presentations of the fundamentals of mathematical thinking for a general audience. The TH was also at the base of numerous research programs for gifted students.

The contents of this book should, and we hope will, initiate further activities of this sort.

Feedback

If you find errors or misleading formulations, please send a note to the authors. Errata, sample implementations of algorithms, and other useful information will appear on the *TH-book* website at http://tohbook.info.

Acknowledgements

We are indebted to many colleagues and students who read parts of the book, gave useful remarks or kept us informed about very recent developments and to those who provided technical support. Especially we thank Jean-Paul Allouche, Jens-P. Bode, Drago Bokal, Christian Clason, Adrian Danek, Yefim Dinitz, Menso Folkerts, Rudolf Fritsch, Florence Gauzy, Katharina A. M. Götz, Andreas Groh, Robert E. Jamison, Marko Petkovšek, Amir Sapir, Marco Schwarz, Walter Spann, Arthur Spitzer, Sebastian Strohhäcker, Karin Wales, and Sara Sabrina Zemljič.

Throughout the years we particularly received input and advice from Simon Aumann, Daniele Parisse, David Singmaster, and Paul Stockmeyer (whose "list" [403] has been a very fruitful source).

Original photos were generously supplied by James Dalgety (The Puzzle Museum) and by Peter Rasmussen and Wei Zhang (Yi Zhi Tang Collection). For the copy of an important historical document we thank Claude Consigny (Cour d'Appel de Lyon). We are grateful to the Cnam – Musée des arts et métiers (Paris) for providing the photos of the original *Tour d'Hanoï*.

Special thanks go to the Birkhäuser/Springer Basel team. In particular, our Publishing Editor Barbara Hellriegel and Managing Director Thomas Hempfling guided us perfectly through all stages of the project for which we are utmost grateful to them, while not forgetting all those whose work in the background has made the book a reality.

A. M. H. wants to express his appreciation of the hospitality during his numerous visits in Maribor.

Last, but not least, we all thank our families and friends for understanding, patience, and support. We are especially grateful to Maja Klavžar, who, as a librarian, suggested to us that it was about time to write a comprehensive and widely accessible book on the Tower of Hanoi.

ANDREAS M. HINZ   *München, Germany*
SANDI KLAVŽAR   *Trzin, Slovenia*
UROŠ MILUTINOVIĆ   *Maribor, Slovenia*
CIRIL PETR   *Hoče, Slovenia*

Preface to the Second Edition

In the preface to the first edition we wrote: "A central theme of our book [...] is the meanwhile notorious Frame-Stewart conjecture, a claim of optimality of a certain solution strategy for what has been called The Reve's puzzle. Despite many attempts and even allegations of proofs, this has been an open problem for more than 70 years." As it happens, in 2014 a historical breakthrough occurred when Thierry Bousch published a solution to The Reve's puzzle! His article is written in French and consequently less accessible to most researchers—especially since Bousch's ingenious proof is rather technical. We believe that this new development alone would have justified a second edition of this book, containing an English rendering of Bousch's approach.

Other significant progress happened since the first edition has been published in 2013. We emphasize here that Stockmeyer's conjecture concerning the smallest number of moves among all procedures that solve the Star puzzle, also listed among the open problems in Chapter 9 of the first edition, has been solved in 2017—again by Tierry Bousch. Some others of these open questions have been settled meanwhile. These solutions are addressed in the present edition. Moreover, extensive computer experiments on the Tower of Hanoi with more than three pegs have been performed in recent years. Other new material includes, e.g., the Tower of Hanoi with unspecified goal peg or with random moves and the Cyclic Tower of Antwerpen.

On our webpage http://tohbook.info twelve reviews of the first edition are referred to. We were pleased by their unanimous appreciation for our book. Specific remarks of the reviewers have been taken into account for the second edition. One desire of the readers was to find additional descriptions of some fundamental mathematical concepts, not to be found easily or satisfactory in the literature but used throughout the book. This guided us to extend Chapter 0 accordingly, e.g., by adding a section on sequences. This will make the book even more suitable as a textbook underlying mathematical seminars and circles which will also appreciate the more than two dozen new exercises.

When it comes to historical matters, an impressive account of the power of myths in math(s) is given in [304] (subtitle not arranged!). During our research for the new edition we found several much more recent legends that do not survive a meticulous reference to original sources which are given whenever we could get hold on them. In citing textbooks, we may not always refer to the most recent editions but to those which were at our disposal.

We extend our thanks to the individuals acknowledged in the preface of the first edition, among whom we want to re-emphasize Jean-Paul Allouche, Daniele Parisse, Sara Sabrina Zemljič, and Paul K. Stockmeyer for their constant support. Paul has been an inspiration for a number of new additions to the book, in particular demonstrating that the classical task can still offer new challenges. By presenting variants like his Star Tower of Hanoi he showed a vision for mathematically significant new topics.

While we were working on the second edition, the following people were of great help: Thierry Bousch, Jasmina Ferme, Brian Hayes, Andreas C. Höck, Caroline Holz auf der Heide, Richard Korf, Borut Lužar, and Michel Mollard.

Again we are grateful for the support from Birkhäuser. Clemens Heine and Luca Sidler guided us with patience and understanding through the jungle of nowadays scientific publishing.

A. M. H. wants to express his gratitude to the people of Slovenia for the opportunity to work on the topics of this book in the frame of the project *Exploration of the intrinsic structure of Tower graphs* supported by the Slovenian Research Agency (code J1-7110). These thanks go in particular to his co-authors and to Boštjan Brešar who stands for the hospitality of the University of Maribor.

Last but not least we thank Uroš Milutinović for inspiring discussions even after he decided, to our great sorrow, not to be a co-author of the second edition.

December 2017

ANDREAS M. HINZ
SANDI KLAVŽAR
CIRIL PETR

München, Germany
Trzin, Slovenia
Hoče, Slovenia

Contents

Contents

Chapter 0

The Beginning of the World

The roots of mathematics go far back in history. To present the origins of the protagonist of this book, we even have to return to the Creation.

0.1 The Legend of the Tower of Brahma

"D'après une vieille légende indienne, les brahmes se succèdent depuis bien longtemps, sur les marches de l'autel, dans le Temple de Bénarès, pour exécuter le déplacement de la *Tour Sacrée de* BRAHMA, aux soixante-quatre étages en or fin, garnis de diamants de Golconde. Quand tout sera fini, la Tour et les brahmes tomberont, et ce sera la fin du monde!"

These are the original words of Professor N. Claus (de Siam) of the Collège Li-Sou-Stian, who reported, in 1883, from Tonkin about the legendary origins of a "true annamite head-breaker", a game which he called LA TOUR D'HANOÏ (see [82]). We do not dare to translate this enchanting story written in a charming language and which developed through the pen of Henri de Parville into an even more fantastic fable [335]. W. W. R. Ball called the latter "a sufficiently pretty conceit to deserve repetition" ([31, p. 79]), so we will follow his view and cite Ball's most popular English translation of de Parville's story:
"In the great temple at Benares, beneath the dome which marks the centre of the world, rests a brass-plate in which are fixed three diamond needles, each a cubit high and as thick as the body of a bee. On one of these needles, at the creation, God placed sixty-four discs of pure gold, the largest disc resting on the brass plate, and the others getting smaller and smaller up to the top one. This is the Tower of Bramah [sic!]. Day and night unceasingly the priests transfer the discs from one diamond needle to another according to the fixed and immutable laws of Bramah [sic!], which require that the priest must not move more than one disc at a time and that he must place this disc on a needle so that there is no smaller disc below it. When the sixty-four discs shall have been thus transferred from the needle on

which at the creation God placed them to one of the other needles, tower, temple, and Brahmins alike will crumble into dust, and with a thunderclap the world will vanish."

Ball adds: "Would that English writers were in the habit of inventing equally interesting origins for the puzzles they produce!", a sentence censored by H. S. M. Coxeter in his revised edition of Ball's classic [32, p. 304].

As with all myths, Claus's legend underwent metamorphoses: the decoration of the discs [étages] with diamonds from Golconda[3] transformed into diamond needles, de Parville put the great temple of Benares[4] to the center of the world and moved "since quite a long time" to "the beginning of the centuries", which Ball interprets as the creation. As if the end of the world would not be dramatic enough, Ball adds a "thunderclap" to it. So the Tower of Brahma became a time-spanning riddle. Apart from his strange spelling of the Hindu god, Ball also re-formulated the rule which de Parville enunciated as "he must not place that disc but on an empty needle or above [au-dessus de] a larger disc". More importantly, while de Parville insists in the task to transport the tower from the first to the third needle, Ball's Bramah did *not* specify the goal needle; we will see later (p. 117) that this makes a difference!

In another early account, the Dutch mathematician P. H. Schoute is more precise by insisting to put the disc (or ring in his diction) "on an empty [needle] or on a larger [disc]" [372, p. 275] and specifying the goal by alluding to the Hindu triad of the gods Brahma, Vishnu, and Shiva: Brahma, the creator, placed the discs on the first needle and when they all reach Shiva's, the world will be destroyed; in between, it is sustained by the presence of Vishnu's needle. The latter god will also watch over the observance of what we will call the (cf. [289, p. 55])

divine rule: you must not place a disc on a smaller one.

Let us hope that Sarasvati, Brahma's consort and the goddess of learning, will guide us through the mathematical exploration of the fascinating story of the sacred tower!

Among the many variants of the story, which would fill a book on its own, let us only mention the reference to "an Oriental temple" by F. Schuh [374, p. 95] where 100 alabaster discs were waiting to be transferred by believers from one of two silver pillars to the only golden one. Quite obviously, Schoute's compatriot was more in favor of the decimal system than the Siamese inventor of the puzzle, who, almost by definition, preferred base two.

De Parville, in his short account of what came to be known as the *Tower of Hanoi (TH)*, was also very keen in identifying this man from Indochina. A mandarin, says he, who invents a game based on combinations, will incessantly think about combinations, see and implement them everywhere. As one is never betrayed but by oneself, permuting the letters of the signatory of the TH, *N. Claus (de*

---

[3]at the time the most important market for diamonds, located near the modern city of Hyderabad
[4]today's Vārānasī

Siam), mandarin of the collège *Li-Sou-Stian* will reveal *Lucas d'Amiens*, teacher at the lycée *Saint-Louis*.

François Édouard Anatole Lucas (see Figure 0.1) was born on 4 April 1842 in the French city of Amiens and worked the later part of his short life at schools in Paris. Apart from being an eminent number theorist, he published, from 1882,

Figure 0.1: Édouard Lucas, 1842–1891

a series of four volumes of *Récréations mathématiques* [283, 284, 289, 290], accomplished posthumously in 1894 and supplemented in the subsequent year by *L'arithmétique amusante* [291].[5] They stand in the tradition of J. Ozanam's popular *Récréations mathématiques et physiques* which saw editions from 1694 until well into the 18th century. The fourth volume of this work contains a plate [327, pl. 16 opposite p. 439][6] showing in its Figure 47 what the author calls "Sigillum Salomonis". It is a mechanical puzzle which Lucas discusses in the first volume of his series under the name of "baguenaudier" (cf. [283, p. 161–186]) and to which his leaflet [82] refers for more details on the TH which only much later enters volume three [289, p. 55–59]. It seems that this more ancient puzzle was the catalyst for Lucas's TH.

---

[5]The editors of [291] point out (p. 210, footnote) that Lucas had the intention to continue his series of *Récréations mathématiques* with two volumes whose chapter headings had already been layed down. Unfortunately, not even summaries of these chapters have come down to us.

[6]Throughout we cite editions which were at our disposal.

0.2   History of the Chinese Rings

The origin of the solitaire game *Chinese rings (CR)*, called jiulianhuan ("Nine linked rings", 九连环) in today's China, seems to be lost in the haze of history. The legends emerging from this long tradition are mostly frivolous in character, reporting from a Chinese hero who gave the puzzle to his wife when he was leaving for war, for the obvious motive of "entertaining" her during his absence. The present might have looked as in Figure 0.2.

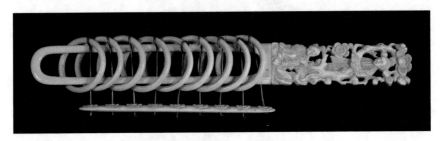

Figure 0.2: The Chinese rings
(courtesy of James Dalgety, http://puzzlemuseum.org)
© 2012 Hordern-Dalgety Collection

The most serious attribution has been made by S. Culin in his ethnological work on Eastern games [86, p. 31f]. According to his "Korean informant", the caring husband was Hung Ming, actually a real person (Zhuge Liang, 诸葛亮, 181–234). However, the material currently available is also compatible with a European origin of the game. The reader may consult [1] for more myths.

The earliest known evidence can be found in Chapter 107 of Luca Pacioli's *De viribus quantitatis* (cf. [328, p. 290–292]) of around 1500, where the physical object, with a certain number of rings, is described and a method to get the rings *onto* the bar is indicated. Pacioli speaks of a "difficult case"; see [187], where also a 7-ring version is discussed which was presented in the middle of the 16th century in Book 15 of G. Cardano's *De subtilitate libri XXI* as a "useless" instrument ("Verùm nullius vsus est instrumétú ex septem annulis...") which embodies a game of "admirable subtlety" ("miræ subtilitatis") [71, p. 492f]. On the other hand, Cardano claimed that the ingenious mechanism was not that useless at all and therefore employed in locks for chests, a claim supported by Lucas in [283, p. 165, footnote].

Many names have been given to the Chinese rings, an expression apparently not used before 1872 (see [73]), over the centuries. They have been called *Delay guest instrument* in Korea, *Cardano's rings* or *tarrying/tiring irons* in England (H. E. Dudeney reports in 1917 that "it is said still to be found in obscure English villages (sometimes deposited in strange places such as a church belfry)"; cf. [119, Problem 417].), *Sinclair's bojor* (Sinclair's shackles) in Sweden, *Vangin* or *Siperian lukko* (prisoner's or Siberian lock) in Finnland, меледа (meleda) in Russia, and *Nürnberger Tand* or *Zankeisen* (quarrel iron) in Germany, but the most puzzling

designation is the French *baguenaudier*. In the note [168] of the Lyonnais barrister
Louis Gros, almost 5 out of 16 text pages are devoted to the etymology with
the conclusion that it should be "baguenodier", deriving from a knot of rings.
However, Lucas did not follow Gros's linguistic arguments and so the French name
of a plant (*Colutea arborescens*) is still attached to the puzzle. But why then has
"baguenauder" in French the meaning of strolling around, waisting time?

The puzzle consists of a system of nine rings, bound together in a sophisticated mechanical arrangement, and a bar (or loop or shuttle as in weaving) with a
handle at one end. At the beginning, all rings are on the bar. They can be moved
off or back onto the bar only at the other end, and the structure allows for just two
kinds of individual ring moves, the details of which we will discuss in Chapter 1.
The task is to move all rings off the bar. Let us assume for the moment that this
and, in fact, all states, i.e. distributions of rings on or off the bar, can be reached.
(Lucas [283, p. 177] actually formulates the generalized problem to find a shortest
possible sequence of moves to get from an arbitrary initial to an arbitrary goal
state.) Then we may view the puzzle as a representation of binary numbers from
0 to $511 = 2^9 - 1$ if we interpret the rings as binary digits, or **bits**, 0 standing for
a ring off the bar and 1 for a ring on the bar. This leads us back even further in
Chinese history, or rather mythology, namely to the legendary Fu Xi (伏羲), who
lived, if at all, some 5000 years ago. To him Leibniz attributes [259, p. 88f] the *ba
gua* (八卦), the eight *trigrams* consisting of three bits each, and usually depicted
in a circular arrangement as in Figure 0.3, where a broken (*yin*, 阴) line stands for
0, a solid (*yang*, 阳) line for 1, and the least significant bit is the outermost one.

Figure 0.3: Fu Xi's arrangement of the trigrams

Legend has it that Fu Xi saw this arrangement on the back of a tortoise.
(Compare also to the Korean flag, the *Taegeukgi*.) Following the yin-yang symbol
in the center of the figure, Fu Xi, Leibniz and we recognize the numbers from
0 to 7 in binary representation. These 8 trigrams therefore provide a supply of
octernary (or *octal*) digits, the *octits*, for a base-8 number system. They have been

used as such in the logo of the *International Congress of Mathematical Education* ICME–14; see Figure 0.4, where the year[7] $(2020)_{10}$ of the event is given as $(3744)_8$ (with Leibniz's *guas* turned upside down, i.e. showing the least significant bit on bottom; this allows us to read the bits from top to bottom and from left to right as $(011111100100)_2$). However, there is no evidence of any ancient (Chinese) tradition for this interpretation.

Figure 0.4: The logo of ICME–14, Shanghai (China) 2020
(designed by Jianpan Wang and Nan Shi)
© 2016–2020 ICME-14 Local Organizing Committee

On the other hand, doubling the number of bits (or combining two octits) will give a collection of 64 *hexagrams* for the I Ching (yi jing, 易经), the famous *Book of Changes*; cf. [451] and [283, p. 149–151]; for its use in magic tricks, see [103, Chapter 8]. It seems, however, that Leibniz went astray with the philosophical and religious implications he drew from yin and yang; see [413]. On the other hand, the mathematical implications of binary thinking can not be over-estimated.

0.3   History of the Tower of Hanoi

The sixth chapter in [283] was devoted by Lucas to *The Binary Numeration*. Here he describes the advantages of the binary system [283, p. 148f], the Yi Jing [283, p. 149–151], and perfect numbers [283, p. 158–160], before he starts his seventh recreation on the baguenaudier, as mentioned before. We do not find, however, the most famous of Lucas's recreations in this first edition, the TH. This is not

---

[7]We write an index "p" for a number represented in a base p system; cf. below. The parentheses may be omitted.

surprising though. In the box containing the original game, preserved today in the *Musée des arts et métiers* in Paris, one can find the following inscription, most probably in Lucas's own hand:

> La tour d'Hanoï, —
>
> Jeu de combinaison pour
> appliquer le système de la numération
> binaire, inventé par M. Edouard Lucas
> (novembre 1883) — donné par l'auteur.

So we have a date of birth for the TH. (In [288, p. xxxii], Lucas claims that the puzzle was published in 1882, but there is no evidence for that.) The idea of the game was immediately pilfered around the world with patents approved, e.g., in the United States (N° 303,946 by A. Ohlert, 1884) and the United Kingdom (N° 20,672 by A. Gartner and G. Talcott, 1890). In 1888, Lucas donated the original puzzle (see Figures 0.5 and 0.6), together with a number of mechanical calculating machines to the *Conservatoire national des arts et métiers* (Cnam) in Paris, where he also gave public lectures for which a larger version of the TH was produced; cf. [287].

The cover of the original box shows the fantastic scenery of Figure 0.7. The picture, also published on 19 January 1884 in [83, p. 128], repeats all the allusions to fancy names of places and persons we already found on the leaflet [82] which accompanied the puzzle. Two details deserve to be looked at more closely. The man supporting the ten-storied pagoda has a tatoo on his belly: A U—Lucas was "agrégé de l'université", entitling him to teach at higher academic institutions. The crane, a symbol for the Far East, holds a sheet of paper on which is written, with bamboo leaves, the name Fo Hi—the former French transliteration of Fu Xi whom we met before.

But why the name "The Tower of *Hanoi*"? We would not think of the capital of today's Vietnam in connection with Brahmins moving 64 golden discs in the great temple of Benares. However, when Lucas started to market the puzzle in its modest version with only eight wooden discs, French newspapers were full of reports from Tonkin. In fact, Hanoi had been seized by the French in 1882, but during the summer of 1883 was under constant siege by troops from the Chinese province of Yunnan on the authority of the local court of Hué, where on 25 August 1883 the Harmand treaty established the rule of France over Annam and Tonkin (cf. [330, Section 11]). In [335], de Parville calls a variant of the TH, where discs of increasing diameter were replaced by hollow pyramids of decreasing size, the "Question of Tonkin" and comments the fact that the discs of Claus's TH were made of wood instead of gold as being more prudent because it concerns Tonkin. So Lucas selected the name of Hanoi because it was in the headlines at the time. Most probably, our book today would sell better had we chosen the title *The Tower of* Kabul!

Figure 0.5: The original Tower of Hanoi
©Musée des arts et métiers–Cnam, Paris / photo: Michèle Favareille
http://www.arts-et-metiers.net/

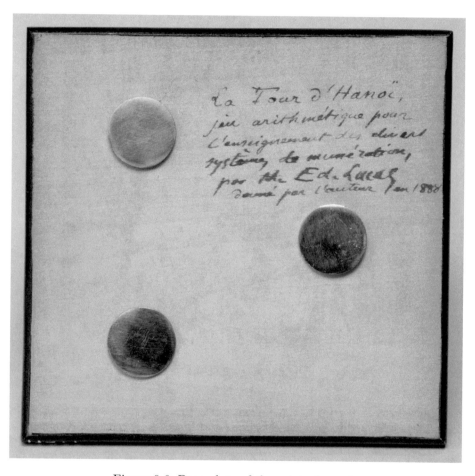

Figure 0.6: Base plate of the original puzzle
©Musée des arts et métiers–Cnam, Paris / photo: Michèle Favareille
http://www.arts-et-metiers.net/

Figure 0.7: The cover plate of the Tower of Hanoi

Lucas never travelled to Hanoi [285, p. 14]. However, he was a member of the commission which edited the collected works of Pierre de Fermat and was sent on a mission to Rome to search in the famous Boncompagni library for unpublished papers of his illustrious compatriot (cf. [290, p. 91]). Was it on this voyage that Lucas invented the TH? The leaflet, which we reproduce here as Figure 0.8, supports this hypothesis when talking in a typical Lucas style about FER-FER-TAM-TAM, thereby transforming the French government into a Chinese one.

Apart from this, the front page of [82] discloses the motive of professor N. Claus (de Siam) for his game, namely the vulgarization of science. He offers enormous amounts of money for the person who solves, by hand, the TH with 64 discs and reveals the necessary number of displacements, namely

$$18\ 446\ 744\ 073\ 709\ 551\ 615 \,, \tag{0.1}$$

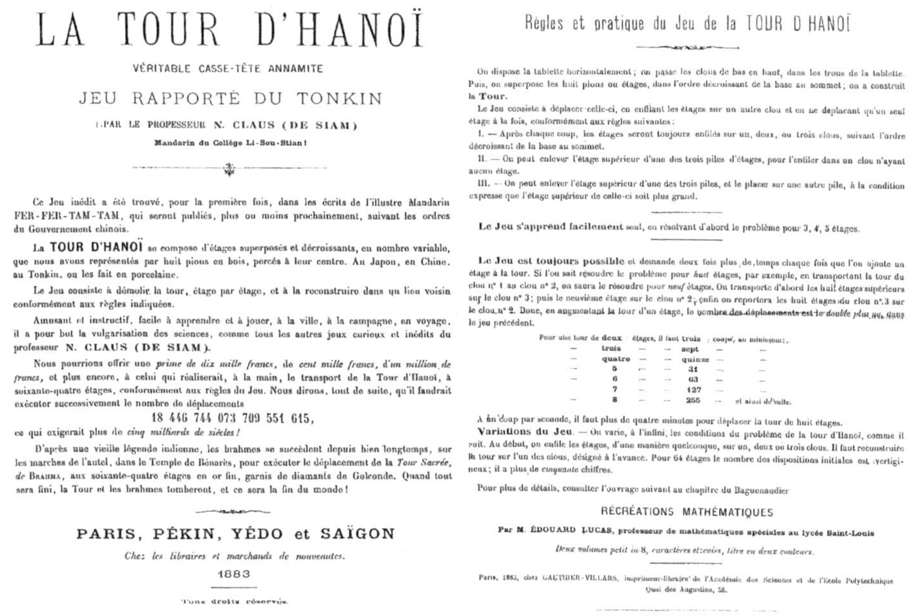

LA TOUR D'HANOÏ

VÉRITABLE CASSE-TÊTE ANNAMITE

JEU RAPPORTÉ DU TONKIN

PAR LE PROFESSEUR N. CLAUS (DE SIAM)

Mandarin du Collège Li-Sou-Stian!

Ce Jeu inédit a été trouvé, pour la première fois, dans les écrits de l'illustre Mandarin FER-FER-TAM-TAM, qui seront publiés, plus ou moins prochainement, suivant les ordres du Gouvernement chinois.

La TOUR D'HANOÏ se compose d'étages superposés et décroissants, en nombre variable, que nous avons représentés par huit pions en bois, percés à leur centre. Au Japon, en Chine, au Tonkin, on les fait en porcelaine.

Le Jeu consiste à démolir la tour, étage par étage, et à la reconstruire dans un lieu voisin conformément aux règles indiquées.

Amusant et instructif, facile à apprendre et à jouer, à la ville, à la campagne, en voyage, il a pour but la vulgarisation des sciences, comme tous les autres jeux curieux et inédits du professeur N. CLAUS (DE SIAM).

Nous pourrions offrir une *prime de dix mille francs*, de cent mille francs, d'un million de francs, et plus encore, à celui qui réaliserait, à la main, le transport de la Tour d'Hanoï, à soixante-quatre étages, conformément aux règles du Jeu. Nous dirons, tout de suite, qu'il faudrait exécuter successivement le nombre de déplacements

18 446 744 073 709 551 615,

ce qui exigerait plus de *cinq milliards de siècles* !

D'après une vieille légende indienne, les brahmes se succèdent depuis bien longtemps, sur les marches de l'autel, dans le Temple de Bénarès, pour exécuter le déplacement de la *Tour Sacrée*, *de* BRAHMA, aux soixante-quatre étages en or fin, garnis de diamants de Golconde. Quand tout sera fini, la Tour et les brahmes tomberont, et ce sera la fin du monde !

PARIS, PÉKIN, YÉDO et SAÏGON

Chez les libraires et marchands de nouveautés.

1883

Tous droits réservés.

Règles et pratique du Jeu de la TOUR D'HANOÏ

On dispose la tablette horizontalement ; on passe les clous de bas en haut, dans les trous de la tablette. Puis, on superpose les huit pions ou étages, dans l'ordre décroissant de la base au sommet ; on a construit la Tour.

Le Jeu consiste à déplacer celle-ci, en enfilant les étages sur un autre clou et en ne déplaçant qu'un seul étage à la fois, conformément aux règles suivantes :

I. — Après chaque coup, les étages seront toujours enfilés sur un, deux, ou trois clous, suivant l'ordre décroissant de la base au sommet.

II. — On peut enlever l'étage supérieur d'une des trois piles d'étages, pour l'enfiler dans un clou n'ayant aucun étage.

III. — On peut enlever l'étage supérieur d'une des trois piles, et le placer sur une autre pile, à la condition expresse que l'étage supérieur de celle-ci soit plus grand.

Le Jeu s'apprend facilement seul, en résolvant d'abord le problème pour 3, 4, 5 étages.

Le Jeu est toujours possible et demande deux fois plus de temps chaque fois que l'on ajoute un étage à la tour. Si l'on sait résoudre le problème pour huit étages, par exemple, en transportant la tour du clou n° 1 au clou n° 3, on saura le résoudre pour neuf étages. On transporte d'abord les huit étages supérieurs sur le clou n° 3 ; puis la neuvième étage sur le clou n° 2 ; enfin on reportera les huit étages du clou n° 3 sur le clou n° 2. Donc, en augmentant la tour d'un étage, le nombre des déplacements est le double plus un, dans le jeu précédent.

Pour une tour de deux étages, il faut trois ; coups, au minimum.

—	trois	—	sept	—
—	quatre	—	quinze	—
—	5	—	31	—
—	6	—	63	—
—	7	—	127	—
—	8	—	255	— et ainsi de suite.

À un coup par seconde, il faut plus de quatre minutes pour déplacer la tour de huit étages.

Variations du Jeu. — On varie, à l'infini, les conditions du problème de la tour d'Hanoï, comme il suit. Au début, on enfile les étages, d'une manière quelconque, sur un, deux ou trois clous. Il faut reconstruire la tour sur l'un des clous, désigné à l'avance. Pour 64 étages le nombre des dispositions initiales est vertigineux ; il a plus de *cinquante* chiffres.

Pour plus de détails, consulter l'ouvrage suivant au chapitre du Baguenaudier

RÉCRÉATIONS MATHÉMATIQUES

Par M. ÉDOUARD LUCAS, professeur de mathématiques spéciales au lycée Saint-Louis

Deux volumes petit in-8, caractères elzévirs, titre en deux couleurs.

Paris, 1883, chez GAUTHIER-VILLARS, Imprimeur-Libraire de l'Académie des Sciences et de l'École Polytechnique Quai des Augustins, 55.

Figure 0.8: Recto and verso of the leaflet accompanying the Tower of Hanoi puzzle

together with the claim that it would take more than *five milliard*[8] *centuries* to carry out the task making one move per second. The number in (0.1) is explained, together with the rules of the game, which are imprecise concerning the goal peg and redundant with respect to the divine rule, on the back of [82]. Here one can find the famous *recursive* solution, stated for an arbitrary number of discs, but demonstrated with an example:[9] if one can solve the puzzle for *eight* discs, one can solve it for *nine* by first transferring the upper eight to the spare peg, then moving the ninth disc to the goal peg and finally the smaller ones to that peg too. So by increasing the number of discs by one, the number of moves for the transfer of the tower doubles plus one move of the largest disc. Now the superiority of the binary number system becomes obvious. We write 2^n for an n-fold product of 2s, $n \in \mathbb{N}_0$, e.g., $2^0 = 1$ (by convention, the product of no factors is 1), $2^1 = 2$, $2^2 = 4$, and so forth. Then every **natural number** $N \neq 0$ can (uniquely) be written as $(N_{K-1} \dots N_1 N_0)_2$, such that $N = \sum_{k=0}^{K-1} N_k \cdot 2^k$ with $K \in \mathbb{N}$ and $N_k \in \{0,1\}$, $N_{K-1} = 1$. (This needs a mathematical proof, but we will not go into it. Later, in Section 0.4.1, we will get a bit more formal about natural numbers.) Clearly, $2^n = 10 \dots 0_2$ with n bits 0 and by binary arithmetic, $1 \dots 1_2 = 2^n - 1$ with n bits 1. Now doubling a number and adding 1 means concatenating a bit 1 to the right

---

[8]billion, i.e. 10^9

[9]This is an instance of the method of *generalizable example*; cf. [227, p. 116ff].

of the number. The recursive solution therefore needs $2^n - 1$ moves to transfer a tower of n discs. Calculating the 64-fold product of 2 and subtracting 1 in decimal representation, we arrive at the number shown in (0.1).

This number evokes another "Indian" myth which comes in even more versions than the Tower of Brahma. The nicest, but not the first, of these legends is by J. F. Montucla [320, p. 379–381], who tells us, in citing the gorgeous number, that the Indian Sessa, son of Daher (Sissa ben Dahir), invented (a prototype of) the chess game which he presented to the Indian king (Shirham). The latter was so pleased that he offered to Sessa whatever he desires. Contrary to our experience with fairy tales, Sessa did not ask for the daughter of the king, but pronounced a "modest" wish: a grain of wheat (rice in other versions) on the first square of the chess board, two on the second, four on the third and so on up to the last, the 64th one. The king's minister, however, found out that it was impossible to amass such an amount of wheat and we are told by Montucla that the king admired Sessa even more for that subtle request than for the invention of the game. (In less romantic versions, Sessa was beheaded for his impertinence.)

In this "arithmetic bagatelle", as Montucla called it, we recognize immediately, that Sessa asked to concatenate a bit 1 to the left of the binary number of grains when adding a new square of the board, all in all $1 \ldots 1_2 = 2^{64} - 1$ grains.

With the *Mersenne numbers*[10] (for M. Mersenne) $M_k = 2^k - 1$, we have an example of a sequence $(M_k)_{k \in \mathbb{N}_0}$, called *Mersenne sequence*, which fulfills two *recurrences*

$$M_0 = 0, \ \forall \, k \in \mathbb{N}_0 : \ M_{k+1} = 2M_k + 1, \tag{0.2}$$

$$M_0 = 0, \ \forall \, k \in \mathbb{N}_0 : \ M_{k+1} = M_k + 2^k. \tag{0.3}$$

We will see later that (0.2) and (0.3) are prototypes of a most fundamental type of recurrences, each of them leading to a uniquely determined sequence.

The most ubiquitous of all sequences defined by a recurrence is even older. In his manuscript *Liber abbaci* of 1202/1228, Leonardo Pisano (see [101]), now commonly called Fibonacci (figlio di Bonaccio), posed the following problem:

Quot paria coniculorum in uno anno ex uno pario germinentur.

The solution of the famous *rabbit problem* invoked the probably most popular integer sequence of all times, which was accordingly named *Fibonacci sequence* by Édouard Lucas (cf. [288, p. 3]). Its members F_k, the *Fibonacci numbers* (cf. [344]), are given by the recurrence

$$F_0 = 0, \ F_1 = 1, \ \forall \, k \in \mathbb{N}_0 : \ F_{k+2} = F_{k+1} + F_k. \tag{0.4}$$

Lucas employed this sequence in a somewhat more serious context in connection with the distribution law for prime numbers in [280], thereby introducing a variant, namely the sequence (for $k \in \mathbb{N}$) $L_k := F_{2k}/F_k$ which fulfils the same recurrence relation as in (0.4), but with the *seeds* replaced by $L_1 = 1$ and $L_2 = 3$ (or by

---

[10]Some authors use this term only if k is prime.

$L_0 = 2$, $L_1 = 1$ to start the sequence at $k = 0$). Lucas still named this sequence for Fibonacci as well in [279, p. 935], but it is now called the *Lucas sequence* (and its members *Lucas numbers*). Fibonacci and Lucas numbers can also be calculated explicitly and they exhibit an interesting relation to the famous *Golden section* as shown in Exercise 0.1; cf. [424]. The methods developed in [280] allowed Lucas to decide whether certain numbers are prime or not without reference to a table of primes, and he announced his discovery that $2^{127} - 1$ is a (Mersenne) prime. (Is there a relation to the TH with 127 discs?) With this he was the last human world record holder for the "largest" prime number, to be beaten only by computers 75 years later, albeit using Lucas's method.

Édouard Lucas's mathematical oeuvre has never been properly recognized in his home country France, neither in his time, nor today. In 1992, M. Schützenberger writes (translated from [375]): "...Édouard Lucas, who has no reputation among professional mathematicians, however, because he is schools inspector and does not publish anything else but books on entertaining mathematics." Lucas's reputation was much higher abroad, in particular in Germany, where, e.g., Edmund Landau declared himself[11] an "ardent admirer of the illustrious Lucas" and called him "immortal". Only in 1998 [93] and with her thesis [95], A.-M. Décaillot put the life and number-theoretical work of Lucas into light. Before that there were just two short biographies in [195, p. 540f] and [435, Section 3.1]. Already in 1907 appeared a collection of biographies and necrologies [292]. As it turns out, Lucas was in the wrong place at the wrong time: with his topics from number theory and his enthusiasm for teaching and popularizing mathematics he put himself outside the infamous main-stream. Being in the wrong place at the wrong time seemed to be Lucas's fate: the story of his death sounds as if it was just a malice of N. Claus (de Siam). Although the source is unknown, we give here a translation of the most moving of all accounts from the journal *La Lanterne* of 6 October 1891 [292, p. 17]:

> The death of this "prince of mathematics", as the young generations of students called him, has been caused by a most vulgar accident. In a banquet at which assisted the members of the [Marseille] congress during an excursion into Provence, a [male] servant, who found himself behind the seat of M. Edouard Lucas, dropped, by unskilfulness, a pile of plates. A broken piece of porcelain came to hit the cheek of M. Lucas and caused him a deep injury from which blood flew in abundance. Forced to suspend his work, he returned to Paris. He took to his bed and soon appeared erysipelas which would take him away.

Édouard Lucas died on 3 October 1891, aged only 49. His tomb, perpetual, but in a deplorable condition (see Figure 0.9), can be found on the Montmartre cemetary of Paris. Lucas was honored by his native city Amiens that named a street after him[12] where one can also find the *Collège Edouard Lucas*; see Figure 0.10.

---

[11] Cited and translated from a letter of 1896 to Henry Delannoy, reproduced in [94, Annexe A].

[12] During its session of 1898–03–30 the municipal council adopted a proposal that the street may "receive the name of a child of Amiens".

Figure 0.9: Detail from Édouard Lucas's grave

© 2015 A. M. Hinz

Upon his untimely death, Lucas left the second volume of his *Théorie des nombres* [288] unfinished. E. T. Bell writes in 1951 [44, p. 230, footnote]: "Some years ago the fantastic price of thirty thousand dollars was being asked for Lucas's manuscripts. In all his life Lucas never had that much money." It is not known where these manuscripts remained; see, however, [94, Section 1]. Only few traces are left of a collection of six *Jeux scientifiques* (cf. [291, Note III]), for which Lucas earned two medals at the World's Fair in Paris in 1889 (the one for which another tower, namely Eiffel's, was erected). Some of these games are presented in Lucas's article [286]. The puzzle collection, dedicated to his children Paul and Madeleine, was advertised in *Cosmos*, a scientific magazine, of 7 December 1889 as a first series, each single game being sold by "Chambon & Baye" of Paris for the price of 10 francs. Among these was a new version of the TH, this time with five pegs and 16 discs in four colors. Lucas says that the number of problems one can pose about the new TH is uncalculable. For the almost original material see Figure 0.11.

(*Bulletin Municipal de la Ville d'Amiens* 24 (1898) 182.)

Figure 0.10: Street sign in Amiens (France)

© 2017 A. M. Hinz

Five of the brochures accompanying the puzzles are preserved, the only missing one is about *Les Pavés Florentins du Père Sébastien*,[13] also described in [30, p. 158]. In [44, p. 230] Bell suggested that "His widely scattered writings should be collected, and his unpublished manuscripts sifted and edited." Despite some efforts (cf. [179, 25]), this goal is still far from being reached. In particular, it would be desirable to review Lucas's œuvre in recreational mathematics just like Décaillot had done it for his work in number theory. A first attempt under sociological aspects has been made by her in [94]. We would also be grateful for any suggestions to locate missing items authored by Lucas; cf. [231, Problem 5].

For now we have to concentrate on Lucas's greatest and most influential invention. At the end of his account on the TH in [289, p. 58f], Lucas writes: "Latterly, the foreign industry has taken possession of the game of our friend [N. Claus (de Siam)] and of his legend; but we can assert that the whole had been

---

[13]Père Sébastien is the French member of the order of the Carmelites Jean Truchet (1657–1729), whose note of 1704 [419] on the *Truchet tiles*, the object of Lucas's game, can be viewed as an early contribution to combinatorics and who was also the inventor of the *typographic point*; see [18] and literature cited therein.

imagined, some time ago already [in 1876 according to [285, p. 14]], in n° 56 of rue Monge, in Paris, in the house built on the site of the one where Pascal died on 19 August 1662."

Figure 0.11: Lucas's new Tower of Hanoi game (slightly complemented)
(courtesy of Francis Lucas, http://edouardlucas.free.fr/)
© 2014 F. Lucas

0.4 Sequences

In the previous section we already met two fundamental integer sequences, the Mersenne sequence and the Fibonacci sequence. Such sequences will abound throughout the book. To make things precise, we have to be a little more formal now.

0.4.1 Integers

We will not give an axiomatic introduction of the natural numbers, for which we refer to [398, p. 160–162]. Readers may rely on whatever they have learned about them. Suffice it here to say that we have a set $\mathbb{N}_0$, elements $0 \in \mathbb{N}_0$, $1 \in \mathbb{N} :=$ $\mathbb{N}_0 \smallsetminus \{0\}$,[14] and an injection (in fact bijection) $\mathbb{N}_0 \to \mathbb{N}$, $n \mapsto n + 1$, such that no proper subset N of $\mathbb{N}_0$ fulfills (we write $N + 1 := \{n + 1 \mid n \in N\}$ for any $N \subseteq \mathbb{N}_0$):

$$0 \in N \wedge N + 1 \subseteq N. \tag{0.5}$$

In other words, if a subset N of $\mathbb{N}_0$ satisfies (0.5), then it must be $\mathbb{N}_0$; this is the *principle of (mathematical* or *complete) induction* and will be the base for virtually all definitions and statements regarding natural numbers.

$\mathbb{N}$ is also called the set of *positive integers*. The set $-\mathbb{N} := \{-n \mid n \in \mathbb{N}\}$ contains the *negative integers*, such that the mutually disjoint union $\mathbb{Z} := -\mathbb{N} \cup \{0\} \cup \mathbb{N}$ is the set of *integers*.[15] For later convenience, we further define with the aid of the *Fundamental Theorem of recursion* (see Remark 9.1 and [398, Theorem 8.3])

$$\forall\, k \in \mathbb{N}_0 : \mathbb{N}_{k+1} = \mathbb{N}_k + 1. \tag{0.6}$$

Obviously, $\mathbb{N}_1 = \mathbb{N}_0 + 1 = \mathbb{N}$ and $\mathbb{N}_0 = \{0\} \cup \mathbb{N}_1$, or more generally

Proposition 0.1.

$$\forall\, k \in \mathbb{N}_0 : \mathbb{N}_k = \{k\} \cup \mathbb{N}_{k+1}. \tag{0.7}$$

Proof. Let $N := \{k \in \mathbb{N}_0 \mid \mathbb{N}_k = \{k\} \cup \mathbb{N}_{k+1}\}$. Then, as seen, $0 \in N$ and for $k \in N$:

$$\mathbb{N}_{k+1} = \mathbb{N}_k + 1 = (\{k\} \cup \mathbb{N}_{k+1}) + 1 = \{k+1\} \cup (\mathbb{N}_{k+1} + 1) = \{k+1\} \cup \mathbb{N}_{(k+1)+1},$$

so that $k + 1 \in N$. □

Of utmost importance in discrete mathematics are the prototype finite sets consisting of $\ell \in \mathbb{N}_0$ consecutive natural numbers, in particular the *initial segments* $[\ell]_0$[16] and $[\ell] := [\ell]_1$ of $\mathbb{N}_0$ and $\mathbb{N}$, respectively:

$$[\ell]_0 = \mathbb{N}_0 \smallsetminus \mathbb{N}_\ell, \ \forall\, k \in \mathbb{N}_0 : [\ell]_{k+1} = [\ell]_k + 1. \tag{0.8}$$

Proposition 0.2. *For every $k \in \mathbb{N}_0$ we have*

$$[0]_k = \varnothing, \ \forall\, \ell \in \mathbb{N}_0 : [\ell+1]_k = \{k\} \cup [\ell]_{k+1}. \tag{0.9}$$

---

[14] According to ISO 31–11:1992 (replaced by ISO 80000–2:2009), our $\mathbb{N}$, which we consider to be the most natural set of natural numbers, should be $\mathbb{N}^*$, while our $\mathbb{N}_0$ should be $\mathbb{N}$, but this is counter-intuitive.

[15] There are mathematical communities, where 0 is positive or even positive *and* negative!

[16] This (finite!) set is denoted by $\mathbb{N}_\ell$ in ISO 31–11:1992, such that $\mathbb{N}_0 = \varnothing$ in their notation.

Proof. Both statements are proved by induction on k. The first induction proof is obvious. For the second put $N = \{k \in \mathbb{N}_0 \mid \forall \ell \in \mathbb{N}_0 : [\ell+1]_k = \{k\} \cup [\ell]_{k+1}\}$. Then for $k = 0$:

$$
\begin{aligned}
\{0\} \cup [\ell]_1 &= \{0\} \cup ([\ell]_0 + 1) \\
&= \{0\} \cup ((\mathbb{N}_0 \setminus \mathbb{N}_\ell) + 1) \\
&= \{0\} \cup ((\mathbb{N}_0 + 1) \setminus (\mathbb{N}_\ell + 1)) \\
&= \mathbb{N}_0 \setminus (\mathbb{N}_\ell + 1) = \mathbb{N}_0 \setminus \mathbb{N}_{\ell+1} = [\ell+1]_0,
\end{aligned}
$$

and if $k \in N$:

$$
\begin{aligned}
[\ell+1]_{k+1} = [\ell+1]_k + 1 &= (\{k\} \cup [\ell]_{k+1}) + 1 \\
&= \{k+1\} \cup ([\ell]_{k+1} + 1) = \{k+1\} \cup [\ell]_{(k+1)+1},
\end{aligned}
$$

whence $k+1 \in N$.   □

In particular, we have $[1]_k = \{k\} \subseteq [\ell+1]_k$ and $[\ell]_k \subseteq \mathbb{N}_k$, the latter again by an easy induction on k. The set $[\ell]_k$ is called the *initial segment* of $\mathbb{N}_k$ with *length* ℓ or ℓ-*segment* of $\mathbb{N}_k$.

As soon as arithmetic and the canonical order relation are established on $\mathbb{N}_0$, we have, of course,

$$
[\ell]_k = \{n \in \mathbb{N}_0 \mid k \le n < k+\ell\} \subseteq \{n \in \mathbb{N}_0 \mid k \le n\} = \mathbb{N}_k.
$$

0.4.2   Integer Sequences

An *integer sequence* is a mapping

$$
a : \mathbb{N}_0 \to \mathbb{Z}, \; n \mapsto a_n \, (:= a(n)).
$$

As a convention we define $a_{-n} := 0$ for $n \in \mathbb{N}$. However, there will be *no* $a(-n)$, which means that $a \in \mathbb{Z}^{\mathbb{N}_0}$.

Elementary examples for integer sequences are $\widehat{0}$, defined by $\widehat{0}(n) = 0^n$ (A000007)[17], and 1 with $1(n) = 1$ (A000012). We also define αa for $\alpha \in \mathbb{Z}$ by $(\alpha a)(n) = \alpha a(n)$ and $a+b$ by $(a+b)(n) = a(n) + b(n)$; in particular, 0 is the *trivial sequence* $0_n = 0$ (A000004).

An utmost important statement about integer sequences is the following

Lemma 0.3. *Let $\alpha \in \mathbb{Z}$, $a, b \in \mathbb{Z}^{\mathbb{N}_0}$. Then*

$$
\forall n \in \mathbb{N}_0 : b_n = a_n - \alpha\, a_{n-1} \Leftrightarrow \forall n \in \mathbb{N}_0 : a_n = \sum_{k=0}^{n} \alpha^{n-k} b_k.
$$

---

[17]This, throughout the book, refers to entries in the *On-Line Encyclopedia of Integer Sequences* (OEIS®, http://oeis.org), initiated by N. J. A. Sloan. In some cases the offset might differ.

Proof. The proof of "⇒" is via a telescoping sum, while "⇐" is trivial. □

If $\alpha = 1$ in Lemma 0.3, then a is the sequence Σb of *partial sums* (or the *integral*) of b, and b is the sequence $\bar{a}$ of *differences* (or the *derivative*) of a. As is desirable, we have the *Fundamental Theorem of integer sequences*:

$$\Sigma \bar{a} = a = \overline{\Sigma a}.$$

As an example let us consider the *identity sequence* (A001477)

$$\mathrm{id} : \mathbb{N}_0 \to \mathbb{N}_0, \ n \mapsto n, \tag{0.10}$$

for which $\overline{\mathrm{id}} = 1 - \widehat{0} = 0, 1, 1, \ldots$ (cf. A057427), i.e. it is the *characteristic function* of $\mathbb{N}$ in $\mathbb{N}_0$. For the sequence $\Delta := \Sigma \mathrm{id} = 0, 1, 3, 6, 10, 15, 21, \ldots$ we obtain, from $\overline{\Delta} = \mathrm{id}$,

$$\Delta_n = \Delta_{n-1} + n; \tag{0.11}$$

these are the *triangular numbers*, known for ages and somehow related to J. F. C. Gauss (1777–1855); see below, p. 50f.

Lemma 0.3 has a useful consequence:

Corollary 0.4. *Let* $\alpha, \beta \in \mathbb{Z}$*; then*

$$\forall \, n \in \mathbb{N}_0 : \ (\beta - \alpha) \sum_{k=0}^{n} \alpha^{n-k} \beta^k = \beta^{n+1} - \alpha^{n+1}.$$

Proof. Put $a_k = \beta^{k+1} - \alpha^{k+1}$ and $b_k = (\beta - \alpha)\beta^k$ in Lemma 0.3. □

0.4.3 The Dyadic Number System

In his (draft of a) letter[18] dated 1697–01–02 (the *New Year's Letter*) to Duke Rudolph August, Prince of Brunswick-Wolfenbüttel, Gottfried Wilhelm Leibni(t)z (1646–1716) presented an explicit list of the first 15 entries of the *dyadic sequence* given by

$$D_n = 2^n \tag{0.12}$$

in both decimal

$$D = 1, 2, 4, 8, 16, 32, 64, \ldots$$

and binary representation

$$D = 1, 10, 100, 1000, 10000, 100000, 1000000 \ldots;$$

they will be identified with each other by writing $D_n = (10^n)_2$, where 0^n means an n-fold repetition of the bit 0.

---

[18]This draft is preserved in the State Library of Lower Saxony in Hanover (LBr II/15); a facsimile can be found between pages 24 and 25 in: R. Loosen, F. Vonessen (Eds.), G. W. Leibniz, Zwei Briefe über das binäre Zahlensystem und die chinesische Philosophie, Belser, Stuttgart, 1968.

Of equal interest is the *Mersenne sequence*

$$M = 0, 1, 3, 7, 15, 31, 63, \ldots$$

defined by $M_n = D_n - 1 = (1^n)_2$ and named for Marin Mersenne (1588–1648). Mersenne, a member of the order of the Minims, had tried, mostly in vain, to devise methods to test the primality of *Mersenne numbers*[19] M_n in order to find (even) *perfect* numbers. The latter are known to be those members of the, also otherwise interesting, sequence $\Delta \circ M$ (A006516) for which M_n is prime (*Euclid-Euler theorem*), namely a *Mersenne prime*; it was Lucas who first came up with a satisfactory test; cf. [435].

For technical reasons it is often useful to employ the sequence (A131577)

$$\overline{M} = 0, 1, 2, 4, 8, 16, 32, \ldots ,$$

where obviously $\overline{M}_n = D_{n-1} = (\omega^{(n)})_2$, with $\omega^{(0)} = \varnothing$ and $\omega^{(n)} = 10^{n-1}$ for $n \in \mathbb{N}$. Moreover, the sequence (A011782)

$$\overline{D} = 1, 1, 2, 4, 8, 16, 32, \ldots$$

fulfills $\overline{D}_n = M_{n-1} + 1$. Finally, from Corollary 0.4 it follows that $\Sigma D_n = M_{n+1} = D_{n+1} - 1$ and $\Sigma M_n = M_{n+1} - (n+1) = 2M_n - n$, the latter because $M_{n+1} = 2M_n + 1$. The sequence

$$E := \Sigma M = 0, 1, 4, 11, 26, 57, 120, \ldots$$

is called (an) *Eulerian sequence* (cf. A000295) for Leonhard Euler (1707–1783), who in 1755 considered a certain list of polynomials whose coefficients have found a modern combinatorial interpretation as follows. Let S_n be the set of **permutations** on $[n]$ and for $\sigma = (\sigma_1, \ldots, \sigma_n) \in S_n$ let $\mathrm{exc}(\sigma) = |\{i \in [n] \mid \sigma_i > i\}|$ ($\in [n]_0$) denote the number of *excedances* of σ; then $\left\langle {n \atop k} \right\rangle := |\{\sigma \in S_n \mid \mathrm{exc}(\sigma) = k\}|$. This opens a whole new world of *eulerian numbers*; cf. [336]. Our sequence is then given by $E_n = \left\langle {n+1 \atop 1} \right\rangle$ (see [336, Chapter 1]). Compare this with $\Delta_n = \binom{n+1}{2}$.

0.4.4   Finite Binary Sequences

Let $\mathbb{B}$ denote the set $B := \{0, 1\}$ equipped with the addition given by $+ = \dot{\vee}$.[20]

A binary sequence $f \in \mathbb{B}^{\mathbb{N}_0}$ is called *finite*, $f \in \mathbb{B}_0^{\mathbb{N}_0}$, if $\exists n_f \in \mathbb{N}_0 \ \forall n \geq n_f : f_n = 0$, i.e. if f is eventually 0; we may assume that n_f is minimal.

For every finite binary sequence f we can define the *derivative* f' by $f'_n = f_{n+1} - f_n = f_{n+1} + f_n$ and the *integral* $\int f$ by $(\int f)_n = \sum_{k=n}^{\infty} f_k = \sum_{k=n}^{n_f-1} f_k$. Both f'

---

[19]So named in [284, p. 230].

[20]The symbol $\dot{\vee}$ represents the *binary exclusive or*, i.e. $\dot{\vee} \colon B^2 \to B$, $(b_1, b_2) \mapsto b_1 \dot{\vee} b_2$, which is 0 iff $b_1 = b_2$; cf. the *binary inclusive or*, defined by $\vee \colon B^2 \to B$, $(b_1, b_2) \mapsto b_1 \vee b_2$, which is 0 iff $b_1 = 0 = b_2$.

and $\int f$ are finite binary sequences (cf. [96]). Then the *Fundamental Theorem for finite binary sequences* holds: $(\int f)' = f = \int (f')$.

Proof. $(\int f)'_n = \sum_{k=n+1}^{\infty} f_k + \sum_{k=n}^{\infty} f_k = f_n$

$$= \sum_{k=n}^{\infty} f_{k+1} + \sum_{k=n}^{\infty} f_k = \sum_{k=n}^{\infty} f'_k = \int (f')_n. \qquad \square$$

Theorem 0.5. *The mapping* $f \mapsto \sum_{k=0}^{\infty} f_k \cdot 2^k$ *is a bijection from* $\mathbb{B}_0^{\mathbb{N}_0}$ *to* $\mathbb{N}_0$.

Proof. Since the sum is finite, it is properly defined. For the injectivity we show that for $N \in \mathbb{N}_0$ and $f \in \{-1, 0, 1\}^{1+N}$ it follows from $0 = \sum_{k=0}^{N} f_k \cdot 2^k$ that $f_N = 0$. Assume that $f_N = 1$; then

$$2^N = f_N \cdot 2^N = \left| \sum_{k=0}^{N-1} f_k \cdot 2^k \right| \leq \sum_{k=0}^{N-1} 2^k = 2^N - 1,$$

a contradiction.

Now let $\phi \in \mathbb{N}_0^{\mathbb{N}_0}$ be defined by $\phi(\nu) = \frac{1}{4}(2\nu - 1 + (-1)^\nu)$; then $\nu - 1 \leq 2\phi(\nu) \leq \nu$. For $n \in \mathbb{N}_0$ let $n_f := \min\{k \in \mathbb{N}_0 \mid \phi^k(n) = 0\}$ and $\forall k \in [n_f]_0 : f_k = \phi^k(n) - 2\phi^{k+1}(n)$. Then

$$\sum_{k=0}^{n_f-1} f_k \cdot 2^k = \sum_{k=0}^{n_f-1} \left(\phi^k(n)2^k - \phi^{k+1}(n)2^{k+1} \right) = n. \qquad \square$$

Remark. Theorem 0.5 shows that $\mathbb{B}_0^{\mathbb{N}_0}$ is countably infinite. Cantor's Second Diagonal method can be used to prove that $\mathbb{B}^{\mathbb{N}_0}$ is uncountable:[21]

Theorem 0.6. *Let* $F : \mathbb{N}_0 \to \mathbb{B}^{\mathbb{N}_0}$ *be a sequence of binary sequences. Then there is a binary sequence* $f \in \mathbb{B}^{\mathbb{N}_0}$ *such that* $\forall n \in \mathbb{N}_0 : f \neq F(n)$.

Proof. Define f by $f_n = 1 - F(n)_n$. $\qquad \square$

We will now identify $d \in \mathbb{N}_0$ with $d \in \mathbb{B}_0^{\mathbb{N}_0}$, $d = \sum_{k=0}^{\infty} d_k \cdot 2^k \left(= \sum_{k=0}^{n_d-1} d_k \cdot 2^k \right)$. Usually, this is written as $d = (d_{n_d-1} \ldots d_0)_2$. For instance, $(10^n)_2 = D_n = 2^n$ and $(1^n)_2 = M_n = 2^n - 1$. Then the finite sequence d', which can also be written as $d'_{n_d-1} \ldots d'_0$, has the following properties.

- For fixed $k \in \mathbb{N}_0$, the infinite sequence (with respect to $d \in \mathbb{N}_0$) d'_k is $(0^{2^k} 1^{2^{k+1}} 0^{2^k})^\infty$, i.e. it has period 2^{k+2}; this is [96, Propriété 1] and follows from the fact that two consecutive bits in $(d)_2$ are either 00, 01, 10, or 11.

---

[21]$\mathbb{B}^{\mathbb{N}_0}$ is, of course, equivalent to the power set $2^{\mathbb{N}_0}$ of $\mathbb{N}_0$, $f_n = 1$ signifying that n belongs to the corresponding subset.

- Going from d' to $(d+1)'$, only one bit is changed! It is d_0', if the number of 1s in d' is even, and otherwise it is bit d_{k+1}', if $d_k' \ldots d_0' = 10^k$; this is [96, Propriété 2] and follows since the bit in position 0 changes every second time when running through the positive natural numbers by the first property and similarly for the other bits.

In Chapter 1 we will show that these facts can be deduced even more easily by recourse to the Chinese rings.

0.5 Indian Verses, Polish Curves, and Italian Pavements

B. Pascal is known, among other things, for the triangle which in his seminal treatise, published posthumously in 1665, he called the *Arithmetical triangle (AT)* (cf. [123]). However, the famous arrangement of numbers was known long before him and can be found implicitly as early as in the tenth century in a commentary on Piṅgala's *Chandaḥśāstra* (ca. -200) by Halāyudha (cf. [225, p. 112f] and [192, Section 3.4]). The *Mount Meru* (Meru-Prastāra) according to this description can be seen in Figure 0.12.

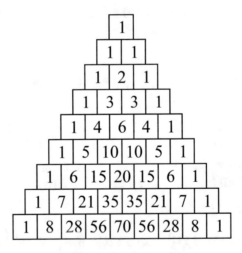

Figure 0.12: The Meru-Prastāra

The rule is to start with a 1 in the single square on top and to fill out each subsequent line by writing the sum of the touching squares of the line immediately above. Piṅgala was investigating poetic meters of short (0) and long (1) syllables, a discipline called *prosody*, cf. [257, p. 42–47]. The problem was to analyze their combinations, and this can be viewed as the birth of *combinatorics*. In how many ways can they be combined to form a word of length k? Quite obviously (for us!), the answer is 2^k, the number of bit strings of length k. But looking closer, how many *combinations* of k syllables are there containing ℓ long and (consequently)

$k - \ell$ short ones? The answer is contained in Mount Meru by looking at the ℓth entry from the left in the kth line from the top, both counting from 0. (It is a puzzle why almost all presentations of AT start with the single 1 on top. Although the modern mathematician is used to such abstruse things, how did the Ancients know that there is precisely one way to choose no long syllable (and no short one either for that matter) to form an empty word?) This number will be pronounced "k *choose* ℓ" and is denoted by $\binom{k}{\ell}$. Summing the entries of a single line in Mount Meru immediately leads to the formula

$$\forall k \in \mathbb{N}_0 : \sum_{\ell=0}^{k} \binom{k}{\ell} = 2^k .$$

Another question that can be solved by looking at Mount Meru (cf. [225, p. 113f]) is the following: if long syllables take 2 beats (or *morae*), short ones only 1 beat (*mora*), how many words (of varying lengths) can be formed from a fixed number of m morae? Since $m = 2\ell + k - \ell = k + \ell$, this number can be found by summing over all $\binom{k}{\ell} = \binom{m-\ell}{\ell}$, i.e. some mounting diagonal in Mount Meru. On the other hand, a word with $m + 2$ morae consists either of $m + 1$ morae plus a single beat or of m morae and a long syllable (see [257, p. 48]) and thus the Fibonacci recurrence relation in (0.4), but with the seeds 1 and 2, is fulfilled, whence we obtain, by the *method of double counting*, the astonishing result that

$$\forall m \in \mathbb{N}_0 : \sum_{\ell=0}^{\lfloor \frac{m}{2} \rfloor} \binom{m - \ell}{\ell} = F_{m+1} , \tag{0.13}$$

where the *floor function* is defined for $x \in \mathbb{R}$ by $\lfloor x \rfloor = \max\{a \in \mathbb{Z} \mid a \le x\}$. (Similarly, the *ceiling function* is characterized by $\lceil x \rceil = \min\{b \in \mathbb{Z} \mid b \ge x\}$.)

The formula in (0.13) is even more surprising because it apparently had been found 50 years before Fibonacci!

During its long history (cf. [437]), the AT reappeared in many disguises. For instance, evaluating the kth power of the *binomial* term $a + b$, we have to sum up products of k factors containing either an a or a b from each of the k individual factors of the power. Identifying a with short and b with long syllables from prosody, we see that the product $a^{k-\ell}b^{\ell}$ shows up precisely $\binom{k}{\ell}$ times, i.e. we arrive at

$$\forall k \in \mathbb{N}_0 : (a + b)^k = \sum_{\ell=0}^{k} \binom{k}{\ell} a^{k-\ell} b^{\ell} .$$

This is the famous *binomial theorem*, and the $\binom{k}{\ell}$ are now commonly known as *binomial coefficients*. However, this denomination addresses only one particular aspect of these numbers and it is a breach of the historical facts; we therefore prefer to call them *combinatorial numbers*.

In more modern terms, the combination of short and long syllables in a word of length k can also be interpreted as choosing a subset from a set K with $|K| = k$

elements. Here 0 means that an element of K does not belong to the subset chosen, while 1 means that it is an element of the subset. In other words, there are exactly $\binom{k}{\ell}$ subsets of K which have precisely ℓ elements. Writing $\binom{K}{\ell}$ for the set of all subsets of K of size ℓ, we get

$$\forall\, \ell \in \{0, \ldots, |K|\} : \left| \binom{K}{\ell} \right| = \binom{|K|}{\ell}.$$

A sound mathematical description of the AT, based on this formula, and some important notation can be found in Exercise 0.2.

The direct application of arguments based on choices is the starting point for proving many interrelations between combinatorial numbers. For instance, if you want to select ℓ balls from a collection of b black and w white balls, you may successively pick λ black ones and then $\ell - \lambda$ white balls. There are $\binom{b}{\lambda}\binom{w}{\ell-\lambda}$ such combinations, whence

$$\forall\, b, w, \ell \in \mathbb{N}_0 : \binom{b+w}{\ell} = \sum_{\lambda=0}^{\ell} \binom{b}{\lambda}\binom{w}{\ell-\lambda}. \tag{0.14}$$

This can be viewed as an extension of the recursive formula in Exercise 0.2 c), which is case $b = 1$ of (0.14) and could therefore be called the *black sheep formula*.

Combinatorics is also the historical starting point for *probability theory*, which has its early roots in discussions about games of chance. One of these was treated mathematically by Euler in 1751 (see [140]) under the name of "Game of encounter" (Jeu de rencontre). It had been studied earlier by Pierre Rémond de Montmort in the first work entirely devoted to probability, his *Essay d'Analyse sur les Jeux de Hazard*, whose second edition appeared in 1713, but Euler's presentation is, as usual, more clearly written.

Two players, A and B, hold identical packs of cards which are shuffled. They start to compare card by card from their respective decks. If during this procedure two cards coincide, then A wins. If all cards have been used up without such an encounter having occurred, then B is the winner. The question is: what is the probability of A (or B for that matter) to win. (In Montmort's version, the number of cards in each deck was 13, and the game was accordingly called "Jeu du Treize".)

Euler solved this question in the following way. Let us assume that a deck has $k \in \mathbb{N}$ cards and that the deck of A is naturally ordered from 1 to k [140, § 3]; this has to be compared with the $k!$ possible decks B can hold [140, § 17]. Some easy cases are done first: for $k = 1$, A is the winner [140, § 6]; for $k = 2$, the chances are $1 : 1$ [140, § 7, 8]. The most instructive case is $k = 3$ [140, § 9–11], where $2! = 2$ decks of B, namely $(1, 2, 3)$ and $(1, 3, 2)$, will make A the winner in the first round. In the second round, the same number of decks $((1, 2, 3)$ and $(3, 2, 1))$, would be favorable for A, however, one obviously has to delete deck $(1, 2, 3)$, from this list. This is because it would have made A the winner in the first round of a game with only 2 cards, namely without card 2 present. This is a prototype application of the **inclusion-exclusion principle**:[22] an element must not be counted twice in the union of two (or more) finite sets. Of the remaining three decks, A will win the third round with deck $(2, 1, 3)$ in B's possession only, so all in all in 4 out of $3! = 6$ cases. Euler makes another full analysis for the case $k = 4$ [140, § 12–14], but thereafter passes to finding the general rule for the number $f_{k,\ell}$ of captures of A in round $\ell \in [k]$ leading to the recurrence [140, § 22, 37]

$$\forall k \in \mathbb{N}: \; f_{k,1} = (k-1)! \;\wedge\; \forall \ell \in [k-1]: \; f_{k,\ell+1} = f_{k,\ell} - f_{k-1,\ell}. \qquad (0.15)$$

From this, Euler deduces (in fact, he considers $f_{k,\ell}/k!$) in [140, § 38–42], by what we would call a double recurrence on k and ℓ,

$$\forall k \in \mathbb{N} \; \forall \ell \in [k]: \; f_{k,\ell} = \sum_{m=0}^{\ell-1} (-1)^m \binom{\ell-1}{m}(k-1-m)!. \qquad (0.16)$$

In fact, formula (0.16) can be proved a posteriori by verifying that it fulfills the recurrence (0.15): $f_{k,1} = (k-1)!$ is obvious and for $\ell \in [k-1]$:

$$
\begin{aligned}
f_{k,l+1} &= \sum_{m=0}^{\ell} (-1)^m \binom{\ell}{m}(k-1-m)! \\
&= (k-1)! + \sum_{m=1}^{\ell} (-1)^m \binom{\ell}{m}(k-1-m)! \\
&= (k-1)! + \sum_{m=1}^{\ell} (-1)^m \left\{ \binom{\ell-1}{m} + \binom{\ell-1}{m-1} \right\}(k-1-m)! \\
&= \sum_{m=0}^{\ell-1} (-1)^m \binom{\ell-1}{m}(k-1-m)! - \sum_{m=0}^{\ell-1} (-1)^m \binom{\ell-1}{m}(k-2-m)! \\
&= f_{k,\ell} - f_{k-1,\ell}.
\end{aligned}
$$

To obtain the number f_k of decks of B favorable for A to win, we have to sum equation (0.16) over ℓ. In [140, § 43, 44] Euler used an ingenious trick based on an observation about the AT: if one adds the entries in a diagonal parallel to the left side of the triangle down to a certain entry, one obtains the entry immediately to the right of it in the next line, i.e.

$$\forall k \in \mathbb{N}, \; m \in [k]_0 : \; \sum_{\ell=m+1}^{k} \binom{\ell-1}{m} = \binom{k}{m+1};$$

---

[22] According to [16, p. 311–314], the inclusion-exclusion principle was used in connection with the problem of derangements by A. De Moivre in 1718. In fact, Problem XXV in [100, p. 59–63] treats the more general task to find the probability that in a permutation of n elements $p \in [n]_0$ remain fixed and $q = n - p$ are out of their place. There is no reference to the Jeu du Treize or the Game of Encounter, nor to Montmort in [100]. The frontispiece [16, p. 308] shows the title page of [100], but the excerpt on [16, p. 313] is taken, without notice, from the second edition of De Moivre's work from 1738; the reference [437, p. 375] gives the impression that both citations stem from one source!

this formula can also easily be proved by induction. We now get for $k \in \mathbb{N}$:

$$
\begin{aligned}
f_k = \sum_{\ell=1}^{k} f_{k,\ell} &= \sum_{\ell=1}^{k} \sum_{m=0}^{\ell-1} (-1)^m \binom{\ell-1}{m} (k-1-m)! \\
&= \sum_{m=0}^{k-1} \sum_{\ell=m+1}^{k} (-1)^m \binom{\ell-1}{m} (k-1-m)! \\
&= \sum_{m=0}^{k-1} (-1)^m (k-1-m)! \sum_{\ell=m+1}^{k} \binom{\ell-1}{m} \\
&= \sum_{m=0}^{k-1} (-1)^m \binom{k}{m+1} (k-1-m)! \quad = \quad k! \sum_{m=0}^{k-1} \frac{(-1)^m}{(m+1)!}.
\end{aligned}
$$

Hence, A's probability to win is $\dfrac{f_k}{k!} = \sum\limits_{m=0}^{k-1} \dfrac{(-1)^m}{(m+1)!}$. Euler ends by remarking that this value tends, for $k \to \infty$, to $1 - \dfrac{1}{e} \approx 0.632$, where e is, very appropriately so, *Euler's number.*

Impressive as this derivation is, there is an easier way to obtain the solution. It is again Euler who, 28 years later, returns to this "curious question" in [141]. In fact, observing that f_k is just the number of permutations σ on $[k]$ which have at least one *fixed point*, i.e. an $i \in [k]$ such that $\sigma(i) = i$, we may equally well consider the complementary number $\overline{f}_k = k! - f_k$ of **derangements** of $[k]$. The reader is invited to follow this approach in Exercise 0.3, where the *subfactorials* $k\mathrm{i}$ are introduced.

Another interesting number scheme can be based on the method of *intercalation*, proposed in connection with a number theoretical function by G. Eisenstein [124]. On his request this has been elaborated by M. Stern in [389], unfortunately only after the former's untimely death. Stern sets out from a pair (a, b) of natural numbers and defines a(n infinite) sequence $((a,b)_n)_{n \in \mathbb{N}_0}$ of (finite) sequences $((a,b)_n(\mu))_{\mu \in [2^n+1]_0}$ in the following way: $(a,b)_0(0) = a$, $(a,b)_0(1) = b$, and the next sequences are obtained by successively intercalating the sum of two neighboring numbers. This can be arranged in an array and will be called *Stern's diatomic array* for the two *atoms* (or *seeds*) a and b. The examples investigated by Stern, namely $(1,1)$ and $(0,1)$ are shown in Figure 0.13.

Contrary to the AT, the method of intercalation preserves all entries from one row to the next and therefore their lengths grow exponentially instead of only linearly in AT. Among the many interesting properties of these arrays which Stern describes, we note that the $(1,1)$ array is, of course, symmetric with respect to the center column:

$$
\forall\, \mu \in [2^n + 1]_0 : \ (1,1)_n(2^n - \mu) = (1,1)_n(\mu),
$$

so it suffices to determine half of the entries, which can be done recursively using the $(0,1)$ scheme:

$$
\forall\, \mu \in [2^n + 1]_0 : \ (1,1)_{n+1}(\mu) = (1,1)_n(\mu) + (0,1)_n(\mu).
$$

1																1
1								2								1
1				3				2				3				1
1		4		3		5		2		5		3		4		1
1	5	4	7	3	8	5	7	2	7	5	8	3	7	4	5	1

0																1
0								1								1
0				1				1				2				1
0		1		1		2		1		3		2		3		1
0	1	1	2	1	3	2	3	1	4	3	5	2	5	3	4	1

Figure 0.13: Stern's diatomic arrays for $(1,1)$ (top) and $(0,1)$ (bottom) and $n \in [5]_0$

This shows that all the information is contained in the $(0,1)$ array which has the property that the beginning segment is preserved in all later rows, i.e. for every $m \geq n$ we have

$$\forall \mu \in [2^n + 1]_0 : (0,1)_n(\mu) = (0,1)_m(\mu).$$

Therefore, the following sequence $(0,1)_\infty$ on $\mathbb{N}_0$, which is called *Stern's diatomic sequence*, is well defined:

$$\forall \mu \in [2^n + 1]_0 : (0,1)_\infty(\mu) = (0,1)_n(\mu).$$

We will meet this sequence on several occasions throughout the book.

The first practical use of Stern's arrays was made implicitly by the French clock maker Louis-Achille Brocot in his seminal account [65], in which he proposed a new method to calculate gearings of pendulum clocks. His contemplations were caused by the problem to repair worn movements and in particular to find the ratios of pinions and angular gears. For practical reasons the fractions involved should have small numerators and denominators, since they are referring to the number of teeth. Brocot also wanted to make his description of the method accessible to those of his colleagues who were not well trained in mathematics. All this led him to consider what generations of pupils assumed to be the most natural method to *add* two fractions, namely by adding the respective numerators and denominators to arrive at the resulting fraction; after all this is the analogue way to *multiply* fractions! However, applying this "addition" to two positive fractions will not lead to a larger one, but to one in between the two; Lucas [288, p. 466–469] therefore called it *mediation* and the resulting fraction

$$\frac{a}{b} \oplus \frac{c}{d} := \frac{a+c}{b+d}$$

the *mediant* of $\dfrac{a}{b}$ and $\dfrac{c}{d}$. It is easy to see that in fact

$$\frac{a}{b} < \frac{c}{d} \Rightarrow \frac{a}{b} < \frac{a}{b} \oplus \frac{c}{d} < \frac{c}{d}.$$

Lucas then [288, p. 469f] defines the *Brocot sequence* $(\beta_n)_{n\in\mathbb{N}_0}$, each element of which consists of a finite sequence of fractions as arranged into the array of Figure 0.14. (Indices have been adapted to our notation.)

$\dfrac{0}{1}$																$\dfrac{1}{1}$
$\dfrac{0}{1}$								$\dfrac{1}{2}$								$\dfrac{1}{1}$
$\dfrac{0}{1}$				$\dfrac{1}{3}$				$\dfrac{1}{2}$				$\dfrac{2}{3}$				$\dfrac{1}{1}$
$\dfrac{0}{1}$		$\dfrac{1}{4}$		$\dfrac{1}{3}$		$\dfrac{2}{5}$		$\dfrac{1}{2}$		$\dfrac{3}{5}$		$\dfrac{2}{3}$		$\dfrac{3}{4}$		$\dfrac{1}{1}$
$\dfrac{0}{1}$	$\dfrac{1}{5}$	$\dfrac{1}{4}$	$\dfrac{2}{7}$	$\dfrac{1}{3}$	$\dfrac{3}{8}$	$\dfrac{2}{5}$	$\dfrac{3}{7}$	$\dfrac{1}{2}$	$\dfrac{4}{7}$	$\dfrac{3}{5}$	$\dfrac{5}{8}$	$\dfrac{2}{3}$	$\dfrac{5}{7}$	$\dfrac{3}{4}$	$\dfrac{4}{5}$	$\dfrac{1}{1}$

Figure 0.14: Brocot array for $n \in [5]_0$

We easily recognize that $\beta_n(\mu) = \dfrac{(0,1)_n(\mu)}{(1,1)_n(\mu)}$ for $\mu \in [2^n+1]_0$. G. Halphén [174, Lemma 1] and Lucas [288, p. 470–474] observed that all rationals from $[0,1]$ can be found as fractions in lowest terms in Brocot's array; in fact, in row n all (proper) fractions with denominator up to $n+1$ can be found (for the first time). This shows that all of Brocot's goals have been reached.

Another interesting consequence ([174, p. 175], [288, p. 474]) is that the number of occurrences of $n \in \mathbb{N}$ in the sequence $((1,1)_{n-1})_{\mu\in[2^{n-1}]}$ is $\varphi(n)$, where φ is *Euler's phi function* that counts the number of positive integers not greater than and relatively prime to n. This and the properties of the Brocot array had already been anticipated in Stern's article [389], but his focus was not on fractions.

During the 19th century, number theorists became interested in the divisibility of combinatorial numbers. In Figure 0.15 squares with even entries have been colored white, those with odd entries black.

The AT mod 2 can, of course, be more easily obtained by using the addition of $\mathbb{B}$ in Halāyudha's construction. This has been interpreted as the action of a cellular automaton by S. Wolfram in [441] (cf. also his *Rule 90* in [442, p. 25f, 610f]) or as the conversion of local interaction to global order by B. Mandelbrot

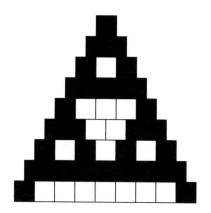

Figure 0.15: The Meru-Prastāra mod 2

[302, p. 328f]; see Figure 0.16, where a single + in the 0th line engenders the familiar pattern in later lines. This is a possible explanation for the development of color patterns in sea shells (Figure 0.17); cf. [309].

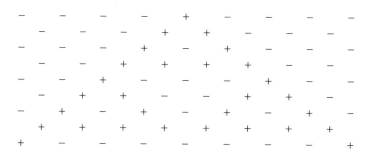

Figure 0.16: AT mod 2 as produced by a cellular automaton

But there is even a way to tell whether $\binom{k}{\ell}$ is odd or even without hunting down Mount Meru, based on a consequence of famous results of E. E. Kummer (cf. [256, p. 115f]) and Lucas (cf. [281, Section XXI]), namely[23]

$$\binom{k}{\ell} \bmod 2 = \prod_{i=0}^{\infty} (\ell_i \le k_i), \qquad (0.17)$$

where k_i and ℓ_i are the bits of k and ℓ, respectively; moreover, we have employed *Iverson's convention* that[24]

$\quad (\mathfrak{S}) = 1$, if statement $\mathfrak{S}$ is true, and $(\mathfrak{S}) = 0$, if $\mathfrak{S}$ is false.

---

[23] By $n \bmod p \in [p]_0$ we denote the remainder of division of n by p.

[24] Some authors use square brackets; see, for instance, [246].

Figure 0.17: Shell of *Lioconcha castrensis* (Linnaeus, 1758), Bohol, The Philippines
© 2017 A. M. Hinz

In practice, the product in (0.17) is, of course, finite because only finitely many
bits of k or ℓ are non-zero.

Equation (0.17) says that $\binom{k}{\ell}$ is odd if and only if every bit of ℓ is less than or
equal to the corresponding bit of k. For instance, since $14 = 1110_2$ and $10 = 1010_2$,
we immediately infer that $\binom{14}{10}$ is odd without calculating its value. If we did, it
would come as no surprise that *One Thousand and One Nights* is full of oddities.
For a more elaborate example, see Exercise 0.4.

A sketch of proof for Lucas's identity (0.17) can be inferred from looking at
Figure 0.16: by induction it can be shown that, starting from $\binom{0}{\ell} = (\ell = 0)$, for
every $n \in \mathbb{N}_0$ rows 2^n to $2^{n+1} - 1$ of AT mod 2 can be obtained from rows 0 to
$2^n - 1$ by taking two copies of these, shifting them 2^n places down and 2^n places to
the left and to the right, respectively, and write + precisely at those places where
+ and − meet. (Actually, the combination ++ does not occur in the two copies.)
But this means that $\binom{2^n+k}{\ell}$ and $\binom{2^n+k}{2^n+\ell}$ both have the same parity as $\binom{k}{\ell}$ for all
$k \in [2^n]_0$ and $\ell \in \mathbb{N}_0$, whereas $\binom{k}{2^n+\ell} = 0$. Again induction proves (0.17).

In the 20th century, people were able to draw ever taller Meru Mounts mod 2
(see Figure 0.18 for one with 65 lines) and discovered an interesting pattern. It
looks like an object introduced by the Polish mathematician W. Sierpiński back
in 1915 ([380]; cf. also [381, p. 99–106]) for a totally different purpose. Sierpiński
wanted to construct an example of a curve all of whose points are ramification
points. He achieved this by repeatedly replacing a line segment by a twofold broken
line segment and rescaling as in the upper row in Figure 0.19.

Figure 0.18: Mount Meru mod 2 with 65 lines

Figure 0.19: Three ways to construct the Sierpiński triangle

The limiting object has been given the disgusting name *Sierpiński gasket* by Mandelbrot [301, p. 56] from the French "jointe de culasse de Sierpiński" in [300]. He was led to this inelegant designation (cf. [302, p. 142]) by the second of three methods to obtain this mathematical object, namely starting from a filled (equilateral) triangle to cut out the open middle (or *medial*) triangle and repeating this for the remaining triangles ad infinitum (see the center row in Figure 0.19).

It is remarkable that 800 years before Sierpiński the artists of the Cosmati family in Italy had already anticipated this construction by producing examples of the third repeating step. One of them is shown in Figure 0.20. This picture, which demonstrates that beauty has a place in mathematics, discloses how tasteless Mandelbrot's naming really is.

Figure 0.20: Cosmatian floor decoration in San Clemente, Rome
© 2012 A. M. Hinz

We will therefore call this most fascinating mathematical object, which is now one of the prototypes of a *fractal*, the *Sierpiński triangle* (ST). A third way to construct it is by successively adding boundary lines of triangles as shown in the bottom row of Figure 0.19 and taking the closure of the emerging set of points in the plane. Each of the three constructions shows that ST is *self-similar*, i.e. each of the three subtriangles has the same mathematical structure as the whole object, but with lengths reduced by the factor 1/2 and "areas" reduced to 1/3 of their original value. (This intuitive notion of area has to be made precise in the sense of a *measure*, which would lead to a *Hausdorff dimension* of $\ln(3)/\ln(2)$ for the ST. More on this below in Section 4.3.1.)

But what, the reader might ask, is the relation with the topic of this book, the TH? For that one had to wait for the article [392] by Ian Stewart (cf. also [394]). To understand this relation, the concept of a **graph** in connection with puzzles has to be introduced in Section 0.7. We begin with some basic facts about elementary types of graphs.

0.6 Elementary Graphs

We first define the notion of a *multigraph* G on a, not necessarily finite, set V of *vertices* (singular *vertex*). Apart from V we need a mapping $e : \binom{V}{2} \to \mathbb{N}_0$. Then $E := \left\{ uv \in \binom{V}{2} \mid e(uv) \neq 0 \right\}$, where we write uv for $\{u, v\}$, is the set of *edges* of G. If necessary, *multiple edges*, where $e(uv) \geq 2$, can be characterized by specifying uv_k for $k \in [e(uv)]$. We do not consider the case where the multiplicity of an edge may be infinite. It is also sometimes useful to allow for *loops*, i.e. edges $vv := \{v\}$, joining one and the same $v \in V$, in which case e has to be defined on $\binom{V}{1} \cup \binom{V}{2}$; we then talk about a *pseudograph*. If $e\left(\binom{V}{1}\right) \subseteq \{0\}$ and $e\left(\binom{V}{2}\right) \subseteq \{0, 1\}$, we are back at *simple graphs* $G = (V, E)$. Finally, we define the *degree* of a vertex v as $\deg(v) = e(vv) + \sum_{u \in V} e(uv) \in \mathbb{N}_0$, if $\{u \in V \mid uv \in E\}$ is finite.

0.6.1 The Handshaking Lemma

A (pseudo/multi)graph G is called *finite*, if $|V| \in \mathbb{N}_0$; this number is then the *order* $|G|$ of G. The *size* of a finite graph is $\|G\| := \sum_{v \in V} e(vv) + \sum_{uv \in \binom{V}{2}} e(uv)$. We then have:

Theorem 0.7. (Handshaking Lemma) *If G is a finite (pseudo/multi)graph, then*

$$\sum_{v \in V} \deg(v) = 2 \|G\| .$$

Proof. Recall Iverson's convention from p. 29. We have

$$
\begin{aligned}
\sum_{v \in V} \deg(v) &= \sum_{v \in V} e(vv) + \sum_{v \in V} \sum_{u \in V} e(uv) \\
&= 2 \sum_{v \in V} e(vv) + \sum_{v \in V} \sum_{u \in V} (u \neq v) \cdot e(uv) \\
&= 2 \sum_{v \in V} e(vv) + 2 \sum_{uv \in \binom{V}{2}} e(uv) = 2 \|G\| . \qquad \square
\end{aligned}
$$

An alternative proof can be based on the method of double counting, namely in this case the elements of the set $X = \{(v, e) \in V \times E \mid v \in e\}$; here edges have to be counted according to their multiplicity and loops can be avoided by replacing them by *lunes*, i.e. subdividing a loop with precisely one extra vertex. Then for a multigraph:

$$|X| = \sum_{v \in V} \sum_{e \in E} (v \in e) = \sum_{v \in V} \deg(v)$$

but also

$$|X| = \sum_{e \in E} \sum_{v \in V} (v \in e) = 2 \|G\| .$$

Remark. The name of the lemma comes from the model of a party, where V is the set of people present and $e(uv)$ is the number of handshakes between participants

u and v (forgetful people may shake hands several times; some pairs of persons may not shake hands at all). Then $\|G\|$ is the total number of handshakes, $\deg(v)$ is the number of handshakes involving person v and $\sum_{v \in V} \deg(v)$ is the number of shaking hands (counted multiply). Theorem 0.7 is an expression of the fact that every handshake shakes two hands.

Corollary 0.8. *In a finite graph the number of* odd *vertices, i.e. vertices of odd degree,*[25] *is even.*

A *walk* of *length* $\ell \in \mathbb{N}_0$ in a graph $G = (V, E)$ is a $v \in V^{1+\ell}$ with $\forall\, k \in [\ell]: v_{k-1}v_k \in E$. If $v_0 = v_\ell$ it is called *closed*, otherwise it is a v_0, v_ℓ-*walk*. Walks of length 0 correspond to vertices, those of length 1 to edges of G. A walk can also be viewed independently of G as a (pseudo/multi)graph $W = (V_W, e_W)$ with $V_W = v([1 + \ell]_0)$ and $e_W : \binom{V_W}{1} \cup \binom{V_W}{2} \to \mathbb{N}_0,\ uv \mapsto |\{k \in [\ell] \mid uv = v_{k-1}v_k\}|$. Then $|E_W| = |V_W| - 1 \le \ell$.

Let $\widetilde{E} := \left\{ xy \in \binom{V}{2} \mid \exists\, x, y\text{-walk in } G \right\}$. A graph G is called *connected* if $\widetilde{E} = \binom{V}{2}$. For $x \in V$, the graph (G_x, e_x) with $V_x = \{x\} \cup \{y \in V \mid xy \in \widetilde{E}\}$ and $e_x = e \upharpoonright \binom{V_x}{2}$ $(e_x = e \upharpoonright \binom{V_x}{1} \cup \binom{V_x}{2}$ for pseudographs$)$ is called a *component* of G.

Proposition 0.9. *Every component of a graph G is connected. In a finite connected graph G we have $|G| \le \|G\| + 1$.*

Proof. The first statement is a consequence of the *transitivity* of $\widetilde{E}$: join an x, z-walk and a y, z-walk at z to obtain an x, y-walk.

The second statement can be proved by induction on $m = \|G\| \in \mathbb{N}_0$. If $m = 0$, then G cannot have more than one vertex because it is connected. Now let $\|G\| = m + 1$ and delete an edge to get G' with $\|G'\| = \|G\| - 1$ and $|G'| = |G|$. If G' is connected, then by induction assumption $|G| = |G'| \le \|G'\| + 1 = \|G\|$; otherwise, G' has two components G'_1 and G'_2, whence $|G| = |G'| = |G'_1| + |G'_2| \le \|G'_1\| + 1 + \|G'_2\| + 1 = \|G'\| + 2 = \|G\| + 1$ by induction assumption.

There is an alternative proof by successively deleting *pendant vertices*, i.e. vertices of degree 1, together with their incident edges. This does not destroy connectedness and neither the inequality. At the end (the graph is finite!) there is either a connected graph with minimal degree strictly larger than 1, whence $\sum_{v \in V} \deg(v) \ge 2|G|$, i.e. $\|G\| \ge |G|$, or (in addition) a single isolated vertex, i.e. $\|G\| \ge |G| - 1$. $\qquad\square$

If $|\{k \in [\ell] \mid uv = v_{k-1}v_k\}| \le e(uv)$ for all $u, v \in V$, then the (closed) walk is called a *(closed) trail*, which is *eulerian*, if equality holds; a closed trail is also called a *circuit*.

---

[25]Similarly, vertices of even degree are called *even*.

0.6.2 Finite Paths and Cycles

Let V be a finite set and $\ell := |V| + 1$.

A connected graph $P_{1+\ell}$ with $V(P_{1+\ell}) = V \cup \{u, w\}$ such that $\deg(u) = 1 = \deg(w)$ and $\deg(v) = 2$ for $v \in V$ is called a *u, w-path* (of *length ℓ*).

Equivalently, $P_{1+\ell}$ is a graph on $V(P_{1+\ell}) = \{v_0, \ldots, v_\ell\}$ with $v_0 = u$, $\{v_1, \ldots, v_{\ell-1}\} = V$, $v_\ell = w$, and $E(P_{1+\ell}) = \{v_{k-1}v_k \mid k \in [\ell]\}$.

Note that $u, w \notin V$ from the requirement on the degrees and that $u \neq w$ by Corollary 0.8; for definiteness we therefore define P_1 with $V(P_1) = \{u\}$ and $E(P_1) = \varnothing$ as the *u, u-path* (of *length* 0). All paths are simple graphs.

A path of length $\ell \in \mathbb{N}_0$ is also a trail P with $|P| = 1 + \ell = 1 + \|P\|$ and vice versa. Whenever there is a *u, v-walk* (of length ℓ) in G, there is also a *u, v-path* (of length $\leq \ell$) as a **subgraph** of G: if a vertex occurs more than once on the walk, just cut out all edges between the first and the last occurance.

A graph $C_{1+\ell}$ is called a *cycle* (of *length* $1 + \ell$), if there is a *u, w-path* $P_{1+\ell}$ such that $V(C_{1+\ell}) = V(P_{1+\ell})$ and $E(C_{1+\ell}) = E(P_{1+\ell}) \cup \{wu\}$. For $\ell = 0$ this means that C_1 is a vertex with a loop, i.e. $V(C_1) = \{u\}$ and $E(C_1) = \{uu\}$, leading to a pseudograph. Although we are dealing with undirected graphs, the definition of a cycle for $\ell = 1$ means that the edge uw is repeated, i.e. we get a lune and C_2 is a multigraph. All other cycles are simple graphs.

For definiteness one may also define the path and cycle of order (and length) 0 by

$$P_0 = (\varnothing, \varnothing) = C_0.$$

Lemma 0.10. *For a connected non-empty finite graph C the following are equivalent:*

(0) *C is a cycle.*

(1) *C is **2-regular**, i.e. $\forall v \in V(C): \deg(v) = 2$.*

(2) *C has no pendant vertex and $|C| = \|C\|$.*

(3) *C is a closed trail and $|C| = \|C\|$.*

Proof. We may assume $C \neq C_1$. (This is actually the reason for the convention that a loop adds 2 to the degree of a vertex.)

The implication "(0) $\Rightarrow$ (1)" follows from the properties of the underlying path; for "(1) $\Rightarrow$ (0)" delete an arbitrary edge from C and obtain an underlying path since connectedness was not destroyed.

The equivalence of (1) and (2) is a consequence of the Handshaking Lemma: if C is 2-regular, then $|C| = \|C\|$ and obviously there is no pendant vertex; if (2) is true, then, since C is without pendant vertices and connected, $\deg(v) \geq 2$ for all vertices and by the Handshaking Lemma $\deg(v) = 2$ for all v.

The equivalence of (3) and (0) follows from the corresponding characterization of paths. $\square$

Proposition 0.11. *A finite graph with all vertices of degree* 2 *except for two pendant vertices is the union of a path with some (maybe no) cycles.*

Proof. By the Handshaking Lemma, the two pendant vertices must belong to the same component, which consequently is a path. The rest, if any, of the graph is 2-regular and therefore a union of cycles.

Instead of using the Handshaking Lemma one may start at one of the pendant vertices, say u, and construct a longest path (exists because u has a neighbor and the graph is finite), which necessarily ends in the other pendant vertex.   □

An application of Proposition 0.11 is in the theory of the Chinese rings; see Chapter 1.

0.6.3   Infinite Cycles and Paths

A connected 2-regular infinite graph is called an *infinite cycle*. A connected graph with maximum degree 2 and exactly one pendant vertex u is called a *u-path*;[26] it is necessarily infinite by the Handshaking Lemma. We now have a complete overview of all graphs with maximum degree at most 2.

Proposition 0.12. *A (pseudo/multi)graph has* maximum degree $\Delta \leq 2$, *if and only if all its components are paths or cycles.*

Proof. It is clear by definition that paths and cycles have $\Delta \leq 2$.

If $\Delta = 0$, then G is the union of isolated vertices, i.e. of P_1s. If $\Delta = 1$, then each component is either an isolated vertex P_1 or an edge, i.e. a path P_2 of length 1.

If $\Delta = 2$, then apart from P_1s every finite component has either no or exactly two pendant vertices by the Handshaking Lemma. (Since $\Delta \leq 2$, we have $2\|G\| = \sum_{v \in V} \deg(v) \leq 2(|G| - k)$, if $2k$ vertices, $k \in \mathbb{N}_0$, have degree 1; but from Proposition 0.9 we know that $\|G\| \geq |G| - 1$, whence $k \leq 1$.) Therefore Lemma 0.10 or Proposition 0.11 apply. An infinite component cannot have more than one pendant vertex: assuming that there are two, consider the connected subgraph consisting of these two, their incident edge(s), their neighbor(s) and a path between the latter; this leads to a finite path which cannot be extended because all vertices are saturated. By definition, the infinite component is either an infinite cycle or an infinite path.   □

A surprising application of Proposition 0.12 is in the proof of the *Cantor-Schröder-Bernstein Theorem* (cf. [41, p. 46–48], where the arguments are rather superficial though):

Theorem 0.13. *For two sets A and B let $f \in B^A$ and $g \in A^B$ be injective. Then there is a bijective $h \in B^A$.*

---

[26]This is compatible with the citation from Lao Tse (fl. -6th century): "a journey of a thousand miles begins with a single step".

Proof. If $A = B$, take $h = \mathrm{id}$. So let $A \neq B$ (then $A \neq \varnothing \neq B$). We first show that we may assume that A and B are disjoint. Define

$$\alpha: A \to A_* := \{(A, a) \mid a \in A\}, \; a \mapsto (A, a)$$

$$\beta: B \to B_* := \{(B, b) \mid b \in B\}, \; b \mapsto (B, b).$$

Obviously, A_* and B_* are disjoint and α and β are bijective. We define $f_* := \beta \circ f \circ \alpha^{-1} \in B_*^{A_*}$ and $g_* := \alpha \circ g \circ \beta^{-1} \in A_*^{B_*}$, which are injective. If we find a bijective $h_* \in B_*^{A_*}$, then $h := \beta^{-1} \circ h_* \circ \alpha \in B^A$ is bijective.

We now define the graph G with $V = A \cup B$ and $E = E_A \cup E_B$ with $E_A := \{\{a, f(a)\} \mid a \in A\}$, $E_B := \{\{b, g(b)\} \mid b \in B\}$; more precisely, $e(uv) = (uv \in E_A \triangle E_B) + 2(uv \in E_A \cap E_B)$ for $uv \in \binom{V}{2}$. This is a, possibly infinite, (multi)graph, and all vertices have degree 1 or 2. So we can apply Proposition 0.12. Since each vertex in a cycle must be incident with one edge from E_A and one from E_B, we can match in pairs the vertices from a cyclic component by only considering edges from E_A, say. A path component of G cannot be finite, because any u, v-path could be extended by $f(v)$ or $g(v)$, depending on whether $v \in A$ or $v \in B$, respectively. The vertices on an infinite u-path can be matched in pairs using only edges from E_A, if $u \in A$, or from E_B, if $u \in B$. □

Remark. For readers who don't like multigraphs, we note that[27] $f' := f \upharpoonright \mathrm{FP}(g \circ f)$ is a bijection to $\mathrm{FP}(f \circ g) \subseteq B$. If $a = g \circ f(a)$, then $f \circ g(f(a)) = f(a)$, so $f(a) \in \mathrm{FP}(f \circ g) \subseteq B$. As a restriction of an injective function, f' is injective. Finally, for $b \in \mathrm{FP}(f \circ g)$ let $a = g(b)$; then $g \circ f(a) = g \circ f(g(b)) = g(b) = a$, i.e. $a \in \mathrm{FP}(g \circ f)$, and $f(a) = f(g(b)) = b$, such that f' is surjective. So one might set $A' := A \smallsetminus \mathrm{FP}(g \circ f)$ and $B' := B \smallsetminus \mathrm{FP}(f \circ g)$ and define the graph G' by $V(G') = A' \cup B'$ and $E(G') = E_A \cup E_B$ with $E_A := \{\{a, f(a)\} \mid a \in A'\}$, $E_B := \{\{b, g(b)\} \mid b \in B'\}$; note that E_A and E_B are disjoint. Then G' is a simple, possibly infinite, graph and we can continue as before. A path cannot end in $v \in A'$ for the reason that $f(v) \in \mathrm{FP}(f \circ g)$ because then $v \in \mathrm{FP}(g \circ f)$ and similarly for $v \in B'$.

In the sequel we will restrict ourselves to finite graphs, because our main intention is to employ them as a mathematical model for puzzles whose ultimate goal is a finite solution.

0.7 Puzzles and Graphs

At an early stage, puzzles have been associated with the mathematical notion of a graph. In fact, a puzzle is considered to be the launch pad for graph theory, namely the famous *Königsberg bridges problem*; cf. [55, Chapter 1] and [438].

---

[27] The set of fixed points of a mapping f is denoted by $\mathrm{FP}(f)$.

0.7.1   The Bridges of Königsberg

The topography of the ancient Prussian city with its seven bridges linking four areas separated by the river Pregel led people to question whether it was possible to make a promenade crossing every bridge exactly once and maybe even to return to the starting point. The solution of this conundrum by L. Euler in 1735 was intriguing for several reasons. Firstly, the answer was in the negative, i.e. such a route is impossible, and secondly, this fact could be proved by a mathematical argument. In modern terminology, the problem is to find a (closed) trail which is eulerian, i.e. it contains all solid lines of the drawing of the multigraph depicted in Figure 0.21, which is a visualization with vertices represented by dots and edges by lines joining them. For instance, the line/edge c = {B, C} stands for a bridge between the quarters of Königsberg represented by dots/vertices B and C; between A and B there are multiple edges a and b.

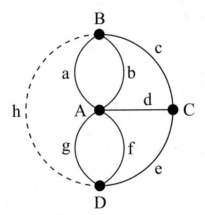

Figure 0.21: Königsberg multigraph drawn according to Euler [142, p. 153]

Quite obviously, an eulerian trail may only be found if the (multi)graph has exactly two vertices of odd degree or none at all if the trail is to be closed. This is because every vertex has to be left as many times as it has been entered, except perhaps for the start and end points of the tour. Comparing the tasks to finding an eulerian circuit or just an eulerian trail in Königsberg, R. J. Wilson once said of the latter that "it's still impossible, but it's easier" [440]. He was right, as can be seen by looking into a letter of Euler to G. Marinoni (see [142, p. 155-157]; cf. also [360] and [439, p. 504]), where Euler drew, apart from the bridges a to g of the original question, an extra passage h between B and D and remarked that this addition would allow for an eulerian trail, A and C then being the only vertices of odd degree. An eulerian circuit would still need an extra bridge between A and C (see Exercise 0.5), but Euler did not consider eulerian circuits. Actually, there *was* a passage in Königsberg in Euler's time at the location of h, called *Holländischer Baum* (Dutch boom),[28] later to become a railroad bridge; cf. also [2, p. 35]. Apparently Euler learned about this passage from a letter by C. L. G. Ehler [360, Figure 2] only *after* he had written his seminal paper on the bridges problem, which can be found in [250, p. 279–288] (for an English translation, see [55, p. 3–8]). Eulerian trails through the ages in Königsberg, or Kaliningrad as it is called today, can be found in [299]. Euler did not consider the question of sufficiency of the obvious necessary condition, which was eventually settled only in 1873 by C. Hierholzer (cf. [55, p. 11f]).

An elegant constructive argument leads to *Fleury's algorithm* (cf. [290, p. 134f], [60, p. 87f]), which we will give here as an example of how algorithms will be presented throughout the book.

Algorithm 1 Fleury's algorithm

Procedure $\mathrm{fl}(G, a)$
Parameter G: non-empty (connected) graph
Parameter a: initial vertex $a \in V(G)$
$\quad t \leftarrow a$ $\hfill$ {trail}
$\quad$ **while** $\deg(a) \neq 0$
$\quad\quad$ **choose** any edge $ab \in E(G)$, but avoid bridges, if possible
$\quad\quad t \leftarrow tb,\ G \leftarrow G - ab,\ a \leftarrow b$
$\quad$ **end while**

We start at any vertex a of the (multi)graph G, repeatedly add an edge to the trail to be constructed and recommence at the other end vertex of that edge, but avoid using a **bridge** of the currently examined graph (not to be confused with the bridges in Königsberg, none of which is a bridge of the Königsberg graph) unless there is no other choice. Edges traversed are deleted from the graph during the process, $G - ab$ being the graph obtained from G by deleting edge $ab := \{a, b\}$, i.e. $V(G - ab) = V(G)$ and $E(G - ab) = E(G) \smallsetminus \{ab\}$.

It is easy to apply Algorithm 1 for the graph in Figure 0.21 (see Exercise 0.6) and in fact it solves the general problem of finding an eulerian trail (circuit) and can, e.g., be employed to show that, just like the well-known pentagram of the Pythagorean school, the *sign manual of Muhammad* of Figure 0.22, showing two opposing crescents, can be drawn in one stroke as according to legend (cf. [283, p. 36]) the prophet had done in the sand with the tip of his scimitar; see Exercise 0.7.

Every algorithm needs a *correctness proof*, i.e. an argument that it produces

---
[28] An inscription (in place in 2016) at the harbor of Stade, a city in the North of Germany which like Königsberg was a member of the *Hanseatic League*, explains the functioning of a toll barrier. The harbor master resided in the *boom house* from 1609 to 1948. The habor entry was closed by a tree stem laid across, whose movements were carried out by the boom keeper from a bascule bridge. By analogy we may assume that there was such a bridge at h in Figure 0.21 blocking the passage of ships mainly from Holland. (A corresponding *Lithuanian boom* was present in the East of Königsberg.)

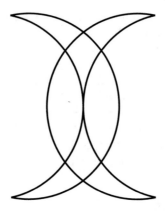

Figure 0.22: The sign manual of Muhammad

after finitely many steps the result it was designed for. For Algorithm 1 we make use of complete induction. The decisive step for an induction proof is the choice of the appropriate statement depending on the induction variable. Here we claim the following for any connected graph G and $a \in V(G)$:

(**A**) *If all vertices of G are even, then $fl(G,a)$ produces an eulerian circuit starting and ending at a.*

(**B**) *If a and $c \neq a$ are the only odd vertices of G, then $fl(G,a)$ produces an eulerian trail from a to c.*

We use the size $m = \|G\| \in \mathbb{N}_0$ of the graph as the induction variable.

If $m = 0$, then G consists of the single vertex a whose degree is 0, such that $fl(G,a)$ will stop with $t = a$, the trivial eulerian circuit.

Now let (A) and (B) be true for all non-empty connected graphs G' with $\|G'\| \leq m \in \mathbb{N}_0$ and let $\|G\| = m + 1$. Then $\deg(a) \neq 0$. In case (A), G has no bridge, because its deletion would lead to two connected components with one odd vertex each, but the number of odd vertices in a graph is always even as a consequence of the Handshaking Lemma, which Euler formulated in his Königsberg note (cf. Paragraph 16 in [55, p. 7]). If edge ab is chosen by the algorithm, then $G' = G - ab$ is connected, has m edges and is in case (B) with a replaced by b and c by a, such that, by induction assumption, the algorithm as applied to G' results in an eulerian trail $b \ldots a$. Hence $fl(G,a)$ produces an eulerian circuit $ab \ldots a$. If in case (B) the chosen edge ab is not a bridge, then G' is connected, has m edges and is in case (A) or (B) for b instead of a, depending on whether $b = c$ or not, such that $fl(G,a)$ produces an eulerian trail $ab \ldots c$ as before. If, however, ab is a bridge, then c lies in the component G_b of G' which contains b, because otherwise b would be its only odd vertex. Since the bridge was unavoidable, the degree of a in G is 1 and $G_b = G - a$, the graph obtained from G by deleting vertex a and its incident edge. But then, depending on whether $b = c$ or not, we are in case

(A) or (B) for G_b and by induction assumption $fl(G, a)$ yields a trail $ab \ldots c$. This completes the induction proof.

Note that one cannot drop the "avoid bridge" condition in Algorithm 1, because a component still containing edges which were not visited can never be entered again after crossing a bridge. Moreover, if neither the condition in (A) nor the one in (B) is fulfilled, the algorithm will nevertheless stop, because in every recurring step one edge is deleted; the resulting trail will be non-eulerian though, because these conditions are necessary for eulericity. The same applies if G is not connected.

For a practical application we need, of course, an efficient method to decide whether an edge is a bridge. This can be based on a *breadth-first search (BFS)*, as demonstrated in Exercise 0.8. The number of steps needed depends polynomially on the number of input data, and since Algorithm 1 can be viewed as a decision tool for eulericity, the latter problem belongs to *complexity class* **P** (cf. [84, p. 87]).

The existence of such an algorithm has not (yet!) been established for another famous problem, namely to find a **hamiltonian path** or a **hamiltonian cycle** in a connected graph. (A graph that contains a hamiltonian cycle is called *hamiltonian*.)

0.7.2 The Icosian Game

W. R. Hamilton's intention was rather more serious when he presented his *Icosian game* (see Figure 0.23 for the original), "of which he was inordinately proud" ([55, p. 31]), in Dublin in 1857, and which was subsequently marketed by Jaques and Son, London. It is said that Hamilton received £ 25 from the dealer, "the only pecuniary reward ever accruing to Hamilton directly from any discovery or publication of his" ([163, p. 55]), and that it was a bad bargain—for the dealer! ([55, p. 31])

One type of tasks given in the leaflet of instructions (to be found, together with examples and hints in [55, p. 32–35]) is to complete a cycle on the *dodeca-hedral graph* (or *Icosagonal* as Lucas has called it in one of his scientific games) in Figure 0.24 when 5 first vertices are prescribed using all other 15 vertices with no vertex occurring more than once. ("Icosian" derives from the Greek εἴκοσι for "twenty".)

This graph is the **1-skeleton** (a drawing of which being obtained by stereographic projection onto the plane) of the dodecahedron, one of the five Platonic bodies. One can imagine a round trip through the 20 vertices of that body/graph along the edges, never visiting a place twice. The use of the words "vertex" and "edge" for the constituting elements of graphs now becomes transparent.

The purpose of the Icosian game was to popularize the *Icosian calculus*, which Hamilton presented in two very short notes [175, 176] in 1856 as an extension to his theory of quaternions. It is based on the *Icosian group* generated by ι, κ, λ

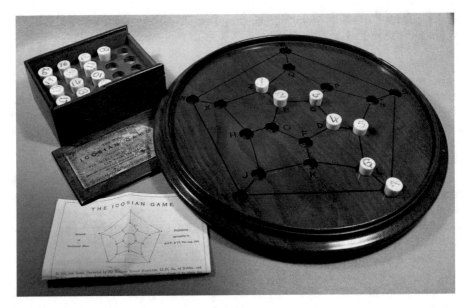

Figure 0.23: Hamilton's Icosian game
(courtesy of James Dalgety, http://puzzlemuseum.org)
© 2012 Hordern-Dalgety Collection

fulfilling

$$1 = \iota^2 = \kappa^3 = \lambda^5 \text{ and } \iota\kappa = \lambda \neq \kappa\iota \, ;$$

obviously, this group is non-commutative. It has order 60 and is isomorphic to the *alternating group* Alt_5, the subgroup of the *symmetric group* Sym_5 (see Exercise 0.9) containing only the *even* permutations of $[5] = \{1, 2, 3, 4, 5\}$, i.e. those which can be expressed as the composition of an even number of **transpositions**; cf. [21, p. 29]. In this interpretation one might put

$$
\begin{aligned}
\iota &\cong 21435 &=& \quad (12) \circ (34) &=& \quad (12)(34) \, , \\
\kappa &\cong 32541 &=& \quad (15) \circ (13) &=& \quad (135) \, , \\
\lambda &\cong 23451 &=& \quad (15) \circ (14) \circ (13) \circ (12) &=& \quad (12345) \, ,
\end{aligned}
$$

where the sign $\cong$ means that the symbols ι, κ, λ are identified with the corresponding permutations σ in Alt_5, given in the form $\sigma = \sigma_1\sigma_2\sigma_3\sigma_4\sigma_5$, and alternatively in cyclic representation showing only the *cycles* of the permutation, i.e. the **cyclic permutations** on disjoint sets which make up the permutation. The group Alt_5 in turn is isomorphic to the rotational symmetry group of the dodecahedron, as can be seen by looking, in Figure 0.25, at the 5 inscribed cubes whose 12 edges are diagonals of the pentagonal faces of the dodecahedron (for details, see [21, p. 40]).

This allowed Hamilton to interpret his principal symbols ι, κ, λ to which he added $\mu := \lambda\kappa \cong 25413 = (12534)$, as operations (executed from left to right in

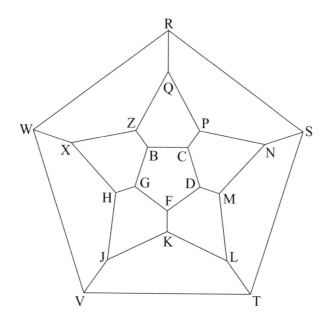

Figure 0.24: Dodecahedral graph with Hamilton's labeling

Hamilton's convention; e.g., $\lambda\kappa \cong 32541 \circ 23451$) on the edges of the dodecahedron: arriving at a vertex of a walk, ι means to go back on the edge just traversed, λ means continuing on the edge to the right, and μ represents the corresponding left turn. This is a complete description of walks on the dodecahedral graph which is 3-regular. For instance, $\omega := \lambda\mu\lambda\mu\lambda = \mu\lambda\mu\lambda\mu$ leads to the antipodal vertex. Cycles being characterized by terms of the form $\gamma = 1$ in the Icosian group, Hamilton was now able to identify the only cycles of order 20 which do *not* contain a strict subsequence of consecutive operations leading to a cycle; they are

$$\left(\lambda^3\mu^3(\lambda\mu)^2\right)^2 \ (=1),\tag{0.18}$$

$$\left(\mu^3\lambda^3(\mu\lambda)^2\right)^2 \ (=1),\tag{0.19}$$

differing only by exchanging left and right. (For an alternative approach to uniqueness of these cycles see [284, p. 210–217].) Thus, starting with the edge from B to C, Equation (0.18) produces the alphabetic path, whereas Equation (0.19) leads to

$$\text{B C D F G H X W R S T V J K L M N P Q Z}.$$

These are the only two hamiltonian cycles starting B C D F G. Now given any path of length 4, i.e. given 5 initial vertices, the last 3 edges correspond to a triple of symbols λ and μ. (The symbol ι cannot occur on a path and in fact, λ and μ also generate the whole group alone, since $\iota = \lambda\mu^4\lambda$ and $\kappa = \lambda^4\mu$.) All these triples

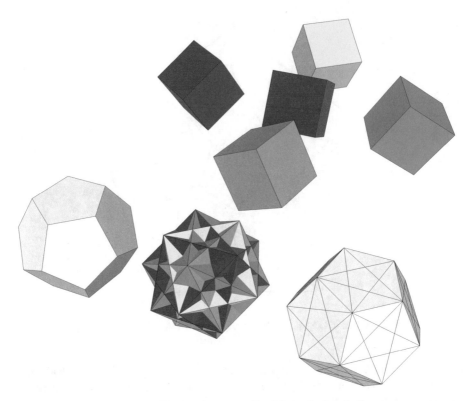

Figure 0.25: Five cubes inscribed in a dodecahedron

do actually occur in the terms of order 20 in (0.18) and (0.19) and can therefore be continued, some of them even in two non-equivalent ways. So there are either two or four solutions for every task; see Exercise 0.10. With 24 different initial paths of length 4 which start in B, say, 6 of which leading to two different completed cycles each, we have 30 hamiltonian cycles on Hamilton's dodecahedral graph. (We are disregarding orientation of the cycles.)

Hamilton's group theoretic approach has been employed, e.g., for symmetry questions (cf. [421, Chapter 6] and [53]), but it can not be used to find hamiltonian cycles in general graphs. As said before, there is as yet no polynomial-time algorithm to decide whether a graph is hamiltonian, although it is, of course, easy to check if a walk in the graph, obtained by any kind of method, is in fact a hamiltonian cycle. This is a typical feature of a problem of complexity class **NP**. It can even be shown that it is **NP**-*complete* (see [432, Corollary B.11]), i.e. finding a polynomial-time algorithm solving the decision would settle the outstanding question "**P**=**NP**?", one of the Millennium Prize Problems worth a million dollars each; cf. [84].

0.7.3 Planar Graphs

Let us go back to old Königsberg and instead of crossing bridges draw a city map with the four quarters *Kneiphof* (A), *Altstadt/Löbenicht* (B), *Lomse* (C), and *Vorstadt* (D). They are separated from each other by the Pregel river as shown in Figure 0.26, where we have added the *dual* graph (cf. [432, p. 236–240]), each quarter being represented by a vertex, two of which are joined by an edge if these quarters have a border in common. You may think of the centers of the areas which are pairwise connected by roads running through the common border.

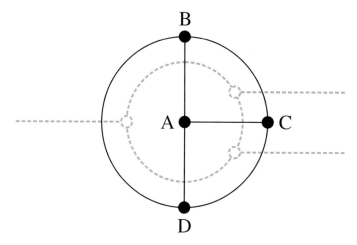

Figure 0.26: City map of Königsberg and its dual graph

The name *dual* for the graph associated with a map derives from the following fact: if we define the center of each face of the dodecahedron to be a vertex and join two such vertices by an edge if the corresponding faces share a common edge, we get the *dual* Platonian body, namely the icosahedron. In other words, the 1-skeleton of the icosahedron is the dual graph of the 1-skeleton of the dodecahedron, i.e. Hamilton's graph in Figure 0.24, and vice versa.

The dual graph of a map is always *planar*, i.e. its edges do not cross outside vertices.[29] This seemingly obvious fact (cf. [384, p. 168]) is actually rather subtle, but we also want to skip the technical details about what a border is mathematically; see [154] for topological details. Suffice it to say that a map can be viewed as a graph as well, namely the dual graph of its dual graph. Its vertices are those points where borders meet as in Figure 0.26, in which you have to think of an extra blue vertex at the antipode of black vertex A, say, where the three open ending blue edges meet. (In this picture, the whole plane or, by stereographic projection again, the entire sphere belongs to Königsberg; if you want to include city limits,

---

[29]We will give a more formal definition of planarity of (simple) graphs later in Section 0.8.1, where we will also specify drawings more carefully.

you have to add three blue vertices and a black one together with the appropriate
edges to the drawing in Figure 0.26.)

As is conventional in cartography, two areas with a common border, in our
case a leg of the river, have to be colored with different colors. This is now equivalent to coloring the vertices of the dual graph in such a way that two adjacent
vertices have different colors. But in the Königsberg graph every pair of the 4 vertices is joined by an edge; this graph, which obviously needs 4 colors to be colored
properly, is denoted by K_4. (K does *not* stand for Königsberg, but for **complete**
graph because C is used for cycle graphs.) The famous question is now: can *every*
map, or every planar graph for that matter, be colored properly with at most
4 colors (cf. [55, Chapter 6], [384, p. 146], [436])? Even before this question was
posed for the first time in 1852, A. F. Möbius asked whether a kingdom could be
divided among 5 heirs, such that each of the 5 new countries has a border with
every other one. If this were possible, then K_5, the complete graph over 5 vertices,
would be a planar graph which is not 4-colorable. So was the *Four color conjecture*
refuted before it was enunciated?

To answer Möbius's question, we need another ingenious observation of Euler's. If in addition to the number of vertices v and the number of edges e we also
count the number of faces f of a convex polyhedron[30] or, which amounts to the
same, the corresponding numbers for its 1-skeleton (where we must not forget the
infinite face if its drawing is in the plane, which corresponds to the (finite) face
on the sphere containing the projection center), we get the amazing formula

$$v - e + f = 2,   \tag{0.20}$$

called *Euler's polyhedral formula* and which is no less than a mathematical gem
(cf. [351, p. 65–73]). Its proof for (connected) planar graphs needs tools not available to Euler though, in particular the *Jordan curve theorem* (which is not named
for a river, but for the 19th century French mathematician C. Jordan), e.g., to
show that an edge separates precisely two faces (cf. [319, p. 25]). Note that the
fact that (0.20) applies to *all* plane drawings of the planar graph G implies that
the number of faces $f = |||G|||$ is the same for all drawings, because the order $v = |G|$
and the size $e = \|G\|$ of G are independent of the drawing.

We can use Equation (0.20) to prove that K_5 is not planar: assume that K_5
is planar; then $3|||K_5||| \le 2\|K_5\|$, because each face is bordered by at least 3 edges
(there are no loops nor multiple edges in the graph), each of which belongs to at
most 2 faces. (The latter again makes use of Jordan's curve theorem!). From (0.20)
it follows that $10 = \|K_5\| \le 3(|K_5| - 2) = 9$, which is obviously a contradiction.

The fact that there are non-planar graphs led to a characterization of connected graphs according to their (topological) *genera*, namely the value of $g \in \mathbb{N}_0$
such that

$$v - e + f = 2(1 - g),$$

---

[30] In Euler's time, mathematicians knew their classical Greek, according to which a polyhedron
has "many seats" and therefore had to be convex, otherwise it could not sit on them. Calling
non-convex solids "polyhedra" is a phantasm of later generations of mathematicians.

where obviously $g = 0$ for planar graphs and, e.g., $g = 1$ for K_5; of course, one has to make precise what a face is for non-planar graphs (cf. [55, Chapter 5]). The quest for the genus $g(G)$ of a graph G turned out to be rather delicate. For instance, a conjecture pronounced implicitly in 1890/1 by P. J. Heawood and L. Heffter (cf. [55, Chapter 7]), namely that for complete graphs we have

$$\forall p \in \mathbb{N}_0 : g(K_p) = \left\lceil \frac{(p-3)(p-4)}{12} \right\rceil ,$$

was proved not earlier than in 1968 by G. Ringel and J. W. Youngs; the complete proof takes almost a whole book; cf. [352].

The genus of a non-planar graph G is related to the task to draw it without a crossing of edges outside vertices on a surface of different topological type than the plane (cf., e.g., [180, Chapter 10]). The other way out, namely allowing for crossings in a drawing in the plane, but trying to keep their number small, is even more intricate. The smallest such number, taken over all (proper) drawings, is called the *crossing number* of G and denoted by $\mathrm{cr}(G)$. Contrary to the genus, it is, in general, easier to obtain an upper bound for the crossing number by "simply" finding a good drawing. It was, e.g., found out in the 1960s (cf. [172, p. 115]) that

$$\forall p \in \mathbb{N}_0 : \mathrm{cr}(K_p) \le \frac{1}{4} \left\lfloor \frac{p}{2} \right\rfloor \left\lfloor \frac{p-1}{2} \right\rfloor \left\lfloor \frac{p-2}{2} \right\rfloor \left\lfloor \frac{p-3}{2} \right\rfloor . \tag{0.21}$$

However, *Guy's conjecture* that equality holds in (0.21), has not been confirmed until today but for a small number of values of p "and in fact the proofs for $7 \le p \le 10$ [cf. [172]] are very uncomfortable" ([180, p. 182]); for a computer-assisted proof of the cases $p = 11$ and 12, see [329].

What does the answer to Möbius's question tell us about the Four color conjecture? Not much; just that colorability and genus are different matters. For the history of the proof of the Four color theorem and the debate (cf. [436, Chapter 11]) that it engendered because it was eventually obtained by the help of computers, see [384, Chapters 18–23]. Many planar graphs are even 3-colorable (the test for (planar) 3-colorability is again **NP**-complete; cf. [432, Theorems B.12,B.7]), and there are 2-colorable (or *bipartite*) graphs which are not planar. (The test for 2-colorability is in complexity class **P**; cf. [306, p. 288].)

For an example of the latter we take recourse to another famous puzzle, the *water, gas and electricity* problem of Dudeney (cf. [55, p. 142f]): three suppliers, for water, gas and electricity, have to be connected directly to three consumers. The corresponding graph is called the **complete bipartite** graph $K_{3,3}$, because neither the suppliers are connected among each other nor are the consumers. In Exercise 0.11 the reader is asked to decide whether the supply lines can be layed out on the surface without crossings. Any graph which contains a **subdivision** of K_5 or $K_{3,3}$ cannot be planar. Note that *Kuratowski's theorem* says that this obvious sufficient condition for non-planarity is also necessary (cf. [432, p. 246–251], [319, Section 2.3]).

The minimum number of colors needed for a proper **vertex coloring** of a graph G is called its *(vertex) chromatic number* $\chi(G)$. It is always at least as large as the *clique number* $\omega(G)$, i.e. the order of a largest **clique** contained in G. For the upper bound we employ a very interesting type of algorithm, called *greedy*, because it always chooses the most profitable next step without thinking further ahead: as long as there are still uncolored vertices, choose any one of these and color it with the smallest possible color (colors being labelled by positive integers), i.e. one that has not yet been assigned to one of its neighbors. This procedure stops after $|G|$ steps, and the largest number used can not be larger than 1 plus the maximum degree $\Delta(G)$ of all vertices of G, because no vertex has more than $\Delta(G)$ neighbors.

Besides the vertex colorings, which can be employed, e.g., in a mathematical model for *Sudoku* (see [354, Section 7.3]; cf. also [98, Chapitre 6]), there are also **edge colorings** (cf. [384, Chapter 16]), where adjacent edges must be colored differently, and **total colorings** (cf. [446]) combining vertex and edge colorings. A theorem by V. G. Vizing says that the *edge chromatic number* (or *chromatic index*) $\chi'(G)$ of a graph G, i.e. the minimum number of colors needed for a proper edge coloring, must take on either the value $\Delta(G)$ or, as already for $G = K_3$, the value $\Delta(G) + 1$ (cf. [432, p. 277f]). For the corresponding question about the *total chromatic number* $\chi''(G)$, the *Total coloring conjecture* claims that it is either $\Delta(G)+1$ or $\Delta(G) + 2$, as for K_2. The conjecture was independently posed in the 1960s by M. Behzad and Vizing; see the book [384] of A. Soifer for a precise history. Specific labelings of (vertices and edges of) graphs lead to other long-standing open questions, like the fanciful *Graceful tree conjecture* (1964); cf. [432, p. 87f].

But let us now come back to problems where graphs provide the solution of a puzzle.

0.7.4   Crossing Rivers without Bridges

The most striking examples of how the modelling by a graph can transform a tricky puzzle into a triviality come from the very old *river crossing problems*. They stem from a time when Königsberg had no bridges and actually did not exist yet and can be found, e.g., in the Latin collection *Propositiones ad acuendos iuvenes*, which appeared around 800 and is generally attributed to Alcuin of York. Item number XVIII is the *propositio de homine et capra et lupo*, which is well known as *the wolf, the goat and the cabbage* problem. It involves a man who wants to cross a river with these three passengers using a boat which can only carry himself and at most one of the three. Moreover, one must not leave the wolf (W) with the goat (G), nor the goat with the (bunch of) cabbage (C) without supervision. The question is: how many one-way trips across the river are necessary to unite all of them on the opposite river bank. The essential mathematical object is the set of passengers who are alone, i.e. without the man, on one side of the river. Of the 8 possible subsets only $\varnothing$, W, G, C, and WC, in obvious notation, are admissible. They form the vertices of the graph in Figure 0.27, whose edges are single passages of the boat and labelled with the symbols of the passengers during the trip.

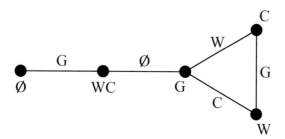

Figure 0.27: The graph of the wolf-goat-cabbage problem

The task is now translated into finding in that graph a closed walk of odd length including the vertex labelled ∅, because every crossing changes the river bank. It is obvious that 7 one-way trips are sufficient and necessary and that there are precisely two optimal solutions ∅–WC–G–C–W–G–WC–∅ and ∅–WC–G–W–C–G–WC–∅.

Much more notorious is problem XVII of the Latin collection, namely the *propositio de tribus frateribus singulas habentibus sorores*, which we call the *careful brothers problem*, avoiding the somewhat lurid and (therefore) more popular name *jealous husbands problem*. It has found coverage over more than a thousand years (cf. [151]), apparently without being properly solved until today; see Exercise 0.12.

Let us mention in passing that another famous riddle can be found in the Latin collection of Alcuin, namely his problem number XLII, *propositio de scala habente gradus centum*: how many pigeons are there on a ladder with 100 steps whose first step is occupied by one pigeon, the second by two birds, and similarly up to the hundredth step holding 100 pigeons; see Figure 0.28 for the example of a four-steps ladder.

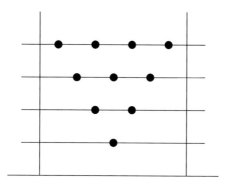

Figure 0.28: A ladder with 4 steps and $\Delta_4 = 10$ "pigeons"

The question is therefore to determine what for reasons obvious from the figure are the triangular numbers Δ_{100} (or Δ_4 in the picture). The general term of the sequence Δ_ν is defined as the sum of the first ν positive integers or, what amounts to the same, of all natural numbers from 0 to $\nu \in \mathbb{N}_0$. The solution given in Alcuin's text, interpreted for general even ν, is to redistribute all entries of the sum into pairs of numbers k and $\nu - k$, $k \in \left[\frac{\nu}{2}\right]_0$, such that only the number $\frac{\nu}{2}$ remains unmatched, whence

$$\Delta_\nu = \frac{\nu}{2} \cdot \nu + \frac{\nu}{2};$$

for odd ν, the pairs $(k, \nu - k)$ for $k \in \left[\frac{\nu+1}{2}\right]_0$ match all entries of the sum, such that altogether

$$\forall\, \nu \in \mathbb{N}_0 : \ \Delta_\nu = \frac{\nu(\nu + 1)}{2}. \tag{0.22}$$

A similar method is said to have been employed by the 9-years-old J. F. C. Gauss when his school class was asked to add up the first 100 numbers. Alas, this is just one of the most outstanding myths in mathematics history and a deplorable example of how facts mutate through thoughtless copying of unreliable sources. B. Hayes presents a list of 70 variants of the story in [183, p. 202]. For an updated and both critical and entertaining version of this account, entitled *Young Gauss Sums It Up*, see [184, Chapter 1]. The only certain root of the anecdote is W. Sartorius von Waltershausen's account [365] written by a confidant shortly after Gauss's death. Sartorius tells us that he learned about the story from the old Gauss himself! In this description [365, p. 12f] there is, e.g., no mention of the sequence from 1 to 100, but of the "summation of an arithmetical series". The question where and when the "100" came about was taken up again recently in [356, p. 45f], but in the subsequent answers [357, p. 41f] differing translations from [183, p. 201] and [366, p. 4] were mixed up. Unfortunately, most authors rely on the latter interpretation which is not particularly faithful as admitted in the "translator's note" [366, p. i].[31] In the original account there is also no reference to the kind of method employed by Gauss to arrive at the solution which the young boy wrote down in "a single number" immediately after the problem was hardly stated, turning in his slate tablet with the words "Ligget se'." ("Here t'lies."). Therefore, the most likely, albeit unromantic, explanation of the historical facts, if there were any, is that the young Gauss already knew formula (0.22), maybe from his own numerical experiments. It is also not true, as claimed in some "popular" accounts, that after the school hour was over all of Gauss's school mates had gotten wrong results, just many of the numbers submitted were false. As a note to modern pupils, and authors of imagined accounts of historical stories for that matter, it should be mentioned that they were "alsbald mit der Karwatsche rectificirt" (immediately rectified with the bullwhip)![32] A much deeper result about triangular numbers was established by Gauss only ten years after the school story. In his scientific diary [157, p. 22] one can find, under the date 10 July [1796] Gött[ingen], the following line

$$* * \text{EΥPHKA. num.} = \Delta + \Delta + \Delta .,$$

---

[31] The translator is a great-granddaughter of Gauss.

[32] The latter subclause has been censored by H. Worthington Gauss!

a very brief, but enthusiastic description of his discovery of a proof for a conjecture which Fermat had announced (as a statement) in a letter to Mersenne 158 years before, namely that every natural number can be written as the sum of three (not necessarily equal) triangular numbers. (The proof can be found in Gauss' *Disqvisitiones arithmeticae* (1801), Section 293.) This shows how long it might take to settle a conjecture!

To emphasize the importance of triangular numbers, we note that every $n \in \mathbb{N}_0$ can be written as $n = \Delta_\nu + x$, where the *index* $\nu \in \mathbb{N}_0$ and the *excess* x are uniquely determined if the latter is chosen in $[\nu + 1]_0$. We then have
$$\nu = \left\lfloor \frac{\sqrt{8n+1} - 1}{2} \right\rfloor,$$
because from $\Delta_\nu \le n < \Delta_{\nu+1}$ it follows that ν is the largest integer with the property $\nu(\nu+1) \le 2n$, and the desired formula follows by completing the square.

With this, the mapping $n \mapsto \dfrac{1+x}{1+\nu-x}$ is a bijective mapping from $\mathbb{N}_0$ to the set of all positive fractions. This is the *first diagonal method* which G. Cantor (cf. [69, § 1]) employed to prove that the set of rational numbers $\mathbb{Q}$ is *countable*. We mention in passing that his famous *second diagonal method* introduced in [70] to show that $2^\mathbb{N}$ is *not* countable, can be adapted for a proof of the uncountability of the set of real numbers $\mathbb{R}$ which even works in binary representation; cf. [199]. To understand what "countable" means, however, one has to understand the notion of equivalence.

0.8 Quotient Sets

The edges of a graph reflect a certain alliance between the corresponding vertices. We may therefore call the edge set an *association* on the vertex set; cf. [201, p. 14]. A graph is then simply a pair $G = (V, E)$ of a set V together with an association E on it. Even if the sets V and E vary, they may lead to the same kind of structure. We therefore want to make the notion of this *isomorphy* of graphs precise in this section, which might be a bit more formal than most of the previous ones.

0.8.1 Equivalence

In a chapter on "the beginning of the world" we must not omit the beginning of the mathematical world, at least with respect to the standards we still use today for writing mathematics. Beyond controversy, this is the *Elements* by Euclid. The first axiom reads

> Things which are equal to the same thing are also equal to one another.

Equality in this sense is the prototype of what we call in mathematics and in every day life an equivalence. We not only apply it to objects which are identical, in Euclid's words from another of his axioms

> Things which coincide with one another are equal to one another.

but also if they have the same "value" agreed upon under fixed circumstances. An example is a *standard* like the International Prototype Kilogram in Saint-Cloud/Sèvres. Euclid's first axiom reflects the property of equality to be "transitive":

Definition 0.14. *Let V be a set. A subset E of $\binom{V}{2}$ is* transitive, *if*

$$\forall \{x,y,z\} \in \binom{V}{3}: \ \{x,z\},\{y,z\} \in E \Rightarrow \{x,y\} \in E.$$

A transitive association is called an equivalence *(on V).*

We agree upon the following

Convention. If E is an equivalence on V, we say that $x \in V$ and $y \in V$ are *equivalent* (in symbols: $x \approx y$) with respect to E, if either $x = y$ or $\{x,y\} \in E$. □

This leads to the

Definition 0.15. *Let E be an equivalence on V. For every $x \in V$, the set*

$$[x] := \{y \in V \mid x \approx y\} = \{x\} \cup \{y \in V \mid \{x,y\} \in E\}$$

is called the equivalence class *(or* equiset *for short) of x in V (with respect to E).*

Equivalence classes are not empty and cover all of V, since $\forall x \in V: x \in [x]$, and two of them are either equal or disjoint because of transitivity. Therefore the quotient set $V/E := [V] = \{[x] \mid x \in V\}$ (sometimes also written as $V/\approx$) is a **partition** of V. In fact, equivalences and partitions are just two sides of the same coin, as can be seen from the following *Fundamental Theorem of equivalence*:

Theorem 0.16. *There is a bijection between the set of partitions of a set V and the set of equivalences on V.*

Proof. Let P be a partition of V. Define the (surjective) mapping $p \in P^V$ by $x \mapsto X$, where the latter set is the unique element of the partition containing x. Then $E = \{\{x,y\} \in \binom{V}{2} \mid p(x) = p(y)\}$ is an equivalence on V. We claim that the mapping $P \mapsto E$ is the desired (canonical) bijection.

It is surjective because every equivalence E is the image of its quotient set. Now assume that P and P' are mapped to the same E. Let $X \in P$ and $x \in X$. (This is the only place where the non-emptiness of elements of P is needed!) Then there is also an $X' \in P'$ such that $x \in X'$. It follows that $X = p(x) = [x] = p'(x) = X'$. This proves that $P \subseteq P'$. In the same way one shows that $P' \subseteq P$, whence $P = P'$. □

For every set V there is the *strict* partition with $E = \varnothing$, where two elements of V are equivalent only if they are equal, as, for example, (hopefully) for the finger-print test on the set of human beings. The other extreme, namely $E = \binom{V}{2}$, leads to the *trivial* partition where all elements of V are equivalent; example: Article 1 of the *Declaration of the Rights of Man and of the Citizen* of the French Revolution: "Men are born and remain free and equal in rights."; here the distinction between equality and equivalence is made precise by the words "in rights". (Men are quite obviously *not* equal!)

The most important mathematical example for an equivalence leads to the *cardinality* of sets. Two sets X and Y are said to be *equipotent*, in symbols $X \approx Y$, if there is a bijection between them. For instance, a set is called *countable*, if it is either finite or *countably infinite*, i.e. equipotent to the set $\mathbb{N}$ of positive integers. We have seen that $[\ell]_k$ and $\mathbb{N}_k$ are prototypes of finite and countably infinite sets, respectively. The notation $[n]$ is now also justified, because $\{1,\dots,n\}$ is a representative of all finite sets containing precisely n elements. In general, if an individual element of an $X \in [V]$ is chosen, it is called a *representative* of X. We have the following

Lemma 0.17. *Let V be a non-empty, finite set. Then*

$$1 \le |[V]| = \sum_{x \in V} \frac{1}{|[x]|} \le |V|.$$

Equality on the left side holds only for the trivial, on the right side only for the strict equivalence.

Proof. $|V| = \sum_{x \in V} 1 \ge \sum_{x \in V} \frac{1}{|[x]|} = \sum_{X \in [V]} \sum_{x \in X} \frac{1}{|[x]|} = \sum_{X \in [V]} 1 = |[V]| \ge 1.$ $\qquad\square$

The strict equivalence is the only *objective* one, all the others are *subjective* in the sense that (a group of) people agree on an equal *value* of non-identical objects. A typical example is given by the Euro coins. The European Central Bank declares all 2 Euro coins, say, issued by the national authorities of participating countries as equal in monetary value. For collectors, however, the 2 Euro coin from Monaco of 2007 showing a portrait of Princess Grace is of totally different value than an ordinary coin of Germany, say. So the individual coins to be found in one and the same pocket of a supermarket cash register, the equivalence class of the monetary value, can be in quite different classes with respect to collectors' values.

A more mathematical example is the following.

Example 0.18. *Let $V = \mathbb{Z}$ and $E = \{\{m,n\} \in \binom{\mathbb{Z}}{2} \mid m \cdot n > 0\}$. Then E forms an equivalence on V, the equivalence classes being the sets of positive and negative integers, respectively, and $\{0\}$.*

This example shows that our definition of an equivalence is more appropriate than the traditional one as a *reflexive, symmetric, transitive relation*, which one can view as the third side of the coin: it is, of course, rather circuitous to regard *ordered* pairs and then ask for symmetry; moreover the symmetric and transitive relation $\{(m,n) \mid \{m,n\} \in E\}$ corresponding to Example 0.18 would *not* be an equivalence relation, because it is lacking reflexivity. (In fact, all symmetric and transitive relations R on a set V have the property $\forall\, x \in V :$ $((x,x) \notin R \Leftrightarrow \forall\, y \in V : (x,y) \notin R)$; therefore, $\overline{R} = R \cup \{(x,x) \mid x \in V\}$ forms an equivalence relation, and $E = \{\{x,y\} \mid (x,y) \in \overline{R},\ x \neq y\}$ is an equivalence.) On the other hand, the partition of $\mathbb{Z}$ into the three equivalence classes according to our definition makes perfect sense. In other words, equivalence (and graph for that matter) is a more fundamental notion than relation.

Example 0.18 can also be used for intractable students, for it has the following consequence.

Corollary 0.19. $\{0\} \neq \varnothing.$[33]

Proof. Equivalence classes are not empty.   □

With respect to graphs, we will now call two graphs $G = (V,E)$ and $G' = (V',E')$ *isomorphic*, in symbols $G \cong G'$, if there is a bijection $\iota : V \to V'$ which is compatible with the graph structure, i.e. $\{v_1,v_2\} \in E \Leftrightarrow \{\iota(v_1),\iota(v_2)\} \in E'$. This, in fact, engenders an equivalence (on the power set of any set of graphs).[34] Similarly, the notion of *isomorphy of groups*, which we encountered in Section 0.7, can now be defined more precisely as an equivalence based on group structure preserving bijections between the underlying sets.

Disposing of a notion of equivalence of graphs, we can now also approach the formal definition of planarity. The usual "dots and lines" representation of a (simple) graph can be viewed as follows. Let $V' \subseteq \mathbb{R}^2$ be the set of dots in the drawing plane and P be the set of polygonal curves in $\mathbb{R}^2$ with two endpoints in V' and otherwise disjoint with V'. A *(graph) drawing* is a pair (V',P'), where P' is a subset of P in which there is at most one x',y'-polygon $p(x',y')$ for any $\{x',y'\} \in \binom{V'}{2}$. Note that $G' := (V',E')$, where $E' = \{\{x',y'\} \in \binom{V'}{2} \mid p(x',y') \in P'\}$, defines a graph. (V',P') is a *drawing of the graph* $G = (V,E)$, if $G \cong G'$.[35] For $\{x',y'\} \neq \{u',v'\}$, an element of $(p(x',y') \cap p(u',v')) \setminus (\{x',y'\} \cap \{u',v'\})$ is called a *crossing* of the drawing. If a drawing has no crossings, it is called a *plane drawing*.

---

[33]This not only shows that zero is not nothing, but it also offers an explanation why the symbol $\varnothing$ is used for the empty set, meaning that not even zero is in it; cf. [226, p. 211]. It was introduced by A. Weil because of its typographical availability from the Norwegian alphabet; cf. [431, p. 119].

[34]The reader not familiar with graph theory should be aware that the word *graph* is used for (at least) two objects: a $G = (V,E)$, sometimes called a *labeled graph*, or an equivalence class of graphs, then called *unlabeled graph*. A *(vertex-)labeling* is a specification of a representative from a class of isomorphic graphs, i.e. an individual graph of that class. Of course, one can always choose a representative with vertex set $[|G|]$.

[35]Sometimes such a drawing is called "the graph", which makes some sense given the word "graph" itself. However, the drawing only represents its equivalence class, i.e. the unlabelled graph. If in the (physical) drawing the dots get labels, namely the names of the corresponding vertices, then it represents an individual graph from the class.

A graph G is *planar*, if it has a plane drawing. Recall that the crossing number $\mathrm{cr}(G)$ of a graph G is the smallest number of crossings, taken over drawings in the plane, which have to fulfil some conditions. For instance, there should be no "hidden" crossings, i.e. not more than two polygons (edges) may pass through one crossing point. Moreover, we will not allow two edges to be tangent and they may only cross each other finitely often, in fact at most once.

As we have seen before, a (finite) graph $G = (V, E)$ is complete, if and only if $E = \binom{V}{2}$ and it is connected, if and only if

$$\widetilde{E} = \left\{\{x,y\} \in \binom{V}{2} \mid \exists\, x, y\text{-walk in } G\right\} \left(= \left\{\{x,y\} \in \binom{V}{2} \mid \exists\, x, y\text{-path in } G\right\}\right) = \binom{V}{2};$$

obviously, $E \subseteq \widetilde{E}$, and if G is complete, then $E = \widetilde{E}$, i.e. we have

$$G \text{ complete } \Leftrightarrow G \text{ connected and } E = \widetilde{E}. \tag{0.23}$$

$\widetilde{E}$ is transitive: join an x, z-walk and a y, z-walk at z to obtain an x, y-walk. So $(V, \widetilde{E})$ is an equivalence. For $x \in V$, the graph G_x induced by the equivalence class V_x of x in V with respect to $\widetilde{E}$, i.e. $G_x = (V_x, E_x)$ with $E_x = \{\{y, z\} \in E;\ y, z \in V_x\}$, is a *component* of G, which by Proposition 0.9 is connected. Moreover, we have:

Proposition 0.20. *If $V \neq \varnothing$, then G is connected, if and only if $|V/\widetilde{E}| = 1$.*

Proof. From $|V/\widetilde{E}| = 1$ it follows that G is connected. If the latter holds, then $V = V_x$ for every $x \in V$, whence $|V/\widetilde{E}| = 1$. □

Proposition 0.21. *E is transitive if and only if every component of G is complete.*

Proof. If E is transitive and there is an x, z-walk in E_x, then this walk can be reduced to the edge $\{x, y\}$ in finitely many steps: $x = x_0 - x_1 - x_2 \to x_0 - x_2$ and so on. Therefore $\widetilde{E}_x \subseteq E_x$, i.e. $\widetilde{E}_x = E_x$, and since G_x is connected we get that G_x is complete from (0.23).

For the converse implication let $\{x, z\}, \{y, z\} \in E$; then $x, y \in V_z$ and since the corresponding component G_z is complete, we get $\{x, y\} \in E_z \subseteq E$. □

0.8.2 Group Actions and Burnside's Lemma

Groups being associated with symmetries, e.g., of geometrical objects or graphs, we will need in later chapters some tools, among which is the so-called Lemma of Burnside. We want to present these statements right now, since they follow in an elementary way from the facts in the previous subsection. Because of its technical nature, this subsection might be skipped and returned to when needed.

Let X be a set and $(\Gamma, \cdot, 1)$ a group. A *group action* of Γ on X is a mapping $\Gamma \times X \to X$, $(g, x) \mapsto g.x$, with the property that for all $x \in X$:

$$1.x = x \ \wedge \ \forall\, g, h \in \Gamma : (g \cdot h).x = g.(h.x).$$

We define

$$\forall\, g \in \Gamma:\; X^g := \{x \in X \mid g.x = x\},$$

$$\forall\, x \in X:\; \Gamma_x := \{g \in \Gamma \mid g.x = x\};$$

X^g is the *fixed point set* of g, and Γ_x is called the *stabilizer* of x. A typical combinatorial technique to count the same set, namely $\{(g,x) \in \Gamma \times X \mid g.x = x\}$, in two different ways immediately leads to

Lemma 0.22. *If Γ acts on X, then*

$$\sum_{g \in \Gamma} |X^g| = \sum_{x \in X} |\Gamma_x|.$$

Calling $\Gamma.x = \{g.x \mid g \in \Gamma\}$ the *orbit* of x, we are able to prove the important *Orbit-Stabilizer theorem.*

Theorem 0.23. *If Γ acts on X, then*

$$\forall\, x \in X:\; |\Gamma| = |\Gamma.x| \cdot |\Gamma_x|.$$

Proof. Let $x \in X$. Obviously, $a \approx b \Leftrightarrow a.x = b.x$ defines an equivalence on the set Γ. Each equivalence class has the form $\widetilde{\Gamma}_y := \{c \in \Gamma \mid c.x = y\}$ for some $y \in \Gamma.x$, and $\widetilde{\Gamma}_y \ne \widetilde{\Gamma}_z$, if $y \ne z \in \Gamma.x$. Therefore, $|\Gamma| = \sum_{y \in \Gamma.x} |\widetilde{\Gamma}_y|$, and it suffices to show that $\forall\, y \in \Gamma.x:\; |\widetilde{\Gamma}_y| = |\Gamma_x|$.

For some $\gamma \in \widetilde{\Gamma}_y$ we define a mapping $\Gamma_x \to \widetilde{\Gamma}_y$ by $g \mapsto \gamma \cdot g$. This mapping is obviously injective. Now let $c \in \widetilde{\Gamma}_y$. Then

$$(\gamma^{-1} \cdot c).x = \gamma^{-1}.(c.x) = \gamma^{-1}.y = \gamma^{-1}.(\gamma.x) = (\gamma^{-1} \cdot \gamma).x = 1.x = x,$$

whence $\gamma^{-1} \cdot c \in \Gamma_x$. So the mapping is also surjective.　　　□

We finally define an equivalence on X by $x \approx y \Leftrightarrow y \in \Gamma.x$. The corresponding equivalence classes are $[x] = \Gamma.x$, and the quotient set $[X]$ is also called the *factor set* X/Γ of X with respect to Γ. We arrive at Burnside's lemma.

Corollary 0.24. *If Γ acts on X, then*

$$\sum_{g \in \Gamma} |X^g| = |\Gamma| \cdot |X/\Gamma|.$$

Proof. $\sum_{g \in \Gamma} |X^g| = \sum_{x \in X} |\Gamma_x| = \sum_{x \in X} \dfrac{|\Gamma|}{|\Gamma.x|} = |\Gamma| \sum_{x \in X} \dfrac{1}{|[x]|} = |\Gamma| \cdot |[X]|$, where use has been made of Lemma 0.22, Theorem 0.23, and Lemma 0.17 in that order.　　　□

We will apply Theorem 0.23 and Corollary 0.24 mainly to the *automorphism group* $(\Gamma, \cdot, 1) = (\mathrm{Aut}(G), \circ, \mathrm{id})$ of a graph $G = (V, E)$, where $\mathrm{Aut}(G)$ is the set of isomorphisms from G to G, $\circ$ stands for the composition of two such mappings, and $\mathrm{id}: V \to V,\; v \mapsto v$, is the identity mapping on V.

0.9 Distance

In order to obtain a *quantitative* measure for the separation of elements of a set, the following is the most natural mathematical model.

Definition 0.25. *Let M be a set. A function* $d : \binom{M}{2} \to \,]0, \infty[$ *which fulfills the* triangle inequality

$$\forall\, m_1, m_2, m \in M : d(m_1, m_2) \leq d(m_1, m) + d(m, m_2), \qquad (0.24)$$

where we write $d(m_1, m_2)$ *for* $d(\{m_1, m_2\})$ *and put* $d(m, m) = 0$, *is called a* dis-tance function *on* M; (M, d) *is called a* metric space *and* $d(m_1, m_2)$ *is the* distance between m_1 and m_2.

Remarks. The word "distance" derives from the Latin "dis-tare", meaning "stand-ing apart" (compare the German "Ab-stand"), such that d should be intrinsically positive. We then put $d(m, m) = 0$ just as a convention.

Customarily, the distance function d is defined on M^2 and then *symmetry* $d((m_1, m_2)) = d((m_2, m_1))$ is required; this is rather absurd!

In the introduction of the axioms of a metric space in his thesis [152, p. 30], M. Fréchet (1878–1973) used the expression "écart" (from the Latin "ex-quartare", meaning cutting into four parts, dismemberment) for distance. Later, however, in [153, p. 61f], he calls a metric space "espace distancié" after criticizing, and rightly so, F. Hausdorff (1868–1942) for the expression "metrischer Raum" [181, p. 211], which Fréchet calls "frappante" and, because of its differing use in geometry, as leading to a "confusion regrettable". Hausdorff used "Ent-fernung" for distance. His "metric space" has become accepted. However, we were not able to find out, when and where the unfortunate term "metric" came into being for the distance, Entfernung or écart function. See [102] for a comprehensive survey on distances.

In the present book we will mainly use the concept of a **graph distance**.

0.10 Early Mathematical Sources

For a serious examination of problems from recreational mathematics there is a BL and an AL—before Lucas and after Lucas. There is apparently no BL for TH and BL sources for CR are rare and mostly obscure.

0.10.1 Chinese Rings

Pacioli's account [328, p. 290–292] contains a description of a solution to get all rings onto the bar from an initial situation where they are all off. Although he sets out from an arbitrary number of rings, his explanations stop as soon as the 7th ring is on the bar, when he writes "e' cosi sucessiue" (and so on). For a recent transcription and translation into English, see [187].

Other early Western sources became known through the pamphlet [168] of Gros from 1872 which in turn became known through the book [283] of Lucas in 1882. These are Cardano's passage mentioned before, an improved translation of which can also be found in [187], and the somewhat more substantial Chapter 111 entitled *De Complicatis Annulis* in J. Wallis's famous treatise on algebra [426, p. 472–478] from 1693. Wallis starts off from the state where the bar and the system of nine rings are separated. As a goal he specifies the state with all rings on the bar, but it is remarkable that in his step by step description of how to put on the rings he ends in the most deranged state ("in implicatissimo statu"), i.e. with the ninth ring on, all others off the bar.

In his account on Wallis's solution, Gros is not quite fair; in fact, in a handwritten annotation from 1884, to be found on the copy of [168] which is at our disposal, the author offers a reason why: "Wallis has been so unjust with respect to French mathematicians that it is well permitted to shut him up [river ses clous] on the subject of the baguenodier." It is true that Wallis complicated his explanations unnecessarily by a dissection of what we call a move into separate movements of rings *and* the bar to the effect that a move of the outer ring needs 2, the move of any other ring 4 movements. On the other hand, this simplifies his calculations. So he finds that in order to get $n \leq 9$ rings onto the bar, one needs $\sum_{k=1}^{n} 2^k = 2^{n+1} - 2$ of his movements. In a last step he adds another $2^9 - 1$ movements to arrive at his most deranged state in altogether 1533 movements. As we will see, this number is correct, but it contains unnecessary movements.

Gros is also unjust to Cardano, a fact to which we will return later. However, the mathematical analysis of Gros is much better, and we will follow his model (with respect to mathematics proper) in Chapter 1, where we also refer to two notes from 1769, one by the German Georg Christoph Lichtenberg (1742–1799) [266], the other by the Japanese Yoriyuki Arima (有马) (1714–1783), who wrote under the pen name Toyota Bunkei (丰田文景) [418]. These notes are discussed in detail in [202] and [107], respectively.

Wei Zhang and Peter Rasmussen have found the (so far) earliest Chinese treatment of the jiulianhuan, which appeared around 1821 in *Xiao hui ji* (*Little wisdoms*, 小慧集) by Zhu Xiang Zhuren (贮香主人). In Figure 0.29 one can see a pictorial representation of the game, and the two tables describe a method to get the nine rings off the bar. We will explain this solution in Chapter 1.

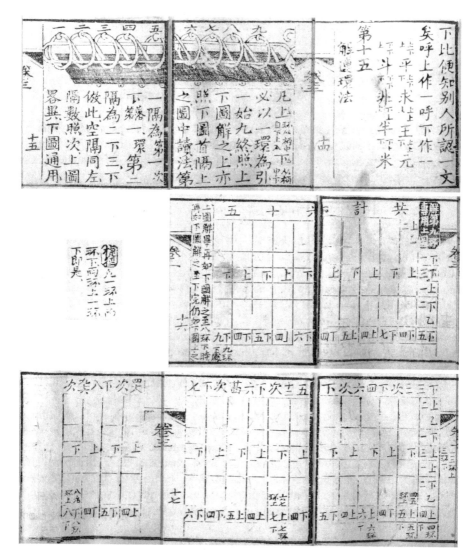

Figure 0.29: Pages 14r to 17v from *Xiao hui ji*, book III (ca. 1821) by Zhu Xiang Zhuren (the insert in the center shows an extra comment on page 14v); from the Yi Zhi Tang collection of traditional Chinese puzzles. For more information: http://ChinesePuzzles.org

0.10.2   Tower of Hanoi

Mathematicians are working on the theory of the TH on the authority of the bible, where in the Gospel according to—whom else?—Luke, we find:

> Which of you, intending to build a tower,
> sitteth not down first, and counteth the cost?
>
> **Luke 14:28**

Following this commandment and attracted by the divine elegance of the TH, mathematicians developed a theory of the puzzle almost instantly after its publication and so it made its first appearances in mathematical literature very early on. Still in 1883, G. de Longchamps gives an account in [275], concentrating on the sequence M_n. Claus de Siam himself, in his 1884 article [83] mentioned earlier, presents the first description of a non-recursive solution, namely to move the smallest disc rotationally over the three pegs, alternating with moves of other discs forced by the divine rule. This strategy, which we will develop in Section 2.1, is attributed to Raoul Maurice Olive (1865–?), a pupil of the Lycée Charlemagne (in Paris) and nephew of Édouard Lucas. Moreover, as already on the leaflet [82], a second problem is proposed in [83], namely to start with an arbitrary distribution of the discs among the three pegs. This will be studied in Section 2.2 and in its most general form in Chapter 3. Finally, [83] points out the intimate relations between the TH and permutations, combinations and the binary number system and anticipates modifications of the game and its rules to represent all number systems and recurrences in general. It is even announced that another game will be published to illustrate the famous theorem of Charles Sturm of 1829 concerning the number of roots of a real algebraic equation in a given interval.

The TH immediately spread outside France. The article [4] by R. E. Allardice and A. Y. Fraser was published in 1884 in Volume II of the *Proceedings of the Edinburgh Mathematical Society*, actually its first volume to appear (cf. [349, p. 140]), and contains a verbal repetition of de Parville's account–in French! Obviously, in those times, Scottish scholars were well-educated. In [4, § 2] the recursive solution is repeated and the observation disclosed that during this solution every disc moves a power of 2 times, according to its size. Also, in [4, § 3], the cyclic pattern of moves is described.

Although these early sources contain the essential mathematical statements about the classical TH problem, they are lacking convincing proofs. This changed with the thoughtful analysis of Schoute in [372]. Stimulated by the publication of Lucas's *Récréations mathématiques* [283], Schoute wrote a series of *Wiskundige Verpoozingen* (mathematical relaxations) in the popular Dutch journal *Eigen Haard* between 1882 and 1884 (cf. [61]); among these are articles on the CR and the binary number system to which the author relates his most original contributions on the TH. Obviously starting off from a German version of the puzzle, which had the discs colored black (the smallest) or red in alternation, he states a number of results, some of them followed by serious proofs. Apart from the facts already known before, he makes use of the coloring of the discs (and the bottoms of pegs, only the intermediate one carrying the same color as the largest disc) to devise a new realization of the classical solution, namely never to place a black disc on another black one (or bottom) and never to undo a move just performed. The most outstanding result, however, is an algorithm [372, p. 286f] which determines the number of steps (in binary) made on the path of the classical recursive solution to arrive at a given configuration. We will present this recipe in Section 2.1.

After that there has been no progress on the classical TH problem for about sixty years. The paper [378] by R. S. Scorer, P. M. Grundy, and C. A. B. Smith was the first to add some mathematical structure to the game by looking at it in the shape of the *state graphs*, which we now call *Hanoi graphs*. They derive typical graph theoretical properties like planarity and edge colorings and even anticipate what we will investigate under the name of *Sierpiński graphs*! Moreover, the authors, like Lucas in the first publication of the TH under his true name [289, p. 55–59], written in 1884, again stressed the value of the TH as representing the formation of numbers in the binary system, and showed the way to more challenging tasks arising from more general starting configurations and the introduction of variations, also including further pegs.

Missing minimality, false assumptions and unproved conjectures

Virtually all early papers on the topic of the TH (re)produce the classical recursive solution together with the number of moves involved. What they do, however, is just provide an *upper* bound for the minimal number by proving that the task can be achieved using that many steps. It is one of the mysteries in the mathematical theory of the TH that apparently nobody felt the necessity to give a *lower* bound, i.e. to prove that there is no better solution, until 1976, when C. Becquaert noticed the gap in the argument in an utmost clear statement in [39, p. 74] and presented the first minimality proof, further elaborated in [40, Section IV]. Due to the inaccessibility of these articles, it is generally D. Wood who is credited with the discovery in his 1981 paper [443, Theorem]. Even today, this proof is missing from most texts and some authors even do not understand where the problem lies (cf. [200]). The crucial point is the assumption that the largest disc will move, in an optimal solution, directly from its source to its target peg or, in other words, it moves precisely once. Even if this seems to be evident, a mathematical proof of minimality can not be based on an unjustified assumption. In fact, the proof is easy and so we will already give it here in a slightly improved rendering of Becquaert's and Wood's arguments.

Theorem 0.26. *An optimal solution for the TH with $n \in \mathbb{N}_0$ discs can be achieved in, but not in less than, M_n moves.*

Proof. The proof is by induction. The statement is trivially true for $n = 0$. It is also clear that if the task can be done for n discs in M_n moves, it can be done for $n+1$ discs in $M_n + 1 + M_n = 2M_n + 1 = M_{n+1}$ moves by transferring the n-tower of smaller discs to the intermediate peg, then disc $n+1$ to the goal peg and finally the n others to the goal too.

Before the first move of disc $n+1$ in any solution, a tower of n discs has to be transferred from the source peg to another one, which takes at least M_n moves by induction assumption. After the last move of disc $n+1$ again an n-tower changes position from some peg to the goal peg, consuming another M_n individual disc moves. Since disc $n+1$ has to move at least once, the solution needs at least M_{n+1} steps.   □

The second half of the argument in the preceding proof shows that a solution with more than one move of the largest disc is longer than the solution in the first part of the proof. That is, in an optimal solution first and last move of the largest disc coincide. This move being mandatory, it follows, again by induction, that the sequence of moves of the optimal solution is unique.

The proof of Theorem 0.26 can be viewed as the beginning of a serious mathematical treatment of the TH almost 100 years after its invention.

The consequence of Theorem 0.26 for the Tower of Bramah (sic!), i.e. the case of M_{64}, the number given in (0.1), has been chosen by W. H. Woodin in [445, p. 344] as a paradigm for the philosophical implications (and those for mathematical logic) of a true sentence whose ultimate effective proof would be coincident with the end of the world.

Unfortunately, mathematical progress on the TH did not come smoothly. The assumption that the largest disc moves only once was thoughtlessly carried over to a generalization of the problem, namely starting from an arbitrary legal initial distribution of the discs, but still heading for a perfect tower on a preassigned peg. Although an argument as for Theorem 0.26 will show that this assumption is still justified, it fails as soon as also the target state may be chosen arbitrarily, because in that case the last move of the largest disc might not be followed by the transfer of a complete tower consisting of the smaller discs. When Wood tried to design rules for adjudicating a two or more person game based on the TH in [444], he failed to recognize this fallacy and claimed in his Theorem 3 the uniqueness of the optimal solution based on the false assumption. A (very simple) counter-example was published by M. C. Er in [138]. However, he himself fails to avoid the error in his algorithm for the general case in [136], where he wrongly claims minimality, which in some cases can only be achieved making two moves of the largest disc. This mistake was repeated over and over again in literature, also by P. Cull and C. Gerety in [88], a paper which we only cite because of its really designative title!

A serious mathematical analysis of this issue was performed by X. Lu in [277], where the author gives a(n equally simple) counter-example (already to be found in 1972 in [224, p. 34]!) to another statement of Wood's [444, Lemma 5] which claims that specific intermediate states, namely with all but the largest disc on the peg different from initial and final position of the latter, are unavoidable in an optimal path, which is essentially equivalent to the assumption that the largest disc moves only once. Lu's approach is also based on unavoidable intermediate states of the puzzle and depends on descriptive arguments deduced from the *Hanoi graphs*, an expression Lu introduces on this occasion.

An example of how dangerous it can be to formulate the rules of the game carelessly led to the *Tower of Stanford*, posed as a problem in *The American Mathematical Monthly* of 2002; see Exercise 0.13.

The Reve's puzzle

The first explicit extension of the classical puzzle to four pegs is given in the book [116] written by Dudeney,[36] who took up G. Chaucer's idea of *The Canterbury Tales*, but changed the tales narrated by pilgrims meeting at night at the old Tabard Inn in Southwark into puzzles. The first of these he called *The Reve's puzzle*.[37] Dudeney writes [116, p. 1]: "There are certainly several far more difficult puzzles extant, but difficulty and interest are two qualities of puzzledom that do not necessarily go together." He erred with respect to The Reve's puzzle! In the story he tells, diamond needles and golden discs are replaced by stools and loaves of cheese, respectively. Although this might be appropriate for an English tavern, it strongly reminds us of the words of Ball (see p. 2). Dudeney proposed (unique) solutions [116, p. 131f] for the cases of 3, 4, and 5 stools for natural, triangular, and *tetrahedral* numbers $\binom{\nu+0}{1}$, $\binom{\nu+1}{2}$, and $\binom{\nu+2}{3}$ of cheeses, respectively, and summarizes the lengths in a table. However, he fails to prove correctness and in particular minimality of his procedure. As it stands, the prize The Reve offered for an optimal solution, namely "a draught of the best that our good host can provide", is still to be won. This situation did not change with the problem being taken up as a challenge by B. M. Stewart to the readers of *The American Mathematical Monthly* of 1939 [390], because his own solution [391] and the one by J. S. Frame [150] were both based on a certain unproven assumption which the editor of the journal, O. Dunkel,[38] framed into a lemma [121]. We will discuss what we now call the *Frame-Stewart conjecture* in Chapter 5.

"No more articles on the Towers of Hanoi for a while."

During the 1970s the TH was (re-)discovered by computer scientists. The classical task to transfer a tower of n discs from one peg to another was considered as a prototype of the *divide-and-conquer* paradigm, namely to cut a problem into smaller

---

[36] In fact, Dudeney first published some instances of the puzzle at the end of the 19th century in various London magazines and reproduced it in the book mentioned.

[37] This was Oswald of Norfolk in Chaucer's work. The re(e)ve was a kind of official for a county; today we meet this word in sheriff=shire-reeve.

[38] Advanced Problem 3918 of the *Monthly* has been selected as one of *The four-hundred 'best' problems (1918–1950)* in [143, p. 30]; there one can also find a portrait and a brief biography [143, p. 1f] (by P. R. Rider) of Dunkel.

pieces and solve the latter successively. Mostly unaware of the mathematical theory of the TH, articles galore appeared whose sole intention was to provide ever
more sophisticated algorithms to produce the list of moves in the solution known
by Lucas and Olive since 1883, which is unique as we know from the remark following the proof of Theorem 0.26. Originally mainly used to demonstrate the power
of recursive approaches in computer science textbooks, a discussion was initiated
by P. J. Hayes concerning the "best" algorithm—recursive or iterative. Although,
as the author explicitly admits in his "disclaimer", there was nothing new about
the TH and its solution, his paper [185] is probably the most cited of all articles
on the TH whose most remarkable innovation was to introduce a plural "s" to
Lucas's tower, which now allows us to distinguish easily between publications on
the topic from computer scientists and those of more respectful mathematicians.
We will ignore this plural form throughout the text except in quotations and in
the references.

Nevertheless, the flood of articles and notes, mainly in *The Computer Journal*, *Information Processing Letters*, and the *ACM SIGPLAN Notices*, produced a
wealth of mostly empirical observations on the properties of the solution. We will
encounter them, together with proofs, in Section 2.1. Striking are the exchanges
of letters to the editors, at times fierce and personal in tone, for instance in *The
Computer Journal* between 1984 and 1986, when the TH was attacked as "a trivial
problem" [186].

This was caused by a series of rather redundant articles, at least two of which
[135, 137] being identical word by word, by Er from 1982 to 1991, in which the
author developed some, mostly algorithmic, properties of the TH and its growing
number of variants.

Finally, in 1985, the editor of the *ACM SIGPLAN Notices* arrived at the end
of his tether and declared [433]: "Towers of Hanoi. Please, no more articles on this
for a while."

The presumed minimal solution for The Reve's puzzle

The same hype occurred in computer science literature about The Reve's puzzle.
The solutions of Stewart and Frame, essentially equivalent and which have hitherto
been called the *presumed minimal solutions* (cf. [193, p. 134]) in view of the lack
of a minimality proof, were studied under algorithmic aspects. Many of these
papers include claims of optimality, like, e.g., [296]. From the mathematical point
of view, however, earnest contributions to clarify the Frame-Stewart conjecture in
the second half of the 20th century have been rare.

The first serious papers

Although the good, the bad and the ugly can be found in all categories of human endeavor, the latter two outnumber the first by far with respect to the TH.
However, a more serious examination of the mathematical theory of the TH began

with the works of Wood [443], Lu [277], T.-H. Chan [74], and the seminal paper by A. M. Hinz [194]. The decade culminated with the most amazing quantitative connection between the TH and the ST in [214]. The majority of (proven) results which we will present in Chapters 2 and 5 originate in subsequent years. The growing scientific interest in the topic became manifest with the first *Workshop on the Tower of Hanoi and Related Problems*, held in Maribor (Slovenia) in 2005. A multidisciplinary meeting *La «Tour d'Hanoï», un casse-tête mathématique d'Édouard Lucas (1842–1891)* in 2009 brought together mathematicians, psychologists and historians of science in Paris. The poster entitled *From London to Hanoi and back—graphs for neuropsychology* by Hinz won a Second Prize at the *International Congress of Mathematicians 2006* in Madrid. It presented the activities of a research project to employ the TH in a computerized test tool for cognitive psychology; cf. [208].

Psychology, variations, open problems

The TH entered the literature of psychology as early as 1897, when E. H. Lindley, wondering "... how myth and legend tend to cluster about puzzles, ...", called the TH, which he groups, together with the CR, into the category of more complex mechanical puzzles, a "familiar mechanical problem" [268, Chapter II]. It turned out to be a valuable tool in cognitive tests because of its attractive aspect, the easy-to-explain rules, the possibility for the experimenter to watch subjects think, and the fact that the average test person does not dispose of a priori special purpose subroutines. The CR with their not so obvious rules were already employed in 1910 in a series of incredibly exhausting (for the test persons, mainly students of Columbia University, New York City) tests with the same task repeated 50 times in a row! In cases it took several hours to solve the CR with only 4 rings, but there was also a solution for 10 rings in about 20 minutes; see [358, p. 74–81]. The TH was used as a test tool in the 1920s by J. Peterson and L. H. Lanier in experimental studies trying to discern the cognitive abilities of black and white children and adults in the United States of America (cf. [337]) in an attempt to decide between genetic and environmental causes for possible differences. At the same time, W. S. Hunter reports [220, p. 331–333] on tests made with college students who were not only asked to develop a solution strategy, but also to "write the equation which will give the least number of moves required to solve the problem for any number of discs." He calls the TH a "very interesting example of the development of scientific thinking". Among the numerous studies of that kind we only mention tests with children of ages between 5 and 12 by J. Piaget and A. Cattin [340] which distinguish between different phases of capability to solve the puzzle and to generalize and explain its solution.

A heuristic analysis of solution strategies was undertaken by H. A. Simon in [382] which influenced both the psychological and the artificial intelligence communities. Simon distinguishes four types of strategies: *goal-recursion, perceptual, sophisticated perceptual*, and *move pattern* based. The first is the classical recursive method, the last is Olive's solution, while the intermediate ones are based on a step-by-step analysis by the problem solver. Memory requirements are forbidding for the recursive solution and low for Olive's. Simon points out that the perceptual strategies have the advantage to allow for a restart, i.e. can be employed even if the initial state is not a perfect tower. In any case, the assumption of one move of the largest disc is made throughout. D. Klahr remarked in 1978 [232] that there might be two solutions for tasks involving *flat* states, i.e. those where discs are lying on mutually different pegs, in the TH with 3 discs. In [233, p. 139] it was even noticed, five years before Lu, that the one-move strategy fails in some tasks to produce a minimal solution. The reason, however, was not well understood, and it must be said that most studies using the TH in psychology lack a mathematical backing. To avoid complications, subjects were even told not to make more than one move with the largest disc; cf. [17]. This is most surprising because the correct state graph for the puzzle was well-known in psychological literature by 1965 (see [208, Section 0.1.1] for the history).

Although some psychological studies did consider more general tasks than the classical perfect-to-perfect game, you hardly find any such task in today's numerous interactive applets on the internet. This misconception of a missing variability of the TH specifications may also have led T. Shallice to invent a variant of the "look-ahead puzzle" which he called *The Tower of London (TL)* [379, p. 205] (see [208, Section 0.1.2] for the history). The game with three pegs of different heights is played with colored, but otherwise identical balls instead of discs, and therefore does not apply anything like the divine rule. It will be explained in detail in Chapter 7 because, although in itself rather simple, its generalizations lead to very interesting mathematics.

Variations of the TH weakening or violating the divine rule or otherwise relaxing the concept of a move have been introduced throughout its history, and we will again only consider those with a strong mathematical context. One of these, the *Switching Tower of Hanoi (Switching TH)*, was in fact introduced by S. Klavžar and U. Milutinović in [235, p. 97f] on a purely mathematical base, namely to illustrate the class of *Sierpiński graphs* to which our Chapter 4 will be devoted. The latter, in turn, are variants of Hanoi graphs which at times allow us to study properties of these in a more straightforward way. Other variations of the TH will be studied in Chapter 6, starting with those already proposed by Lucas in [285]. Finally, a last group of variants, initiated in [378], will be addressed in Chapter 8. Here the possibility to move between pegs is restricted, but usually with the divine rule in force.

To get an impression how varied the applications of the divine rule are, let us only mention that it might serve, as pointed out by H. Hering in [188, p. 411], as a model for shunting railway cars on three tracks merging into one without changing the order of the cars. Another instance is the question whether a given permutation can be sorted by a passage through a single stack. Permutations that allow such a sorting were described by D. Knuth and are sometimes called *Catalan permutations*. In this problem no integer can be placed on top of a smaller integer in the stack, which is just the divine rule! We refer to [171] for more information on this (and related) problem(s).

The principal practical application of the TH remains, however, the testing in cognitive psychology. In order to put these experiments onto a firm and reliable fundament, the mathematical theory of *tower tasks* has been presented in [208] together with the description of a computerized test tool based on it. The special role played by the corresponding state graphs has been stressed in [201], where also a misconception to be found in some psychology papers has been addressed, namely to mistake these graphs for what A. Newell and Simon have called in [322] the *problem space* of a puzzle. The latter concept takes in more than the sheer mathematical structure of the problem manifest in the graph. The crucial point is the distinction between isomorphy and equivalence. As pointed out in Section 0.8.1, the concept of equivalence requires a subjective agreement on some "value". The isomorphy of graphs is just one of those possible values, and two puzzles will be considered to be mathematically equivalent if their corresponding state graphs are isomorphic. This means, e.g., that whenever a solution has been found for one of the isomorphs, a solution for the other can be deduced mathematically. This does *not* imply, however, that both tasks are equally easy or difficult to solve for a human being. For instance, it has been felt that color coding of differences of discs is easier to perceive than sizes and so the *Tower of Toronto* in [362] replaces diameter by shading, disallowing darker discs to be placed on lighter ones. A more striking example is *Monsters and Globes* from [254, p. 251] as discussed in [201], where the TH problem is administered in a way to switch the roles of static and dynamic components of the puzzle to the effect that even those familiar with the solution to the TH are hardly able to solve the variant. We will meet a similar *problem isomorph* in connection with the CR. Therefore, these variants, although mathematically equivalent, are *not* equivalent psychologically. If one would take into account only mathematical equivalence, even (Argentine) ants would be able to solve the TH; cf. [350]!

Other problems in this book will be too difficult for them, but suitable for readers on all levels of ambition. The most challenging ones, namely those which are still open to be resolved, are summarized in the final chapter which may last until the end of the world brought about by the Brahmins.

0.11 Exercises

0.1. Solve the recurrence equation in (0.4), i.e. $x_{n+2} = x_{n+1} + x_n$, by the ansatz $x_n = \xi^n$ for some $\xi \in \mathbb{R}$ and write F_n and L_n as linear combinations of these particular solutions. Show that the ratios F_{n+1}/F_n and L_{n+1}/L_n tend to the Golden section $(1 + \sqrt{5})/2$ as $n \to \infty$.

0.2. Recall that for $k \in \mathbb{N}_0$ we write $[k] = \{1, \ldots, k\}$, $[k]_0 = \{0, \ldots, k-1\}$ for the k-*segments* of $\mathbb{N}$ and $\mathbb{N}_0$, respectively. This leads to a proper definition of the size of a set M, denoted by $|M|$: we say that M is *finite* and has *size*

$|M| := m \in \mathbb{N}_0$, if there is a bijection from M to $[m]$. That this notion is well-defined follows from the **Pigeonhole principle**.

We define
$$\forall\, k, \ell \in \mathbb{N}_0 : \binom{k}{\ell} = \left| \binom{[k]}{\ell} \right|.$$

Moreover, let $k!$ (*k factorial*) be defined by the recurrence
$$0! = 1, \quad \forall\, k \in \mathbb{N}_0 : (k+1)! = (k+1) \cdot k!.$$

This is well-defined by the Fundamental Theorem of recursion (cf. the solution of Exercise 0.1), where we put $M = \mathbb{N}$, $\eta_0 = 1$, and $\varphi(k,m) = (k+1) \cdot m$. An alternative representation is $k! = \prod_{\kappa=1}^{k} \kappa$.

Prove the following.

a) For $\ell \in [k+1]_0$, there are $\dfrac{k!}{(k-\ell)!}$ injective mappings from $[\ell]$ to $[k]$ (*arrangements* of ℓ out of k objects); in particular, $k!$ is the number of permutations on $[k]$, i.e. $|S_k|$.

b) $\forall\, k \in [\ell]_0 : \binom{k}{\ell} = 0$; otherwise $\binom{k}{\ell} = \dfrac{k!}{\ell!(k-\ell)!}$.

c) $\forall\, k, \ell \in \mathbb{N}_0 : \binom{k+1}{\ell+1} = \binom{k}{\ell} + \binom{k}{\ell+1}$.

0.3. (To be solved *without* recourse to the sequence $(f_k)_{k \in \mathbb{N}}$ of p. 25!)

a) Show that the number $\overline{f}_k$ of derangements of $[k]$, $k \in \mathbb{N}_0$, fulfills the recurrence
$$x_0 = 1, \; x_1 = 0, \; \forall\, k \in \mathbb{N} : x_{k+1} = k\,(x_k + x_{k-1}). \tag{0.25}$$

b) Prove that (0.25), recurrence
$$x_0 = 1, \; \forall\, k \in \mathbb{N} : x_k = k\,x_{k-1} + (-1)^k \tag{0.26}$$
and formula
$$\forall\, k \in \mathbb{N}_0 : x_k = k! \sum_{\ell=0}^{k} \frac{(-1)^\ell}{\ell!} \tag{0.27}$$
are equivalent.

Remark 0.27. *From the Fundamental Theorem of recursion (put $M = \mathbb{Z}$, $\eta_0 = 1$, $\varphi(k,n) = (k+1)n - (-1)^k$) it follows that the sequence $(x_k)_{k \in \mathbb{N}_0}$ is properly and uniquely defined by (0.26). The term x_k is denoted by $k\mathrm{i}$ and called k subfactorial. From (a) it follows that $k\mathrm{i}$ is the number of derangements on*

[k]. *The series in* (0.27) *converges very rapidly to* $1/e$; *in fact,* $k\textsf{i}$ *is the closest integer to* $k!/e$ *for every* $k \in \mathbb{N}$: *Taylor's formula tells us that*

$$\forall\, k \in \mathbb{N}_0 \;\exists\, \xi \in\,] -1, 0[\,: \; e^{-1} - \sum_{\ell=0}^{k} \frac{(-1)^\ell}{\ell!} = \frac{(-1)^{k+1}}{(k+1)!} e^\xi \,.$$

Now

$$\frac{k!}{e} - k\textsf{i} = k! \left(e^{-1} - \frac{k\textsf{i}}{k!} \right) = k! \left(e^{-1} - \sum_{\ell=0}^{k} \frac{(-1)^\ell}{\ell!} \right) = \frac{(-1)^{k+1}}{k+1} e^\xi \,.$$

For $k \in \mathbb{N}$, *the absolute value of the right-hand term is strictly less than* $\frac{1}{2}$.

0.4. Determine the parity of $\binom{1039}{11}$.

0.5. Prove the following:

   a) Every connected (multi-)graph [with $2k$ odd vertices, $k \in \mathbb{N}_0$] can be made **eulerian** by adding [k] edges.

   b) By adding only $k - 1$ edges (if $k \in \mathbb{N}$), the graph becomes **semi-eulerian**.

0.6. Find an eulerian trail in Euler's multigraph of Figure 0.21, including passage h.

0.7. Show that the symbol in Figure 0.22 can be drawn in one line.

0.8. Construct an algorithm which decides whether an edge in a given graph is a bridge or not.

0.9. Let $n \in \mathbb{N}_0$. Show that $\mathrm{Sym}_n := (S_n, \iota, \circ)$, where $\iota \in S_n$, $k \mapsto k$ for $k \in [n]$, forms a group which is commutative if and only if $n \leq 2$.

0.10. Find all hamiltonian cycles on Hamilton's dodecahedral graph which start

   a) L T S R Q,

   b) J V T S R.

0.11. Decide whether $K_{3,3}$ is planar or not.

0.12. Three men and their corresponding three sisters want to cross a river using a boat which can take one or two persons. No brother wants to leave his sister in the presence of another man. How many single river crossings are necessary and how many solutions to the task can be found?

0.13. Voluntarily misinterpreting the, admittedly very terse, description of the rules for the TH given by D. Knuth in [244, p. 321][39] (and actually citing it (involuntarily?) imprecisely), J. McCarthy replaced the divine rule by requiring that during the transfer of a regularly ordered tower of discs from one peg to another the largest disc on any peg must lie on bottom; see [307]. Find the minimum number of moves necessary and sufficient to solve the task under the relaxed rule.

---

[39]in answering a question posed by A. M. Hinz

Chapter 1

The Chinese Rings

In this chapter, we will discuss the mathematical theory of the CR which goes back to the booklet by Gros of 1872 [168]. This mathematical model may serve as a prototype for the approach to analyze other puzzles in later chapters. In Section 1.1 we develop the theory based on binary coding leading to a remarkable sequence to be discussed in Section 1.2. Some applications will be presented in Section 1.3.

1.1 Theory of the Chinese Rings

Recall the initial appearance of the puzzle with all rings on the bar (cf. Figure 1.1).

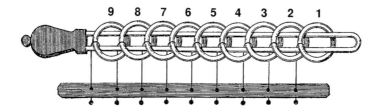

Figure 1.1: Chinese rings

In the introduction we assumed that all configurations of the system of rings can be reached from this initial state using just two kinds of individual ring moves, which we will now specify as:

- the rightmost ring can always be moved (*move type* 0),

- the ring after the first ring on the bar (from the right) can be moved (*move type* 1);

© Springer International Publishing AG, part of Springer Nature 2018
A.M. Hinz et al., *The Tower of Hanoi – Myths and Maths*,
https://doi.org/10.1007/978-3-319-73779-9_2

we have chosen the handle to be on the left for mathematical reasons, the rightmost ring having the least "place value". From a practical point of view, a move of type 0 consists of two steps, or movements in the sense of Wallis (cf. Section 0.10.1): to move the rightmost ring off the bar, one has to pull the bar back and then the ring through the loop; for every other ring, the bar has to be pulled back through its right-hand neighbor and itself (two movements), then the moving ring goes through the loop and finally the bar forward through the neigboring ring, all in all four movements. Since all the movements of the bar are "forced" by the material, we will only count moves of rings, i.e. movements through the loop. It is not easy to imagine the manipulation of the contraption without feeling it in one's hands. But there are movies on the Web where you can find people solving the puzzle in a sometimes incredible speed.

The original task was to get all rings off the bar, and we have to show that our assumption was correct. The questions arising are

- Is there a solution? (If the answer is "yes", then there is also a shortest solution with respect to the number of moves needed.)

- Is there only one (shortest) solution?

- Is there an *efficient* solution, i.e. an algorithm realizing the shortest solution?

The same types of questions will be asked for other puzzles in later chapters.

For every ring there are two conditions, namely to be off or on the bar, represented by 0 and 1, respectively. As before, $B = [2]_0 = \{0, 1\}$. Then every state of the CR with $n \in \mathbb{N}_0$ rings can be represented by an $s = s_n \ldots s_1 \in B^n$, where $s_r = 0$ (1) means that ring $r \in [n]$ (numbered from right to left) is off (on) the bar. (We include the case $n = 0$ for technical reasons; $s \in B^0$ is the empty string then by convention.) According to the rules, a bit s_{k+1} can be switched, if either $k = 0$ (move type 0) or $s_k = 1$ and $\forall l \in [k-1] : s_l = 0$ (move type 1). That is to say, with $b \in B$,

- any state $x \ldots yb$ can be transformed into $x \ldots y(1-b)$ (move type 0),

- any state $x \ldots yb10 \ldots 0$ into $x \ldots y(1-b)10 \ldots 0$ (move type 1).

For $n \in \mathbb{N}_0$ the task translates to finding a (shortest) path from 1^n to 0^n in the graph R^n whose vertex set is B^n and whose edges are formed by pairs of states differing by the legal switch of one bit. (Here and in the sequel the string $b \ldots b$ of length $k \in \mathbb{N}_0 \cup \{\infty\}$ will be written as b^k; similarly, for strings s and t, st will denote the concatenated string.) Since this graph is undirected, we may as well start in $\alpha^{(n)} := 0^n$, a vertex of degree 1, if $n \in \mathbb{N}$. Its only neighbor is $0^{n-1}1$. If $n > 1$, the next move is either to go back to 0^n, which in view of a *shortest* path to 1^n is certainly not a good idea, or to the only other neighbor $0^{n-2}11$. Continuing this way we will never return to a vertex already visited, because their degrees (at most 2) have been used up already. Therefore, we finally arrive on this path graph at the only other vertex of degree 1, namely $\omega^{(n)} := 10^{n-1}$. (In addition, we define $\omega^{(0)}$ as the empty string for definiteness as before.) This does, however, not guarantee that we passed the goal state 1^n on our way! A graph with vertices of degree 2 except for exactly two pendant vertices consists of a path and a certain number of cycles (cf. Proposition 0.11). So what we have to show is that R^n is connected.

Theorem 1.1. *The graph R^n is connected for every $n \in \mathbb{N}_0$. More precisely, R^n is the path on 2^n vertices from $\alpha^{(n)}$ to $\omega^{(n)}$.*

Proof. The proof is by induction on n. The case $n = 0$ is trivial. For $n \in \mathbb{N}_0$ we know by induction assumption that R^n is the path from $\alpha^{(n)}$ to $\omega^{(n)}$. Attaching 0 at the left of each of its vertices we get a path from $\alpha^{(1+n)}$ to $0\omega^{(n)}$ in R^{1+n} which passes through all states that start with 0. Similarly, attaching 1 to the vertices of the same path but taken in reverse order, gives a path on 2^n vertices in R^{1+n} between $1\omega^{(n)}$ and $\omega^{(1+n)}$. Since these two paths are linked in R^{1+n} by precisely one edge, namely the edge between $0\omega^{(n)}$ and $1\omega^{(n)}$, the argument is complete. $\square$

Remark. Readers with a *horror vacui* are advised to base induction on $n = 1$.

Remark 1.2. *Combining the move types we get the more formal definition of the graphs R^n by*

$$V(R^n) = B^n, \ E(R^n) = \left\{ \left\{ \underline{s}0\omega^{(r-1)}, \underline{s}1\omega^{(r-1)} \right\} \mid r \in [n], \ \underline{s} \in B^{n-r} \right\},$$

where for each edge r is the moving ring. Note that the distribution $\underline{s}$ of rings $r+1$ to n is arbitrary.

Here is an alternative proof for Theorem 1.1. The path $P^n \subseteq R^n$ leading from $\alpha^{(n)}$ to $\omega^{(n)}$ must contain the edge $e := \{0\omega^{(n-1)}, 1\omega^{(n-1)}\}$, because this is the only legal way to move ring n onto the bar. This means that P^n contains, in obvious notation, $0P^{n-1}$, e, and $1P^{n-1}$, the latter traversed in inverse sense. Hence, $|P^n| \geq 2|P^{n-1}|$, such that, with $|P^0| = 1$, we get $|P^n| \geq 2^n$ and consequently $P^n = R^n$. In other words, R^n is obtained by taking two copies of R^{n-1}, reflecting the second, and joining them by an edge. As an example, R^3 is the path graph depicted in Figure 1.2. The reflection is indicated with the dashed line and the digits that were added to the graphs R^2 are in red color.

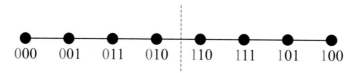

Figure 1.2: The graph R^3

From this it is obvious that the edge sets of the *Chinese rings graphs* can also be defined recursively.

Remark 1.3. *The edge sets of* R^n *are given by*

$$E(R^0) = \varnothing,$$

$$\forall n \in \mathbb{N}_0 : E(R^{1+n}) = \{\{ir, is\} \mid i \in B, \{r, s\} \in E(R^n)\} \cup \{\{0\omega^{(n)}, 1\omega^{(n)}\}\}.$$
(1.1)

The reader is invited (Exercise 1.1) to deduce the recurrence

$$\forall n \in \mathbb{N}_0 : \ell_n + \ell_{n-1} = M_n$$
(1.2)

for the number ℓ_n of moves needed to take off $n \in \mathbb{N}_0$ rings from the bar, i.e. to get from 1^n to 0^n in R^n. Obviously, $\ell = 0_2, 1_2, 10_2, 101_2, 1010_2, 10101_2, \ldots$ in binary representation, such that $\ell_n = \lceil \frac{2}{3}(2^n - 1) \rceil = \lfloor \frac{2}{3} 2^n \rfloor$ (A000975). This sequence is denoted by ℓ for Lichtenberg who derived it as early as 1769 in [266] and already used n to denote the number of rings! The sequence of differences $J := \overline{\ell}$ is the *Jacobsthal sequence* (A001045). Also in 1769, Arima came up with an investigation into the CR (cf. the *Historical note* in [248, p. 679]) resulting in his sequence $A_n = \frac{1}{2}(J_n + 1)$ (for $n \in \mathbb{N}$) which gives the number of down moves of ring 1 in the solution for n rings; see [107]. The history and mathematics of the Lichtenberg sequence can be found in [202]; Paul K. Stockmeyer provided some more properties in [406]. J. Huang, M. Mickey, and J. Xu discovered a connection of the sequence with the Catalan numbers (cf. below, Exercise 8.10) by interpreting the Lichtenberg sequence as the so-called non-associativity measurement of the double minus operation [219, Theorem 9].

The classical CR with $n = 9$ rings can therefore be solved in a minimum of 341 moves. In order to find out whether we have to move ring 1 or ring 2 first when we begin with all rings *on* the bar, we make the following observation.

Proposition 1.4. *The function* $B^n \ni s \mapsto (\sum_{r=1}^n s_r) \bmod 2 \in B$ *defines a vertex coloring of* R^n*; the type of the move associated with an edge defines an edge coloring.*

The proof is left as an exercise (Exercise 1.2).

Let us break to look back at early theories about the number of moves for a solution of the CR.

Since, starting in 0^n, the first and then every second move on the path graph R^n is of type 0, ring 1 moves 2^{n-1} times on it. This makes up for 2^n Wallis movements (cf. p. 58). The other $2^{n-1} - 1$ moves produce $2^{n+1} - 4$ movements. For $n = 9$, this leads to 1532 movements; to this one has to add the movement of the bar right to the left end to arrive at Wallis's value in Section 0.10.1.

Another counting was used by Cardano (cf. [187]), where a simultaneous move up or down of rings 1 and 2 is allowed for $n \in \mathbb{N}_2$. Lucas called this the *accelerated Chinese rings* in [283, p. 183–186]. The **diameter** of the corresponding graph reduces to $3 \cdot 2^{n-2} - 1$, and the standard task can then be solved in $p_n = 2^{n-1} - (n \text{ even})$ moves; see Exercise 1.3. The sequence p is called *Purkiss sequence* (essentially A051049) for Henry John Purkiss (1842–1865), who seems to have been the first to publish it in 1865 ([347]; cf. the *Historical note* in [248, p. 679]). (Lucas attributes the solution to Théodore Parmentier (1821–1910) ([282, p. 42]), from whom he learned about Gros's note ([282, p. 37]), so that the letter "p" is justified anyway.) The three CR sequences are related by the formula (see [202, (PLA)])

$$\forall\, n \in \mathbb{N}_2 : p_n = \ell_n - 2A_{n-1} + 1\,.$$

It might be true that the Latin in Cardano's passage [71, p. 492f] is difficult to understand; possibly, as Gros [168] assumes, because written by a second hand. But it is clear from the beginning that he is dealing with the accelerated Chinese rings for $n = 7$. Therefore, the (only) numbers 31, 64, 95, and 190 occurring in the text are neither misprinted nor mathematical errors, as Gros and others claim, but are, respectively, p_6, p_7, $3 \cdot 2^5 - 1$, and twice the latter number. So Cardano gave the lengths of the solutions to get from 10^6 to 1^7, from 1^7 to 0^7, for the whole path and for the "circulus".

We are now also able[40] to interpret the two tables in the book by Zhu Xiang Zhuren mentioned in the introduction. We refer to Figure 0.29 whose contents have to be read from top to bottom and from right to left. The passage is entitled *Untangling Linked Rings Method* (解连环法). In the top row the rings are numbered above the puzzle (which has its handle on the right) as in Figure 1.1; instructions are given how to operate it and how to use the tables, the second of which "many times". The insert in the center row says "When possible, move the first two rings at the same time in one move.", that is, Cardano's counting is employed. Table I in the same row consists of 11 columns and 6 rows. The first case (top right) does not count, but has the overlaid inscription "This table takes the 9th ring off." And that is what it does! It starts (below the latter inscription) with "1 off, 3 off, 1 on, 2 and 1 off" and reaches the end of the first column with "5 off". (The Chinese characters for "on" (up) and "off" (down) are 上 (shang) and 下 (xia), respectively; the number symbols can easily be taken from the very top of the figure.) Then the second column starts with "2 and 1 on". Below the bottom left case, we find the inscription "The 9th ring is now off". In our notation this table solves the task $1^9 \to 010^7$, and the number of moves needed can be calculated from the dimensions of Table I as $11 \times 6 - 1 = 65$ and read from the line above the table. But for the last move, this is the solution for the 7-ring puzzle in $p_7 = 64$ moves! The task is therefore reduced to solving $10^7 \to 0^8$, i.e. to traverse the whole path graph R^8, which should take $3 \cdot 2^6 - 1 = 191$ further moves. This is done with the aid of Table II in the bottom row of Figure 0.29. A first run leads to ring 8 being taken off, i.e. to state $0^2 10^6$, after $16 \times 6 = 96$ moves. (The entry in the penultimate case of the table wants to indicate the state *before* ring 8 is taken off; given that ring 9 was taken off by Table I, "rings 8 and 9 are on" is an obvious misprint for "rings 7 and 8 are on".) Instead of making use of the reflective structure of the state graph and going back Table II switching "on" and "off", the author now employs an iterative method, using the table once again

---

[40] only with the generous aid by Wei Zhang and Peter Rasmussen

from the beginning, but stopping in the middle, when ring 7 can be taken off. This is mathematically correct, because the first half of the CR graph is the same if the leading 0 is deleted. The kth application of Table II therefore solves the task $0^k 10^{8-k} \to 0^k 1^2 0^{7-k}$, i.e. $0^{8-k} \to 10^{7-k}$, in $3 \cdot 2^{6-k} - 1$ moves and then puts ring $9 - k$ down. So the 6th run ends already halfway through the rightmost column. This is indicated by the two lines on the margin saying that "rings 2 and 3 are on" and then "ring 3 is now off." Finally, the last two moves are in cases 4 and 5 of the right-hand column. The move numbers to get rings 3 to 8 off are again given above Table II. Adding up these values together with the two last moves gives the correct sum 191 indicated above.

Coming back to the standard style of counting individual ring moves, we have:

Proposition 1.5. *The CR with $n \in \mathbb{N}$ rings have a unique minimal solution of length ℓ_n. It can be realized by alternating moves of types 0 and 1, starting with type 0 if n is odd and type 1 otherwise.*

Remark 1.6. *If one finds the CR abandoned in state $s \in B^n$, there is an easy way to decide about the best first move to get to $\alpha^{(n)}$ (the next moves being obvious on a path graph). As neighboring states differ by one bit only, the sum of bits of s will be odd if the last move on the path from $\alpha^{(n)}$ to s is made by ring 1 and even otherwise. So the parity of the sum of bits of s will tell you which move to make first on the reverse path.*

Every connected graph has a canonical metric, namely its *graph distance* $\mathrm{d}(s,t)$ between two vertices s and t being given by the length of a shortest s,t-path. A direct application is the existence of *perfect codes* for R^n, see Exercise 1.4. The proof of Theorem 1.1 shows that the diameter $\mathrm{diam}(R^n)$ of R^n is $\mathrm{d}(\alpha^{(n)}, \omega^{(n)}) = 2^n - 1$ and that there is a unique shortest path between any two states of the CR. Their distance is given by $\mathrm{d}(s,t) = |\mathrm{d}(s) - \mathrm{d}(t)|$, where $\mathrm{d}(s) := \mathrm{d}(s, \alpha^{(n)})$ can be determined by a finite automaton (cf. Figure 1.3). It consists of two states A and B. The input of a bit a in A results in printing a and moving to state B if $a = 1$; the input of b in B leads to printing of $1 - b$ and moving to A if $b = 1$.

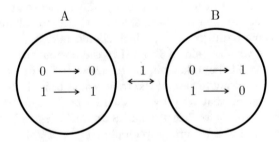

Figure 1.3: Automaton for the Gros code

Proposition 1.7. *If we enter the components s_r of $s = s_n \ldots s_1 \in B^n$, $n \in \mathbb{N}$, from left to right, starting in state* A *of the automaton, the resulting output, read from left to right and interpreted as a binary number, gives the value of* $\mathrm{d}(s)$.

Proof. We prove by induction on n that input of s starting in A gives $\mathrm{d}(s)$ and input of s starting in B gives $2^n - 1 - \mathrm{d}(s)$. This is obviously true for $n = 1$. Let $n \in \mathbb{N}$ and $s \in B^{1+n}$. If $s = 0\bar{s}$, $\bar{s} \in B^n$, then starting in A leads to $\mathrm{d}(\bar{s}) = \mathrm{d}(s)$, because $\mathrm{d}(s) < 2^n$; starting in B gives $2^n + 2^n - 1 - \mathrm{d}(\bar{s}) = 2^{1+n} - 1 - \mathrm{d}(s)$. If $s = 1\bar{s}$, then starting in A leads to $2^n + 2^n - 1 - \mathrm{d}(\bar{s}) = \mathrm{d}(s)$, because $\mathrm{d}(s) = 2^n + \mathrm{d}(\bar{s}, \omega^{(n)})$; starting in B gives $\mathrm{d}(\bar{s}) = 2^{1+n} - 1 - \mathrm{d}(s)$. □

Remark 1.8. *It follows immediately from Proposition 1.4 that the best first move from state s to state t is of type $\mathrm{d}(s) \bmod 2$, if $\mathrm{d}(s) < \mathrm{d}(t)$ and of type $1 - (\mathrm{d}(s) \bmod 2)$ otherwise. The rest of the shortest path is then again obtained by alternating the types of moves.*

The bijection from B^n to $[2^n]_0$ provided by the automaton above is a coding of the states of the CR by the distance from the state $\alpha^{(n)}$. This code goes back to Gros from his work [168] on the baguenaudier. Its inverse is called *Gray code*, after F. Gray, who got a patent [165] for it in 1953. See Figure 1.4 for the case $n = 3$.

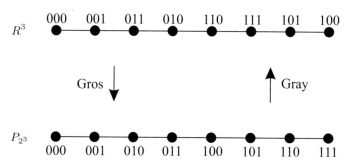

Figure 1.4: The graph R^3 and a path graph of length 7

In fact, if d_r are the bits of $d \in \mathbb{N}_0$, we may put

$$\forall\, r \in \mathbb{N}: \; s_r = d_r \,\dot{\vee}\, d_{r-1}. \tag{1.3}$$

In other words, the bit string $(s_r)_{r \in \mathbb{N}}$ is obtained as the *Nim-sum* of d and $\lfloor \frac{d}{2} \rfloor$. Then $\mathrm{d}(s) = d$, as can be seen by applying the automaton or either Gros's formula $d_{r-1} = d_r \,\dot{\vee}\, s_r$ (cf. [168, p. 13]). The corresponding automaton for the Gray code is shown in Figure 1.5. Here moves between the two states of the automaton are performed according to the one-sided arrows. For an application, see Exercise 1.5.

Because of the construction in the proof of Theorem 1.1, the Gray code is also called *reflected binary code* (cf. Figure 1.2). Its main advantage is that neighboring code numbers differ by exactly one bit. More on Gray codes can be found in [248,

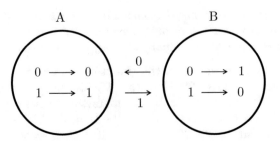

Figure 1.5: Automaton for the Gray code

Section 7.2.1.1]. For instance, it can also be produced in an iterative way with the aid of the Gros sequence; see Section 1.2 below.

A state $s \in B^n$ being uniquely determined by its distance $\mathrm{d}(s)$ from $\alpha^{(n)}$ and with all values from 0 to $2^n - 1$ occurring, it is obvious that the average distance to $\alpha^{(n)}$ (or to $\omega^{(n)}$ for that matter) in R^n is $2^{-n}\Delta_{2^n - 1} = (2^n - 1)/2$. This is not too surprising given that R^n is isomorphic to the path graph on 2^n vertices P_{2^n}.

The **eccentricity** $\varepsilon(v)$ of a vertex v in a path graph is always the maximum of its distances to the two end vertices. Therefore, in R^{1+n}, $n \in \mathbb{N}_0$, we have

$$\forall\, i \in B,\ s \in B^n : 2^n \le \varepsilon(is) = 2^{n+1} - 1 - \mathrm{d}(s) \le 2^{n+1} - 1 \,,$$

and every value in that range appears precisely once for each i. Hence, the **total eccentricity** of R^{1+n} is (cf. A010036)

$$\mathrm{E}(R^{1+n}) = 2 \sum_{k=2^n}^{2^{n+1}-1} k = (2^{n+1} - 1)2^{n+1} - (2^n - 1)2^n = 2^n(3 \cdot 2^n - 1)\,,$$

such that the *average eccentricity* on R^{1+n} turns out to be (cf. [212, Proposition 4.1])

$$\bar{\varepsilon}(R^{1+n}) = \frac{1}{2}(3 \cdot 2^n - 1)\,, \tag{1.4}$$

i.e. asymptotically (for $n \to \infty$) 3/4 of the diameter.

To find the average distance on R^n, we can start off from the notion of the *Wiener index* $\mathrm{W}(G)$ of a connected graph G, namely the total sum $\mathrm{W}(G) = \displaystyle\sum_{\{s,t\}\in\binom{V(G)}{2}} \mathrm{d}(s,t)$ of distances between any two vertices of G. The *average distance* on G is then[41]

$$\bar{\mathrm{d}}(G) = \frac{2\,\mathrm{W}(G)}{|V(G)|^2}\,.$$

Knowing the Wiener indices of path graphs from Exercise 1.6, we get

$$\bar{\mathrm{d}}(R^n) = \frac{1}{3}(2^n - 2^{-n})\,, \tag{1.5}$$

---

[41]Some authors prefer to divide by $|V(G)|(|V(G)| - 1)$ only; cf. [67, Equation (9.4)].

or approximately 1/3 of the diameter.

Finally, it would be interesting to know whether it is likely to arrive at a solution for the CR by chance, that is if the player makes moves at random choosing move types 0 or 1 with equal likelihood 1/2 (except in states $\alpha^{(n)}$ and $\omega^{(n)}$). Based on the method of Markov chains (cf. [276] and below, Section 4.1), H. L. Wiesenberger has found [434, p. 66f] that the expected number of moves to reach state t from s on such a random walk in R^n is

$$\mathrm{d}_e(s,t) = \begin{cases} \mathrm{d}(t)^2 - \mathrm{d}(s), & \text{if } \mathrm{d}(t) > \mathrm{d}(s); \\ 0, & \text{if } \mathrm{d}(t) = \mathrm{d}(s); \\ (2^n - 1 - \mathrm{d}(t))^2 - (2^n - 1 - \mathrm{d}(s)), & \text{if } \mathrm{d}(t) < \mathrm{d}(s). \end{cases}$$

In particular, $\mathrm{d}_e(0^n, 1^n) = \ell_n^2$, but $\mathrm{d}_e(1^n, 0^n) = (2^n - 1)^2 - \ell_{n-1}$ for $n \in \mathbb{N}$. So it was a good idea of Hung Ming to give the jiulianhuan to his wife with all rings on the bar, because she would need 260951 moves to get them all off, in contrast to 116281 moves the other way round, if she wanted to solve it without thinking. These huge numbers demonstrate the advantage of a good mathematical model!

1.2 The Gros Sequence

CR yields an interesting integer sequence—the Gros sequence. It will become clear later that this is just the tip of an iceberg because many additional interesting integer sequences will appear in due course.

An analysis of the solution for the CR reveals the following: assuming that there are infinitely many rings and starting with all of them off the bar, the sequence of rings moved starts

$$g = 1, 2, 1, 3, 1, 2, 1, 4, 1, 2, 1, 3, 1, 2, 1, 5, 1, 2, 1, 3, 1, \ldots$$

Note that we have paused at the moment when the first 5 rings are on the bar. This sequence is, of course, already anticipated in Zhu Xiang Zhuren's solution: the ring numbers in the bottom line of his Table II (cf. the bottom row in Figure 0.29) reading $4, 5, 4, 6, 4, 5, 4, 7, 4, 5, 4, 6, 4, 5, 4, 8$. P. Rasmussen came up with a nice mnemonic:

one two one three one two one four one two one three one two one more .

To the best of our knowledge, the sequence g was studied for the first time by Gros in his 1872 pamphlet [168], hence we named it the *Gros sequence*. It is the sequence A001511, referred to as the *ruler function*; cf. the markings on an imperial ruler, just like the heights of the columns in Figure 0.13. The Gros sequence obeys the following appealing recurrence.

Proposition 1.9. *For any* $k \in \mathbb{N}$,

$$g_k = \begin{cases} 1, & k \text{ odd}; \\ g_{k/2} + 1, & k \text{ even}. \end{cases}$$

Proof. Recall from Proposition 1.4 that in the solution of the CR, the moves alternate between type 0 and type 1. So every odd numbered move (or odd move for short) is of type 0 which implies that $g_k = 1$ when k is odd. Now consider even moves (that is, moves of rings $2, 3, 2, 4, \ldots$) and observe that they are identical to the solution of the puzzle with rings $2, 3, \ldots$. Since these moves appear in even steps, the second assertion follows.   □

Repeated application of the recursion in Proposition 1.9 implies:

Corollary 1.10. *Let* $k \in \mathbb{N}$ *and write* $k = 2^r(2s+1)$, $r, s \in \mathbb{N}_0$. *Then* $g_k = r + 1$.

An easy consequence (see Exercise 1.7) of this corollary is yet another characterization of the Gros sequence reflecting the reflectedness of the Gray code, namely

$$\forall n \in \mathbb{N}_0 : \left(g_{2^n} = n + 1 \wedge \forall k \in [M_n] : g_{2^n - k} = g_k = g_{2^n + k} \right). \tag{1.6}$$

In this form the Gros sequence has been employed as a "naming sequence" in a "silent symmetric space-optimal [population] protocol" in [37, Section 4].

Corollary 1.10 can be rephrased by saying that g_k is the number of times the factor 2 appears in $2k$. Already in 1808, A. M. Legendre posed the question, how many factors 2 the number $n!$ has [258, p. 8–10] (cf. [247, p. 51, Exercises 11/12]). This amounts to summing the *binary carry sequence* $\widetilde{g}$ (A007814), defined in Exercise 1.8, where it is shown that $\widetilde{g}_k = g_k - 1$. He obtained the remarkable formula

$$\sum_{k=1}^{n} \widetilde{g}_k = n - q(n), \tag{1.7}$$

where $q(n)$ denotes the number of 1s in the binary representation of n; see Exercise 1.9 for the proof.

Gros makes use of Corollary 1.10 for his "practical rule" [168, p. 14f] for a move in a state s with $\mathrm{d}(s) = k$: the move which led to s from $\alpha^{(n)}$ is by ring g_k, which is the position (counted from 1) of the rightmost bit 1 in k. Therefore, a move of this ring *from* s will lead to $\alpha^{(n)}$. By symmetry, moving away from $\alpha^{(n)}$ involves the ring with the position number of the rightmost 0 in the binary representation of k. Compared to our recipe in Remark 1.6, this procedure has the practical disadvantage that the state s is visible for the problem solver, but k has to be calculated. For the general task to get from s to t, solved by Lucas [283, p. 179], this is unavoidable though; cf. Remark 1.8.

We may now come back to the two properties of the derivative d' of the binary representation d of a natural number, as derived after Theorem 0.6. In (1.3) a state $s \in B^n$ was constructed from its *Gros weight*, i.e. its distance $(d_{n-1} \ldots d_0)_2$ to state

0^n on R^n, in such a way that $\forall r \in \mathbb{N}: d'_{r-1} = s_r$. The finite binary sequence $\sigma = (s_{k+1})_{k \in \mathbb{N}_0}$ is the Gray code of the number (or finite binary sequence) d. Of course, we may identify $s \in B^n$ and the corresponding $\sigma \in B^{\mathbb{N}_0}$ with

$$\sum_{k=1}^{n} s_k \cdot 2^{k-1} = \sum_{k=0}^{n-1} \sigma_k \cdot 2^k \in [D_n]_0,$$

respectively. Then, the edge $\{\underline{s}0\omega^{(r-1)}, \underline{s}1\omega^{(r-1)}\} \in E(R^n)$ joins the natural numbers $\overline{M}_{r-1} + \underline{s} \cdot 2^r$ and $\overline{M}_{r-1} + 2^{r-1} + \underline{s} \cdot 2^r$. The function b in [96] is $(\cdot)_2^{-1}$. Therefore, if $b(d)' = \underline{s}s_r\omega^{(r-1)}$ and $b(d+1)' = \underline{s}(1 - s_r)\omega^{(r-1)}$, we have $|(b(d+1)')_2 - (b(d)')_2| = 2^{r-1}$ and r follows the Gros sequence when d runs through $\mathbb{N}_0$.

On the other hand, for each state $s \in B^n$, its Gros weight $(d_{n-1} \ldots d_0)_2$ can be found by $d_{r-1} = d_r \dot{\vee} s_r$, beginning with $r = n$ (where $d_n = 0$) and down to $r = 1$. We have $(\int \sigma)_r = \sum_{k=r}^{n-1} \sigma_k = \sum_{k=r}^{n-1} s_{k+1} =: \delta_r$, so that $\delta_n = 0$ and $\delta_{r-1} - \delta_r = s_r$, whence $d = \int \sigma$, which corresponds to Gros's algorithm in [168, p. 13]. We therefore call d the *Gros code* of σ.

Another use can be made of the binary representation to find out whether in move $k = \sum_{\nu=0}^{\infty} k_\nu \cdot 2^\nu$ ring $r = g_k$ goes up (1) onto the bar or down (0), still assuming an infinite supply of rings and starting with all of them off the bar. Let us denote the resulting binary sequence by $f \in B^{\mathbb{N}}$. Then $\forall k \in \mathbb{N}: f_k = 1 - k_{g_k}$, because this is the position (off or on the bar) of ring g_k after move number k according to (1.3). From Proposition 1.9 it is clear that f_k can be obtained recursively by

$$\forall \ell \in \mathbb{N}_0: f_{2\ell+1} = 1 - \ell \bmod 2, \quad f_{2\ell+2} = f_{\ell+1}.$$

Let us look at the sequence f from a different prospect. Have you ever wondered why it is so difficult to fold a package insert of some medicine back after it had been unfolded? Try the inverse: take a lengthy strip of paper, fold it on the shorter center line and keep on doing this, always in the same direction, as long as it is physically feasible, the mth bending producing 2^{m-1} new folds. In the left picture of Figure 1.6 four foldings have been performed and colored according to their appearance, such that altogether 15 edges occur on the paper strip. Now unfold with approximately right angles at these fold edges. Despite the straightforward, symmetric procedure, the paper strip, viewed from the edge, will look surprisingly erratic as in the right-hand picture of Figure 1.6. Rotating at the center (red) fold, then at the secondary (green) one, then at blue and finally violet, you can solve the package insert problem.

The arising polygon has been called a *dragon curve* (of order 4) because its shape tends (for higher orders) to the silhouette of a sea dragon; it is one of the favorites of fractal people (cf. [302, p. 66]) because of its interesting properties. For instance, in the top left picture of Figure 1.7 four copies of the curve of order 4 from Figure 1.6 meet at right angles in the center point. In the subsequent pictures

Figure 1.6: Folding and unfolding a paper strip

the order grows up to 11, and at each step a scaling by the factor $1/\sqrt{2}$ has been performed such that the maximal squares inside which all gridpoints are covered have equal side-length. Figure 1.7 can be viewed as a proof without words for the space-filling property of dragon curves; cf. [361, p. 163f].

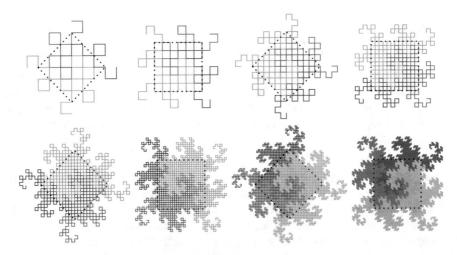

Figure 1.7: Space-filling property of the dragon curve

The folding is completely described by the sequence of orientations of the turns at the fold edges, i.e. the corners of the curve, either right (0) or left (1). For an $(m + 1)$-fold folding, $m \in \mathbb{N}_0$, this *paper-folding sequence* $\varphi \in B^{\mathbb{N}}$ fulfills, if we start with a left bending,

$$\forall\, m \in \mathbb{N}_0: \ \varphi_{2^m} = 1, \ \forall\, \mu \in [2^m - 1]: \ \varphi_{2^m + \mu} = 1 - \varphi_{2^m - \mu}$$

because of the reflection at the center fold 2^m. This is similar to the behavior of the Gros sequence as noticed by A. Sainte-Laguë [363, p. 40]. (The Gros sequence is misprinted there.) More precisely, as observed in [110], φ represents the pattern of ups and downs in the CR! Ring $r \in \mathbb{N}$ is moved up for the first time in move number 2^{r-1} after a sequence of moves leading from $\alpha^{(r-1)}$ to $\omega^{(r-1)}$ has been performed. This is followed by a complete transformation from $\omega^{(r-1)}$ to $\alpha^{(r-1)}$, i.e. the original subsequence in reflected order. So the sequence

$$f = 1, 1, 0, 1, 1, 0, 0, 1, 1, 1, 0, 0, 1, 0, 0, \ldots$$

is the paper-folding sequence φ (cf A014577), and our Chinese lady may amuse herself with either the CR or the folding of paper strips.

The Gros sequence can be found all over mathematics, for instance in connection with hamiltonian cycles on the edges of n-dimensional cubes; see Exercise 1.10. If in Figure 0.3 we start on bottom at 0 and change bits going around counter-clockwise according to the Gros sequence, i.e. $1, 2, 1, 3, 1, 2, 1$, we obtain the numbers from 0 to 7 in the order of the Gray code as in Figure 1.8.

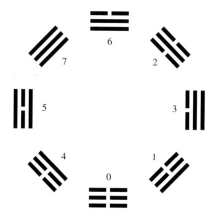

Figure 1.8: Gray's arrangement of the trigrams

Before we turn to some applications of the CR in Section 1.3, we show the presence of the Gros sequence in the fascinating theory of square-free sequences.

The greedy square-free sequence

A sequence $a = (a_n)_{n \in \mathbb{N}}$ of symbols a_n from an *alphabet* A is called *non-repetitive* or *square-free* (over A) if it does not contain a subsequence of the form xx (a *square*), where x is a non-empty subsequence of consecutive symbols of a. Clearly, $1, 2, 3, 4, \ldots$ is a square-free sequence over the alphabet $\mathbb{N}$. It is not very exciting though and expensive in the sense that it uses large numbers at early stages. So let us try to find a cheaper one by one of the most popular strategies in the theory of algorithms, namely the greedy approach; cf. p. 48. Roughly speaking, given a problem, a solution to the problem is built step by step, where in each step a partial solution is selected that optimizes certain greediness criteria.

In our case, the most obvious greediness criterion is to select the next term of the sequence as the smallest positive integer that does not produce a square. Let $a = (a_n)_{n \in \mathbb{N}}$ be the sequence obtained by this procedure; i.e.,

$$\forall\, n \in \mathbb{N} : a_n = \min \{\alpha \in \mathbb{N} \mid a_1, \ldots, a_{n-1}, \alpha \text{ contains no square}\}.$$

The sequence a is square-free, because any square would already occur at some finite stage of its construction. Clearly, $a_1 = 1$, $a_2 = 2$, and $a_3 = 1$. Then $a_4 \neq 1$ and $a_4 \neq 2$, hence $a_4 = 3$. Continuing such a reasoning it is easy to see that the sequence a begins as follows:

$$1, 2, 1, 3, 1, 2, 1, 4, 1, 2, 1, 3, 1, 2, 1, 5, 1, \ldots$$

From here we can guess the following result, a formal proof of which is left for Exercise 1.11:

Proposition 1.11. *The sequences g and a are the same.*

In particular, the Gros sequence is square-free ([197, Theorem 0]); it is even strongly square-free; cf. Exercise 1.12. We will return to square-free sequences in Chapter 2.

1.3   Two Applications

In this section we demonstrate how the theory of the CR and the Gros sequence can be helpful in the analysis of other problems.

Topological variations

We may turn our viewpoint around and ask whether there are other puzzles whose state graphs are isomorphic to R^n. In psychology literature they are called *problem isomorphs*. An example is the study [255] by Kenneth Kotovsky and Herbert A. Simon. They introduce a computerized version of the CR which they call *digital*, the original CR being considered as analog. Instead of moving rings on and off a bar, balls are switched between two boxes each. Figure 1.9 shows the move of ball 1, corresponding to the edge $\{1010000, 1010001\}$ in R^7. In each state balls that can move are colored green, the others red. As we know from the CR ball 1 is always movable and, except for states $\alpha^{(n]}$ and $\omega^{(n]}$, exactly one other ball, to the right of which there is the first ball in box 1 and all others in their respective box 0, can be moved.

Although the state graphs of the classical and the digital CR are isomorphic, the human test person will face different difficulties in solving the tasks. In fact, the classical CR with just 5 rings "proved to be virtually impossible to solve" in 90 minutes [255, p. 152]; this is quite alarming when compared to the performance of

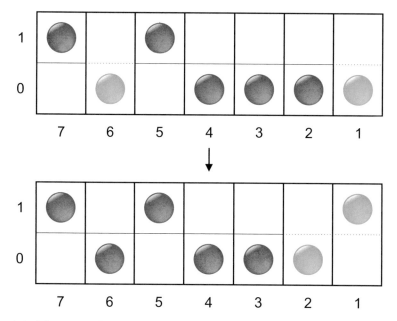

Figure 1.9: The move from 1010000 to 1010001 in the digital Chinese rings with 7 balls

students 80 years before in Ruger's experiments [358]! The difficulty in several experiments with additional information turned out to vary and the effect of *transfer* was studied, i.e. whether the order in which different isomorphs were administered has an influence on performance.

It has been regarded as a deficiency that the classical realization of R^9 in form of the CR as shown in Figure 1.1 physically allows us to move rings 1 and 2 simultaneously if they are either both on or both off the bar. This is why we insisted on individual ring moves and counted them accordingly (see, however, Exercise 1.3). A variant which does not display this ambiguity was patented under the title *Locking disc puzzle* (LD) by W. Keister (cf. [229]). Here the intertwined system of rings of the CR is replaced by an arrangement of circular discs on a slide with three of the 90° sectors of the circle cut out to allow for rotation about their center (see Figure 1.10 for the 6-disc version). (To avoid a trivial solution, the first disc has only two sectors cut out and the one opposite to the convex side just flattened.) The bar of the CR corresponds to a frame in LD which is designed in a way to realize the same kind of individual moves (i.e. rotation of discs and moving the slide back and forth) as for the CR. Thus the conditions "off" and "on" the bar of the rings in CR is translated into an orientation of the convex side of a disc parallel to the slide or perpendicular to it: as soon as all discs are positioned horizontally, the slide can be pulled out of the frame. However, this puzzle is *not* equivalent to the CR, because there are two horizontal orientations of the discs (a second vertical orientation is debarred by the frame), such that the corresponding state graph has more vertices than R^6 and therefore can not be isomorphic to it.

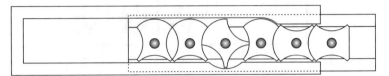

Figure 1.10: The Locking disc puzzle

In a second attempt to improve the CR hardware, Keister came forward with the *Pattern-matching puzzle* (PM) (cf. [230]), where in addition to a slide and a frame there is a rack attached to the frame which holds 4 pattern bars. The latter constitute a 4-bit code which is fixed at the beginning. Due to the mechanical arrangement of PM only one of the 8 teeter bars on the slide can be moved at a time and this move of bar $b \in [8]$ between two positions "up" (1) and "down" (0) is possible only if the bars with smaller labels coincide with the given code (filled up to the right by 0s; we adapted the labelings to our CR notations). Again the task starts with all teeter bars up, i.e. in state 1^8, and the slide can be detached from the frame only in state 0^8. It turns out that one can reach each of the 2^8 states of the teeter bars from every other one in a unique path, such that the corresponding state graph for each of the 16 codings of the pattern bars is a **tree** on 2^8 vertices. Therefore, the task to free the slide has different (lengths of) solutions depending on the code, 0000 leading to the shortest one with just 8 moves. Only in the special case 1000 of the pattern code the corresponding tree is actually a path graph and the puzzle isomorphic to the CR with 8 rings, such that the task needs $\ell_8 = 10101010_2 = 170$ moves.

A much more challenging variant of the CR is depicted in Figure 1.11, and following E. R. Berlekamp, J. H. Conway and R. K. Guy [52, p. 858] we will call it the *Chinese string (CS)*. Here the system of $n \in \mathbb{N}$ rings in the original puzzle is arranged in a rigid, but otherwise topologically equivalent manner in a *frame*, e.g., made of wood. The shuttle of the CS is replaced by a flexible *rope* (or string) of sufficient length. At the beginning, the rope, initially separated from the frame, is somehow entangled with it and the task for the player is to disentangle the rope from the frame. (The dotted arc in Figure 1.11 does not belong to the puzzle and will be explained later.)

Because of the flexibility of the rope it is clear that there are many more possible states of the CS puzzle, but that any distribution $s \in B^n$ of the CR can also be realized in the CS by inserting the rope into the frame in the same geometric fashion as the shuttle is moved into the rings in the CR. For an example, see Figure 1.12 showing $s = 11001$.

Since there is no obvious notion of a *move* in the CS puzzle, the complexity of a solution has to be defined in a topological way. In [228], L. H. Kauffman

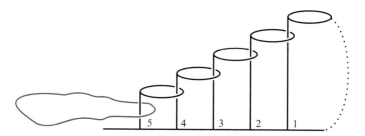

Figure 1.11: Chinese string puzzle

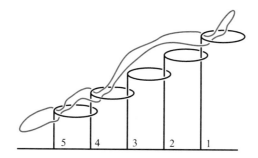

Figure 1.12: State 11001 of the Chinese string

suggests to add an imaginary arc running from the tip of ring 1 to the base of the frame as in Figure 1.11 and to count the number of crossings of this arc with the loop during the execution of a solution; the minimal number of crossings taken over all solutions is called the *topological exchange number* $E_{\text{top}}(s)$ of s. Similarly, we will define $E_{\text{top}}(s,t)$ as the *topological exchange number* to get from s to t; i.e. $E_{\text{top}}(s) = E_{\text{top}}(s, \alpha^{(n)})$. Every solution to get from s to t in CR can also be performed on the CS. Herein only the moves of ring 1 will correspond to crossing the arc. Therefore the topological exchange number is bounded from above by the number of moves of ring 1 in the optimal solution in CR, which will be called the *mechanical exchange number*; it is

$$E_{\text{mec}}(s,t) = \left\| \left\lceil \frac{\text{d}(s)}{2} \right\rceil - \left\lceil \frac{\text{d}(t)}{2} \right\rceil \right\| .$$

Kauffman's *Ring conjecture* [228, p. 8] says that this is also a lower bound for the topological exchange number of s; in particular, $E_{\text{mec}}(s) = E_{\text{top}}(s)$.

This conjecture has been confirmed, for a topologically equivalent version of the CS, by J. H. Przytycki and A. S. Sikora in [346, Theorem 1.1] in the special case $s = \omega^{(n)}$ (cf. Figure 1.11 for $n = 5$), where the exchange number is 2^{n-1}. A. C. Höck showed in [216, Chapter 4], using some basic algebraic topology as in [346], that several variants are indeed equivalent with respect to Kauffman's question and provided for [216, Theorem 5.4] an alternative, mainly combinatorial proof of

$$E_{\text{top}}(\omega^{(n)}) = 2^{n-1} = E_{\text{mec}}(\omega^{(n)}). \tag{1.8}$$

The general case of the Ring conjecture had been posed as an open problem in Chapter 9 of the first edition of this book and has now also been verified by Höck; in fact he got [216, Theorem 5.5]:

Theorem 1.12. *For all s and t in B^n we have $E_{\text{mec}}(s,t) = E_{\text{top}}(s,t)$.*

Proof. Suppose that $E_{\text{top}}(s,t) < E_{\text{mec}}(s,t)$ for some s,t; since $E_{\text{top}}(s,t) = E_{\text{top}}(t,s)$ and $E_{\text{mec}}(s,t) = E_{\text{mec}}(t,s)$, we may assume that $E_{\text{mec}}(s) \leq E_{\text{mec}}(t)$. Then, since $E_{\text{top}} \leq E_{\text{mec}}$, we get

$$\begin{aligned}
E_{\text{top}}(\alpha^{(n)}, \omega^{(n)}) &\leq E_{\text{top}}(\alpha^{(n)}, s) + E_{\text{top}}(s,t) + E_{\text{top}}(t, \omega^{(n)}) \\
&< E_{\text{mec}}(s) + E_{\text{mec}}(s,t) + E_{\text{mec}}(t, \omega^{(n)}) \\
&= E_{\text{mec}}(s) + E_{\text{mec}}(t) - E_{\text{mec}}(s) + 2^{n-1} - E_{\text{mec}}(t) \\
&= 2^{n-1}.
\end{aligned}$$

This is a contradiction to (1.8).   □

Remark. Although the exchange number only counts moves in CS corresponding to moves of ring 1 in CR, there can not be an altogether shorter solution, because consecutive crossings in CS cancel out and ring 1 is moved in every second move in CR. On the other hand, the solution for CS engendered by the unique shortest solution for CR will not be the only optimal solution for CS; cf. [216, Section 2.4].

For psychological tests it might be interesting to compare the performance of subjects who are first confronted with the CR and thereafter with the CS or vice versa, because it seems to be very confusing that fixed and moving parts of the puzzles are interchanged in the two versions.

Tower of Hanoi networks

We next briefly describe an application of the sequence g from physics. S. Boettcher, B. Gonçalves, and H. Guclu [57, 58] introduced two infinite graphs which they named *Tower of Hanoi networks*. They introduced them to explore aspects of a small-world behavior and demonstrated that they possess appealing properties.

The network/graph HN4 is defined on the vertex set $\mathbb{Z}$. Connect each vertex k to $k+1$ and $k-1$, so that the two-way infinite path, or infinite cycle, is constructed. Next write $k \in \mathbb{Z} \setminus \{0\}$ as $k = 2^r(2s+1)$, $r \in \mathbb{N}_0$, $s \in \mathbb{Z}$, and connect k to $2^r(2s+3)$, and $2^r(2s-1)$. Finally add a loop at the vertex 0 so that the constructed graph HN4 is 4-regular. It is shown in Figure 1.13.

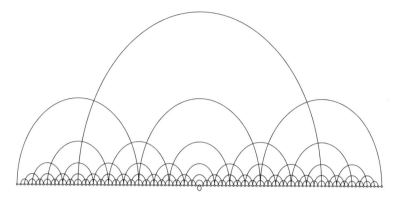

Figure 1.13: Network HN4

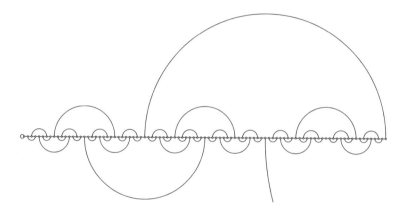

Figure 1.14: Network HN3

The graph HN3 is defined similarly, except that it is built on the basis of the one-way infinite path, or 0-path, to which only edges that form "forward jumps" are added. Instead of giving its formal definition we refer to Figure 1.14.

To see the connection between the networks HN3/HN4 and the sequence g, consider the vertices of HN4 that belong to $\mathbb{N}$. By the definition of the network, a vertex $k = 2^r(2s+1)$ has four neighbors, the largest of them being $2^r(2s+3)$. The jump from k to $2^r(2s+3)$ is

$$2^r(2s+3) - 2^r(2s+1) = 2^{r+1} = 2^{g_k},$$

the latter equality following from Corollary 1.10. A similar conclusion holds for HN3. Hence a historically better justified name for the HN3/HN4 networks would be *Chinese rings networks*.

As already mentioned, the sequence g appears in many situations, too many to be even listed here. The biggest surprise, however, is that the Gros sequence solved the Tower of Hanoi before the latter was at all invented!

1.4   Exercises

1.1. Derive recurrence (1.2).

1.2. Prove Proposition 1.4.

1.3. Consider the accelerated Chinese rings for $n \in \mathbb{N}_2$.

a) What is the corresponding state graph?

b) Calculate the value of its diameter.

c) Determine the minimal number of moves p_n needed to solve the accelerated Chinese rings task $1^n \to 0^n$ (or vice versa).

1.4. A *perfect code* in a connected graph $G = (V, E)$ is a subset C of the vertex set V with the property

$$\forall\, v \in V\ \exists_1\, c \in C:\ d(v, c) \le 1\,.$$

The elements of C are called *codewords*.

Show that R^n, $n \in \mathbb{N}_0$, contains precisely two perfect codes C, if n is odd, and precisely one, if n is even. How large is $|C|$?

1.5. [119, Problem 417] Suppose there are altogether fourteen rings on the tiring irons [Chinese rings] and we proceed to take them all off in the correct way so as not to waste any moves. What will be the position of the rings after the 9999th move has been made?

1.6. Determine the Wiener index of the path graph on $k \in \mathbb{N}$ vertices P_k.

1.7. Prove (1.6).

1.8. Let $k = (b_n b_{n-1} \dots b_1 b_0)_2$ be the binary representation of an integer $k \in \mathbb{N}$. Define $\widetilde{g}_k = i$, where $b_i = 1$ and $b_j = 0$ for $j \in [i]_0$. (That is, $\widetilde{g}_k$ is the index of the right-most non-zero bit in the binary representation of k. For instance, $\widetilde{g}_1 = 0$ and $\widetilde{g}_{24} = 3$.) Show that

$$\widetilde{g}_k + 1 = g_k\,.$$

1.9. Show that for $k \in \mathbb{N}$, the function q fulfills

$$q(k) = \begin{cases} q(k-1) + 1\,, & k \text{ odd}; \\ q\left(\frac{k}{2}\right)\,, & k \text{ even} \end{cases}$$

and deduce Legendre's formula (1.7) from this.

1.10. The n-cube is the graph with vertex set $\{b_1 b_2 \ldots b_n \mid b_i \in B\}$, two vertices being adjacent if their labels differ in exactly one position. Show that the n-cube contains a hamiltonian cycle for any $n \in \mathbb{N}_2$.

1.11. Show that the Gros sequence and the greedy square-free sequence are the same.

1.12. [197, p. 259] A sequence of symbols $a = (a_n)_{n \in \mathbb{N}}$ is called *strongly square-free* if it does not contain a subsequence of the form xy (an *abelian square*), where x and y are non-empty subsequences of consecutive symbols of a such that the symbols from y form a permutation of the symbols from x. Show that the Gros sequence is strongly square-free.

Chapter 2

The Classical Tower of Hanoi

This chapter describes the classical TH with three pegs. In the first section, the original task to transfer a tower from one peg to another is studied in detail. We then extend our considerations to tasks that transfer discs from an arbitrary regular state to a selected peg. We further broaden our view in Section 2.4 to tasks transforming an arbitrary regular state into another regular state. For this purpose it will be useful to introduce Hanoi graphs in Section 2.3.

2.1 Perfect to Perfect

In this section we analyze the classical task to transfer a tower from one peg to another in the minimum number of moves. We will call this problem *type* P0. After introducing a mathematical representation of the problem, its unique solution is given as an iterative algorithm. Olive's algorithm is much more practical than the classical recursive algorithm which we restate anyway in order to make the picture complete. (In fact, in the majority of books that come in contact with the TH recursion is the central if not the only approach eventually presented!) We begin with the concept of regular states.

Regular states

The TH consists of three vertical pegs, anchored in a horizontal base plate, and a certain number of discs of mutually different diameters. Each disc is pierced in its center so that it can be stacked onto one of the pegs. Any distribution of all discs on the three pegs with no larger disc lying on a smaller one is called a *regular state* of the puzzle; a *perfect state* is a regular state with all discs arranged on one and the same peg. See Figure 2.1 for examples of a perfect and a regular state.

Following the model of the theory for the Chinese rings in Chapter 1, we define $T = \{0, 1, 2\}$ (*T* for ternary) and

© Springer International Publishing AG, part of Springer Nature 2018
A.M. Hinz et al., *The Tower of Hanoi – Myths and Maths*,
https://doi.org/10.1007/978-3-319-73779-9_3

- label the pegs 0, 1, and 2.

We denote the number of discs by n and

- label the discs from 1 to n in increasing order of diameter.

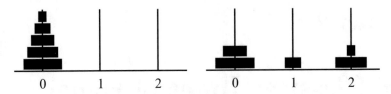

Figure 2.1: A perfect and a non-perfect regular state

Then every regular state is represented uniquely by an element $s = s_n \ldots s_1 \in T^n$, where s_d is the peg on which disc d is lying. For instance, the perfect state and the non-perfect regular state of Figure 2.1 are represented by 00000 (or 0^5) and 02012, respectively. (Mathematically it makes no difference how the pegs are physically distributed. For simplicity, most of our pictures show them arranged in line, but the reader should still imagine them as in the original TH of Figure 0.5.)

For future purposes the following observation is useful. For $m \in [n]$ and $s \in T^m$ let $sT^{n-m} = \{st \mid t \in T^{n-m}\}$. Obviously, $|sT^{n-m}| = |T^{n-m}|$. Moreover,[42]

$$T^n = \bigcup_{s \in T^m}^{\cdot} sT^{n-m}. \tag{2.1}$$

This fact is proved in Exercise 2.1. The most important case is $m = 1$. It shows that T^n decomposes naturally into the three (disjoint) sets $0T^{n-1}$, $1T^{n-1}$, and $2T^{n-1}$, characterized by the position of the largest disc.

Legal moves

A disc can be moved from the top of a stack on one peg to the top of the (possibly empty) stack on another peg, provided one obeys the divine rule. The classical task is to get from a perfect state to another perfect state by a sequence of such *legal moves*. The ultimate goal is to find a shortest solution, that is, a solution with the minimum number of moves. In Figure 2.2 a solution for the TH with 4 discs is presented. (We will see soon that this solution is **the** solution of the puzzle.) The middle column of the figure gives the labels of the discs moved.

The optimal solution

As in the case of the CR it is clear that in an optimal solution disc 1 has to be moved in the first step and then in every second move. Hence it will move

---

[42]By $\dot{\cup}$ we mean the union of pairwise disjoint sets.

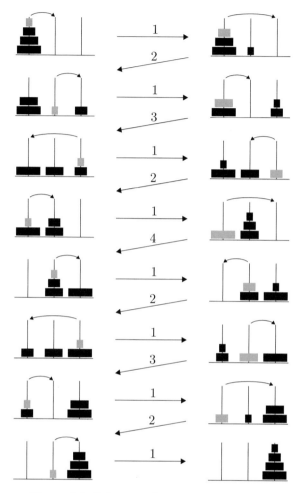

Figure 2.2: Solution of the TH with 4 discs

in every odd numbered move (cf. the moves from the left to the right column in Figure 2.2). The even numbered moves are then imposed by the divine rule because the smallest top disc that is not disc 1 has to be moved. Moreover, the moves of discs 2 to n form an optimal solution with respect to the same task but with only $n-1$ discs (cf. the left column in Figure 2.2). This is because disc 1 can always be moved out of the way for another disc to be moved. Recalling the Gros sequence g and its basic property from Proposition 1.9, we are therefore led to the following alternative to Theorem 0.26.

Theorem 2.1. *The classical TH task for $n \in \mathbb{N}_0$ discs has a unique optimal solution of length $2^n - 1$. The k-th move in this solution is by disc g_k, $k \in [2^n - 1]$.*

Proof. We proceed by induction on n, where in the induction step from n to $n+1$ we employ the unique optimal solution of length $2^n - 1$ which we get by induction assumption for discs 2 to $n+1$ in even moves. The first $2^n - 1$ odd moves are made by disc 1 in the unique way such that the even moves can be performed. Finally, disc 1 makes the last of altogether $2(2^n - 1) + 1 = 2^{n+1} - 1$ moves.

The second statement of the theorem can also be proved by induction: in the induction step, odd moves k are made, as we have seen, by disc $1 = g_k$ and even moves k by disc $g_{k/2} + 1 = g_k$.   □

An alternative for the induction step in the preceding proof can be based on the following observation.

Proposition 2.2. *In the optimal solution to transfer $n \in \mathbb{N}_0$ discs from one peg to another, disc $d \in [n]$ moves for the first time in step 2^{d-1} and for the last time in step $2^n - 2^{d-1}$; in particular, the largest disc n moves exactly once, namely in the middle of the solution.*

The proof is left as Exercise 2.2.

2.1.1   Olive's Algorithm

The discussion before the proof of Theorem 2.1 shows that similarly to the CR, disc 1 is moved in every odd move and the even moves are dictated by the divine rule. However, now we need an additional information about the direction of the moves of disc 1. It can be observed by considering small numbers of discs that in the optimal solution disc 1 moves cyclically among the pegs. When n is odd, it moves from source peg via goal peg to intermediate peg; if n is even, the cycle is from source peg via intermediate peg to goal peg. This observation has been made by Lucas's nephew Olive and leads to Algorithm 2. (Note that $3 - i - j$ is the peg different from both i and j.)

We leave the correctness proof for Olive's algorithm for Exercise 2.3. The formally even simpler Algorithm 3 (cf. [386]) can be obtained by looking at the *idle* peg, i.e. the peg not involved in a move (cf. [109, p. 55], where the name "idle" (пустой) is attributed to M. Fyodorov). It can be thought of as a virtual disc 0 which always makes room for the real discs. To distinguish it from the latter let us visualize it by a thimble which is placed at the beginning on top of peg i where the n-tower initially rests. Take the thimble in your left hand and move it cyclically on top of the pegs. In between, with your right hand, you make the only legal move of a disc avoiding the peg sealed by the thimble. Then eventually all discs will arrive at peg j if the direction in the cycle for the thimble had been chosen correctly according to parity of the number of discs.

Assuming that we go from peg $i = 0$ to peg $j = 2$, the idle peg in move number $\ell \in [2^n - 1]$ is $((2 - (n \bmod 2))\,\ell) \bmod 3$. If ℓ is given in binary representation $\sum_{\nu=0}^{n-1} \ell_\nu 2^\nu$, we can determine the idle peg as the final state of the automaton in

---

Algorithm 2 Olive's algorithm

Procedure $p0(n,i,j)$
Parameter n: number of discs $\{n \in \mathbb{N}_0\}$
Parameter i: source peg $\{i \in T\}$
Parameter j: goal peg $\{j \in T\}$
    if $n = 0$ or $i = j$ **then** STOP
    if n is odd **then**
        move disc 1 from peg i to peg j
    else
        move disc 1 from peg i to peg $3 - i - j$
    end if
    remember move direction of disc 1
    while not all discs are on peg j
        make legal move of disc not equal 1
        make one move of disc 1 cyclically in its proper direction
    end while

Algorithm 3 Idle peg algorithm

Procedure $p0i(n,i,j)$
Parameter n: number of discs $\{n \in \mathbb{N}_0\}$
Parameter i: source peg $\{i \in T\}$
Parameter j: goal peg $\{j \in T\}$
    $idle \leftarrow i,\ dir \leftarrow (-1)^n (j - i)$ {idle peg and its cyclic direction}
    while not all discs are on peg j
        $idle \leftarrow (idle + dir) \bmod 3$
        make legal move between pegs different from $idle$
    end while

Figure 2.3 (cf. [339, Fig. 2]) after the successive input of ℓ_ν, starting in state 0 and going from $\nu = n - 1$ to 0. (A transition between states of the automaton only occurs if the input matches the label on an arrow.) The proof of correctness for this procedure is by induction, where one has to take thorough care of the parity of n.

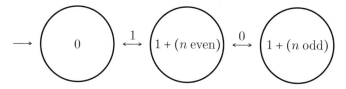

Figure 2.3: Automaton for the idle peg

It comes as a surprise that the highly repetitive sequence of positions of the thimble leads to another square-free sequence (cf. Section 1.2).

More square-free sequences

Indeed, let us assume that we have an infinite supply of discs. (A bold assumption given that the original ones were made of pure gold!) Let the position of the thimble follow the sequence $0, 1, 2, 0, 1, 2, \ldots$. For instance, position 1 of the idle peg means a move of a disc between pegs 0 and 2. Taking into account the direction of that move, we leave the entry 1 if this move is from 0 to 2, but write $\bar{1}$ if it is from 2 to 0. Similarly, 0 stands for a move from 2 to 1, $\bar{0}$ for 1 to 2, 2 for a move from 1 to 0, $\bar{2}$ for 0 to 1. This encoding is summarized in Table 2.1.

move between pegs	$0 \to 2$	$2 \to 1$	$1 \to 0$	$0 \to 1$	$1 \to 2$	$2 \to 0$
encoding	1	0	2	$\bar{2}$	$\bar{0}$	$\bar{1}$

Table 2.1: Encoding of move types

Then our infinite sequence of moves, called *Olive sequence*, starts

$$o = 1, \bar{2}, 0, 1, 2, \bar{0}, 1, \bar{2}, 0, 1, \bar{1}, 2, 0, 1, \bar{2}, 0, \ldots ,$$

when a 4-tower has been moved from peg 0 to peg 1. Note that the unbarred version of o provides yet another practical solution: follow this sequence without regarding the direction of the move, the latter being implied by the divine rule anyway, cf. [378, p. 97].

The orientation respecting sequence o has the further property of being square-free. This was proved for the first time by J.-P. Allouche, D. Astoorian, J. Randall, and J. Shallit in [6, Theorem 9]; for an alternative approach, see [197, Theorem 1].

The sequence o uses six symbols. As noticed by Hinz [197, Theorem 2], this can be reduced by combining its terms into triples. It turns out that only five of these actually occur, namely

$$A = (1, \bar{2}, 0),\ B = (1, 2, \bar{0}),\ \Gamma = (\bar{1}, 2, 0),\ \Delta = (1, 2, 0),\ E = (\bar{1}, 2, \bar{0}).$$

This leads to a new sequence

$$h = A, B, A, \Gamma, A, B, \Delta, E, A, B, A, \Gamma, A, E, \Delta, \ldots ,$$

which is square-free, for otherwise the sequence o would contain a square as well. Contrary to the Gros sequence g, however, both h and o are *not* strongly square-free (Exercise 2.4).

2.1.2 Other Algorithms

It is obvious that Olive's algorithm can be carried out by a human being without further thinking. It is therefore much more practical than the classical recursive algorithm mentioned in the introduction: imagine, for the latter, brahmin $d \in [64] \setminus \{1\}$, responsible for disc d, passing on duties to brahmin $d - 1$. For the general case, the recursive algorithm is realized as Algorithm 4.

Algorithm 4 Recursive algorithm

Procedure $p0r(n, i, j)$
Parameter n: number of discs $\quad \{n \in \mathbb{N}_0\}$
Parameter i: source peg $\quad \{i \in T\}$
Parameter j: goal peg $\quad \{j \in T\}$
 if $n \neq 0$ and $i \neq j$ **then**
  $k \leftarrow 3 - i - j$ {the auxiliary peg different from i and j}
  $p0r(n-1, i, k)$ {transfers $n-1$ smallest discs to auxiliary peg}
  move disc n from i to j {moves largest disc to goal peg}
  $p0r(n-1, k, j)$ {transfers $n-1$ smallest discs to goal peg}
 end if

The recursive argument, however, can be used for a wealth of useful statements about the optimal solution. For instance, let us try to characterize those states $s \in T^n$ which belong to the shortest path from i^n to j^n, say. We interpret the bottom of pegs i, $3-i-j$, j as (immovable) discs $n+1$, $n+2$, $n+3$, respectively. Moreover, we call a legal arrangement (i.e. with no larger disc on a smaller one) of at most n (movable) discs on a fixed peg *admissible* (with respect to i and j), if the discs (including the immovable one) on that peg appear in alternating parity. Then the following holds (cf. [427, Theorem A]).

Proposition 2.3. *A state $s \in T^n$ belongs to the optimal solution path from i^n to j^n if and only if s is* admissible, *i.e. the corresponding arrangements on all three pegs are admissible.*

Proof. We prove by induction that there are 2^n admissible states in T^n with respect to i and j and that all (2^n) states on the shortest path from i^n to j^n are admissible.

 For $n = 0$, we just note that the empty arrangement on a peg (i.e. only the immovable disc is present) is admissible.

 Let $\bar{s} := s_{n-1} \ldots s_1$. Then $s = s_n \bar{s}$ is admissible with respect to i and j if and only if $s_n \in \{i, j\}$ and $\bar{s}$ is admissible in T^{n-1} with respect to s_n and $3 - i - j$. Therefore, by induction assumption, there are $2 \cdot 2^{n-1} = 2^n$ admissible states in T^n. The first 2^{n-1} states on the optimal path from i^n to j^n are those where $s_n = i$, i.e. the largest disc n is lying on a(n immovable) disc of different parity, and $\bar{s}$ belongs to the shortest path from i^{n-1} to $(3 - i - j)^{n-1}$ such that s is admissible with respect to i and j because disc n replaces the immovable disc $n + 1$. The same argument applies for the last 2^{n-1} states, where now disc n is on disc $n + 3$ of opposite parity.   □

The characterization of states on the optimal path can be used for another "human" algorithm; see Exercise 2.5. Moreover, there is a fascinating relation between admissible arrangements on a peg and Fibonacci numbers; see Exercise 2.6.

By a recursive argument, we also obtain the following information about the number and direction of moves of individual discs.

Proposition 2.4. *In the optimal solution to move an n-tower from peg i to peg j, disc $d \in [n]$ is moved in moves $(2k + 1)\,2^{d-1}$, $k \in [2^{n-d}]_0$, i.e. 2^{n-d} times, and in every move*

$$[(j - i)\,((n - d) \bmod 2 + 1)] \bmod 3$$

is added (modulo 3) to the position of d.

The proof of this result is left for Exercise 2.7. It shows that every odd disc moves cyclically like disc 1, the others in reversed cycles. This statement has been used as an example to demonstrate the associativity of the equivalence of boolean values in [28].

Olive's sequence can now easily be calculated from the index (move number) $\ell \in \mathbb{N}$ in binary representation $\sum_{\nu=0}^{\infty} \ell_\nu 2^\nu$. In fact, we may identify o with the sequence in $B \times T$, where (b, i) stands for $\bar{i}$, if $b = 0$ and for i itself if $b = 1$. Then

$$o_\ell = (g_\ell \bmod 2, \ell \bmod 3) \tag{2.2}$$

because by the definition of o, which assumes that disc 1 moves to peg 2 first, we are in the case $i = 0$, $j = 2$ with n odd of Proposition 2.4 such that odd moving discs correspond to $b = 1$ and even discs to $b = 0$; by Theorem 2.1 the moving disc is g_ℓ. The formula (2.2) is truly remarkable! It demonstrates in one line the genius of Lucas when he extended the binary-based CR to the ternary TH and of his nephew Raoul who unveiled the interplay of the two number systems in the TH, essentially engendered by the identity $2^n \bmod 3 = 2^{n \bmod 2}$ which we already encountered in the solution of Exercise 1.4. The calculation of $\ell \bmod 3$ can be done very efficiently with the aid of the automaton in Figure 2.3 (case n is odd), and we know from Corollary 1.10 that $\bar{g}_\ell := g_\ell \bmod 2 = 0$ if and only if the binary representation of ℓ ends with an odd number of bits 0. The automaton of Allouche and F. Dress in [9, Figure 2] also determines the Olive sequence, but entering the bits of ℓ from the right it needs 14 states. As remarked in [9, p. 13], the barred idle pegs are in a $1:2$ minority with respect to the unbarred ones in Olive's sequence; see also [197, p. 261]. Allouche and co-workers emphasize the automatic generation of sequences like Olive's by means of the *Toeplitz transform*: if in the formal infinite sequence $(1\bar{2}0 \diamond 2\bar{0}1 \diamond 0\bar{1}2\diamond)^\infty$ the diamonds are successively replaced by the sequence itself, then the limit is o [7, Section 4]. This offers a unified approach covering also the paper-folding sequence f we met in Chapter 1 if the Toeplitz transform is applied to $(1 \diamond 0 \diamond)^\infty$ [7, Section 2]. But what would Lucas and Olive have thought about their post-bourbakian compatriots and associates who in [6, Corollary 5] characterize Olive's sequence as "a fixed point of a 2-uniform homomorphism" on $(B \times T)^{\mathbb{N}}$?[43]

The sequence $(\bar{g}_\ell)_{\ell \in \mathbb{N}}$ (cf. A035263) is interesting in its own right; $(1 - \bar{g}_{\ell+1})_{\ell \in \mathbb{N}_0}$ is the *period-doubling sequence* of [12, Example 6.3.4] (cf. A096268). From Proposition 1.9 we deduce:

$$\bar{g}_\ell = \begin{cases} 1, & \ell \text{ odd}; \\ 1 - \bar{g}_{\ell/2}, & \ell \text{ even}. \end{cases}$$

Moreover, if we define $t_k = \left(\sum_{\ell=1}^{k} g_\ell \right) \bmod 2$ for $k \in \mathbb{N}_0$, we obtain from (1.7) that $t_k = q(k) \bmod 2$, whence the sequence t fulfills the recurrence

$$t_0 = 0, \ \forall\, k \in \mathbb{N}_0 : t_{2k+1} = 1 - t_k, \ t_{2k+2} = t_{k+1}.$$

So $t = 0, 1, 1, 0, 1, 0, 0, 1, 1, 0, 0, 1, 0, 1, 1, 0, \ldots$ is the well-known *Prouhet-Thue-Morse sequence* (cf. [11]). Subsequent substitution of a for $(0, 1, 1)$, b for $(0, 1)$, and c for (0) leads to the *Thue sequence* $a, b, c, a, c, b, a, b, c, b, a, \ldots$, which is square-free over the three-letter alphabet $\{a, b, c\}$. (A square-free sequence over a two-element alphabet cannot be infinite; it ends already after three entries.)

Coming back to the general $i^n \to j^n$ task, we can also provide a recipe (cf. [385, p. 103f]) to specify the move with number $\ell = \sum_{\nu=0}^{n-1} \ell_\nu 2^\nu$ making use of only some portion of the bits of ℓ. If $\ell = 2^{d-1}$, then disc d moves to an empty peg; this is unique but for the very first move which goes to the goal peg if and only if n is odd. Otherwise, consider the smallest discs $d_1 < d_2$ with $\ell_{d_1-1} = 1 = \ell_{d_2-1}$ and move d_1 onto disc d_2 if of different parity or else to the peg not occupied by d_2. This works because ℓ is the first move of any disc in $[d_2]_0 \setminus [d_1]_0$ after d_2 has made a move and therefore the tower of discs smaller than d_1 either rests on top of disc d_2 or d_1 can move there, which by Proposition 2.3 is only possible if they are of different parity.

Proposition 2.4 also leads to Algorithm 5, which provides initial peg i_ℓ and goal peg j_ℓ for any individual move $\ell \in [2^n - 1]$ explicitly (cf. [194, Proposition 1]).

An immediate consequence is the *parallel* Algorithm 6, containing a formula which gives the position s_d of disc $d \in [n]$ after move $\ell \in [2^n - 1]$ (cf. [188, Folgerung 3]), this position being equal to the last goal of disc d up to move ℓ (cf. [194, Proposition 2]).

In particular, this algorithm allows us to determine the state s on the shortest path from 0^n to 2^n, say, whose distance from 0^n is equal to an $\ell \in [2^n]_0$ given by

---

[43] The homomorphism is generated by $(b, i) \mapsto ((0, 1 \vartriangle i), (1 - b, 0 \vartriangle i))$; for the definition of $\vartriangle$ see Table 2.2 below.

Algorithm 5 Determination of move

Procedure $p0m(n, i, j, \ell)$
Parameter n: number of discs $\{n \in \mathbb{N}_0\}$
Parameter i: source peg $\{i \in T\}$
Parameter j: goal peg $\{j \in T\}$
Parameter ℓ: move number $\{\ell \in [2^n - 1]\}$

$\quad d \leftarrow g_\ell$ $\{$disc moved in move number $\ell\}$
$\quad k \leftarrow \dfrac{\ell}{2^d} - \dfrac{1}{2}$ $\{\ell = 2^{d-1}(2k + 1)\}$
$\quad i_\ell \leftarrow \big(k(j - i)\,((n - d) \bmod 2 + 1) + i\big) \bmod 3$
$\quad j_\ell \leftarrow \big((k + 1)(j - i)\,((n - d) \bmod 2 + 1) + i\big) \bmod 3$

Algorithm 6 Parallel algorithm

Procedure $p0p(n, i, j, \ell, d)$
Parameter n: number of discs $\{n \in \mathbb{N}_0\}$
Parameter i: source peg $\{i \in T\}$
Parameter j: goal peg $\{j \in T\}$
Parameter ℓ: move number $\{\ell \in [2^n - 1]\}$
Parameter d: disc number $\{d \in [n]\}$

$\quad s_d \leftarrow \left(\left\lfloor \dfrac{\ell}{2^d} + \dfrac{1}{2} \right\rfloor (j - i)\,((n - d) \bmod 2 + 1) + i \right) \bmod 3$

---

its binary representation $\sum_{\nu=0}^{n-1} \ell_\nu 2^\nu$. This corresponds to applying Gray coding to ℓ to obtain the state with that distance from $\alpha^{(n)}$ in the CR (cf. Section 1.1). In fact, putting $i = 0$ and $j = 2$ in Algorithm 6, we get, after some straightforward manipulation,

$$s_d = (\ell_{d-1} + \ell_d + 2\ell_{d+1} + \ell_{d+2} + \cdots + 2\ell_{n-2} + \ell_{n-1}) \bmod 3, \quad \text{if } n - d \text{ is odd,} \quad (2.3)$$
$$s_d = (2\ell_{d-1} + 2\ell_d + \ell_{d+1} + \cdots + 2\ell_{n-2} + \ell_{n-1}) \bmod 3, \quad\quad \text{if } n - d \text{ is even.} \quad (2.4)$$

This compares to the recipe given without proof in [52, p. 862]. There the example of a 7-disc tower after 13 steps is shown, i.e. the case $n = 7$ and $\ell = 13_{10} = 0001101_2$; from Equations (2.3) and (2.4) one gets $s = 0001102_2$.

The parallel algorithm is also a special case of a *spreadsheet* solution in the sense of B. Hayes in [182]. The position of disc d after $\ell \in [2^n]_0$ moves in the shortest path from 0^n to 2^n can be determined simultaneously in each cell of an $n \times 2^n$-grid by one and the same formula; see Figure 2.4 for the example $n = 4$. This shows how time can be traded for memory.

Another elegant algorithm to determine the state after ℓ moves in the optimal path from 0^n to 2^n, and hence by relabeling for any initial and terminal pegs, can be based on an observation by Er [133, p. 150f]. It is shown here as Algorithm 7. The correctness proof is by induction. Note that the sign in front of ℓ_{d-1} in the

d \ ℓ	0	1	2	3	4	5	6	7	8	9	10	11	12	13	14	15
1	0	1	1	2	2	0	0	1	1	2	2	0	0	1	1	2
2	0	0	2	2	2	2	1	1	1	1	0	0	0	0	2	2
3	0	0	0	0	1	1	1	1	1	1	1	1	2	2	2	2
4	0	0	0	0	0	0	0	0	2	2	2	2	2	2	2	2

Figure 2.4: Spreadsheet solution for 4 discs

allocation of the value of s_d determines the move direction and that the initial peg i is modified after a move of disc d. The reader is invited to test the algorithm with the example preceding the previous paragraph.

Algorithm 7 Alternative parallel algorithm

Procedure p0pa(n, ℓ)
Parameter n: number of discs $\{n \in \mathbb{N}\}$
Parameter ℓ: move number $\{\ell \in [2^n]_0$ in binary representation$\}$
    $i \leftarrow 0$
    for $d = n$ **downto** 2
        $s_d \leftarrow (i - \ell_{d-1}((n - d) \bmod 2 + 1)) \bmod 3$
        $i \leftarrow (i + \ell_{d-1}(i - s_d)) \bmod 3$
    end for
    $s_1 \leftarrow (i + \ell_0(n \bmod 2 + 1)) \bmod 3$

It is now natural, in particular in view of the activities of the Brahmins in Benares, to ask the *inverse* question, namely given a distribution of discs somewhere on the optimal path from 0^n to 2^n, say, to determine how many moves have led to this state. We can use the fact that by Theorem 2.1 this path P^n has length $2^n - 1$ and is therefore, by Theorem 1.1, isomorphic to the CR graph R^n. Moreover, starting P^n in 0^n we can in parallel traverse R^n, beginning in state $\alpha^{(n)}$, by replacing a move of disc d in P^n by a move of ring d in R^n. Since the disc/ring number follows the Gros sequence in both path graphs, every state s on P^n is uniquely specified by the values, for all $d \in [n]$, of $\tilde{s}_d \in B$, being defined as the number of moves, taken modulo 2, made by disc, and hence ring, d to arrive at s, respectively $\tilde{s}$. The mapping $s \mapsto \tilde{s}$ is therefore an isomorphism between P^n and R^n. In particular, $\mathrm{d}(s, 0^n) = \mathrm{d}(\tilde{s})$, and the latter can easily be calculated using the Gros automaton in Figure 1.3. If we put $s_{n+1} = 0$, $\tilde{s}$ can be obtained from s as $\tilde{s}_d = (s_d \neq s_{d+1})$, a formula which can be proved by induction on n making use of the symmetry with respect to the move of the largest disc as reflected in Proposition 2.2. Combined, we obtain an algorithm to determine the bits of

$d(s, 0^n) = \sum_{\nu=0}^{n} \ell_\nu \cdot 2^\nu$ with $\ell_n = 0$ without recourse to the CR as

$$\forall\, d \in [n]: \; \ell_{d-1} = \ell_d \,\dot\vee\, (s_d \neq s_{d+1}). \tag{2.5}$$

This is the recipe which has already been given by P. H. Schoute [372, p. 286f].

We are now in the position to determine the age of the universe and consequently how much time the Brahmins will leave us to stay alive; see Exercise 2.8.

The isomorphism between the optimal path in the TH and the CR graph has been used implicitly, without mentioning the CR, by D. W. Crowe in [85] to establish a connection between a hamiltonian cycle on the n-dimensional cube and the TH. We have seen in Chapter 1 that the connection is actually with the CR; cf. Exercise 1.10. For the same reason many of the early workers on the TH considered the game to be equivalent to the CR, an opinion still shared by some more recent authors (cf. the "TH networks" in Chapter 1). However, contrary to the Brahmins, mere mortals *do* make mistakes and go astray from the optimal path. We will now see that this little flaw in human nature will turn the TH from an isomorph of the trivial CR to a mathematically challenging object.

2.2 Regular to Perfect

In the previous section we showed how to transfer the whole tower of discs from one peg to another. Now we move to the question how to reach a perfect state from an *arbitrary* regular state; we call this problem *type* P1.

We first consider the special case in which the optimal solution has been abandoned at some stage, in state s say. (Imagine the situation when one monk transfers his duties to another one.) Then it is still possible to continue with the following insight from the considerations in Section 2.1: odd numbered discs move cyclically in the same direction as disc 1, and even discs move cyclically the other way round. This will be called the *right direction* for the disc. Algorithm 8 provides the optimal first move after which one may continue with Olive's procedure.

Algorithm 8 Optimal first move from abandoned state

Procedure $\mathrm{pla}(n, i, j, s)$
Parameter n: number of discs $\{n \in \mathbb{N}\}$
Parameter i: source peg $\{i \in T\}$
Parameter j: goal peg $\{j \in T\}$
Parameter s: regular state on the optimal path $\{s \in T^n\}$
    let d be the smallest top disc of s different from 1
    if the legal move of d is in the right direction **then**
      move disc d
    else
      move disc 1 in the right direction
    end if

The algorithm is illustrated in Figure 2.5 with two states. In both cases we have six discs, hence the right direction of the even discs is the same in both cases. In state 012002, the legal move of disc 2 is in its wrong direction, therefore the next optimal move is the move of disc 1 in its right direction. In the state 012000, the legal move of disc 4 is in its right direction, hence this move is the optimal next move.

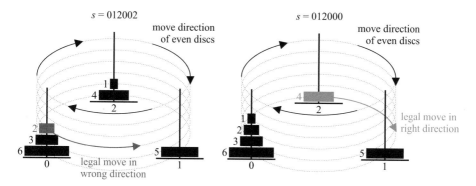

Figure 2.5: Move directions of discs

Before we can employ Algorithm 8 to resume an abandoned solution, we have to check whether s indeed lies on the optimal path from i to j. This can be done in Algorithm 9 by building an automaton from Figure 2.6 that we will call the *P1-automaton*.

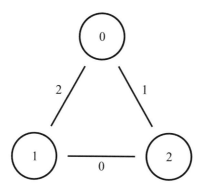

Figure 2.6: P1-automaton

On a more formal level, this automaton realizes the binary operation

$$i \vartriangle j = 2(i + j) \bmod 3, \ i, j \in T,$$

namely starting in state j of the automaton, the change of state according to i will end in state $i \vartriangle j$. Note that $i \vartriangle j = 3 - i - j$, if $i \neq j$ and $i \vartriangle j = j$ otherwise.

Therefore, a missing label means no change of state of the automaton, as it is the case for all automata presented in this book. The idea to use this kind of automaton for TH problems goes back to Stockmeyer (cf. [407, Table 4.2]). Let us mention, however, that on an even more abstract level $(T, \vartriangle)$ forms a *quasigroup*, i.e. the *Cayley table* of $\vartriangle$ on T is a *latin square* where each of the three elements of T occurs once in every row and once in every column; see Table 2.2. In a latin square on numbers, the rows and columns obviously add up to the same sum. In our case also the diagonals do. Viewed modulo 3, our table even gives 0 for the sum on the broken diagonals. It's a kind of magic! The binary operation $\vartriangle$ on T is commutative, but *not* associative: $0 \vartriangle (1 \vartriangle 2) = 0 \neq 2 = (0 \vartriangle 1) \vartriangle 2$. Moreover, the equations $i \vartriangle x = j$ and $y \vartriangle i = j$ have unique solutions $i \backslash j = x = i \vartriangle j$ and $j/i = y = j \vartriangle i$, respectively; that is, left and right division are the same. The operation appears already in [373, p. 248], and $(T, \vartriangle)$ can also be viewed as the smallest non-trivial *Steiner quasigroup*.

$\vartriangle$	0	1	2
0	0	2	1
1	2	1	0
2	1	0	2

Table 2.2: Cayley table for $\vartriangle$ on T

Algorithm 9 Detection of deviation from optimal path

Procedure $\mathrm{pld}(n, i, j, s)$
Parameter n: number of discs $\{n \in \mathbb{N}\}$
Parameter i: source peg $\{i \in T\}$
Parameter j: goal peg $\{j \in T\}$
Parameter s: regular state $\{s \in T^n\}$
   $k \leftarrow 3 - i - j$ {state of the P1-automaton}
   for $d = n$ **downto** 2
     if $s_d = k$ **then**
       s is *not* on the optimal path, STOP
     else
       $k \leftarrow 3 - k - s_d$ {updated state of the P1-automaton}
     end if
   end for
   if $s_1 = k$ **then**
     s is *not* on the optimal path, STOP
   end if
   s is on the optimal path

Theorem 2.5. *Algorithm 9 detects whether the state s is on the optimal path from the source peg i to the goal peg j.*

Proof. Let $k = 3 - i - j$. We proceed by induction on n. Obviously, the algorithm works correctly for $n = 1$. Let $n \in \mathbb{N}$ and $s \in T^{1+n}$. If $s_{n+1} = k$, then s is not on the right track, because the largest disc just moves from i to j. If $s_{n+1} = i$, then we move to state j of the P1-automaton. The n-tower has to be moved from i to k and therefore the roles of i, j, k have changed to i, k, j; but then the P1-automaton does the right thing by induction assumption. Similarly for $s_{n+1} = j$.   □

In Figure 2.7 we apply Algorithm 9 to the state $s = 212210$ where $i = 0$ is the source peg and $j = 2$ is the goal.

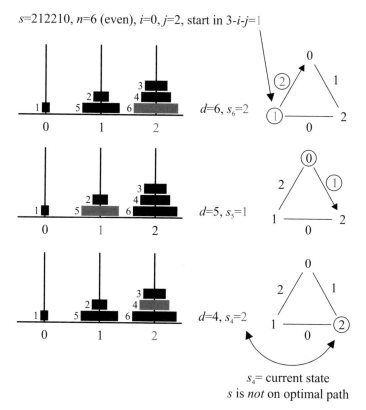

$s{=}212210, n{=}6$ (even), $i{=}0, j{=}2$, start in $3{-}i{-}j{=}1$

$d{=}6, s_6{=}2$

$d{=}5, s_5{=}1$

$d{=}4, s_4{=}2$

$s_4 =$ current state

s is *not* on optimal path

Figure 2.7: Algorithm 9 stops after checking 3 tits.

Algorithm 9 stopped already after 3 **tits** for $s = 212210$. This is in fact the expected value of checks! To state this result rigorously, let us introduce the notation $\bar{\bar{3}}$ for $2/3$. This was a special fraction in Ancient Egypt; cf. [159, p. 21]. Note that in binary representation $\bar{\bar{3}} = (.101010\ldots)_2$, a fact which relates to the Lichtenberg sequence; cf. [202, p. 6] and [406, p. 175]. Moreover, the area of a parabolic segment is $\overline{\overline{3}}$ times the area of the underlying rectangle. Likewise, the surface and volume of a ball are $\overline{\overline{3}}$ of the respective measures of the circumscribed cylinder; cf. [146, p. 156–158]. It seems that Archimedes was so proud of these discoveries that he asked a cylinder and inscribed ball to be installed on his grave stone. The International Prototype Kilogram is such an *archimedian cylinder*.

Theorem 2.6. *The number of tits checked on average by Algorithm 9 is*

$$3\left(1 - \overline{\overline{3}}^{\,n}\right).$$

Proof. The algorithm stops after the input of $\ell \in [n-1]$ tits if and only if each s_{n-m}, $m \in [\ell-1]_0$, is different from the current state of the P1-automaton and $s_{n-\ell+1}$ is equal to the current state. In this case, the tits s_d, $d \in [n-\ell]$, are arbitrary. Hence the algorithm stops after the input of ℓ tits for exactly $2^{\ell-1} \cdot 3^{n-\ell}$ states. Consequently, it needs the full n tits just for the remaining $2^{n-1} \cdot 3$ states. (Equivalently, it checks n tits if and only if s_2 to s_n are different from the current state and s_1 is arbitrary.) So for all states together only

$$\frac{1}{2}\sum_{\ell=1}^{n-1} \ell\, 2^\ell\, 3^{n-\ell} + 3n2^{n-1} = \frac{1}{2}\sum_{\ell=1}^{n} \ell\, 2^\ell\, 3^{n-\ell} + n\, 2^n$$

tits are used, i.e. on the average

$$\frac{1}{2}\sum_{\ell=1}^{n} \ell\, \overline{\overline{3}}^{\,\ell} + n\, \overline{\overline{3}}^{\,n}.$$

By formula [162, (2.26)] this is equal to

$$\frac{1}{2}\,\frac{\overline{\overline{3}} - (n+1)\overline{\overline{3}}^{\,n+1} + n\overline{\overline{3}}^{\,n+2}}{\left(1 - \overline{\overline{3}}\right)^2} + n\,\overline{\overline{3}}^{\,n} = 3\left(1 - \overline{\overline{3}}^{\,n}\right)$$

as claimed. □

In Figure 2.8 we apply Algorithm 9 to the state $s = 012002$. It turns out that s lies on the optimal path between perfect states on pegs 0 and 2, hence we can resume this path with the aid of Algorithm 8.

In view of the example from Figure 2.7, namely $s = 212210$, it is now natural to ask whether the game can still be continued, if s is found to be *off* the optimal path, a situation quite typical for human problem solvers. In fact, when the classical task to transfer a tower from one peg to another one is administered to subjects in psychological tests, the majority of them start moving discs around and do not wait until the optimal solution has been found by planning ahead. After this "chaotic starting phase" the test person finds himself in some more or less arbitrary state $s \in T^n$ and usually does not remember how he got there. The goal, a perfect tower on peg j, say, is still present in his mind and he is facing a

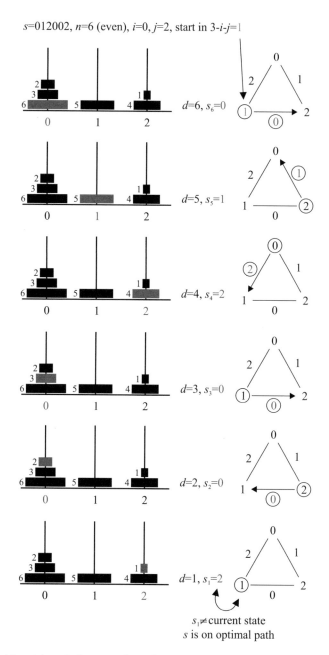

Figure 2.8: Algorithm 9 detects that the state on the left is on the optimal path from peg $i = 0$ to peg $j = 2$.

problem of type P1, namely to get from a given regular to a preassigned perfect
state with the least possible number of moves (cf. [277, Lemma 4, Theorem 1] and
[194, Theorem 3]).

Theorem 2.7. *The task to get from a regular state $s \in T^n$ to the perfect one on peg
$j \in T$ has a unique optimal solution of length $\leq 2^n - 1$. Moreover, if $s \neq j^n$ and δ is
the largest disc not on peg j in state s, i.e. $\delta = \max\{d \in [n] \mid s_d \neq j\}$, then in the
course of this solution δ is moved precisely once, and larger discs are not moved
at all.*

Proof. It is clear that the larger discs are not moved at all because the moves of
those discs could just be deleted to arrive at a shorter solution. We may therefore
assume that the largest disc is not on peg j, i.e. $s_n \neq j$.

We proceed by induction. Let $s = s_{n+1}\bar{s}$ with $s_{n+1} \neq j$ and $\bar{s} \in T^n$. If disc $n+1$
were to move more than once, this would involve at least $2(2^n - 1)$ moves of discs
smaller than $n+1$ between the first two moves of disc $n+1$ and after the last one,
resulting in at least 2^{n+1} moves. However, the combination of the unique optimal
solution from $\bar{s}$ to peg $3 - s_{n+1} - j$, the move of disc $n+1$ from s_{n+1} to j and finally
the unique optimal transfer of the n-tower from $3 - s_{n+1} - j$ to j results in a unique
optimal solution of length less than 2^{n+1}.   □

In the sequel we will denote the length of an optimal solution to get from a
state s to the perfect state on peg j by $\mathrm{d}(s, j^n)$. In this notation, Theorem 2.7 in
particular says that $\mathrm{d}(s, j^n) \leq 2^n - 1$ holds for any state s. An easy but important
consequence of this is the following.

Lemma 2.8. *State $s \in T^n$, $n \in \mathbb{N}$, is uniquely determined by the three values $\mathrm{d}(s, k^n)$,
$k \in T$.*

Proof. Induction on n. The case $n = 1$ is trivial. Let $i \in T$ and $t \in T^n$, $n \in \mathbb{N}$. Then,
by Theorem 2.7,

$$\mathrm{d}(it, i^{1+n}) = \mathrm{d}(t, i^n) < 2^n, \tag{2.6}$$

and, together with Theorem 2.1,

$$\mathrm{d}(it, j^{1+n}) = \mathrm{d}(t, (3 - i - j)^n) + 2^n \geq 2^n, \; j \in T \setminus \{i\}. \tag{2.7}$$

Therefore, given $s = s_{n+1}\bar{s}$ with $s_{n+1} \in T$ and $\bar{s} \in T^n$, tit s_{n+1} is given as the $k \in T$
for which the value $\mathrm{d}(s, k^{1+n})$ is less than 2^n, i.e. the smallest. Moreover, the three
values $\mathrm{d}(\bar{s}, k^n)$ are given by the equalities in (2.6) and (2.7), so that by induction
assumption also $\bar{s}$ is uniquely determined.   □

The formula in (2.5) for states s lying on the optimal path from 0^n to 2^n
does not hold for other s in general; for instance, $\mathrm{d}(20, 00) = 3$, where (2.5) would
give 2. But although the easy recipe of Exercise 2.5 cannot be carried over for
a practical solution to a P1 task, we nevertheless observe the following (cf. [428,
p. 412]).

Proposition 2.9. *Suppose that during the optimal solution of a P1 task, disc $d \neq 1$ is moved. Then the next step is a move of disc 1, and it moves onto d if and only if d is even.*

Proof. Again, this is done by induction on the number of discs $n \in \mathbb{N}_2$. For $n = 2$, disc 1 has to move atop disc 2 after the latter's only move. For the induction step we may assume that disc $n + 1$ is originally not on its goal peg, in which case it moves exactly once. Before this move the induction assumption applies. The move of disc $n + 1$ is followed by a transfer of a perfect n-tower onto disc $n + 1$ and here, according to Olive's solution, the first move, necessarily by disc 1, is to the goal peg if and only if n is odd, i.e. $n + 1$ is even. □

Since obviously no two moves of disc 1 follow each other immediately, we are left with finding the best first move. This is achieved in Algorithm 10 (cf. [196, p. 178]), again by recourse to the P1-automaton.

Algorithm 10 Best first move to a perfect state

Procedure p1(n, s, j)
Parameter n: number of discs $\{n \in \mathbb{N}\}$
Parameter s: regular state $\{s \in T^n\}$
Parameter j: goal peg $\{j \in T\}$
   $\mu \leftarrow 0$ {length of path}
   $\delta \leftarrow n + 1$ {active disc}
   $k \leftarrow j$ {state of the P1-automaton}
   for $d = n$ **downto** 1
     if $s_d \neq k$ **then**
       $\mu \leftarrow \mu + 2^{d-1}$
       $\delta \leftarrow d$
       $k \leftarrow 3 - k - s_d$ {updated state of the P1-automaton}
     end if
   end for

Theorem 2.10. *Algorithm 10 returns the length μ of the optimal path from the regular state s to the goal peg j and, for $s \neq j^n$, the disc δ to be moved in the first move of that path. Moreover, the idle peg, i.e. the peg not involved in that move, is given by the final state k of the P1-automaton, i.e. the move is from s_δ to $3 - k - s_\delta$.*

Remark 2.11. *For $n = 0$ or $s = j^n$, Algorithm 10 returns $\mu = 0$, $\delta = n + 1$, and $k = j$. The first value is, of course, trivial, the last two are void, but will be needed in a future algorithm.*

Proof of Theorem 2.10. We again proceed by induction. Let $s = s_{n+1}\bar{s}$, $\bar{s} \in T^n$. If $s_{n+1} = j$, then, by Theorem 2.7, the largest disc does not move, and the induction hypothesis can be applied to $\bar{s}$. Otherwise, $\bar{s}$ has to be brought to peg $3 - s_{n+1} - j$ to allow for the single move of disc $n + 1$ from s_{n+1} to j. In this case another

$1 + 2^n - 1 = 2^{(1+n)-1}$ moves have to be carried out, and disc $n + 1$ moves first only if $\bar{s}$ is already perfect on $3 - s_{n+1} - j$. □

For an early example of a P1 task, see Exercise 2.9.

Remark 2.12. *In formal terms, the application of Algorithm 10 amounts to the expression $k = s_1 \vartriangle \cdots \vartriangle s_n \vartriangle j$. (Recall that the operation $\vartriangle$, although commutative, is* not *associative; therefore, concatenate expressions have always to be read from the right.) Since $k \vartriangle i \neq k \vartriangle j$, if $i \neq j$, we deduce that applications of the algorithm to the same s, but different goal pegs, will lead to different resulting final states of the P1-automaton, i.e. different idle pegs for the best first move. Moreover, except for $s = (3 - i - j)^n$, the distances $\mathrm{d}(s, i^n)$ and $\mathrm{d}(s, j^n)$ will not be equal; see Exercise 2.10.*

A comparison of Algorithms 9 and 10 shows that s is on the optimal path from i^n to j^n if and only if $\mathrm{d}(s, k^n) = 2^n - 1$, $\{i, j, k\} = T$. More generally, we get the following nice invariant.

Proposition 2.13. *For any $s \in T^n$,*

$$\mathrm{d}(s, 0^n) + \mathrm{d}(s, 1^n) + \mathrm{d}(s, 2^n) = 2 \cdot (2^n - 1).$$

Proof. If we run Algorithm 10 in parallel for all three goal pegs, then the if-clause will add each value 2^{d-1} precisely twice. Therefore,

$$\mathrm{d}(s, 0^n) + \mathrm{d}(s, 1^n) + \mathrm{d}(s, 2^n) = 2 \cdot \sum_{d=1}^{n} 2^{d-1} = 2 \cdot (2^n - 1).$$

□

Remark 2.14. *From Proposition 2.13 it follows that in Lemma 2.8 in fact two of the values $\mathrm{d}(s, k^n)$ suffice to identify $s \in T^n$. This can be compared to the Gray code in Section 1.1. Moreover, suppose that we make a single legal move away from s. Then we have approached one perfect state by one step, moved away from another one by one step, and stay at the same distance from the third one. Indeed, this follows from Lemma 2.8 and Proposition 2.13, as elaborated in Exercise 2.11.*

The following immediate consequence of Proposition 2.13 has been found, without minimality considerations though, by Er [129, Equation (6)] and by F. Scarioni and M. G. Speranza [367], who in their Theorem also give the variance $2(4^n - 1)/27$.

Corollary 2.15. *The average distance of states in T^n, $n \in \mathbb{N}_0$, from a fixed perfect state is $\overline{3} \, (2^n - 1)$.*

Proof. If we sum the formula in Proposition 2.13 for all $s \in T^n$ and divide by the number 3^n of these states, we get, by symmetry, three times the average distance from a specific perfect state. □

An alternative proof of Corollary 2.15 can be based on symmetry and the fact, again following from Proposition 2.13, that $\bar{\bar{3}} \, (2^n - 1)$ is also the average distance from a fixed state s to the three perfect states.

Another immediate implication of Proposition 2.13 is the number of states realizing the worst-case distance from a specific perfect state; see Exercise 2.12. This is a special case of the following result [194, Proposition 5] (also implicitly contained in the exceeding formula of [368, Theorem 4.6]); recall that $q(m)$ is the number of 1s in the binary representation of m as on p. 80.

Proposition 2.16. *Let j^n be a fixed perfect state. Then for $\mu \in [2^n]_0$:*

$$|\{s \in T^n \mid \mathrm{d}(s, j^n) = \mu\}| = 2^{q(\mu)} \, .$$

Proof. According to Algorithm 10, μ is the sum of 2^{d-1} for those discs d where the P1-automaton changed its state because s_d was different from the previous one. There are two possibilities for this to happen.   □

An immediate consequence is the following.

Corollary 2.17. *For all $n \in \mathbb{N}_0$:* $\displaystyle\sum_{\mu=0}^{2^n-1} 2^{q(\mu)} = 3^n$.

Occasionally, we will be interested in the asymptotic behavior, for large n, of (integer) sequences. To investigate this behavior we first note that Lemma 0.3 and Corollary 0.4 and their proofs remain valid for $\alpha, \beta \in \mathbb{R}$ and $a, b \in \mathbb{R}^{\mathbb{N}_0}$. We then have the following elementary, but extremely useful lemma which can hardly be found in textbooks (cf., however, [263, p. 151–153]).

Lemma 2.18. *Let $\alpha, \beta, \gamma \in \mathbb{R}$ and $a, b \in \mathbb{R}^{\mathbb{N}_0}$. Then*

a) $\forall n \in \mathbb{N}_0 : a_{n+1} = \alpha a_n + b_n \Leftrightarrow \forall n \in \mathbb{N}_0 : a_n = \alpha^n a_0 + \displaystyle\sum_{k=0}^{n-1} \alpha^k b_{n-1-k}$,

b) $\forall n \in \mathbb{N}_0 : (\beta - \alpha) \displaystyle\sum_{k=0}^{n-1} \alpha^k \beta^{n-1-k} = \beta^n - \alpha^n$.

c) *If $|\alpha| < |\gamma|$ and a fulfills the recurrence in (a) with* $\dfrac{b_n}{\gamma^n} \to b_\infty \in \mathbb{R}$, *then*

$$\frac{a_n}{\gamma^n} \to \frac{b_\infty}{\gamma - \alpha}, \text{ as } n \to \infty \, .$$

Proof. We get (a) by replacing b_0 by a_0 and b_n by b_{n-1} for $n \in \mathbb{N}$ in Lemma 0.3. Similarly, (b) follows from Corollary 0.4 by writing $n - 1$ for n.

For (c) we remark that if $\dfrac{a_n}{\gamma^n}$ converges at all, it is necessarily to $\dfrac{b_\infty}{\gamma - \alpha}$.

Replacing a_n with $\dfrac{a_n}{\gamma^n} - \dfrac{b_\infty}{\gamma - \alpha}$, we may therefore assume that $b_\infty = 0$ and $\gamma = 1$.

From (a) we know that

$$a_n = \alpha^n a_0 + \sum_{k=0}^{n-1} \alpha^k b_{n-1-k} = \alpha^n a_0 + \sum_{k=0}^{n-1} \alpha^{n-1-k} b_k \,.$$

As $|\alpha| < |\gamma| = 1$, it suffices to show that the last sum tends to 0 as n goes to infinity. Let $\varepsilon > 0$ and define $\mu := \frac{1}{2}(1 - |\alpha|)\varepsilon$ (> 0). Then $\exists\, n_\mu \; \forall\, k \geq n_\mu : |b_k| \leq \mu$, such that for all $n \geq n_\mu$ and making use of (b):

$$
\begin{aligned}
\left| \sum_{k=n_\mu}^{n-1} \alpha^{n-1-k} b_k \right|
&\leq \mu \sum_{k=n_\mu}^{n-1} |\alpha|^{n-1-k} \\
&= \mu \left(\frac{1 - |\alpha|^n}{1 - |\alpha|} - \frac{1 - |\alpha|^{n_\mu}}{1 - |\alpha|} \right) \\
&\leq \frac{|\alpha|^{n_\mu}}{1 - |\alpha|} \mu \leq \frac{1}{1 - |\alpha|} \mu = \frac{\varepsilon}{2} \,.
\end{aligned}
$$

Now $\exists\, B > 0 \; \forall\, k \in \mathbb{N}_0 : |b_k| \leq B$ and

$$\exists\, n_\varepsilon \geq n_\mu \; \forall\, n \geq n_\varepsilon \; |\alpha|^n \leq \varepsilon \frac{(1 - |\alpha|)|\alpha|^{n_\mu}}{2B} \,,$$

whence, again with (b),

$$
\begin{aligned}
\left| \sum_{k=0}^{n_\mu-1} \alpha^{n-1-k} b_k \right|
&\leq B|\alpha|^{n-n_\mu} \sum_{k=0}^{n_\mu-1} |\alpha|^{n_\mu-1-k} \\
&= B|\alpha|^{n-n_\mu} \frac{1 - |\alpha|^{n_\mu}}{1 - |\alpha|} \\
&\leq B \frac{1}{(1 - |\alpha|)|\alpha|^{n_\mu}} |\alpha|^n < \frac{\varepsilon}{2} \,. \qquad \square
\end{aligned}
$$

Lemma 2.18 c) and its proof even extend to matrices $A \in \mathbb{R}^{\ell \times \ell}$ and sequences of vectors $a, b \in (\mathbb{R}^\ell)^{\mathbb{N}_0}$, $\ell \in \mathbb{N}$, without any further change. This leads to

Lemma 2.19. *If $\|A\| < |\gamma|$ and a fulfills the recurrence in Lemma 2.18 a) with $\gamma^{-n} b_n \to b_\infty \in \mathbb{R}^\ell$, then $\gamma^{-n} a_n \to (\gamma I - A)^{-1} b_\infty$, where I is the unit matrix in $\mathbb{R}^{\ell \times \ell}$.*

The proof is the same as for Lemma 2.18 c); the only extra ingredient we need is

$$\|A\| < 1 \Rightarrow I - A \text{ is invertible}.$$

This is easy: if $I - A$ were not invertible, then we would have an $x \in \mathbb{R}^\ell$ with $\|x\| = 1$ and $(I - A)x = 0$, i.e. $Ix = Ax$, but then $1 = \|x\| = \|Ix\| = \|Ax\| \leq \|A\| \, \|x\| = \|A\|$.

2.2.1 Noland's Problem

We can now approach a somewhat more complex task of type P1.

Proposition 2.20. *Let* $s \in T^N$, $N \in \mathbb{N}_0$, *with* $s_d \equiv d \bmod 2$ *for all* $d \in [N]$. *Then*

$$d\left(s, s_N^N\right) = \left\lfloor \frac{3}{7} 2^N \right\rfloor, \quad d\left(s, 2^N\right) = \left\lfloor \frac{5}{7} 2^N \right\rfloor, \quad d\left(s, (1 - s_N)^N\right) = \left\lfloor \frac{6}{7} 2^N \right\rfloor.$$

Remark. The three distances in Proposition 2.20 are from a state where the discs are ordered according to parity to the perfect state on the peg where the largest disc initially lies, to the peg where no disc lies, and to the peg where the second largest starts, respectively. This type of tasks has been posed several times in literature, mostly in the case of even $N = 2n$, where n even numbered discs on peg 0 and n odd numbered discs on peg 1 have to be reunited on either peg 1 [108, p. 76] or the third peg 2. The latter problem was formulated by H. Noland in [323]. It was solved by A. J. van Zanten in [448] for odd N as a lemma for yet another task, the Twin-Tower problem to be discussed later in Section 6.3. S. Obara and Y. Hirayama asked in [326] to sort a perfect tower into odd and even discs on the two other pegs; see Figure 2.9. Although this is, of course, mathematically equivalent to the inverse task, it seems that the human problem solver has more difficulties to reach a non-perfect state from a perfect one than vice versa. Without the requirement that the sorted discs end up on the pegs different from the starting position, the sorting task can be fulfilled in $d(s, s_N^N)$ moves.

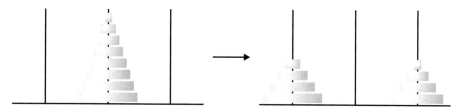

Figure 2.9: Sorting golden (even) and silver (odd) discs

Proof of Proposition 2.20. Let

$$x_N := d\left(s, s_N^N\right), \quad y_N := d\left(s, 2^N\right), \quad z_N := d\left(s, (1 - s_N)^N\right).$$

Since all tasks are of type P1, the largest disc will move at most once and we have

$$x_N = y_{N+1} - 2^N, \quad x_{N+1} = z_N, \quad x_{N+2} = y_N + 2^N,$$

and consequently

$$x_0 = 0, \quad x_1 = 0, \quad x_2 = 1, \quad x_{N+3} = x_N + 3 \cdot 2^N;$$

$$y_0 = 0, \quad y_1 = 1, \quad y_2 = 2, \quad y_{N+3} = y_N + 5 \cdot 2^N.$$

We can solve these recurrences in six steps using Lemma 2.18. For example, let $N = 3n$, $n \in \mathbb{N}_0$. Then, with $\alpha = 1$, $\beta = 8$, $a_n = x_{3n}$, and $b_n = 3 \cdot 8^n$ we get

$$x_{3n} = 3 \sum_{k=0}^{n-1} 8^{n-1-k} = \frac{3}{7}(8^n - 1).$$

Similarly, we obtain

$$x_{3n+1} = \frac{6}{7}(8^n - 1), \ x_{3n+2} = \frac{12}{7}8^n - \frac{5}{7},$$

altogether $x_N = \left\lfloor \frac{3}{7}2^N \right\rfloor$, $z_N = \left\lfloor \frac{6}{7}2^N \right\rfloor$.

In the same way we get

$$y_{3n} = \frac{5}{7}(8^n - 1), \ y_{3n+1} = \frac{10}{7}8^n - \frac{3}{7}, \ y_{3n+2} = \frac{20}{7}8^n - \frac{6}{7},$$

whence $y_N = \left\lfloor \frac{5}{7}2^N \right\rfloor$.   □

A solution of this type to Noland's problem was given by D. G. Poole in [342]. N. F. Lindquist [269] observed that for $t_n := y_{2n}$ the following recurrence holds:

$$t_0 = 0, \ t_1 = 2, \ t_2 = 11, \ t_{n+3} = t_n + 45 \cdot 4^n.$$

Yet another way to solve the tasks of Proposition 2.20 can, of course, be based on the P1-automaton with the advantage that the best first move, and consequently the complete optimal solution will be obtained. For instance, in the task to reunite all discs on the peg which is originally empty (corresponding to the sequence y_N) one finds that the best first move has idle peg $(N + 2) \bmod 3$. The sorting problem of Figure 2.9 is slightly more subtle; see Exercise 2.13.

In general, the state of the P1-automaton immediately *before* the evaluation of disc $d \in [n]$ in Algorithm 10 can be interpreted as the *first goal* of disc d (cf. [128]), denoted by

$$(s \vartriangle j)_d,$$

to make the best first move possible in the course of the optimal solution to get from state s to the perfect state on peg j. Note that $(s \vartriangle j)_d = s_{d+1} \vartriangle \cdots \vartriangle s_n \vartriangle j$ for $d \in [n]$, and $(s \vartriangle j)_0 := s_1 \vartriangle \cdots \vartriangle s_n \vartriangle j$ is the idle peg in the best first move, which can be interpreted as the first goal of the "thimble" (cf. p. 96). Moreover (cf. [194, Theorem 3]),

$$\mathrm{d}(s, j^n) = \sum_{d=1}^{n} (s_d \neq (s \vartriangle j)_d) \cdot 2^{d-1}. \tag{2.8}$$

An immediate consequence is another characterization of perfect states.

Lemma 2.21. *Let $s \in T^n$, $n \in \mathbb{N}_0$, and $\{i, j\} \in \binom{T}{2}$. Then*

$$\mathrm{d}(s, i^n) = \mathrm{d}(s, j^n) \Leftrightarrow s = (i \vartriangle j)^n.$$

Proof by induction on n. The case $n = 0$ is clear. So let $s = s_{n+1}\overline{s} \in T^{1+n}$. Then, by (2.8) and the uniqueness of finite binary representations,

$$d(s, i^n) = d(s, j^n) \Leftrightarrow \forall d \in [n+1] : (s_d \neq (s \vartriangle i)_d) = (s_d \neq (s \vartriangle j)_d) .$$

This, in turn, is equivalent by induction assumption, to $\overline{s} = (i \vartriangle j)^n$ and $(s_{n+1} \neq (s \vartriangle i)_{n+1}) = (s_{n+1} \neq (s \vartriangle j)_{n+1})$. The proof is completed by the fact that $(s \vartriangle i)_{n+1} = i$ and $(s \vartriangle j)_{n+1} = j$. □

Another application of (2.8) is made in Exercise 2.14. Employing (2.8) in Noland's problem with $2n$ discs, we easily see that

$$\forall d \in [2n] : s_d = (s \vartriangle 2)_d \Leftrightarrow (2n - d) \bmod 3 = 1 ,$$

such that

$$y_{2n} = \sum_{d=1}^{2n} (s_d \neq (s \vartriangle 2)_d) \cdot 2^{d-1}$$

$$= 2^{2n} - 1 - \sum_{d=1}^{2n} ((2n - d) \bmod 3 = 1) \cdot 2^{d-1}$$

$$= 2^{2n} \left(1 - \frac{1}{4} \sum_{k=1}^{\lfloor 2(n-1)/3 \rfloor} 8^{-k}\right) - 1 = \left\lfloor \frac{5}{7} 4^n \right\rfloor ,$$

where use has been made of Lemma 2.18 again.

We now come back to P1-type tasks $s \to j^n$, $n \in \mathbb{N}$, where the goal peg j is *not* specified. For instance, starting a P0-type task from 0^n, the target might be either 1^n or 2^n. If a player gets lost in $s \in T^n$, the question arises for which $j \in [2]$ we have $d(s, j^n) < d(s, (3 - j)^n)$. (The case of equality can only occur for $s = 0^n$; cf. Exercise 2.10 a.) We make the following observation for $s \neq 0^n$: if the number of leading 0s is even, then j is the first (from the left) non-zero tit i in s; otherwise $j = 3 - i$. (In particular, $j = s_n$, if $s_n \neq 0$; this is already a consequence of (2.6) and (2.7).) More formally,

Proposition 2.22. *Let $n \in \mathbb{N}$ and $s = 0^{n-d}s_d\overline{s}$ with $d \in [n]$, $s_d \in [2]$, $\overline{s} \in T^{d-1}$. For $j \in [2]$ with $d(s, j^n) < d(s, (3 - j)^n)$ we have: $j = s_d$, if $n - d$ is even, and $j = 3 - s_d$ otherwise.*

Proof. Entering s into two P1-automata with initial states $j = 1$ and $j = 2$, respectively, will find these automata after $n - d$ inputs in states 1 and 2, or 2 and 1, depending on whether $n - d$ is even or odd. In the first case, entering $s_d \in [2]$ will leave either the first or the second state unchanged depending on whether $s_d = 1$ or $s_d = 2$, such that the corresponding j^n is the closer perfect state from s; cf. (2.8). Similarly for the odd case. □

The decision about the nearest perfect state can be realized in Algorithm 11.

Algorithm 11 Closest perfect state

Procedure cp(n, s)
Parameter n: number of discs $\{n \in \mathbb{N}\}$
Parameter s: regular state, not perfect on 0 $\{s \in T^n \setminus \{0^n\}\}$
    $d \leftarrow n$ {index of tit}, $b \leftarrow 0$ {parity of $n - d$}
    while $s_d = 0$
       $d \leftarrow d - 1, b \leftarrow 1 - b$
    end while
    $j = 1 + b \ \dot{\vee} \ (s_d - 1)$

On the average, this algorithm performs much better than evaluating (2.8) twice; see Exercise 2.15.

Let us finally mention that Lu proposed in [277, Section 4] an algorithm to get from a perfect to an arbitrary regular state, by which means he also arrives at Proposition 2.13 ([277, Corollary 1]).

2.2.2   Tower of Hanoi with Random Moves

At the end of Section 1.1 results on the expected number of moves when the Chinese Rings puzzle is played randomly were presented. We now consider related tasks on the Tower of Hanoi. More precisely, denoting by r a random regular state from T^n and by i a random peg from T, we are interested in the expected number of moves d_e between the state 0^n and one of the states i^n and 2^n, and between r and one of the states i^n and 0^n. For instance, in the task to reach an arbitrary perfect state i^n from the perfect state 0^n, the game ends as soon as a perfect state is reached. To make this task non-trivial, it is required that at least one disc is moved. In all the tasks it is assumed that the next move is chosen uniformly from the set of current legal moves. M. A. Alekseyev and T. Berger [3] solved these problems as follows.

Theorem 2.23. *For every* $n \in \mathbb{N}$,

$$d_e(0^n, i^n) = \frac{3^n - 1}{2},$$

$$d_e(0^n, 2^n) = \frac{(3^n - 1)(5^n - 3^n)}{2 \cdot 3^{n-1}},$$

$$d_e(r, i^n) = \frac{5^n - 2 \cdot 3^n + 1}{4},$$

$$d_e(r, 0^n) = \frac{(3^n - 1)(5^{n+1} - 2 \cdot 3^{n+1}) + 5^n - 3^n}{4 \cdot 3^n}.$$

Among the four tasks of Theorem 2.23, the first and the third are more symmetric and also seem more natural than the other two because we finish the game as soon as a perfect state is reached (and not when a *specific* perfect state is reached). This is reflected in the formulas for $d_e(0^n, i^n)$ and $d_e(r, i^n)$ which are significantly simpler than the formulas for $d_e(0^n, 2^n)$ and $d_e(r, 0^n)$, respectively.

The formula for $d_e(0^n, 2^n)$ asserts that a thoughtless person would need an expected number of approximately $576\,008$ moves already for the original 8-disc version of the TH! (For 4 discs, the number is about 806; compare this with the performance of an ape, albeit on drugs, in the movie *Rise of the Planet of the Apes* (2011).)

In [3] it is also proved that $d_e(10^{n-1}, i^n) = 3(5^{n-1} - 3^{n-1})/2$ holds for every $n \in \mathbb{N}$. The task to reach an arbitrary perfect state randomly from the state 10^{n-1} is not an important one, but it turned out to be utmost useful in deriving the results of Theorem 2.23. In the rest of the subsection we demonstrate this by proving the first formula of Theorem 2.23.

So consider the task to get randomly from 0^n to an arbitrary perfect state i^n, where at least one move is made. Let $p_j(n)$, $j \in [3]_0$, be the probabilities that the final state reached is the perfect state on peg j. It is clear that $p_1(n) = p_2(n)$, so that $p_0(n) + 2p_1(n) = 1$. Using this notation we will first prove that:

$$d_e(0^n, i^n) = d_e(0^{n-1}, i^{n-1}) + 2p_1(n-1)d_e(10^{n-1}, i^n). \tag{2.9}$$

After the game starts with a move of disc 1, at some point we get for the first time to a state in which all discs from $[n-1]$ lie on the same peg. The expected number of moves to reach this state is $d_e(0^{n-1}, i^{n-1})$. In the case that these discs lie on peg 0, the game is over. Otherwise we are in a state where disc n can (but not necessarily will) move for the first time. We have reached this state with the probability $1 - p_0(n-1) = 2p_1(n-1)$. The expected number of moves to reach the final state of the game (a perfect state) from such a state is, by definition, $d_e(10^{n-1}, i^n)$. In conclusion, $d_e(0^n, i^n) = d_e(0^{n-1}, i^{n-1}) + 2p_1(n-1)d_e(10^{n-1}, i^n)$, whence (2.9) is proved.

In view of Equation (2.9) we next have a closer look at $d_e(10^{n-1}, i^n)$ and prove:

$$d_e(10^{n-1}, i^n) = \frac{1}{2} + d_e(0^{n-1}, i^{n-1}) + (p_0(n-1) + p_1(n-1))d_e(10^{n-1}, i^n). \tag{2.10}$$

Being in the state 10^{n-1}, disc n is moved with the probability $1/3$. The probability that it is moved twice in a row is $1/3^2$. Continuing in this manner we infer that the expected number of moves of disc n at the start of the game is $\sum_{k=1}^{\infty} 3^{-k} = \frac{1}{2}$. After this initial stage of the game, disc 1 is moved and then the game continues until all discs from $[n-1]$ lie for the first time on the same peg. The expected number of moves for this part of the game is, by definition, $d_e(0^{n-1}, i^{n-1})$. At this stage, the discs from $[n-1]$ are on peg 0 with the probability $p_0(n-1)$, and on pegs 1 and 2 with the same probability $p_1(n-1)$. It follows that no matter where disc n lies (either on peg 1 or on peg 2), the game is over with the probability $p_1(n-1)$. Otherwise, with the probability $1 - p_1(n-1) = p_0(n-1) + p_1(n-1)$, the game continues. Since this is just an instance of the initial game, the expected number of moves to finish it is $d_e(10^{n-1}, i^n)$. This proves Equation (2.10).

To conclude the argument, using $p_0(n-1) + p_1(n-1) = 1 - p_1(n-1)$, Equation (2.10) can be rewritten as

$$p_1(n-1) \, d_e(10^{n-1}, i^n) = \frac{1}{2} + d_e(0^{n-1}, i^{n-1}).$$

Inserting this into Equation (2.9) yields:

$$d_e(0^n, i^n) = 3d_e(0^{n-1}, i^{n-1}) + 1.$$

The solution of this recurrence, together with the initial condition $d_e(0, i) = 1$, gives the claimed result: $d_e(0^n, i^n) = (3^n - 1)/2$.

The second task of Theorem 2.23 was analyzed in [3, Section 3] also based on networks of electrical resistors.

2.3   Hanoi Graphs

In Section 2.2 we introduced the set T^n in order to accommodate all regular states of the Tower of Hanoi with n discs. A graph structure on T^n is now imposed by the prototype R^n in Chapter 1. Theorem 2.7 suggests to unite all shortest paths ending in the perfect state 0^n, say. This would result in a **binary tree** with vertex set T^n, **root** 0^n and **height** $2^n - 1$; cf. Er [129]. However, although all states of the Tower of Hanoi with n discs will be represented in this tree, its edge set does not comprise all admissible moves. Their inclusion will allow us to discuss more general tasks in the rest of this chapter as well as new views on facts already presented.

Two vertices from the vertex set T^n (that is, regular states) are adjacent if they are obtained from each other by a legal move of one disc. We call this graph on 3^n vertices, first considered by Scorer et al. [378], the *Hanoi graph* (cf. [277, p. 24]) for n discs and 3 pegs and denote it by H_3^n. The degree sequence of H_3^n is very simple; cf. Exercise 2.16. In analogy with Remark 1.2 we can state formally

$$V(H_3^n) = T^n,$$
$$E(H_3^n) = \left\{ \{\underline{s}i(3-i-j)^{d-1}, \underline{s}j(3-i-j)^{d-1}\} \mid i,j \in T,\ i \neq j,\ d \in [n],\ \underline{s} \in T^{n-d} \right\};$$
$$\tag{2.11}$$

here each edge represents a move of disc d between pegs i and j, which is independent of the distribution $\underline{s}$ of larger discs, i.e. the bottom $n - d$ discs.

This immediately leads to colorings of Hanoi graphs.

Proposition 2.24. *For any* $n \in \mathbb{N}$, $\chi(H_3^n) = 3 = \chi'(H_3^n)$. *Moreover, the function* $s \mapsto (\sum_{d=1}^n s_d) \bmod 3$ *defines a vertex coloring of* H_3^n; *the label of the idle peg of the move associated with an edge defines an edge coloring.*

Proof. The second part of the statement follows from the fact that every edge represents the change of precisely one tit in the adjacent vertices. Moreover, all edges coming together in a fixed state correspond to moves with different idle pegs, two for the moves of disc 1, and the position of disc 1 in case another disc is moving. See Figure 2.10 for an illustration.

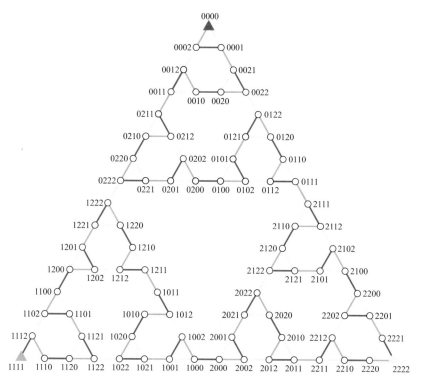

Figure 2.10: The graph H_3^4 with 3-edge coloring

By the above, $\chi(H_3^n) \leq 3$ and $\chi'(H_3^n) \leq 3$. To complete the argument note that for $\underline{s} \in T^{n-1}$, the states $\underline{s}0$, $\underline{s}1$, and $\underline{s}2$ induce a complete graph on 3 vertices.

□

The question to find total colorings for Hanoi graphs is somewhat more subtle and therefore left as Exercise 2.17.

The size of H_3^n is the topic of Exercise 2.18. To illustrate the recursive structure of Hanoi graphs, Figure 2.11 shows the standard drawings of the first four Hanoi graphs. The trees mentioned above are obtained from these graphs by deleting all horizontal edges.

We observe that H_3^{1+n} is composed of three subgraphs iH_3^n, $i \in T$, induced by the vertex sets iT^n, any two of these subgraphs, say iH_3^n and jH_3^n, being joined by precisely one edge, the edge between ik^n and jk^n, $k = 3 - i - j$. See Figure 2.12.

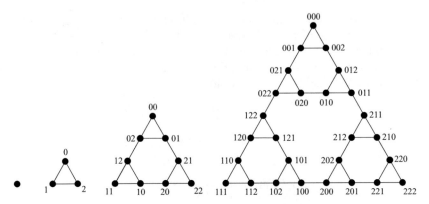

Figure 2.11: Hanoi graphs H_3^0, H_3^1, H_3^2, and H_3^3

Planarity of H_3^n is then obvious; cf. [378, p. 97]. Since all edges can be drawn with the same length, H_3^n is a *matchstick graph*; cf. [98, Chapitre 12].

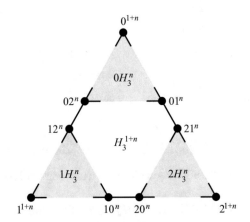

Figure 2.12: Recursive structure of Hanoi graphs

Note that the order of the perfect states in the graphical representation of the subgraphs iH_3^n have to be chosen appropriately. More precisely, these subgraphs are reflected at the vertical axis through $i0^n$, and $1H_3^n$ and $2H_3^n$ are rotated by 120° clockwise or counter-clockwise, respectively. This recursive labeling procedure is schematically explained in Figure 2.13. For an alternative visualization, see [396, Figure 4].

The formal recursive definition of the edge sets of the Hanoi graphs H_3^n is (cf. Equation (1.1))

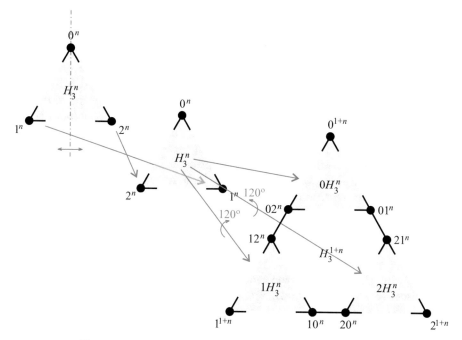

Figure 2.13: Recursive composition of labels of H_3^{1+n}

$$E(H_3^0) = \varnothing,$$
$$\forall\, n \in \mathbb{N}_0 : E(H_3^{1+n}) = \{\{ir, is\} \mid i \in T,\ \{r, s\} \in E(H_3^n)\}$$
$$\cup \{\{i(3-i-j)^n, j(3-i-j)^n\} \mid i, j \in T,\ i \neq j\}. \tag{2.12}$$

Many properties of Hanoi graphs can be deduced from their recursive structure, like obviously connectedness or, more precise, 2-connectedness.

Proposition 2.25. *The **connectivity** of* H_3^n, $n \in \mathbb{N}$, *is* $\kappa\left(H_3^n\right) = 2$.

Proof. Deleting the two neighbors of a perfect state will separate the latter from the rest of the graph. To show that the deletion of only one vertex is not sufficient, we employ induction: for $H_3^1 \cong K_3$ this is clear; otherwise, by induction assumption, H_3^{1+n} could only be disconnected by deleting a vertex ik^n and an edge $\{ik^n, jk^n\}$ with $|\{i, j, k\}| = 3$, but then two vertices from iH_3^n and jH_3^n, respectively, are still linked by a path through edges $\{ij^n, kj^n\}$ and $\{ki^n, ji^n\}$. □

Before moving on we remark that the Hanoi graphs, together with some of their structural properties and applications to the TH (say Olive's algorithm), were rediscovered by R. Hinze in [215] within the frame of the so-called wholemeal and projective programming.

2.3.1   The Linear Tower of Hanoi

Another quality of Hanoi graphs is hamiltonicity; see Exercise 2.19. As a consequence of that exercise, one gets a hamiltonian $1^4, 2^4$-path in the graph H_3^4 of Figure 2.10 by avoiding all edges colored in red. They correspond to moves between pegs 1 and 2, i.e. with idle peg 0. So this path is the state graph of a variant of the TH, where these moves are prohibited. In fact, most versions of the TH available on the market today have the three pegs arranged in line and not, as the original, in a triangle. This demonstrates the ignorance of modern manufacturers because it *does* make a difference for human problem solvers who may be reluctant to make (long) moves between exterior pegs or may even think that they are not allowed. This latter restriction leads to what has been called the *Linear Tower of Hanoi* (or *Three-in-a-row Tower of Hanoi*). It was first briefly mentioned back in 1944 by Scorer, Grundy, and Smith [378, p. 99] and studied in detail by Hering in [190]. It is the prototype of TH variants with restricted disc moves which will be addressed in Chapter 8.

Choosing the label 0 for the central peg, we will denote the modelling graph for n discs by $H_{3,\mathrm{lin}}^n$. As shown above, it is a path graph on 3^n vertices; see Figure 2.14 for the case $n = 2$.

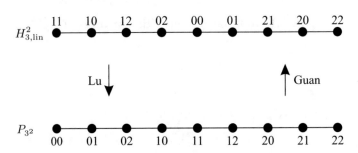

Figure 2.14: The graph $H_{3,\mathrm{lin}}^2$ and a path graph with length 8

This path is realized by the following procedure starting in 1^n: move disc 1 twice in a row (in the same direction), make the only legal move of a larger disc, and keep on doing this. Even simpler is the recipe always to make the only legal move not undoing the previous one (because we are on a path graph). However, both strategies do not allow a human problem solver to continue an abandoned puzzle. Here another easy human strategy is given in Exercise 2.20; cf. also the remark on the sequence of discs moved $1, 1, 2, 1, 1, 2, 1, 1, 3, \ldots$ (A051064) in [378, p. 99]. The number of moves made by individual discs is determined in Exercise 2.21.

In analogy to the Gros code of Chapter 1 (cf. Figure 1.4), let us introduce the *Lu code* (or *Lu weight*) of $s \in T^n$ as $\mathrm{d}(s) := \mathrm{d}(s, 1^n)$ with the graph distance in $H_{3,\mathrm{lin}}^n$. Again we then have $\mathrm{d}(s, t) = |\mathrm{d}(s) - \mathrm{d}(t)|$ for $s, t \in T^n$. So distances in $H_{3,\mathrm{lin}}^n$

can easily be obtained from the Lu weights. For their calculation let

$$\forall r \in [n] : \widetilde{s}_r = \sum_{\rho=r+1}^{n} (s_\rho = 0),$$

where the sum is the binary sum $+ = \dot{\vee}$. (This means that $\widetilde{s}_r$ is the parity of the number of 0s in s strictly to the left of s_r.) Then we have (cf. [277, Theorem 5]):

Proposition 2.26. *Let $s \in T^n$. We write the ternary representation of* $\mathrm{d}(s)$ *as* $\sum_{r=1}^{n} \mathrm{d}(s)_r \cdot 3^{r-1}$, $\mathrm{d}(s)_r \in T$, *i.e.* $\mathrm{d}(s) = (\mathrm{d}(s)_n \ldots \mathrm{d}(s)_1)_3$. *Then*

$$\forall r \in [n] : \mathrm{d}(s)_r = (1 + (2\widetilde{s}_r - 1)s_r) \bmod 3. \tag{2.13}$$

The proof is left to the reader as Exercise 2.22 a). Part (b) of that exercise demonstrates in what sense the inverse *Guan code* (see Figure 2.14), which assigns the state $s(d)$ to a distance d such that $\mathrm{d}(s(d)) = d$, can be viewed as a *reflected ternary code*. It corresponds to the reflected binary or Gray code, characterized as a special sequence of words from B^n in which any two consecutive terms differ by exactly one bit. This idea can be generalized by saying that a listing of $[p]_0^n$ is a (p,n)-*Gray code* if any two consecutive terms differ in exactly one position. D.-J. Guan [170] constructed (p,n)-Gray codes and showed that the corresponding $(3,n)$-Gray code encodes the traverse of the state graph of the Linear TH between its end vertices.

2.3.2 Perfect Codes and Domination

A somewhat more subtle recursive argument, stressing the role of parity of n in some features of Hanoi graphs, leads to perfect codes on H_3^n. Figure 2.15 shows the iterative construction of two families $(A_n)_{n\in\mathbb{N}}$ and $(C_n)_{n\in\mathbb{N}}$ of subsets of the vertex sets T^n of H_3^n as they appear in the standard drawings of these graphs.

The seeds are $A_0 = \varnothing$ and $C_0 = T^0$. In $0H_3^n$, we include all those vertices into A_{1+n} and C_{1+n} which are in the corresponding positions of the elements of A_n and C_n, respectively, in H_3^n. For $1H_3^n$ and $2H_3^n$ we have to take parity of n into account. For even n we add to C_{1+n} those vertices $1s$, $2s$ for which $s \in A_n$; the additional elements of A_{1+n} are obtained from those in corresponding positions in H_3^n, but with the latter rotated by $120°$ counter-clockwise for $1H_3^n$ and clockwise for $2H_3^n$. Similarly, for odd n, C_{1+n} will receive the corresponding elements contained in the positions of C_n in the copies of H_3^n rotated as before, and A_{1+n} those of the not rotated one. The reason for the necessity of the auxiliary sequence of sets A is that all perfect states have to be codewords for even n, so that we cannot link three of these together because then codewords would be neighbors. The set A_n does not cover perfect states, whereas C_n is a perfect code on H_3^n. See Figure 2.16 for the first few cases.

Because of the imposing character of the construction, the perfect codes are unique for even n and there are precisely three of them for odd n, depending

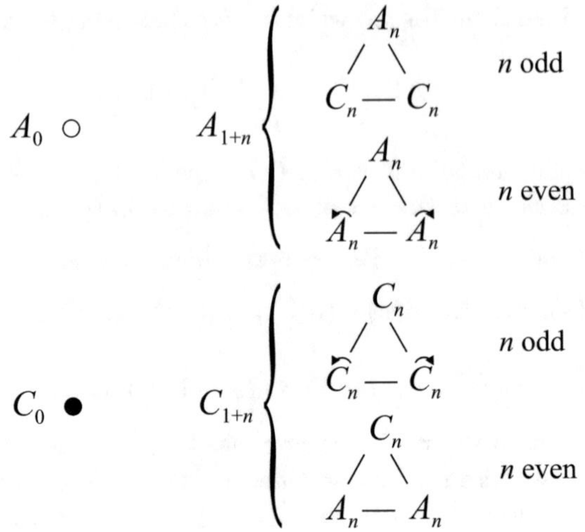

$$A_0 \; \circ \qquad\qquad A_{1+n} \begin{cases} \begin{array}{c} A_n \\ / \; \backslash \\ C_n - C_n \end{array} & n \text{ odd} \\[2em] \begin{array}{c} A_n \\ / \; \backslash \\ \overleftarrow{A}_n - \overrightarrow{A}_n \end{array} & n \text{ even} \end{cases}$$

$$C_0 \; \bullet \qquad\qquad C_{1+n} \begin{cases} \begin{array}{c} C_n \\ / \; \backslash \\ \overleftarrow{C}_n - \overrightarrow{C}_n \end{array} & n \text{ odd} \\[2em] \begin{array}{c} C_n \\ / \; \backslash \\ A_n - A_n \end{array} & n \text{ even} \end{cases}$$

Figure 2.15: Iterative construction of sets A_n and C_n

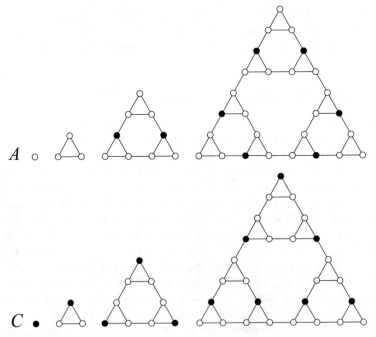

Figure 2.16: The vertices in A_n and C_n (represented by filled dots) in H_3^n for $n = 0, 1, 2, 3$

on which perfect state is a codeword and therefore obtained from each other by rotations of 120°.

The existence and uniqueness of perfect codes in Hanoi graphs H_3^n were first proved by Cull and I. Nelson in [89]. C.-K. Li and Nelson [264] followed with a short proof for the existence of these perfect codes. They also considered the more general problem of perfect k-codes for $k \in \mathbb{N}_2$ and proved that no such codes exist in H_3^n except the trivial ones, namely the codes with one vertex $(3 \cdot 2^{n-2} \le k \le 2^n - 1)$ and the code containing the three perfect states $(k = 2^{n-1} - 1)$.

Perfect codes in a graph G are special cases of G-dominating sets. To define these, we introduce the *(closed) neighborhood* of a vertex $u \in V(G)$ by

$$N[u] = \{u\} \cup \{v \in V(G) \mid \{u, v\} \in E(G)\} .$$

Moreover, the *(closed) neighborhood* of a subset U of $V(G)$ is

$$N[U] = \bigcup_{u \in U} N[u].$$

We say that $v \in V(G)$ is *dominated* by $d \in V(G)$ (or by $D \subseteq V(G)$), if $v \in N[d]$ (or $v \in N[D]$). A $D \subseteq V(G)$ is called (a G-)*dominating (set)* if $N[D] = V(G)$, i.e. if all $v \in V(G)$ are dominated by D, or in other words, if every vertex $v \in V(G) \setminus D$ is adjacent to some vertex $d \in D$. Since obviously $V(G)$ is dominating, it makes sense to define the *domination number* $\gamma(G)$ of G as the size of a smallest G-dominating set, that is,

$$\gamma(G) = \min \{|D| \mid D \subseteq V(G), \, N[D] = V(G)\} .$$

An obvious upper bound is $\gamma(G) \le |G|$ with equality if and only if $\Delta(G) = 0$; see Exercise 2.23.

To get a lower bound, we call a subset C of $V(G)$ 1-*error correcting*, if the neighborhoods of elements of C do not overlap, i.e. if

$$\forall \{c, c'\} \in \binom{C}{2}: \; N[c] \cap N[c'] = \varnothing .$$

We then have:

Proposition 2.27. *If $C \subseteq V(G)$ is 1-error correcting, then $\gamma(G) \ge |C|$.*

Proof. Let $|C| = k$ and $C = \{c_1, \ldots, c_k\}$ and let $D = \{d_1, \ldots, d_\ell\}$ be a G-dominating set with $|D| = \ell$. Then for every $i \in [k]$ there is a $j \in [\ell]$ such that $d_j \in N[c_i]$. By taking, e.g., the minimal such j, we get a mapping from $[k]$ to $[\ell]$, which is injective because $N[c] \cap N[c'] = \varnothing$ if $\{c, c'\} \in \binom{C}{2}$. Therefore $k \le \ell$ by the Pigeonhole principle. $\square$

Since a subset C of $V(G)$ is a perfect code if and only if it is dominating and 1-error correcting at the same time, we immediately get:

Corollary 2.28. *If C is a perfect code of G, then $\gamma(G) = |C|$. In particular, all perfect codes of G have the same cardinality.*

For an application to some easy cases we refer to Exercise 2.24. A perfect code of H_3^n contains $\frac{1}{4}(3^n + 2 + (-1)^n)$ codewords (A122983); see Exercise 2.25. Therefore:

$$\gamma(H_3^n) = \tfrac{1}{4}\left(3^n + 2 + (-1)^n\right).$$

2.3.3 Symmetries

Apart from the rotations there are three additional obvious symmetries of Hanoi graphs given by reflections at axes through perfect states and perpendicular to the opposite side. These symmetries of the drawing of the graph correspond to permutations of the peg labels. To show that there are no additional symmetries hidden, we look at the automorphism group $\mathrm{Aut}(H_3^n)$ of a Hanoi graph, i.e. the set of isomorphisms from H_3^n to H_3^n together with the canonical operation of composition, and show that it is isomorphic to the permutation group on T, which we will denote by $\mathrm{Sym}(T)$ ($\cong \mathrm{Sym}_3$). Note that this is the smallest non-abelian group, in turn isomorphic to the dihedral group of order 6 (cf. [343, p. 327]).

Theorem 2.29. *For any $n \in \mathbb{N}$, $\mathrm{Aut}(H_3^n) \cong \mathrm{Sym}(T)$.*

Proof. The six automorphisms of H_3^n described above, namely

$$g_\sigma : T^n \to T^n,\ s \mapsto \sigma(s_n)\dots\sigma(s_1)$$

for $\sigma \in \mathrm{Sym}(T)$, form a subgroup of $\mathrm{Aut}(H_3^n)$ isomorphic to $\mathrm{Sym}(T)$. Hence it remains to prove that they comprise already the complete automorphism group.

Let $g \in \mathrm{Aut}(H_3^n)$. Since g preserves degrees, we necessarily have $g(k^n) = \sigma(k)^n$ for some $\sigma \in \mathrm{Sym}(T)$ and all $k \in T$. But then $g = g_\sigma$ by Lemma 2.8, because automorphisms preserve distances.   □

A perfect state can thus be mapped to any other perfect state by some automorphism of H_3^n. On the other hand, a perfect state can not be mapped to a non-perfect one by an automorphism. This means that the three perfect states form an orbit under the action $g_\sigma.s = \sigma(s_n)\dots\sigma(s_1)$ of $\mathrm{Aut}(H_3^n)$ on $V(H_3^n) = T^n$; cf. Section 0.8.2. Any other orbit contains six states. These facts, rather obvious from the graphs, follow formally from the Orbit-Stabilizer Theorem 0.23, because perfect states are invariant under two of the six symmetries, namely the identity and the reflection on the line through that vertex, whereas every other vertex is only fixed by the identity. For the number of orbits, i.e. the size of the factor set $[T^n]$ of T^n with respect to the automorphism group $\mathrm{Aut}(H_3^n)$, we thus have $3 + 6([T^n] - 1) = 3^n$, such that there is a total of

$$|[T^n]| = \frac{1}{2}\left(3^{n-1} + 1\right) \tag{2.14}$$

orbits under the action of $\mathrm{Aut}(H_3^n)$ on T^n, $n \in \mathbb{N}$. (This is sequence A007051.) In other words, there exist that many essentially different states of the TH with n discs. Compared with the number of symmetry classes of Rubik's cube, namely $901\,083\,404\,981\,813\,616$ (cf. [223, Section 8.7]), this means that a TH with 40 or more discs has more configurations than the famous Hungarian puzzle.

The identity (2.14) is, of course, also a consequence of Corollary 0.24, because, apart from the trivial fixed point set $(T^n)^{\mathrm{id}}$ containing 3^n elements, only three other automorphisms, namely those corresponding to reflections, have fixed points, one each. Therefore, the sum in Burnside's lemma is $3(3^{n-1} + 1)$, which then has to be divided by the number of automorphisms.

A slightly more involved application of factor sets is obtained if we let $\mathrm{Aut}(H_3^n)$ act on $X = T^n \dot{\times} T^n$, where $A \dot{\times} A := \{(a,b) \in A \times A \mid a \neq b\}$; X is the set of non-trivial tasks (s,t) to get from a given state s to another one t. We will come to the metric aspects of this problem in the next section, but want to point out here that these tasks also come in sets of six equivalent copies, differing only by the labeling of the pegs. Psychologists, who use the TH in test tools (cf. [208]), do *not* regard (s,s) to be a task; therefore, for better comparison, we only consider the $3^n(3^n - 1)$ non-trivial tasks as well. The group action is now given by $g_\sigma.(s,t) = (\sigma(s_n)\ldots\sigma(s_1), \sigma(t_n)\ldots\sigma(t_1))$, and the only automorphism having fixed points is the identity. Therefore, from Corollary 0.24, the number of equisets is (cf. A016142)

$$\left|[X]\right| = \frac{1}{2}3^{n-1}(3^n - 1).$$

Moreover, since $\{\mathrm{id}\}$ is the stabilizer for every $(s,t) \in X$, we deduce from Theorem 0.23 that every equiset $[(s,t)]$ contains exactly 6 equivalent tasks which psychologists called *iso-problems*; cf. [201, p. 20].

At this point we should mention again that mathematical and psychological equivalence are not equivalent! A human problem solver might not realize the mathematical equivalence of two tasks where just two pegs have been switched because of the position of these pegs: one of them might be closer to the test person or they come in a version with the pegs in a row, making the middle one special. More seriously, the set of tasks X has another symmetry, namely tasks (s,t) and (t,s) are mathematically equivalent. They are not, in general, from the psychological point of view: it seems to be much harder to get from a perfect to a regular state than vice versa. It might be contrary to our everyday experience that disorder can be achieved much more easily, but here it is a prescribed, deterministic disorder! We therefore did not include reversion of tasks into the symmetry group acting on the set of tasks. We will come back to symmetry considerations in connection with the other so-called *tower tasks*.

2.3.4 Spanning Trees

In the 19th century, G. R. Kirchhoff considered the flow of electricity in a network of wires which can be modelled as circuits in a corresponding graph; cf. [55,

Chapter 8]. In the hands of mathematicians (cf. [421, p. 7–11]), this study of the complexity of a network turned into the question for the number of *spanning trees* of a connected graph G, i.e. **spanning subgraphs** of G which are trees (cf. [377]); this number is denoted by $\tau(G)$ and called, following W. T. Tutte, the *complexity* of G. Culminating point of this development is the *Matrix-Tree theorem* [422, Theorem VI.29], which states that $\tau(G)$ can be calculated from the so-called *Kirchhoff matrix* $\mathrm{K}(G)$ in a rather simple way: $\mathrm{K}(G)$ is obtained by subtracting the **adjacency matrix** of G from the diagonal matrix whose non-trivial entries are the degrees of the respective vertices; choose a vertex from G and delete the corresponding row and column from $\mathrm{K}(G)$, then $\tau(G)$ is equal to the determinant of the remaining matrix. As an example,

$$\mathrm{K}(H_3^1) = \begin{pmatrix} 2 & -1 & -1 \\ -1 & 2 & -1 \\ -1 & -1 & 2 \end{pmatrix},$$

and

$$\tau(H_3^1) = \begin{vmatrix} 2 & -1 \\ -1 & 2 \end{vmatrix} = 3.$$

The number of vertices of H_3^n growing like 3^n, it seems not to be feasible, however, to apply the Matrix-Tree theorem for general n. Nevertheless, we have the following impressive result.

Theorem 2.30. *For every $n \in \mathbb{N}_0$, the complexity of H_3^n is given by*

$$\tau(H_3^n) = 3^{\frac{1}{4}(3^n-1)+\frac{1}{2}n} \cdot 5^{\frac{1}{4}(3^n-1)-\frac{1}{2}n}$$

$$= \left(\sqrt{\frac{3}{5}} \right)^n \cdot \left(\sqrt[4]{15} \right)^{3^n-1}.$$

The first proof is due to E. Teufl and S. Wagner [417, p. 892], who made use of a resistance scaling factor on what they call a self-similar lattice, i.e. exploiting the recursive structure of H_3^n. The authors also determined the asymptotic behavior of the number m_n of **matchings** of H_3^n, namely $0.6971213284 \cdot 1.77973468825^{3^n}$ in [416, Example 6.3]. Compatible with this result is the calculation by H. Chen, R. Wu, G. Huang, and H. Deng in [78, Proposition 5] of the average entropy $\mu(H_3^n) = \lim_{n\to\infty} 3^{-n}\ln(m_n)) \approx 0.5764643\ldots$ The exact formula for m_n, however, is not yet known, cf. Exercise 2.26.

An entirely combinatorial proof of Theorem 2.30 was given by Z. Zhang, S. Wu, M. Li, and F. Comellas[44] in [450] following the model of S.-C. Chang, L.-C. Chen, and W.-S. Yang [75] who did the same kind of calculation for the approximating lattices of the Sierpiński triangle; cf. also [92, Section 3.2]. Because of its intrinsic challenges, we will sketch this argument, thereby essentially following J. Brendel [64].

---

[44]who kindly provided a draft version by Zhang, Wu and Comellas in 2011

Proof of Theorem 2.30. We will first try to establish a recurrence for $u_n := \tau(H_3^n)$. A spanning tree U of H_3^{1+n} decomposes into three spanning **forests** U_i of iH_3^n, respectively, together with a subset E of the connecting edges $e_k = \{ik^n, jk^n\}$, $\{i,j,k\} = T$. So we have $|U| - 1 = \|U\| = \|U_0\| + \|U_1\| + \|U_2\| + |E| = |U_0| - c_0 + |U_1| - c_1 + |U_2| - c_2 + |E|$, where $c_i \in \mathbb{N}$ is the number of components of U_i; hence $3 \geq |E| = c_0 + c_1 + c_2 - 1 \geq 2$. If $|E| = 2$, all U_is are trees, such that we get $3u_n^3$ spanning trees, depending on which of the three connecting edges are missing in E; cf. the upper row in Figure 2.17.

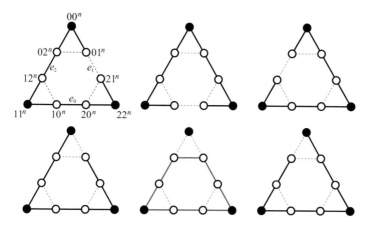

Figure 2.17: Composition of spanning trees of H_3^{1+n} (large triangles) from spanning forests of H_3^n (small triangles) with the number of respective components of the latter given by the number of dashed lines; filled dots are perfect states, unfilled ones are connecting vertices

If, however, all connecting edges are used, one of the subgraphs, say U_0, has to have two components, while the other two are trees. Since U has to be connected, each of the two components of U_0 has to contain at least one of the connecting vertices $0k^n$, $k \neq 0$. If we call b_n the number of spanning forests of H_3^n with two components, each of which containing at least one perfect state, we have to exclude those where 0^n is isolated (cf. the lower row in Figure 2.17) and will get $3 \cdot \frac{2}{3} b_n u_n^2$ spanning trees of H_3^{1+n} which are distinct from the ones constructed earlier. It is also clear that the counting is complete. We therefore arrive at

$$u_{1+n} = 3u_n^3 + 2b_n u_n^2. \qquad (2.15)$$

Unfortunately, we were not able to produce a recurrence for u_n directly, but had to introduce the counting of specific two-component spanning forests as well, such that we will now have to find a recurrence for the sequence $(b_n)_{n \in \mathbb{N}_0}$ too. With B such a subgraph of H_3^{1+n} decomposed as before into B_i and E, we get $3 \geq |E| = c_0 + c_1 + c_2 - 2 \geq 1$ and have to distinguish three cases for the size of E. This can be analyzed case by case as before; we omit the details. However, if $|E| = 3$, we are facing the possibility that, say, $c_0 = 3$, $c_1 = 1 = c_2$, i.e. B_0 is a spanning forest of $0H_3^n$ with three components, each of them containing a vertex $0k^n$; cf. the center picture in Figure 2.18.

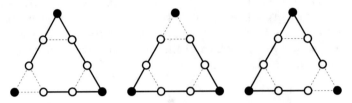

Figure 2.18: Composition of two-component spanning forests of H_3^{1+n} from two spanning trees and one three-component spanning forest of H_3^n; cf. the caption of Figure 2.17

Denoting the number of spanning forests of H_3^n with three components each containing a perfect state by t_n, we clearly get an extra additive term $3t_n u_n^2$ for b_{1+n}. All in all we obtain

$$b_{1+n} = 3u_n^3 + 7b_n u_n^2 + \frac{7}{3}b_n^2 u_n + 3t_n u_n^2 . \qquad (2.16)$$

Fortunately, this is the end of the chain, because spanning forests with more than three components must have a component without a perfect state (Pigeon hole principle!). A final analysis for the sequence $(t_n)_{n \in \mathbb{N}_0}$, the details of which we omit again, yields

$$t_{1+n} = u_n^3 + 4b_n u_n^2 + 4b_n^2 u_n + \frac{14}{27}b_n^3 + 3t_n u_n^2 + 4t_n b_n u_n . \qquad (2.17)$$

We are left with the formidable task to solve the non-linear recurrence consisting of Equations (2.15), (2.16) and (2.17), with the obvious seeds $u_0 = 1$, $b_0 = 0 = t_0$. There seems to be no general method, so one has to do some handicraft! First of all one realizes that b_n is linear in (2.15) and since $u_n \neq 0$ (every connected graph has a spanning tree: take a minimal (with respect to size) spanning subgraph), we can solve for b_n and insert this into the other two equations. Now the same is true for t_n in (2.16), such that from (2.17) we get a recurrence of order 3 for u_n (cf. [64, p. 19]):

$$u_{3+n} = \frac{u_{2+n}^2 (36 u_{2+n} u_{1+n}^4 u_n^3 + 21 u_{2+n}^2 u_n^6 - 35 u_{1+n}^8)}{18 u_{1+n}^3 u_n^6} . \qquad (2.18)$$

This does not look more promising, even though it is easy to calculate the first few terms of the sequence: $u_0 = 1$, $u_1 = 3$, $u_2 = 135$, $u_3 = 20\,503\,125$. But a good idea is to factor these values, and observing that $u_0 = 3^0 \cdot 5^0$, $u_1 = 3^1 \cdot 5^0$, $u_2 = 3^3 \cdot 5^1$, $u_3 = 3^8 \cdot 5^5$, we are led to the "educated guess" that $u_n = 3^{\alpha_n} \cdot 5^{\alpha_n - n}$. Plugging this into (2.18), we get the following *linear* third-order recurrence for α:

$$\alpha_{n+3} = 5\alpha_{n+2} - 7\alpha_{n+1} + 3\alpha_n$$

with the seeds

$$\alpha_0 = 0, \ \alpha_1 = 1, \ \alpha_2 = 3,$$

and as a check $\alpha_3 = 8$. Although this homogeneous recurrence with constant coefficients is a little harder than the ones we met before, it can be solved by standard methods (cf. [139, Theorem 7.2]): the *characteristic polynomial* is $x^3 - 5x^2 + 7x - 3$ and has the double root 1 and the simple root 3, such that $\alpha_n = a + bn + c3^n$, and the seeds can be employed to specify the parameters a, b and c to obtain $\alpha_n = \frac{1}{4}(3^n - 1) + \frac{1}{2}n$.

It should be emphasized that this is *not* the end of the proof, but that we have to check that $u_n = 3^{\alpha_n} \cdot 5^{\alpha_n - n}$ really solves (2.18), because our ansatz could have been wrong!

Although we do not need them anymore, we give the solutions for b_n and t_n for the readers interested in difficult recurrences:

$$b_n = \frac{3}{2}\left(\left(\frac{5}{3}\right)^n - 1\right) u_n, \ t_n = \frac{3}{4}\left(\left(\frac{5}{3}\right)^n - 1\right)^2 u_n. \qquad \Box$$

To readers not yet exhausted from such arguments we recommend Exercise 2.26.

We add in passing that in [59] Boettcher and S. Li investigated the number of spanning trees for the so-called Tower of Hanoi networks described in Section 1.3 as well as for some additional Hanoi networks. Their complexity is determined asymptotically.

Apollonian networks $A(n)$, $n \in \mathbb{N}_0$, named for Apollonios of Perge (fl. -3rd to -2nd c.), are *maximal* (with respect to planarity, which is destroyed when adding an edge) planar graphs constructed as follows. $A(0)$ is a plane drawing of K_3; $A(n + 1)$ is obtained from $A(n)$ by adding one new dot/vertex into each of the *inner* faces (those which are not the infinite face) of $A(n)$ and connecting it by (non-crossing) lines/edges to all three vertices of the face. Note that $A(1) \cong K_4$. Then the *inner dual*, i.e. ignoring the infinite face, of $A(n)$ is isomorphic to the Hanoi graph H_3^n. This appealing fact seems to be implicitly around for a while and explicitly stated and used in [449] to enumerate spanning trees of Apollonian networks and in [265] to calculate the Tutte polynomial of these graphs.

Despite rapidly increasing sequences like the above $\tau(H_3^n)$, we finally notice that in connection with the Olive sequence

$$o = 1, \bar{2}, 0, 1, 2, \bar{0}, 1, \bar{2}, 0, \bar{1}, 2, 0, 1, \bar{2}, 0, \dots ,$$

it is worthwhile to introduce a limit graph in the following way. To any state $s \in H_3^n$ we add infinitely many zeros to the left and put:

$$V(H_3^\infty) = \bigcup_{n \in \mathbb{N}_0} \{0^\infty s \mid s \in T^n\}.$$

For any $\bar{s}, \bar{t} \in V(H_3^\infty)$, there is an $n \in \mathbb{N}_0$, such that $\bar{s} = 0^\infty s$ and $\bar{t} = 0^\infty t$, where $s, t \in H_3^n$; then $\bar{s}$ and $\bar{t}$ are adjacent in H_3^∞ if and only if s and t are adjacent in H_3^n. Because of the recursive definition of H_3^n this does not depend on the n chosen. Note that the graph H_3^∞, called *Sisyphean Hanoi graph*, is connected, see Exercise 2.27, and represents all regular states of the Tower of Hanoi with pegs unlimited in height and an infinite supply of discs reachable from the state where all discs start on peg 0. That is to say that no matter how many discs have been transferred to another peg after finitely many moves, there will always be (infinitely many) discs left on peg 0. Among the infinite paths in this graph we rediscover the Olive sequence as can be descried in Figure 2.19.

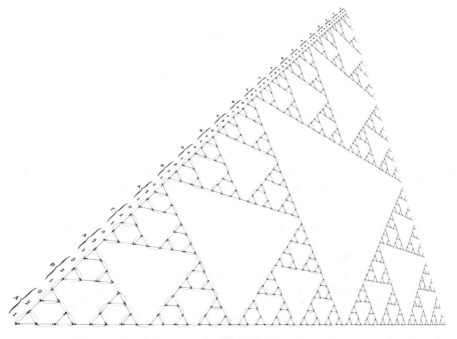

Figure 2.19: Olive's sequence in the Sisyphean Hanoi graph

The representation of Olive's sequence on H_3^∞ reveals yet another connection to special sequences. We observe again (cf. Proposition 2.16) that the numbers of states at fixed distances $\mu \in \mathbb{N}_0$ from a perfect state are always powers of two; see the second numerical column in Figure 2.20. This also easily follows from the fact that this number is doubled whenever we go from $\mu - 2^i$ to μ for $2^i \le \mu < 2^{i+1}$, $i \in \mathbb{N}_0$. We recall from Proposition 2.16 that the exponents of these powers are $q(\mu)$, the number of 1s in the binary representation of μ; cf. the third numerical column in Figure 2.20. In the fourth column we list the parity of these numbers, i.e. the sequence t given by $t_n = q(n) \bmod 2$, which we recall from Section 2.1 as the Prouhet-Thue-Morse sequence, leading to the square-free Thue sequence.

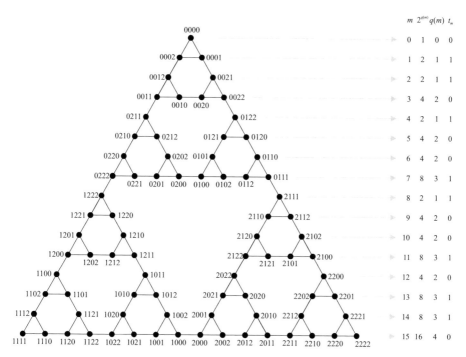

Figure 2.20: Prouhet-Thue-Morse sequence in H_3^4

2.4 Regular to Regular

Given the connectedness of the graphs H_3^n and the corresponding distance function d, it is natural to impose a new problem *type* P2, namely tasks to get from an arbitrary regular initial state s to an arbitrary regular goal state t or, in other words, to find (shortest) paths between any two vertices of H_3^n and their distance $d(s,t)$. It turns out that the worst case is still the classical perfect-to-perfect task:

Theorem 2.31. $\forall\, n \in \mathbb{N}_0 :\ \mathrm{diam}\,(H_3^n) = 2^n - 1$.

Proof. By Theorem 2.1 we only have to prove $\mathrm{diam}\left(H_3^{1+n}\right) \le 2^{n+1}-1$ for all $n \in \mathbb{N}_0$. Let $s, t \in T^n$. Then, for $i \ne k \ne j$,

$$d(is, jt) \le d(is, ik^n) + 1 + d(jk^n, jt) = d(s, k^n) + 1 + d(k^n, t) \le 2^n - 1 + 1 + 2^n - 1 = 2^{n+1} - 1,$$

by Theorem 2.7. □

The path implicit in the preceding proof is far from being of minimal length in general. For instance, if $i = j$, then $d(is, jt) = d(s,t) \le 2^n - 1 < 2^{n+1} - 1$. This is a direct consequence of the *boxer rule*:

Lemma 2.32 (They never come back). *If on a **geodesic** of H_3^{1+n}, $n \in \mathbb{N}_0$, the largest disc is moved away from a peg, it will not return to the same peg.*

Proof. If on a path in H_3^{1+n} disc $n+1$ moves away from and back to some peg, one may leave out these two moves and all moves of disc $n+1$ in between, because the moves of the smaller discs are not effected by the largest disc. This will result in a strictly shorter path between the end vertices such that the original path can not be a geodesic.   □

For what follows, let us define two notations given a vertex $is \in T^{1+n}$ of H_3^{1+n} and with $\{i,j,k\} = T$, namely

$$\mathrm{d}(s;j,k) := \mathrm{d}(s,j^n) - \mathrm{d}(s,k^n) \tag{2.19}$$

and $\mathrm{d}(is) := |\mathrm{d}(s;j,k)| \in [2^n]_0$; the latter is well defined since $|\mathrm{d}(s;k,j)| = |\mathrm{d}(s;j,k)|$. The meaning of these notions can be sensed from Figure 2.21.

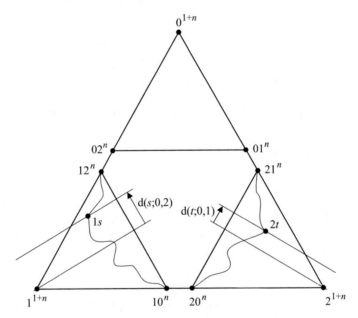

Figure 2.21: The meaning of $\mathrm{d}(s;j,k)$

We can now derive the following analogue of the result we got for the CR (cf. p. 78).

Proposition 2.33. *Let $s \in T^n$, $n \in \mathbb{N}_0$, and $\{i,j,k\} = T$. Then*

$$\varepsilon(is) = \max\left\{\mathrm{d}(is,j^{1+n}), \mathrm{d}(is,k^{1+n})\right\} = 2^{n+1} - 1 - \frac{1}{2}\left(\mathrm{d}(s,i^n) - \mathrm{d}(is)\right).$$

Proof. We may assume that $\varepsilon(is) = d(is, jt)$ for some $t \in T^n$, because from Theorem 2.31 and by virtue of (2.8) we know that $d(is, it) < 2^n \le d(is, j^{1+n})$. Then

$$\varepsilon(is) \le d(is, ik^n) + 1 + d(jk^n, jt) \le d(is, ik^n) + 2^n = d(is, j^{1+n}).$$

We now apply the very useful formula

$$\max\{a, b\} = \frac{1}{2}\left(a + b + |a - b|\right),$$

which can easily be checked for $a, b \in \mathbb{R}$, to get

$$\varepsilon(is) = \frac{1}{2}\left(d(is, j^{1+n}) + d(is, k^{1+n}) + \left|d(is, j^{1+n}) - d(is, k^{1+n})\right|\right).$$

Now, by Proposition 2.13,

$$d(is, j^{1+n}) + d(is, k^{1+n}) = 2(2^{n+1} - 1) - d(is, i^{1+n}) = 2(2^{n+1} - 1) - d(s, i^n).$$

Moreover,

$$\begin{aligned}
d(is, j^{1+n}) - d(is, k^{1+n}) &= d(is, ik^n) + 1 + d(jk^n, j^{1+n}) \\
&\quad - (d(is, ij^n) + 1 + d(kj^n, k^{1+n})) \\
&= d(s, k^n) + d(k^n, j^n) - d(s, j^n) - d(j^n, k^n) = d(s; k, j).
\end{aligned}$$

□

The range of eccentricities in H_3^{1+n}, $n \in \mathbb{N}$, is from $3 \cdot 2^{n-1}$ to $2^{n+1} - 1$ (see Exercise 2.28), and we can now approach the goal to determine the average eccentricity of H_3^{1+n}. Adding eccentricities we get from Proposition 2.33 for a fixed $i \in T$:

$$\sum_{s \in T^n} \varepsilon(is) = 3^n(2^{n+1} - 1) + \frac{1}{2}\sum_{s \in T^n}|d(s; k, j)| - \frac{1}{2}\sum_{s \in T^n}d(s, i^n). \qquad (2.20)$$

We know from Proposition 2.13 that the last term is equal to $3^{n-1}(2^n - 1)$. To obtain the value of the penultimate term, we note that $-2^n < d(s; k, j) < 2^n$, and a good idea would be to count, for each value μ in this range, the number of vertices s such that $d(s; k, j) = \mu$. So let us define

$$\forall n \in \mathbb{N}_0 \ \forall \mu \in \mathbb{Z}: \ z_n(\mu) = |\{s \in T^n \mid d(s; k, j) = \mu\}|; \qquad (2.21)$$

this is independent of k and j as long as they are different.

The sequence of functions z_n on $\mathbb{Z}$ (cf. [194, Definition 10]) is extremely interesting and useful. In relation to Hanoi graphs, each $z_n(\mu)$ counts the vertices on a vertical line superposed to the standard drawing of H_3^n; see Figure 2.22 for

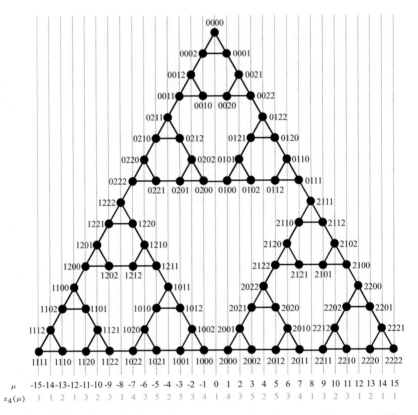

Figure 2.22: Values for $z_4(\mu)$ as the numbers of vertices on the vertical lines at distance μ from the center, representing the difference in distances from the bottom corners

the case $n = 4$ with $k = 1, j = 2$. It is also the number of ways μ can be written as a binary expansion, but with coefficients from the set $\widetilde{T} := \{-1, 0, 1\}$:

$$z_n(\mu) = \left|\left\{\widetilde{s} \in \widetilde{T}^n \;\middle|\; \mu = \sum_{d=1}^{n} \widetilde{s}_d \cdot 2^{d-1}\right\}\right|. \tag{2.22}$$

This follows from (2.8) as soon as one can convince oneself that the mapping from T^n to $\widetilde{T}^n$, $s \mapsto \widetilde{s}$, where $\widetilde{s}_d = (s_d \neq (s \vartriangle 0)_d) - (s_d \neq (s \vartriangle 2)_d)$ for $d \in [n]$, is bijective. Since $|T^n| = 3^n = |\widetilde{T}^n|$ and by the Pigeonhole principle, it suffices to show that the mapping is injective, i.e. that s can be uniquely recovered from $\widetilde{s}$, which is possible because $s_d = (s \vartriangle (\widetilde{s}_d + 1))_d$, and this can be solved for s.

From (2.22) one obtains the following recurrence:

$$\forall\, \mu \in \mathbb{Z}: \; z_0(\mu) = (\mu = 0),$$

$$\forall\, n \in \mathbb{N}_0 \; \forall\, \mu \in \mathbb{Z}: \; z_{n+1}(\mu) = z_n(\mu - 2^n) + z_n(\mu) + z_n(\mu + 2^n); \tag{2.23}$$

this three-term sum arises from the three values $1, 0$, and -1 which $\widetilde{s}_{n+1}$ can adopt.

The values of $z_n(\mu)$ can be arranged in an array as in Figure 2.23, where the rows correspond to n, starting with $n = 0$, and the columns to μ, with $\mu = 0$ in the center.

$$0\,0\,0\,0\,0\,0\,0\,0\,0\,0\,0\,0\,0\,0\,0\,0\,1\,0\,0\,0\,0\,0\,0\,0\,0\,0\,0\,0\,0\,0\,0\,0\,0$$
$$0\,0\,0\,0\,0\,0\,0\,0\,0\,0\,0\,0\,0\,0\,0\,1\,1\,1\,0\,0\,0\,0\,0\,0\,0\,0\,0\,0\,0\,0\,0\,0\,0$$
$$0\,0\,0\,0\,0\,0\,0\,0\,0\,0\,0\,0\,1\,1\,2\,1\,2\,1\,1\,0\,0\,0\,0\,0\,0\,0\,0\,0\,0\,0\,0\,0\,0$$
$$0\,0\,0\,0\,0\,0\,0\,0\,1\,1\,2\,1\,3\,2\,3\,1\,3\,2\,3\,1\,2\,1\,1\,0\,0\,0\,0\,0\,0\,0\,0\,0\,0$$
$$0\,1\,1\,2\,1\,3\,2\,3\,1\,4\,3\,5\,2\,5\,3\,4\,1\,4\,3\,5\,2\,5\,3\,4\,1\,3\,2\,3\,1\,2\,1\,1\,0$$

Figure 2.23: The $z_n(\mu)$ array for $n \in [5]_0$ and $|\mu| \in [17]_0$

Restricting ourselves to non-negative arguments, we recognize a Stern array (cf. p. 26), namely $((1,0)_n)_{n\in\mathbb{N}_0}$ for the atoms 1 and 0, as in Figure 2.24.

1									0
1				1					0
1		2		1		1			0
1	3	2	3	1	2	1	1		0
1	4	3	5	2	5	3	4	1	3	2	3	1	2	1	1	0

Figure 2.24: Stern's diatomic array for $(1,0)$ and $n \in [5]_0$

This shows that

$$\forall\, \mu \in [2^n + 1]_0 : \ z_n(\mu) = (1,0)_n(\mu) = (0,1)_\infty(2^n - \mu), \tag{2.24}$$

and

$$\forall\, \mu \in [2^{n-1} + 1]_0 : \ z_n(\mu) = (1,1)_{n-1}(\mu) = (1,1)_{n-1}(2^{n-1} - \mu). \tag{2.25}$$

The identity (2.25) has two immediate consequences. The Brocot sequence can be written as (cf. p. 28) $\beta_n(\mu) = \dfrac{z_n(2^n - \mu)}{z_{n+1}(2^n - \mu)}$ on $[2^n + 1]_0$, and we have the following identity (cf. [331, Theorem 7]):

$$\forall\, n \in \mathbb{N} : \ \left|\{\mu \in [2^{n-1}] \mid z_n(\mu) = n\}\right| = \varphi(n)$$

with Euler's phi function (cf. p. 28).

By induction on n we get from (2.23) (see Exercise 2.29):

$$z_n(-\mu) = z_n(\mu), \ z_n(\mu) = 0 \Leftrightarrow |\mu| \geq 2^n, \ z_n(\mu) = 1 \Leftrightarrow \exists\, k \in [n+1]_0 : |\mu| = 2^n - 2^k. \tag{2.26}$$

The most exciting special values of $z_n(\mu)$ can be found when μ runs through the Jacobsthal sequence (for E. Jacobsthal), which we encountered briefly in Chapter 1 (cf. p. 74). This sequence is explicitly given by $J_k = \frac{1}{3}\left(2^k - (-1)^k\right)$ and fulfills the two recurrence relations

$$\forall\, k \in \mathbb{N}_0 : 2^k - J_k = J_{k+1} = 2J_k + (-1)^k, \tag{2.27}$$

with the seed $J_0 = 0$. We now get

Theorem 2.34. *For every $n \in \mathbb{N}_0$:*

$$
\begin{aligned}
F_{1+n} &= z_n(J_n) &= \max\{z_n(\mu) \mid \mu \in \mathbb{Z}\}, \\
L_{1+n} &= z_{2+n}(J_n).
\end{aligned}
$$

Proof. We first show that $f_n := \max\{z_n(\mu) \mid \mu \in \mathbb{Z}\}$ fulfills the Fibonacci recurrence relation in (0.4); the seeds are obviously $f_0 = 1 = f_1$. We proceed by induction. For $n = 0$ we get $f_2 = f_1 + f_0$ by inspection. It is clear that $f_{n+3} \leq f_{n+2} + f_{n+1}$, because one of two neighbors in the sequence z_{n+2} appears already in z_{n+1}. Moreover, by induction assumption $f_{n+2} = f_{n+1} + f_n$ such that, from intercalation, f_{n+2} and f_{n+1} are neighbors in z_{n+2}, whence $f_{n+3} \geq f_{n+2} + f_{n+1}$.

We now put $g_n := z_n(J_n)$ and show again that the Fibonacci recurrence holds with $g_0 = 1 = g_1$ as before:

$$
\begin{aligned}
g_{n+2} = z_{n+2}(J_{n+2}) &= z_{n+2}(2^{n+1} - J_{n+1}) \\
&= z_{n+1}(J_{n+1}) + z_{n+1}(2^n + J_n) \\
&= z_{n+1}(J_{n+1}) + z_n(J_n) \\
&= g_{n+1} + g_n.
\end{aligned}
$$

Similarly, one shows, this time based on the second recurrence relation for the *Jacobsthal numbers* in (2.27), that the sequence with the general term $(1,2)_n(J_n)$ fulfills the Fibonacci recurrence with seeds 1 and 3, such that this sequence represents the Lucas numbers L_{1+n}. Finally, we have to note that $(1,2)_n(\mu) = z_{n+2}(\mu)$. □

In the sequel we will need some further properties of the sequences $z_n(\mu)$ (cf. [194, Lemma 2ii]).

Lemma 2.35. *For all $n \in \mathbb{N}_0$:*

$$\sum_{\mu \in \mathbb{Z}} z_n(\mu) = 3^n, \quad \sum_{\mu \in \mathbb{N}} z_n(\mu) = \frac{1}{2}(3^n - 1), \quad \sum_{\mu \in \mathbb{N}} \mu z_n(\mu) = \frac{1}{5}(6^n - 1).$$

The proof will be done in Exercise 2.30.

From (2.26) and Lemma 2.35 we see immediately that

$$\frac{1}{2}\sum_{s \in T^n} |d(s;k,j)| = \sum_{\mu=1}^{2^n-1} \mu z_n(\mu) = \frac{1}{5}(6^n - 1).$$

Putting everything into (2.20), we arrive at

$$\sum_{s \in T^n} \varepsilon(is) = 3^n(2^{n+1} - 1) + \frac{1}{5}(6^n - 1) - 3^{n-1}(2^n - 1),$$

whence

$$E(H_3^{1+n}) = \frac{14}{15}6^{n+1} - \frac{2}{3}3^{n+1} - \frac{3}{5}.$$

This leads to (cf. [212, Proposition 4.4]):

Theorem 2.36. *For $n \in \mathbb{N}_0$, the average eccentricity of H_3^{1+n} is*

$$\bar{\varepsilon}(H_3^{1+n}) = \frac{14}{15}2^{n+1} - \frac{2}{3} - \frac{3}{5}3^{-(n+1)}. \qquad (2.28)$$

An immediate consequence is that asymptotically the average eccentricity is $\frac{14}{15}$ of the diameter. This has some practical significance for psychological tests based on TH, because it means that a randomly chosen starting configuration will normally lead to sufficiently challenging tasks. We will say more about the *normalized average eccentricity* $\bar{\varepsilon}/\text{diam}$ of other graphs in Section 5.7.1.

Another important consequence of the boxer rule is (cf. [388, p. 62f] and [194, Lemma 1ii]):

Corollary 2.37. *On a geodesic of H_3^n, the largest disc moves at most twice.*

Proof. By Lemma 2.32 (and induction), $m \in \mathbb{N}_0$ moves of the largest disc involve $m + 1$ pegs. By the Pigeonhole principle, m must be smaller than 3, the number of pegs available. □

A long standing myth was that the largest disc will not move more than once on a geodesic—after all, its moves are the most restricted and therefore costly ones. This assumption led (via induction) to the belief that the shortest path between two given states is unique (cf., e.g., [444, Theorem 3]). The most easy example in Figure 2.25 shows that this is false already for $n = 2$: to interchange two discs on different pegs in an optimal 3-move path, one can move the larger disc just once or either twice.

But the situation is even worse: Figure 2.26 shows that two moves of the largest disc (*largest disc moves, LDMs*) are *necessary* on a shortest path between two states like in this example the transposition of the largest disc 3 and the 2-tower consisting of discs 1 and 2. Here the shortest path with one LDM has length 7 (dashed arrow), whereas the optimal path using 2 LDMs needs only 5 moves (solid arrow).

On the other hand, nothing else could go wrong (cf. [194, Theorem 4]):

Theorem 2.38. *Let $s, t \in T^n$. Then there are at most two shortest s,t-paths. If there are two, they differ by the number (one or two) of moves of the largest disc $d \in [n]$ for which $s_d \neq t_d$.*

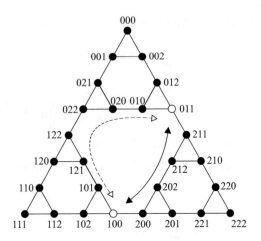

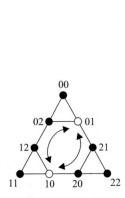

Figure   2.25:   Non-unique   Figure 2.26: Shortest path needs two LDMs
shortest paths in H_3^2 in H_3^3

Proof. We may clearly assume that $s \neq t$. Then there exists the largest disc d for which $s_d \neq t_d$. By Lemma 2.32, discs $d+1,\dots,n$ do not move on an optimal path. In addition, Corollary 2.37 implies that disc d

  - either moves only once: from peg s_d to peg t_d,

  - or it moves first from peg s_d to peg $3 - s_d - t_d$, and then from $3 - s_d - t_d$ to t_d.

In both cases, the moves of discs $1,\dots,d-1$ are unique by the uniqueness of the optimal solutions for P1 tasks (Theorem 2.7). $\square$

As a consequence of Theorem 2.38 an interesting metric property of Hanoi graphs can be considered. A pair of vertices $\{u,v\} \in \binom{V(G)}{2}$ of a connected graph G is called *diametrical*, if $d(u,v) = \mathrm{diam}(G)$. We call the number of diametrical pairs of G its *diametrical number*, denoted by $\mathrm{Diam}(G)$. If $|G| \geq 2$, then $\mathrm{Diam}(G) \in \left[\binom{|G|}{2}\right]$ with the extreme values taken by path graphs and complete graphs, respectively. Stockmeyer[45] came up with the following statement:

Proposition 2.39. *For all $n \in \mathbb{N}_0$:*

$$\mathrm{Diam}(H_3^{1+n}) = \tfrac{3}{2}(4^n + 3 \cdot 2^n - 2).$$

Among the diametrical pairs of vertices $3(2^n-1)$ have two shortest paths connecting them. The remaining $3 \cdot 2^{n-1}(2^n + 1)$ pairs have unique shortest paths involving a single move of the largest disc $n + 1$.

We will give our proof of this proposition as Exercise 2.31.

---

[45] P. K. Stockmeyer, private communication, 2017

Theorem 2.38 and its proof immediately lead to Algorithm 12 which solves a P2 task $s \to t$. Recall that the call $\mathrm{p1}(n, s, j)$ of Algorithm 10 returns the disc δ active and the peg i passive in the best first move to get from an n-disc state s to the perfect state on peg j, as well as the overall distance $\mu = \mathrm{d}(s, j^n)$.

Algorithm 12 Best first move to a regular state

Procedure $\mathrm{p2}(n, s, t)$
Parameter n: number of discs $\quad \{n \in \mathbb{N}\}$
Parameter s: regular initial state $\quad \{s \in T^n\}$
Parameter t: regular goal state $\quad \{t \in T^n\}$

$\quad \mu \leftarrow 0$ $\hfill$ {length of path}
$\quad N \leftarrow n$ $\hfill$ {largest disc to be moved}
$\quad$ **while** $s_N = t_N$
$\quad\quad N \leftarrow N - 1$
$\quad$ **end while**
$\quad$ **if** $N = 0$ **then**
$\quad\quad$ STOP
$\quad$ **end if**
$\quad C \leftarrow 1$ $\hfill$ {case of P2 decision}
$\quad \bar{s} \leftarrow s \upharpoonright [N-1], \bar{t} \leftarrow t \upharpoonright [N-1]$
$\quad \mathrm{p1}(N-1, \bar{s}, 3 - s_N - t_N)$
$\quad\quad \mu_1 \leftarrow \mu, \delta_1 \leftarrow \delta, i_1 \leftarrow i$ $\hfill$ {μ, δ, i: output of p1}
$\quad \mathrm{p1}(N-1, \bar{t}, 3 - s_N - t_N)$
$\quad\quad \mu_1 \leftarrow \mu_1 + 1 + \mu$ $\hfill$ {length of one-move path}
$\quad \mathrm{p1}(N-1, \bar{s}, t_N)$
$\quad\quad \mu_2 \leftarrow \mu, \delta_2 \leftarrow \delta, i_2 \leftarrow i$
$\quad \mathrm{p1}(N-1, \bar{t}, s_N)$
$\quad\quad \mu_2 \leftarrow \mu_2 + 2^{N-1} + 1 + \mu$ $\hfill$ {length of two-move path}
$\quad \mu \leftarrow \mu_1$
$\quad$ **if** $\mu_1 \geq \mu_2$ **then**
$\quad\quad C \leftarrow C + 1, \mu \leftarrow \mu_2$
$\quad$ **end if**
$\quad$ **if** $\mu_1 = \mu_2$ **then**
$\quad\quad C \leftarrow C + 1$
$\quad$ **end if**

By the above, it is not necessary to give a formal correctness proof for Algorithm 12 which returns the length μ of an optimal solution. In addition, it specifies the case C the task leads to, namely if $C = 1$, then the largest moving disc moves once only, if $C = 2$, then it moves twice, and if $C = 3$, we have a draw, i.e. both alternatives lead to a shortest path. The disc active in the best first move is δ_C and the idle peg is i_C in the first two cases; if $C = 3$, then we may either move disc δ_1 avoiding peg i_1 or use disc δ_2 with idle peg i_2.

Although Algorithm 12 can easily be implemented on a computer and there-

fore be used in a test tool, it has two serious drawbacks. First of all, it is not *human*
in the sense that it needs too much memory to be employed by a human problem
solver faced with a P2-type task; we will come back to this in Section 6.3 when dealing with the so-called Twin-Tower problem, a proto-type of which had been posed
already by Lucas in [285, Deuxième problème], namely the task $(01)^4 \to (10)^4$ in
H_3^8 (cf. Section 6.1), thus being the earliest concrete example for P2. Secondly, the
four-fold call of procedure p1 and the comparison of the two alternative solutions
is not particularly elegant. The challenge to devise a more efficient solution to this
so-called *P2 decision problem* was pronounced in [196, p. 179] and successfully
taken up by D. Romik in [353]. Before we describe his approach, based again on
employing an appropriate automaton, in Section 2.4.3, we will show how statistical
results can be obtained without solving the decision problem for every single task.

2.4.1   The Average Distance on H_3^n

Let us consider a P2-type task $is \to jt$ in H_3^{1+n} with $s, t \in T^n$, $i, j \in T$, $i \neq j$. By
Theorem 2.38, there are exactly two candidates for an optimal solution, namely,
with $k = 3 - i - j$, the paths $is \to ik^n \to jk^n \to jt$ and $is \to ij^n \to kj^n \to ki^n \to ji^n \to$
jt of lengths

$$\mathrm{d}_1(is, jt) := \mathrm{d}(s, k^n) + 1 + \mathrm{d}(t, k^n)$$

and

$$\mathrm{d}_2(is, jt) := \mathrm{d}(s, j^n) + 1 + 2^n + \mathrm{d}(t, i^n),$$

respectively. In order to compare these two values, we notice that

$$\mathrm{d}_1(is, jt) \,\Box\, \mathrm{d}_2(is, jt) \Leftrightarrow \mathrm{d}(s; k, j) + \mathrm{d}(t; k, i) \,\Box\, 2^n, \qquad (2.29)$$

where we make use of the *box notation*, which the reader may recall from school
when asked to compare two algebraic expressions A and B. Here $\Box \in \{<, =, >\}$ and
the comparison $A \,\Box\, B$ can be transformed into $A' \,\Box\, B'$ as long as all operations,
performed simultaneously on both sides of the box, do not change signs. For in-
stance, $A \,\Box\, B \Leftrightarrow 2A \,\Box\, 2B$. If at the end of such a chain of equivalent expressions
stands an expression $A' \,\Box\, B'$ which can be identified to be true for one of the
admissible symbols $\Box$ is replacing, then this is also the appropriate symbol in the
original expression $A \,\Box\, B$. If also operations are employed which *do* change signs,
one may use $\bar{\Box} \in \{<, =, >\}$ with $\bar{\Box} \,==$, iff $\Box \,==$, and $\bar{\Box} \,=>$, iff $\Box \,=<$. For instance,
$A \,\Box\, B \Leftrightarrow -A\,\bar{\Box}\, -B \Leftrightarrow B\,\bar{\Box}\, A$.

Let us now ask how many vertices there are in T^{1+n} which can be reached
from a fixed vertex is on two different optimal paths; we call the set of these
vertices $X(is)$. Since $X(i^{1+n}) = \varnothing$ by Theorem 2.7, we may assume that $n \neq 0$
and $s \neq i^n$. We first search for them in jT^n, $j \in T \setminus \{i\}$. From (2.29) it is clear
that $jt \in X(is)$ if and only if $\mathrm{d}(s; k, j) + \mathrm{d}(t; k, i) = 2^n$ for $k = i \wedge j$; in particular,
both summands must be strictly positive. This means that j and k are uniquely

determined by $d(s, k^n) > d(s, j^n)$ and that

$$X(is) \smallsetminus iT^n = X(is) \cap jT^n = \{jt \mid d(t, k^n) > d(t, i^n),\ d(jt) = 2^n - d(is)\}\ .$$

The number of elements in the latter set is just $z_n(2^n - d(is))$, which is positive since $s \neq i^n$. So we arrived at a converse of Theorem 2.7, namely that perfect states are the only ones which are linked to every other state by a unique optimal path using at most one move of the largest disc (cf. [205, Corollary 3.7], [204, Proposition 2.2]).

Proposition 2.40. *For every* $is \in T^{1+n} \smallsetminus \{0^{1+n}, 1^{1+n}, 2^{1+n}\}$, $n \in \mathbb{N}$, *there is a* $jt \in T^{1+n}$ *with* $j \in T \smallsetminus \{i\}$ *such that there are two shortest paths between these vertices in* H_3^{1+n}.

The value of $z_n(2^n - \nu)$ for $\nu \in [2^n]_0$ is actually independent of n; more precisely,

$$\forall\, n \in \mathbb{N}_0\ \forall\, \nu \in [2^n]_0\ \forall\, m \in \mathbb{N}_n :\ z_n(2^n - \nu) = z_m(2^m - \nu)\,.$$

This can easily be seen by induction on m making use of (2.23) and (2.26). We can therefore define the sequence b by $b(\nu) = z_n(2^n - \nu)$ for $\nu \in [2^n]_0$. Note that it also follows from (2.26) that $b(\nu) = 0 \Leftrightarrow \nu = 0$ and $b(\nu) = 1 \Leftrightarrow \nu = 2^k$ for some $k \in \mathbb{N}_0$. We then obtain

Proposition 2.41. *For* $is \in T^{1+n}$, $n \in \mathbb{N}_0$,

$$|X(is) \smallsetminus iT^n| = b\,(d(is))\,.$$

Putting $\mu = 2^{n+1} - \nu$ in (2.23), we get

$$\forall\, n \in \mathbb{N}_0\ \forall\, \nu \in [2^{n+1}] \smallsetminus [2^n] :\ b(\nu) = b(2^{n+1} - \nu) + b(\nu - 2^n)\,. \tag{2.30}$$

Together with the two seeds (or "atoms") $b(0) = 0$ and $b(1) = 1$, (2.30) constitutes a recurrence for what we recognize, by the identity in (2.24), as Stern's diatomic sequence: $b(\nu) = (0, 1)_\infty(\nu)$. This integer sequence turns up in many mathematical disguises. It is most commonly defined using the recurrence relation

$$\forall\, \nu \in \mathbb{N} :\ b(2\nu) = b(\nu),\ b(2\nu + 1) = b(\nu) + b(\nu + 1)\,. \tag{2.31}$$

The equivalence of (2.30) and (2.31) for the chosen seeds is proved in Exercise 2.32.

But there can be even more elements of $X(is)$, namely those in iT^n. By Lemma 2.32, the two shortest paths must then lie entirely in iH_3^n, such that we may apply the above argument to $s_n s^{(n-1)}$, with $s^{(n-1)} := s_{n-1} \ldots s_1$. The condition with $\{s_n, j_n, k\} = T$ is now that $d(s^{(n-1)}; k, j_n) + d(t; k, s_n) = 2^{n-1}$, and j_n has to be chosen such that the first summand is positive. This leads to new elements in $X(is)$ of the form $ij_n t$ with $t \in T^{n-1}$ fulfilling $d(t, k^{n-1}) > d(t, s_n^{n-1})$

and $d(j_n t) = 2^{n-1} - d(s_n s^{(n-1)})$. We may continue this procedure in T^m for further decreasing $m \in [n]$, where we put $s^{(m)} := s_m \dots s_1$, $j_{1+m} \in T \setminus \{s_{1+m}\}$ is chosen such that $d(s^{(m)}; s_{1+m} \vartriangle j_{1+m}, j_{1+m}) > 0$, and we define $s_{(n-m)} := s_{1+n} \dots s_{2+m}$ with the understanding that $s_{1+n} := i$. Then we arrive at

Theorem 2.42. *For is $\in T^{1+n}$, $n \in \mathbb{N}_0$, the set of vertices in H_3^{1+n} to which two optimal paths lead is*

$$X(is) \quad = \quad \bigcup_{m=1}^{n} \left\{ s_{(n-m)} j_{1+m} t \mid t \in T^m, \right.$$

$$\left. d(t, (s_{1+m} \vartriangle j_{1+m})^m) > d(t, s_{1+m}^m), \ d(j_{1+m} t) = 2^m - d(s_{1+m} s^{(m)}) \right\} .$$

For each $m \in [n]$ the corresponding individual set in this union has size $z_m(2^m - d(s_{1+m} s^{(m)}))$. Hence we get

Corollary 2.43. *For is $\in T^{1+n}$, $n \in \mathbb{N}_0$,*

$$|X(is)| = \sum_{m=1}^{n} b\left(d(s_{1+m} s^{(m)}) \right) .$$

As Stockmeyer pointed out,[46] the sum in this corollary is not quite suited to establish the fact that $|X(is)| \geq n$. His cute observation will be addressed in Exercise 2.33.

By similar arguments as those leading to Proposition 2.41 and Corollary 2.43, just making use of $\square \Rightarrow$ in (2.29), we can count those vertices jt of H_3^{1+n} which are linked to is by a unique shortest path, but where a second move of the largest disc not initially in goal position is necessary. Calling this set $Y(is)$ and defining

$$B(N) = \sum_{\nu=0}^{N-1} b(\nu) \text{ for } N \in \mathbb{N}_0, \text{ we obtain}$$

Theorem 2.44. *Let is $\in T^{1+n}$, $n \in \mathbb{N}_0$. Then*

$$|Y(is) \setminus iT^n| = B(d(is))$$

and

$$|Y(is)| = \sum_{m=1}^{n} B\left(d(s_{1+m} s^{(m)}) \right) .$$

We now turn to the question how many of *all* tasks $is \to jt$, $i \neq j$, have two optimal solutions. We refer back to (2.29) in the case where $\square$ is equality. By symmetry it is clear that their number is equal to

$$6 \left| \{ (s,t) \in (T^n)^2 \mid d(s; 1, 2) + d(t; 1, 0) = 2^n \} \right| =: 6x_n .$$

---

[46]P. K. Stockmeyer, private communication, 2017.

Since $d(0s) = |d(s; 1, 2)| < 2^n$,

$$x_n = \sum_{\mu=1}^{2^n-1} \left|\{(s, t) \in (T^n)^2 \mid d(s; 1, 2) = \mu,\ d(t; 1, 0) = 2^n - \mu\}\right| = \sum_{\mu \in \mathbb{N}} z_n(\mu) z_n(2^n - \mu).$$

Introducing the auxiliary sequence given by $y_n = \sum_{\mu \in \mathbb{N}} z_n(\mu)^2$, (2.23) and (2.26) yield the following recurrence relations.

$$\forall n \in \mathbb{N}_0 : \quad x_{n+1} = 2x_n + 2y_n + 1,\ y_{n+1} = 2x_n + 3y_n + 1, \tag{2.32}$$

whence

$$x_{n+2} = 2x_{n+1} + 2y_{n+1} + 1 = 2x_{n+1} + 4x_n + 6y_n + 3 = 5x_{n+1} - 2x_n.$$

To solve the recurrence

$$x_0 = 0,\ x_1 = 1,\ \forall n \in \mathbb{N}_0 : x_{n+2} = 5x_{n+1} - 2x_n, \tag{2.33}$$

we recall the idea of Exercise 0.1 and try the ansatz $\xi_n = \Theta^n$ for the recurrence equation, leading to $\Theta^2 = 5\Theta - 2$, such that $\Theta_\pm = (5 \pm \sqrt{17})/2$. The solution of (2.33) is therefore $x_n = \dfrac{\Theta_+^n - \Theta_-^n}{\sqrt{17}}$ (cf. [194, Proposition 6i]). (Inserting this into the first equation in (2.32), we further get $y_n = \dfrac{1}{4} \left(\dfrac{\sqrt{17}+1}{\sqrt{17}} \Theta_+^n - 2 + \dfrac{\sqrt{17}-1}{\sqrt{17}} \Theta_-^n \right)$.)

Remark. Although it is not too surprising, in particular in view of the abundance of applications of the Fibonacci sequence, that the same recurrence might occur in different settings, let us mention that x_{n+1} (cf. A107839) has been identified as the number of so-called Kekulé structures of a specific class of benzenoid hydrocarbons with the molecular formula $C_{12n+2}H_{6n+4}$; cf. [90, p. 75–78]. This is a nice example of the mathematical microcosmos the TH embraces.

We summarize the discussion on TH tasks with two optimal solutions (cf. [205, Proposition 3.9]).

Proposition 2.45. *Among the 9^n tasks for the TH with $n \in \mathbb{N}_0$ discs,*

$$\frac{3}{4} \left(\frac{\sqrt{17}+1}{\sqrt{17}} \Theta_+^n - 2 \cdot 3^n + \frac{\sqrt{17}-1}{\sqrt{17}} \Theta_-^n \right) = 3 \left(\sum_{\mu \in \mathbb{N}} z_n(\mu)^2 - \sum_{\mu \in \mathbb{N}} z_n(\mu) \right)$$

have two optimal solutions.

Proof. We only have to remark that the desired number is equal to $6 \sum_{k=0}^{n-1} x_k 3^{n-1-k}$, because there are 3^{n-1-k} states of larger discs if $k+1$ is the largest disc not initially on its goal peg. An application of Lemma 2.18 then yields the stated formula on the left. For the right-hand one have a look at Lemma 2.35.   □

The first few values of the sequence in Proposition 2.45 are

$$0, 0, 6, 48, 282, 1\,476, 7\,302, 35\,016, 164\,850, \ldots .$$

Note that, since $\Theta_+ \approx 4.56 < 9$, the portion of tasks with non-unique solutions is asymptotically, for $n \to \infty$, vanishing. This is not the case for those tasks, where two moves of the largest disc are *necessary*. To see this, let us define w_n as the number of tasks $0s \to 2t$, $s, t \in T^n$ where disc $n + 1$ necessarily moves twice in an optimal solution; in other words, we consider the case $□ => $ in (2.29). By a similar argument as before (just put $\mu = d(s; 1, 2)$ and $\nu = 2^n - d(t; 1, 0)$), we see that

$$w_n = \sum_{\mu \in \mathbb{N}} \sum_{\nu \in \mathbb{N}} (\nu < \mu) z_n(\mu) z_n(2^n - \nu).$$

In analogy to the solution of Exercise 2.30 we decompose the set of pairs (μ, ν) to be considered in evaluating w_{n+1} according to

$$(\nu < \mu) = (\nu < \mu < 2^n) + (\nu < \mu = 2^n) + (\nu < 2^n < \mu) + (\nu = 2^n < \mu) + (2^n < \nu < \mu)$$

and arrive at the recurrence

$$w_0 = 0, \ \forall\, n \in \mathbb{N}_0 : \ w_{n+1} = 2w_n - y_n + \frac{1}{2}(9^n - 1),$$

which can be solved with the aid of Lemma 2.18 to yield $w_n = \dfrac{1}{14}(9^n - 2^n) - \dfrac{\Theta_+^n - \Theta_-^n}{2\sqrt{17}}$, so that we obtain the following result (cf. [194, Proposition 6ii]).

Proposition 2.46. *Among the* $6 \cdot 9^n$ *tasks for the TH with* $n + 1$ *discs,* $n \in \mathbb{N}_0$, *where the largest disc* $n + 1$ *is originally not on its goal peg,*

$$\frac{3}{7}(9^n - 2^n) - \frac{3}{\sqrt{17}}(\Theta_+^n - \Theta_-^n)$$

need two moves of disc $n + 1$ *in the only optimal solution.*

This means that if TH tasks (with a large number of discs are chosen randomly, then, in an optimal solution, the largest disc will not move at all in $1/3$, exactly once in $13/21$, and exactly twice in $1/21$ of all cases, the possibility of two solutions being negligible (cf. [194, Corollary 2]).

For the CR the average distance of R^n was given in (1.5). We will now approach the formidable task to find $\overline{d}(H_3^n)$. As with the CR, we will calculate the Wiener index $W(H_3^n)$ or rather, for symmetry reasons, the total number of moves over all tasks, namely $D_n := 2W(H_3^n) = \sum\limits_{(s,t)\in(T^n)^2} d(s,t)$. Quite obviously, $D_0 = 0$. To obtain a recurrence relation, we observe that

$$D_{n+1} = \sum_{(s,t)\in(T^{n+1})^2} (s_{n+1} = t_{n+1})\, d(s,t) + \sum_{(s,t)\in(T^{n+1})^2} (s_{n+1} \neq t_{n+1})\, d(s,t)$$

$$= 3D_n + 6 \sum_{(s,t)\in(T^n)^2} d(0s, 2t)$$

$$= 3D_n + 6 \sum_{(s,t)\in(T^n)^2} d_1(0s, 2t) - 6u_n,$$

where (cf. the expression for w_n)

$$u_n = \sum_{\mu\in\mathbb{N}} \sum_{\nu\in\mathbb{N}} (\nu < \mu)\,(\mu - \nu) z_n(\mu) z_n(2^n - \nu).$$

With the aid of Corollary 2.15, we get

$$\sum_{(s,t)\in(T^n)^2} d_1(0s, 2t) = 9^n \left(\frac{2}{3}(2^n - 1) + 1 + \frac{2}{3}(2^n - 1)\right) = \frac{1}{3}9^n (2^{n+2} - 1),$$

whence

$$D_{n+1} = 3D_n + 2\cdot 9^n (2^{n+2} - 1) - 6u_n. \tag{2.34}$$

To evaluate the sequence u_n, we proceed as before and introduce the auxiliary sequence

$$v_n = \sum_{\mu\in\mathbb{N}} \sum_{\nu\in\mathbb{N}} (\nu < \mu)\,(\mu - \nu) z_n(\mu) z_n(\nu).$$

With the same decomposition as in the case of w_n and making extensive use of (2.23), (2.26), and Lemma 2.35, we arrive at the recurrence relations

$$u_{n+1} = 2u_n + 2v_n + \frac{1}{5}(3^n + 1)(6^n - 1), \quad v_{n+1} = 2u_n + 3v_n + \frac{1}{5}(6^n - 1) + \frac{1}{2}6^n(3^n - 1).$$

Combining these, we arrive at a linear non-homogeneous recurrence of second order (cf. [303, Chapter 6]) for the u_n, namely

$$u_0 = 0 = u_1, \quad \forall n \in \mathbb{N}_0: \; u_{n+2} = 5u_{n+1} - 2u_n + 4\cdot 18^n. \tag{2.35}$$

Since the homogeneous part of the recurrence relation in (2.35) is the same as for x_n, we already know that its general solution is $a\Theta_+^n + b\Theta_-^n$. Observing that any two solutions of the inhomogeneous equation differ by a solution of the homogeneous one (this is a consequence of the linearity), we are left with finding a particular solution of the inhomogeneous equation. The ansatz $c\cdot 18^n$ is a reasonable guess and indeed leads to a solution if $c = 1/59$. So $u_n = 18^n/59 + a\Theta_+^n + b\Theta_-^n$. Using the two seeds we arrive at

$$u_n = \frac{1}{59}18^n - \frac{1}{118}\left(1 + \frac{31}{17}\sqrt{17}\right)\Theta_+^n - \frac{1}{118}\left(1 - \frac{31}{17}\sqrt{17}\right)\Theta_-^n.$$

Putting everything together, the recurrence relation for D_n reads

$$\forall\, n \in \mathbb{N}_0 : \ D_{n+1} = 3D_n + \frac{466}{59}18^n - 2 \cdot 9^n + \frac{3}{59}\left(1 + \frac{31}{17}\sqrt{17}\right)\Theta_+^n + \frac{3}{59}\left(1 - \frac{31}{17}\sqrt{17}\right)\Theta_-^n.$$

Complicated as it might look, this is just a non-homogeneous linear recurrence of first order with constant coefficient and can therefore be solved for the seed $D_0 = 0$ by Lemma 2.18 to yield (cf. [194, Proposition 7])

$$D_n = \frac{466}{885}18^n - \frac{1}{3}9^n + \frac{6}{59}\left(2 + \frac{3}{17}\sqrt{17}\right)\Theta_+^n - \frac{3}{5}3^n + \frac{6}{59}\left(2 - \frac{3}{17}\sqrt{17}\right)\Theta_-^n. \quad (2.36)$$

We thus arrive at:

Theorem 2.47. *The average distance on Hanoi graph H_3^n, $n \in \mathbb{N}_0$, is $\overline{d}(H_3^n) = D_n/9^n$ with D_n given by (2.36). Asymptotically, for large n, it is $466/885$ of the diameter of H_3^n.*

Independently, this result has been given, albeit in somewhat cryptic formulas, by Chan [74, Theorem 1 and Corollary], who even determined the variance [74, Theorem 2 and Corollary]; its asymptotic value is $904808318/14448151575$ (cf. [214, p. 136]). Chan's approach is essentially to consider distances between vertices $0s$ and $2t$, $s, t \in T^n$, in a weighted graph obtained from H_3^{1+n} by reducing the whole subgraph $1H_3^n$ to a single edge of weight 2^n.

If we had been interested only in the asymptotic behavior of D_n for large n, we might have put $\widetilde{D}_n = 18^{-n}D_n$ and $\widetilde{u}_n = 18^{-n}u_n$ and divided (2.34) by 18^{n+1} to get

$$\widetilde{D}_{n+1} = \frac{1}{6}\widetilde{D}_n + \frac{4}{9} - \frac{1}{3}\widetilde{u}_n + o(1), \quad (2.37)$$

where $o(1)$ stands for a term which goes to 0 when $n \to \infty$. As before, we can calculate u_n explicitly (or asymptotically, i.e. $\widetilde{u}_n$) to arrive at

$$\widetilde{D}_{n+1} = \frac{1}{6}\widetilde{D}_n + \frac{233}{531} + o(1). \quad (2.38)$$

If we *assume* that $\widetilde{D}_n \to \widetilde{D}_\infty \in \mathbb{R}$, as $n \to \infty$, we would obtain $\widetilde{D}_\infty = \dfrac{466}{885}$. But (2.38) does *not* yield convergence immediately! (As an example take $a_{n+1} = a_n + \frac{1}{n}$.) However, recall our Lemma 2.18 c)! If we put $a = \widetilde{D}$, $\alpha = \frac{1}{6}$, $\gamma = 1$, and $b = \frac{233}{531} + o(1)$, i.e. $b_\infty = \frac{233}{531}$, we are done.

Of course, we may also employ (the asymptotics of) u_n directly in (2.34) which then becomes

$$D_{n+1} = 3D_n + \frac{466}{59}18^n + o(18^n), \quad (2.39)$$

with the term $o(18^n)$ going to 0 when divided by 18^n. Again we make use of Lemma 2.18 c), this time with $a = D$, $\alpha = 3$, $\gamma = 18$, and $b_n = \frac{466}{59}18^n + o(18^n)$, i.e. $b_\infty = \frac{466}{59}$.

For those who do not like the guesswork associated with the above explicit determination of the sequence u, there is a way to obtain the asymptotic behavior of u_n without calculating the individual members of the sequence explicitly. Again, taking convergence of $\tilde{u}_n$ for granted, we get the limit $\frac{1}{59}$ if we divide (2.35) by 18^{n+2}. However, to *guarantee* convergence, we make use of Lemma 2.19. We start directly from the recurrence for u and v and put $a_n = (u_n, v_n) \in \mathbb{R}^2$, $A = \begin{pmatrix} 2 & 2 \\ 2 & 3 \end{pmatrix}$, $\gamma = 18$, and $b_n = 18^n(\frac{1}{5}, \frac{1}{2}) + o(18^n)$, i.e. $b_\infty = (\frac{1}{5}, \frac{1}{2})$. Since Θ_+ is the largest eigenvalue of A, we have $\|A\| = \Theta_+ < 18 = \gamma$ and we can employ Lemma 2.19 with $(\gamma I - A)^{-1} = \frac{1}{236} \begin{pmatrix} 15 & 2 \\ 2 & 16 \end{pmatrix}$ to arrive at

$$18^{-n}(u_n, v_n) \to \left(\frac{1}{59}, \frac{21}{590} \right), \text{ as } n \to \infty.$$

An impressive formula like (2.36) cannot be without significance! And indeed, as indicated in the introductory chapter, it has found an application to the seemingly unrelated Sierpiński triangle. Stewart called this "yet another demonstration of the remarkable unity of mathematics" [392, p. 106] (translated in [393, p. 13]) and in fact it was him who pointed out in that article that "as the number [n] of discs becomes larger and larger, the graph [H_3^n] becomes more and more intricate, looking more and more like the Sierpiński gasket". (In the French original, Stewart used the much more romantic cognomen "napperon de Sierpiński"—Sierpiński's doily.) Based on the observation of Stewart's, Hinz and A. Schief were able to make the notion of this limit precise; see [214]. In fact, it can be viewed as the object in the plane which one obtains by rescaling the diameter $2^n - 1$ of the canonical drawing of H_3^n to the side length of the Sierpiński triangle, say 1, and letting n tend to infinity. Note the difference to the construction of the Sisyphean Hanoi graph H_3^∞ whose drawing is not bounded.

The distance on the Sierpiński triangle can be derived from one of the three equivalent constructions of that object, namely to take the closure of the set of lines obtained by successively adding triangles as in the bottom part of Figure 0.19. It can be shown (cf. [214, Theorem 1]) that any point x on ST can be joined by a rectifiable curve (of length at most 1) lying completely on ST to one of the corners of ST, and consequently x can be joined to any other point y on ST; the infimum (in fact, minimum) of the lengths of such curves is then $d(x, y)$. While it is obvious that $d(x, y) \geq \|x - y\|$ for the euclidean norm, with equality for any two points lying on a side of one of the constituting triangles, the (sharp) estimate $d(x, y) \leq 2\|x - y\|$ has to be proved by induction; see Exercise 2.34.

For other metric properties of ST it is tempting to use self-similarity arguments. Let us, for instance, ask for the average distance γ to a corner point of ST, the top one, say. Again we treat the question of measure in a naive way; see [214, Definition 2] for details. With probability 1/3 the randomly chosen point x lies in the upper subtriangle of ST, such that its expected distance to the top is

$\gamma/2$. With probability $1/3$, the point x lies in the bottom left subtriangle and the shortest curve linking it to the top will go through the only common point of the two subtriangles. Hence the expected distance of x from the top is $(\gamma + 1)/2$. The same applies to x in the bottom right subtriangle. All in all we get

$$\gamma = \frac{1}{3} \cdot \frac{\gamma}{2} + \frac{2}{3} \cdot \frac{\gamma + 1}{2} = \frac{\gamma}{2} + \frac{1}{3},$$

such that $\gamma = 2/3$.

Let us now employ the same type of argument to determine the average distance δ on ST. With probability $1/3$, the two random points lie in the same subtriangle, such that their expected distance is $\delta/2$. For the other pairs the shortest path passing through the shared point of their respective subtriangles has expected length $2 \cdot (\gamma/2) = \gamma$, such that this time

$$\delta = \frac{1}{3} \cdot \frac{\delta}{2} + \frac{2}{3} \cdot \gamma = \frac{1}{3} \cdot \frac{\delta}{2} + \frac{4}{9},$$

whence $\delta = 8/15$. This is about as easy as it is false!

To see why, let us return to the TH. If we can show that the metric properties of H_3^n and the nth order approximation of ST are asymptotically, for large n, the same, we may let n go to infinity in (2.36) and obtain the following surprising result [214, Theorem 2].

Theorem 2.48. *The average distance on the Sierpiński triangle is* $\dfrac{466}{885}$ *(of the diameter).*

Ian Stewart honored the number $\frac{466}{885}$ by including it into his collection of *incredible numbers* and devoting a section to it in [397, p. 152–159].

The way the limit process leading to Theorem 2.48 is justified is intimately connected with the introduction of another class of graphs, the so-called *Sierpiński graphs* S_3^n. In [214] they are defined (Definition 6) and it is shown that they are isomorphic to the corresponding H_3^n (Lemma 2). These graphs, which can, of course, be viewed as produced by an alternative labeling of the Hanoi graphs, turned out to be so useful for the theory of the TH and most interesting in their own, that we will devote Chapter 4 to them. Readers who are curious about the limit process just employed may have a quick look at Figure 4.11 to be convinced of its legitimacy. It shows essentially that any curve in ST can be approximated by a path on a renormalized H_3^n. In particular, it follows that diam(ST) = 1 and that Theorem 2.48 is true.

Another interpretation of ST as a sequence of graphs has been given by M. T. Barlow and E. A. Perkins in [36]. Their graphs, drawings of which are essentially the unions of triangles we used in one of the constructions of ST and which have later been called *Sierpiński triangle graphs* (cf. [207]), were used to study a diffusion process on ST arising as the limit of random walks on these graphs. This opened the new field of analysis on fractals; cf. [410]. Hanoi graphs, in fact a twofold Sisyphean Hanoi graph, also entered the business, albeit under the name of *Pascal graph* in [348]. To understand the reason for this choice of appellation, we come back to Pascal's Arithmetical triangle.

2.4.2 Pascal's Triangle and Stern's Diatomic Sequence

In Chapter 0 we presented the morphogenesis of ST from AT mod 2. The latter was viewed in [302, p. 329] as a(n infinite) graph by joining closest (odd) neighbors in AT by edges; cf. Figure 0.16. Since there is a connection between ST and Hanoi graphs, by transitivity there must be a relation between the latter and the graph AT mod 2, namely $H_3^n \cong \mathrm{AT}_{2^n} \bmod 2$ (see [195, Theorem 1]), where $\mathrm{AT}_k \bmod 2$ is the graph obtained from the first k rows of the odd entries of AT; see the black part of Figure 2.27.

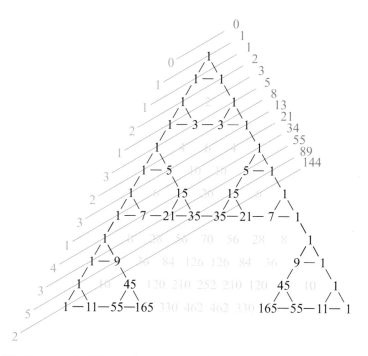

Figure 2.27: The graph $\mathrm{AT}_k \bmod 2$ for $k = 12$ (black), its subdiagonals (red), even combinatorial numbers (light green), Fibonacci numbers (dark green), and Stern numbers (blue)

An isomorphism between H_3^n and $\mathrm{AT}_{2^n} \bmod 2$, which Poole has called *Lucas correspondence* in [343], can be given by

$$s \mapsto \left(\sum_{d=1}^{n} \left(s_d \neq (s \vartriangle 0)_d \right) \cdot 2^{d-1}, \; \sum_{d=1}^{n} \left(s_d = (s \vartriangle 2^{n \bmod 2})_d \right) \cdot 2^{d-1} \right), \qquad (2.40)$$

where the vertex of AT mod 2 representing combinatorial number $\binom{k}{\ell}$ is given by the ordered pair (k, ℓ). In fact, by (2.8), the two sums in (2.40) are $\mathrm{d}(s, 0^n)$, i.e. the distance of s to the top corner of H_3^n, and $2^n - 1 - \mathrm{d}(s, 2^{n \bmod 2})$, which is the complement of the distance to the bottom right corner according to our convention, respectively. That the image refers to an odd combinatorial number follows from (0.17). The number of images is 3^n by Corollary 2.17; injectivity and compatibility with the graph structures can be deduced from Remark 2.14. This construction actually produces an isomorphism between the Sisyphean Hanoi graph H_3^∞ and the graph AT mod 2, identifying 0^∞ with $\binom{0}{0}$.

One of the most fruitful techniques in mathematics is to transfer properties from one object to an isomorphic one. So everything we learned about the graphs H_3^n is now also true for AT_{2^n} mod 2. This approach was taken in [195]. As a prototype of such a derivation of statements about combinatorial numbers, we let the reader prove a result of J. W. L. Glaisher in Exercise 2.35.

Here is a more striking example. In (0.13) we have seen that the entries (odd and even) on the νth *subdiagonal* of AT (the red lines in Figure 2.27), $\nu \in \mathbb{N}$, i.e. all $\binom{\nu-1-\ell}{\ell}$ with $\ell \in \left[\left[\frac{\nu+1}{2} \right] \right]_0$, add up to the Fibonacci number F_ν; see Figure 2.27. For $\nu \in [2^n]$ and every odd combinatorial number on the νth subdiagonal, the difference of the distances to the bottom right and to the top corner of AT_{2^n} mod 2 is $2^n - \nu$. So their number is $z_n(2^n - \nu) = b(\nu)$ [195, Proposition 3]; compare the red lines in Figure 2.27 with those of Figure 2.22.

Proposition 2.49. *On the νth subdiagonal of AT, $\nu \in \mathbb{N}$, there are $b(\nu)$ odd entries.*

With the recurrence from (2.31), the proof of Proposition 2.49 is implicit in [72, p. 18f]. For an easy statement about *Stern numbers* $b(\nu)$ see Exercise 2.36.

Stern's sequence has been thoroughly studied over the years; rich sources are [331] and the survey paper of I. Urbiha [423] including information on the history of the sequence and many of its properties.

For instance, it has been shown by D. Parisse [331, Theorem 2] (cf. also [198, Theorem]) that for fixed $\mu \in \mathbb{N}_0$, the sequence $(z_n(\mu))_{n \in \mathbb{N}_0}$ is eventually in arithmetic progression, which led him to an efficient method to calculate its elements with the aid of continued fractions. Since Stern numbers can easily be computed from (2.31), we may use

$$\forall\, n \in [n_\mu]_0 : z_n(\mu) = 0, \; \forall\, n \in \mathbb{N}_0 \smallsetminus [n_\mu]_0 : z_n(\mu) = (n - n_\mu) b(\mu) + b\left(2^{n_\mu} - \mu \right),$$
$$(2.41)$$

where n_μ is the smallest integer such that $\mu < 2^n$, i.e. $n_0 = 0$ and $n_\mu = \lfloor \mathrm{lb}(\mu) \rfloor + 1$[47]

---

[47] $\mathrm{lb}(x) = \ln(x)/\ln(2)$ is the *binary logarithm* of x.

otherwise, to determine $z_n(\mu)$ efficiently for large n; (2.41) is an easy consequence of (2.23).

As a curiosity we note that from Theorem 2.34 we get the impressive formula uniting Fibonacci, Stern, and Jacobsthal numbers

$$\forall\, n \in \mathbb{N}_0 : F_n = b(J_n).$$

Let us finally mention just one more property of Stern numbers, namely a specific interpretation of the sequence.

A *hyperbinary representation* of a non-negative integer ν is a representation of ν as a sum of powers of 2, each power being used at most twice. We will employ the notation $(s_n \ldots s_1)_{[2]}$ to describe the hyperbinary sum $\sum_{d=1}^{n} s_d \cdot 2^{d-1}$, $s_d \in T$. Let $\mathcal{H}(\nu)$ denote the set of all hyperbinary representations of ν, where any two representations of the same integer differing only by zeros on the left-hand side are identified. For instance, $(1)_{[2]}$ is the same representation of 1 as $(01)_{[2]}$.

Clearly, $|\mathcal{H}(0)| = 1 = |\mathcal{H}(1)|$. When $x = (s_n \ldots s_1)_{[2]}$ is odd, then s_1 must be 1, hence

$$|\mathcal{H}(2\nu + 1)| = |\mathcal{H}(\nu)|, \ \nu \in \mathbb{N}.$$

When x is even, s_1 may be 0 or 2 which in turn implies that

$$|\mathcal{H}(2\nu)| = |\mathcal{H}(\nu)| + |\mathcal{H}(\nu - 1)|, \nu \in \mathbb{N}.$$

Comparing this recurrence with the one that characterizes Stern's sequence, namely (2.31), we obtain:

Theorem 2.50. *For every $\nu \in \mathbb{N}_0$, $|\mathcal{H}(\nu)| = b(\nu + 1)$.*

Theorem 2.50 is, of course, also an immediate consequence of the identity in (2.22): for $\nu \in [2^n]_0$, there is a one-to-one correspondence between the hyperbinary representations $\nu = \sum_{d=1}^{n} s_d \cdot 2^{d-1}$, $s_d \in T$, and the representations of $\mu := 2^n - 1 - \nu \in [2^n]_0$ as $\mu = \sum_{d=1}^{n} \tilde{s}_d \cdot 2^{d-1}$ with $\tilde{s}_d = 1 - s_d \in \tilde{T} = \{-1, 0, 1\}$, because there cannot be a hyperbinary sum for ν with $s_m \neq 0$ for some $m > n$; it follows that $|\mathcal{H}(\nu)| = z_n(\mu) = b(2^n - \mu) = b(\nu + 1)$.

N. Calkin and H. S. Wilf [68] constructed a binary tree which contains every positive fraction in lowest terms precisely once and such that reading the successive rows of this tree results in the sequence $\left(\dfrac{\mathcal{H}(n)}{\mathcal{H}(n+1)}\right)_{n \in \mathbb{N}_0}$. Since $\mathbb{Q}_+ := \{x \in \mathbb{Q} \mid x > 0\}$ is equivalent to the set of these fractions, we obtain a bijection from $\mathbb{N}$ to $\mathbb{Q}_+$, namely

$$n \mapsto \frac{b(n)}{b(n+1)}.$$

Compared to Cantor's diagonalization (cf. p. 51) or the surjection onto $\mathbb{Q} \cap [0, 1]$ obtained from Brocot's array (cf. p. 28) this has the advantage that no fraction is repeated in the sequence.

The relations between Stern's diatomic sequence and graphs of Hanoi type become a little more lucid in the context of Sierpiński graphs to which we will turn in Chapter 4. The same applies to some extent to the notorious P2 decision problem, to which we will nevertheless come back right now.

2.4.3   Romik's Solution to the P2 Decision Problem

We warn the leisurely reader that the contents of this subsection may become somewhat technical and encourage those who lack patience to skip it.

Instead of averaging over many optimal solutions, we will now look at individual tasks $is \to jt$, where $s, t \in T^n$, $n \in \mathbb{N}$, and with $i, j, k \in T$ such that $|\{i, j, k\}| = 3$. We want to construct an algorithm which decides efficiently about the number of moves the largest disc $n + 1$ has to make in a shortest path. Since $i \neq j$, we know from Theorem 2.38 that there are three possible cases: "I" will mean that $n + 1$ moves only once, "II" that it *necessarily* moves twice, and "I/II" that both strategies lead to an optimal solution. We can exclude $n = 0$, because there is no choice in H_3^1. We only consider the standard case that $i \neq j$; tasks with some largest discs initially already in their goal position can be reduced to the standard case by appropriate pre-processing, i.e. ignoring these discs.

From (2.29) we know that

$$d_1(is, jt) \,\Box\, d_2(is, jt) \Leftrightarrow \rho := d(s; k, j) + d(t; k, i) \,\Box\, 2^n, \quad \Box \in \{<, =, >\},$$

so that we may now concentrate on the right side of this equivalence. By definition and with (2.8),

$$\rho = d(s, k^n) - d(s, j^n) + d(t, k^n) - d(t, i^n) = \sum_{d=1}^{n} \beta_d \cdot 2^{d-1},$$

where

$$\beta_d = (s_d \neq (s \vartriangle k)_d) - (s_d \neq (s \vartriangle j)_d) + (t_d \neq (t \vartriangle k)_d) - (t_d \neq (t \vartriangle i)_d)$$
$$= (s_d \neq (is \vartriangle j)_d) - (s_d \neq (is \vartriangle k)_d) + (t_d \neq (jt \vartriangle i)_d) - (t_d \neq (jt \vartriangle k)_d) .$$

There are 9 different pairs (s_d, t_d), leading to only 5 values of β_d, namely

$$\beta_d = \begin{cases} -2, & \text{if } (s_d, t_d) = ((is \vartriangle j)_d, (jt \vartriangle i)_d) ; \\ -1, & \text{if } (s_d, t_d) = ((is \vartriangle j)_d, (jt \vartriangle j)_d) \quad \text{or} \quad ((is \vartriangle i)_d, (jt \vartriangle i)_d) ; \\ 0, & \text{if } (s_d, t_d) = ((is \vartriangle j)_d, (jt \vartriangle k)_d) \quad \text{or} \quad ((is \vartriangle i)_d, (jt \vartriangle j)_d) \\ & \qquad\qquad\qquad\qquad\qquad\qquad\quad \text{or} \quad ((is \vartriangle k)_d, (jt \vartriangle i)_d) ; \\ 1, & \text{if } (s_d, t_d) = ((is \vartriangle i)_d, (jt \vartriangle k)_d) \quad \text{or} \quad ((is \vartriangle k)_d, (jt \vartriangle j)_d) ; \\ 2, & \text{if } (s_d, t_d) = ((is \vartriangle k)_d, (jt \vartriangle k)_d) . \end{cases}$$

For $\delta \in [n+1]_0$ let $\rho_\delta = \sum_{d=1}^{\delta} \beta_d \cdot 2^{d-1}$. Then $|\rho_\delta| < 2^{\delta+1}$ and $\rho_\delta - \rho_{\delta-1} = \beta_\delta \cdot 2^{\delta-1}$.

for $\delta \in [n]$. Let us look at three cases.

$$
\begin{aligned}
\text{A}: \quad & \rho \,\square\, 2^n && \Leftrightarrow && \rho_\delta \,\square\, 2^\delta, \\
\text{B}: \quad & \rho \,\square\, 2^n && \Leftrightarrow && \rho_\delta \,\square\, 0, \\
\text{C}: \quad & \rho \,\square\, 2^n && \Leftrightarrow && \rho_\delta \,\square -2^\delta.
\end{aligned}
$$

For $\delta = n$ we are in case A, because $\rho = \rho_n$. Now let us assume that for some $\delta \in [n]$ we are in case A. Then

$$
\rho_\delta \,\square\, 2^\delta \Leftrightarrow \rho_{\delta-1} \,\square\, (2 - \beta_\delta)2^{\delta-1}.
$$

For $\beta_\delta \leq 0$ it follows that $\square \, = <$, i.e. we detect a type I task and stop; for later purpose, we will call this the terminating case D. For $\beta_\delta = 1$ we have $\rho \,\square\, 2^n \Leftrightarrow \rho_{\delta-1} \,\square\, 2^{\delta-1}$, and we stay in case A, but with δ replaced with $\delta - 1$. Similarly, if $\beta_\delta = 2$, then we have to continue with case B and $\delta - 1$ instead of δ.

If we are in case B for some $\delta \in [n-1]$, then $\rho_\delta \,\square\, 0 \Leftrightarrow \rho_{\delta-1} \,\square -\beta_\delta \cdot 2^{\delta-1}$. For $\beta_\delta = -2$ we end in D. For $\beta_\delta = -1$ we go to A with δ replaced with $\delta - 1$. For $\beta_\delta = 0$ we stay in B for $\delta - 1$ and for $\beta_\delta = 1$ we move to case C for $\delta - 1$. Finally, if $\beta_\delta = 2$, then two moves of disc $n + 1$ are necessary and we end in what we call terminating case E.

Finally, in case C for some $\delta \in [n-2]$, $n \neq 1$, we have $\rho_\delta \,\square -2^\delta \Leftrightarrow \rho_{\delta-1} \,\square -(2 + \beta_\delta)2^{\delta-1}$. Here, for $\beta_\delta = -2$ we have to move to B and change δ to $\delta - 1$. Similarly for $\beta_\delta = -1$, where we stay in C. Finally for $\beta_\delta \geq 0$ we have $\square \, = >$ and we end in E.

This analysis shows that all possible cases are covered and that we can base Algorithm 13 on an automaton which is essentially due to Romik (cf. [353, Figure 2]) and shown in Figure 2.28.

Algorithm 13 P2 decision algorithm for H_3^{1+n}

Procedure p2H(n, s, t)
Parameter n: number of discs minus 1 $\{n \in \mathbb{N}\}$
Parameter s: initial configuration $\{s \in T^{1+n}\}$
Parameter t: goal configuration $\{t \in T^{1+n}, t_{n+1} \neq s_{n+1}\}$
   $i \leftarrow s_{n+1}$, $j \leftarrow t_{n+1}$
   start in state A of P2-automaton
   $\delta \leftarrow n$
   while $\delta > 0$
     replace pairs (ι, κ) on the arcs by $(s_{\delta+1} \vartriangle \iota, t_{\delta+1} \vartriangle \kappa)$
     apply automaton to pair (s_δ, t_δ)
       {algorithm STOPs if automaton reaches terminating state D or E}
     $\delta \leftarrow \delta - 1$
   end while

Starting with $\delta = n$ and in state A of the automaton, the labels (ι, κ) on the arcs of the automaton have to be updated to obtain $((is \vartriangle \iota)_\delta, (jt \vartriangle \kappa)_\delta)$ which can

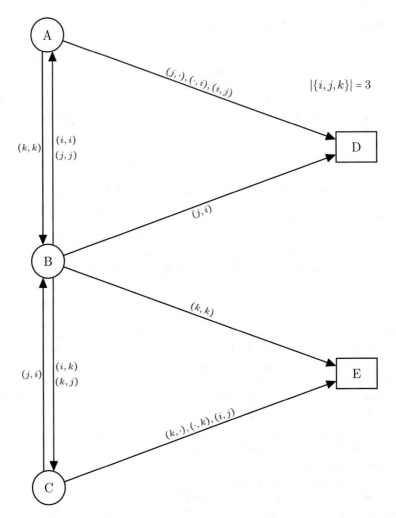

Figure 2.28: Romik's automaton for the P2 decision problem

then be compared with the pair (s_δ, t_δ) by the automaton which will accordingly adapt its state. The automaton either stops, if reaching states D (one move of largest disc; class I task) or E (two moves; class II task) or ends in one of the other states when input data have run out, i.e. after n steps for $\delta = 0$. In the latter case, ending in A means $\rho \,\square\, 2^n \Leftrightarrow \rho_0 \,\square\, 1$; since $\rho_0 = 0$, we have $\square\, = <$, such that disc $n + 1$ makes one move only. In B we have $\rho \,\square\, 2^n \Leftrightarrow \rho_0 \,\square\, 0$, such that $\square$ is equality and we have a draw, that is there are two shortest paths using one or two largest disc moves, respectively (class I/II task). If the algorithm ends in state C, then $\rho \,\square\, 2^n \Leftrightarrow \rho_0 \,\square\, -1$, i.e. $\square\, => $ and we necessarily have two moves of disc $n + 1$.

Note that in many cases the algorithm stops after the input of less than $n+1$ pairs (s_δ, t_δ). However, the update of labels in the automaton is clumsy and in fact Romik applied his automaton unchanged to the P2 decision problem for Sierpiński graphs S_3^n. We will therefore postpone examples and the discussion of complexity to Chapter 4.

2.4.4   The Double P2 Problem

Z. Šunić [412] studied the problem of simultaneously solving two P2 tasks. More precisely, recall first that for a given regular state, a legal (possibly empty) move is uniquely determined if the idle peg of the move is prescribed. Now, we are given two initial states $s, s' \in T^n$ and two goal states $t, t' \in T^n$. The task is to find a sequence of idle pegs such that the moves dictated by the idle pegs simultaneously transfer state s to t and state s' to t'. We will use the notation

$$ s \to t \parallel s' \to t' $$

for this task and call it the *double P2 problem*. In Figure 2.29 a solution for the task $00 \to 01 \parallel 01 \to 02$ is shown. The color of the arrow of a move indicates the idle peg of that move.

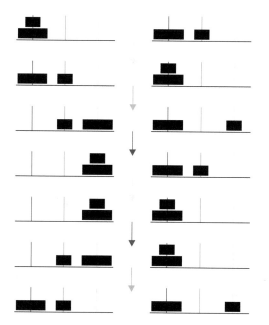

Figure 2.29: A solution for a double P2 task

The task $s \to t \parallel s' \to t'$ is called *solvable* if there exists a sequence of idle pegs that transfers state s to t and state s' to t'. Applying the theory of Hanoi

Towers groups (cf. below, p. 265), D. D'Angeli and A. Donno [91] were able to characterize solvable tasks as follows (see [412] for an alternative approach):

Theorem 2.51. *Let $n \in \mathbb{N}$ and $s, s', t, t' \in T^n$. Then the task $s \to t \parallel s' \to t'$ is solvable if and only if the length of the longest common* **suffix** *of s and t is the same as the length of the longest common suffix of s' and t'.*

Theorem 2.51 in particular implies that the tasks $s \to t \parallel s' \to t'$, for which $s_1 \neq t_1$ and $s'_1 \neq t'_1$ hold, are solvable. Such tasks are called *basic double tasks* and for them Šunić [412] proved:

Theorem 2.52. *Let $n \in \mathbb{N}_3$. Then the basic double tasks induce a connected subgraph X_n of the state graph of the double P2 problem on n discs. Moreover, the diameter of X_n is at most $\frac{11}{3} 2^n$.*

The diameter of X_n can also be bounded from below, it is at least $2 \cdot 2^n$. This follows from another result of Šunić asserting that an optimal solution of the task

$$0^n \to 0^{n-1}1 \parallel 0^{n-1}1 \to 0^{n-1}2$$

makes exactly $2 \cdot 2^n$ moves for any $n \in \mathbb{N}_3$. (Note that this is not true for $n = 2$ because the solution for $n = 2$ presented in Figure 2.29 makes 6 moves.) We also mention that for the task

$$0^n \to 2^n \parallel 2^n \to 0^n$$

a solution is constructed in [412] that makes $\frac{1}{3}\left(2^{n+2} - (-1)^n\right)$ moves for any $n \in \mathbb{N}_2$ and it is conjectured that this number of moves is optimal.

2.5   Exercises

2.1. Verify Equation (2.1).

2.2. Prove Proposition 2.2 without referring to the Gros sequence.

2.3. Prove that Olive's algorithm is correct.

2.4. Show that neither of the sequences h and o is strongly square-free.

2.5. (Schoute [372]; cf. [52, p. 862]) Use Proposition 2.3 to describe a human algorithm for the shortest path from a tower on peg 0 to a tower on peg 2, i.e. one which does not need large long-term memory.

2.6. (J. Bennish [46]) Recall from Equation (0.4) that $F_0 = 0$, $F_1 = 1$, and $F_k = F_{k-1} + F_{k-2}$, $k \in \mathbb{N}_2$, are the Fibonacci numbers. Let a_n be the number of legal arrangements of discs on the intermediate peg in the optimal solution of the classical TH task with n discs. Show that $a_n = F_{n+1}$ for any $n \in \mathbb{N}_0$.

2.7. Prove Proposition 2.4.

2.8. A snapshot from Benares shows the Brahmin on duty moving disc $\delta \in [64]$ and the Tower of Brahma with the other discs. How can we determine the age of the universe?

It has been observed (by Schoute already; cf. [372, p. 275]) that the Brahmins make precisely one move per second without deviating from the optimal solution. Until March 2013, the space observatory *Planck* of the *European Space Agency (ESA)* has sent back data to Earth which can be interpreted as a snapshot of the Tower in Benares showing the Brahmin on duty moving disc 51; disc 52 lies alone on the goal needle, discs 58 and 59 on the intermediate needle, and all the others on the initial needle. How old is the universe accordingly?

2.9. (A. P. Domoryad [108, p. 76]) Let 8 discs initially be distributed as in Figure 2.30. The goal is the 8-tower on the middle peg. Describe the optimal solution and determine its length.

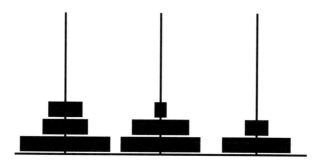

Figure 2.30: Initial state of Domoryad's task

2.10. **a)** Let $\{i, j, k\} = T$ and $n \in \mathbb{N}$.
Show that $d(s, i^n) \neq d(s, j^n)$ for all $s \in T^n \setminus \{k^n\}$.

b) (Domoryad [108, p. 76]) Assume that we have $2n$ discs, $n \in \mathbb{N}$, and let initially the n largest discs lie on peg 0, the others on peg 1. Calculate the distances of this state to the perfect states 1^{2n} and 2^{2n}, respectively.

2.11. Complete the argument for Remark 2.14.

2.12. Let $j \in T$ be a fixed peg in a TH with n discs. Determine the number of states with $d(s, j^n) = 2^n - 1$ and characterize them.

2.13. Calculate the number of moves in the optimal solution for the task in Figure 2.9 and determine the best first move.

2.14. (Stockmeyer [404]) For $n \in \mathbb{N}_0$ and $E \in 2^{[n]}$ define

$$\Psi(E) = \sum_{d \in E} 2^{d-1} ;$$

note that $0 = \Psi(\varnothing) \le \Psi(E) \le \Psi([n]) = 2^n - 1$. Let $\{i,j\} \in \binom{T}{2}$.

Show that

$$\forall\, s \in T^n : \mathrm{d}(s, j^n) \ge \Psi\left(s^{-1}(\{i\})\right) \qquad (2.42)$$

and

$$\forall\, E \in 2^{[n]} \,\exists_1\, s \in [p]_0^n : s^{-1}(\{i\}) = E \wedge \mathrm{d}(s, j^n) = \Psi(E). \qquad (2.43)$$

2.15. Find the average performance of Algorithm 11.

2.16. Determine degrees of vertices of H_3^n.

2.17. Show that there is a total coloring of H_3^n using at most 4 colors.

2.18. Show that $\|H_3^n\| = \frac{3}{2}(3^n - 1)$.

2.19. Let $n \in \mathbb{N}$.

a) Show that in H_3^n any pair of perfect states i^n and j^n is linked by a unique hamiltonian path (cf. [277, Lemma 6]). Its edge set consists of all $\{sik^{d-1}, sjk^{d-1}\} \in E(H_3^n)$, i.e. those edges corresponding to moves whose idle peg is $k = 3 - i - j$.

b) Show that H_3^n is hamiltonian.

c) Are the graphs H_3^n (semi-)eulerian?

2.20. (Stockmeyer [407, Section 4.1]) Consider the Linear TH and the task to transfer a tower of discs from peg 1 to peg 2. Show that the following strategy returns the optimal sequence of moves: if the number of discs on peg 0 is m, then the idle peg of the next move is $2 - m \bmod 2$.

2.21. Consider the optimal solution for the Linear TH that transfers the tower of n discs from peg 1 to peg 2. For any disc $d \in [n]$ determine the number of its moves during the optimal solution.

2.22. a) Give a proof of Proposition 2.26.

b) Show that (cf. [277, p. 37])

$$\forall\, d \in \left[\frac{3^n + 1}{2}\right]_0, \ \forall\, r \in [n] : s(3^n - 1 - d)_r = s(d)_r \mathbin{\vartriangle} 0.$$

2.23. Show that $\gamma(G) = |G|$ if and only if $\Delta(G) = 0$.

2.24. Find the domination numbers for all connected graphs G with maximal degree $\Delta(G) \le 2$.

2.25. Prove that the number of codewords in a perfect code on H_3^n is $\frac{1}{4}(3^n + 2 + (-1)^n)$.

2.26. a) Find a recurrence for the number of matchings of H_3^n, $n \in \mathbb{N}_0$.

b) Determine the number of **perfect matchings** for the graphs obtained from H_3^n by deleting one or three perfect states.

2.27. Show that the Sisyphean Hanoi graph H_3^∞ is connected.

2.28. Show that the **radius** of H_3^{1+n}, $n \in \mathbb{N}$, is $\mathrm{rad}(H_3^{1+n}) = 3 \cdot 2^{n-1}$ and that for $n \in \mathbb{N}_2$ its **center** consists of the 6 vertices iji^{n-1}, $i, j \in T$, $i \neq j$.

2.29. Prove the formulas in (2.26).

2.30. Prove Lemma 2.35.

2.31. Prove Proposition 2.39.

2.32. Show that (2.30) and (2.31) are equivalent if $b(0) = 0$ and $b(1) = 1$.

2.33. [48] Let $n \in \mathbb{N}$, $s \in T^n$, and $i \in T$.

a) Show that $|X(is)| = n$ if $s = j^n$, $j \in T \setminus \{i\}$.

b) Show that $|X(is)| \geq n$ for $s \neq i^n$.

c) Find a "small" example for $|X(is)| > n$.

d) Show that $X(01^n) = \{1^\nu(1 + (-1)^\nu)1^{n-\nu} \mid \nu \in [n]\}$.

2.34. Show that $\mathrm{d}(x, y) \leq 2\|x - y\|$ for $x, y \in \mathrm{ST}$ and that this bound is sharp.

2.35. Prove that the number of odd combinatorial numbers in row $\mu \in \mathbb{N}_0$ of the AT is $2^{q(\mu)}$.

2.36. Find out about the parity of Stern numbers $b(\nu)$, $\nu \in \mathbb{N}_0$.

---

[48] P. K. Stockmeyer, private communication, 2017

Chapter 3

Lucas's Second Problem

In his early descriptions of the TH (see, e.g., [83]), Lucas pointed out the possibility of starting with an arbitrary distribution of n discs among three pegs, i.e. allowing for discs lying on a smaller one. The task is again to arrive at a perfect state on a preassigned peg, while still obeying the divine rule. This, in Lucas's opinion [82], will vary the conditions of the problem of the TH "to infinity". We may even go beyond by prescribing an arbitrary, albeit regular, state as the goal. We will approach this problem in the next section and round this chapter off with a section on an algorithmic solution to *Lucas's second problem*.

3.1 Irregular to Regular

For the sake of this chapter (only), a distribution of discs among pegs in which larger discs are allowed to lie above smaller ones will simply be called a *state*. An *irregular state* is a state which is not regular in the classical meaning; see Figure 3.1 for an example.

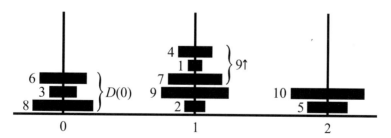

Figure 3.1: An irregular state

We consider an arbitrary state s. Let $d_1, \ldots, d_{i_1}$ be the discs, listed from top to bottom, that are stacked onto peg 0, and let $d_{i_1+1}, \ldots, d_{i_2}$ and $d_{i_2+1}, \ldots, d_n$ be the

corresponding discs, also listed from top to bottom, on pegs 1 and 2, respectively.
Then s can be described by the string

$$d_1 \ldots d_{i_1} \,|\, d_{i_1+1} \ldots d_{i_2} \,|\, d_{i_2+1} \ldots d_n \,,$$

where the symbol "$|$" is used to delimit discs among pegs; this notation goes
back to [378]. For instance, the state shown in Figure 3.1 is represented by
6 3 8 | 4 1 7 9 2 | 10 5. Clearly, every state, regular or irregular, can uniquely be
described in such a way, and different strings correspond to different states. Since
the two symbols "$|$" can appear anywhere, but are interchangeable, we thus infer:

Proposition 3.1. *The number of states in the puzzle with $n \in \mathbb{N}_0$ discs is $\frac{1}{2}(n+2)!$.*

Lucas calls this number, for $n = 64$, "vertiginous" and claims that it has more
than fifty digits. Although this is true, it underestimates the actual size of the
figure by more than 40 digits! Lacking modern computing equipment he might
just have observed that the number has more than 50 factors greater than 10.

In the previous chapter we have considered the graph H_3^n as a natural mathematical model for the TH with 3 pegs. We now extend this idea to the more
general situation in which irregular states are allowed. In this case, however, some
of the edges must be oriented since it is possible that one state can be obtained
from another by a legal move but not the other way round. To be more precise,
let $\overrightarrow{H}_3^n$ be the **digraph** whose vertex set is the set $\mathfrak{T}^n$ of states and where there is
an arc (σ, τ) from state σ to state τ, if τ results from σ by a legal move of one
disc. If arcs (σ, τ) and (τ, σ) are simultaneously present, it is customary to identify
them with an edge between σ and τ. Hence $\overrightarrow{H}_3^n$ is an example of a **mixed graph**,
that is a graph containing both edges and arcs. Clearly, $\overrightarrow{H}_3^n$ contains H_3^n as a
subgraph. In this sense the regular states form a subset $T^n \subseteq \mathfrak{T}^n$, proper for $n \geq 2$.
In Figure 3.2 the digraphs $\overrightarrow{H}_3^2$ and $\overrightarrow{H}_3^3$ are shown, where the respective subgraphs
H_3^2 and H_3^3 are emphasised. Note that in the drawing of $\overrightarrow{H}_3^3$ (cf. [147, p. 34]) six
outer triangles are drawn twice, that is, 18 vertices are drawn in duplicates, so
that the mixed graph contains 60 vertices in agreement with Proposition 3.1.

The problem to transfer discs from a(n irregular) state to a perfect state
obeying the classical rules is called *type* P3 problem. This problem has been independently considered and solved by Hinz in [194, Chapter 2], and C. S. Klein and
S. Minsker in [243]. Hinz approached the theory in the framework of the even more
general *type* P4 problem, namely to reach a regular state from an irregular one.
Although there *are* arcs and even edges between irregular states, some (ir)regular
to irregular tasks are not solvable, as, e.g., $21|| \rightarrow ||21$ in $\overrightarrow{H}_3^2$. Therefore, we will
not consider irregular goals.

A state will be addressed by a Greek letter such as σ, and the corresponding
Latin character, like s, will be used to designate its *regularization*, i.e. the regular
state where all discs are on the same peg as in σ. Our first result [194, Lemma 3]

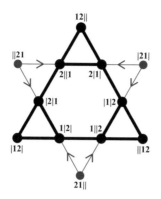

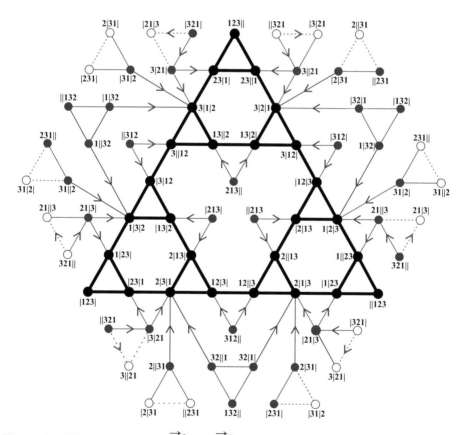

Figure 3.2: The mixed graphs $\overrightarrow{H}_3^2$ and $\overrightarrow{H}_3^3$ (Unfilled dots and dotted lines represent repeated occurrences of a vertex or edge, respectively.)

leads to goal states which are *semi-perfect*, i.e. states whose regularizations are perfect.

Lemma 3.2. *For every $\sigma \in \mathfrak{T}^n$, $n \in \mathbb{N}$, and every $j \in T$, there is a semi-perfect state on j that can be reached from σ in at most $2^n - 1$ moves; if $s_n \neq j$, then the goal state is perfect.*

Proof. Induction on n. The case $n = 1$ is trivial. If $s_{n+1} = j$, then consider the state $\bar{\sigma}$, obtained from σ by ignoring disc $n + 1$ and all discs underneath that disc and relabeling the remaining discs. This state is either empty or, by induction assumption, at most $2^n - 1$ moves are needed to transfer its discs to peg j. If $s_{n+1} \neq j$, then $\bar{\sigma}$ is empty or can be transferred to a semi-perfect state on peg $3 - j - s_{n+1}$ in at most $2^n - 1$ moves by induction assumption. Then disc $n + 1$ moves to peg j and finally the n smaller discs to j as well, again in at most $2^n - 1$ moves. Altogether, at most $2^{n+1} - 1$ moves are made, and since disc n started from a peg different from j in the last part of the procedure, the final state is perfect by induction assumption. $\square$

The upper bound in Lemma 3.2 is sharp because it applies to the classical P0 problem as well. Since we know to solve any P2 task in H_3^n in at most $2^n - 1$ moves (cf. Theorem 2.31), we can try to reduce a P4 to a P2 task by going from the irregular initial state to some regular one (cf. [194, Proposition 8]). We may assume that $n \in \mathbb{N}_2$, since otherwise there are no irregular states.

Proposition 3.3. *For every $\sigma \in \mathfrak{T}^n$, $n \in \mathbb{N}_2$, there is a regular state $t \in T^n$ that can be reached from σ in at most 2^{n-2} moves.*

Remark 3.4. *The upper bound is sharp. Take the state $\sigma = 1 \ldots (n-2)\, n\, (n-1) || $. As long as disc n has not been moved, the state will remain irregular. This move can only happen after a P0 task for $n-2$ discs has been solved, which takes another $2^{n-2} - 1$ moves.*

Proof of Proposition 3.3 by induction on n. For $n = 2$ take a look at Figure 3.2. If disc $n + 1$ lies at the bottom of peg s_{n+1} we may apply the induction assumption on $\bar{\sigma}$. If there is precisely one disc $d \in [n]$ underneath $n + 1$, we can transfer $\bar{\sigma}$ to a perfect tower on a peg j different from s_{n+1} and $s_{\max([n]\setminus\{d\})}$ in at most $2^{n-1} - 1$ moves by Lemma 3.2, such that after the move of disc $n + 1$ to peg $3 - s_{n+1} - j$ a regular state is reached. Finally, if there is more than one disc underneath $n + 1$, the latter disc can move at the latest in the 2^{n-2}th move according to Lemma 3.2, and by induction assumption the resulting state can be made regular in at most another 2^{n-2} moves. $\square$

The main statement [194, Theorem 6] on the P4 problem is now a direct consequence of Proposition 3.3 and Theorem 2.31:

Theorem 3.5. *For every $(\sigma, t) \in \mathfrak{T}^n \times T^n$, $n \in \mathbb{N}_2$, there is a **path** from σ to t in $\overrightarrow{H}_3^n$ of length less than or equal to $2^{n-2} + 2^n - 1$.*

We leave it as an exercise to the reader to show that the upper bound can not be improved, even for a P3-type task; see Exercise 3.1.

For practical purposes, it will be useful to single out a class of tasks $\sigma \to t$ in $\overrightarrow{H}_3^n$, namely the *special case* where disc n is *not* on the bottom of peg s_n in state σ and $s_n = t_n$; all other tasks will be called the *standard case*. We then have [194, Proposition 9]:

Proposition 3.6. *All standard case tasks in $\overrightarrow{H}_3^n$, $n \in \mathbb{N}_2$, can be solved in at most $2^n - 1$ moves.*

The proof is left as Exercise 3.2. Based on the arguments for Theorem 3.5 (or Proposition 3.6) a recursive algorithm could be constructed which solves every task in $\overrightarrow{H}_3^n$ with at most $2^{n-2} + 2^n - 1$ (or $2^n - 1$) moves. However, the resulting solution needs not to be minimal! One reason is, as for P2 tasks which after all will also be solved by the algorithm, that there might be an option to move the largest disc more than once; for special case tasks this is, of course, mandatory. But there are even tasks where disc n has to move three times for an optimal solution; see Exercise 3.3! This is, however, the worst case.

Lemma 3.7. *In a shortest path from $\sigma \in \mathfrak{T}^n$ to $t \in T^n$ in $\overrightarrow{H}_3^n$, $n \in \mathbb{N}_2$, disc n does not move twice to the same peg; in particular, it moves at most three times. If $s_n = t_n$ in a standard case task, then disc n does not move at all.*

Proof. As soon as disc n has been moved to some peg, it lies on the bottom of that peg, where it does not obstruct the moves of other discs. Therefore, all further moves of disc n eventually leading to the same peg would be a waste. An analogous argument holds for the last statement. □

For special case tasks three moves of the largest disc cannot appear in an optimal solution.

Lemma 3.8. *In a shortest solution for a special case task in $\overrightarrow{H}_3^n$, $n \in \mathbb{N}_2$, disc n moves precisely twice.*

Proof. Assume that disc n moves three times in an optimal σ, t-path P, i.e., by Lemma 3.7, from $i := s_n$ to j to k and back to $i = t_n$, where $\{i, j, k\} = T$. We will construct a σ, t-path P' which is strictly shorter than P and employs two moves of disc n only. By D we will denote the (non-empty) set of discs lying underneath disc n in state σ.

The moves before the first move of n in P are carried over to P', but that move of n is skipped. All discs from $[n-1] \setminus D$ are now on peg k, and they are then transferred in P to peg i to be united with those from D in preparation for the second move of disc n. Instead, in P', we move the discs on k to peg j, which can be done by switching pegs i and j in the moves of P and ignoring moves of discs from D (which are hidden underneath disc n in P' so far). Now the move of n from j to k in P is replaced by one from i to k in P'.

We call the states reached by now σ_P and $\sigma_{P'}$, respectively. Let $d \in D$ be the largest disc such that the arrangement of d and discs above it on peg i is regular in σ_P. Then the regular, albeit possibly incomplete, tower above d has to be transferred in P to some other peg. The same set of discs forms a regular arrangement on pegs i and j in $\sigma_{P'}$, because all discs on j have been moved already. Therefore, the same state as in P can be reached from $\sigma_{P'}$ in at most as many moves.

Now P' can be continued just like P. All in all we have saved at least one move (of disc n).   □

Remark. The task $n(n-1)||1\ldots(n-2) \to 01^{n-1}$ shows that even for large n a path with three moves of disc n may just be one move longer than the optimal path.

Although the number of moves of disc n is fixed by Lemma 3.8, optimal solutions for special case tasks may still vary depending on the choice of the intermediate peg. This can lead to non-uniqueness even if the initial state is not semi-perfect as demonstrated by the example of the task $32||1 \to 13|2|$ in $\overrightarrow{H}_3^3$.

For standard case tasks, things are more subtle: the largest disc may move not at all, once, twice, or three times on an optimal path for a P4-type task. Moreover, there are also some surprises for the number of optimal solutions. For instance, there are no less than six shortest paths (of length 6) from $43|21|$ to $|12|34 = 2211$ in $\overrightarrow{H}_3^4$ [147, Beispiel 3.4], depending on whether disc 2 moves before disc 4, immediately after it, or after disc 3 and whether this move is to peg 0 or 2.

We therefore give up the hope for an easy recursive algorithm which finds optimal solutions for P4 and turn to the more comfortable situation of P3 tasks.

3.2   Irregular to Perfect

After all this it may come as a surprise that with a perfect goal state j^n we are (almost) back to normal. We only need to look at the standard case; cf. [194, Lemma 7].

Lemma 3.9. *In a shortest solution for a standard type task $\sigma \to j^n$ in $\overrightarrow{H}_3^n$, $n \in \mathbb{N}_2$, with $s_n \neq j$, disc n moves precisely once.*

The proof proceeds by constructing a σ, j^n-path with only one move of n from any assumed optimal path employing two or three, as in the proof of Lemma 3.8. For the details we refer to [194, p. 316f].

Although we now know the itinerary of the largest disc from Lemmas 3.8, 3.7, and 3.9, we still have choices to make for an optimal path. For instance, in $\overrightarrow{H}_3^3$, we may move disc 3 from peg 0 to peg 2 with the smaller discs in regular or irregular order on peg 1. In special case tasks we also have to decide on the intermediate peg to which the largest disc must move to liberate the discs underneath. This leads to the idea of best buffer discs and special pegs; the following notation will be used for a fixed state $\sigma \in \mathfrak{T}^n$:

- For a given peg $i \in T$, let $D(i) = \{d \in [n] \mid s_d = i\}$, i.e. the set of discs lying on i.

- For a given disc $d \in [n]$, let $d{\uparrow}$ be the set of discs lying atop d.

We refer back to Figure 3.1 for an illustration of these concepts. We will assume throughout this section that $\{i, j, k\} = T$.

In a special case task $\sigma \to j^n$, the *special peg* for σ, denoted by $\mathrm{sp}(\sigma)$, is defined as follows. If $D(i) = \emptyset = D(k)$, set $\mathrm{sp}(\sigma) = i$. Otherwise, let n' be the largest disc in $D(i) \cup D(k)$, say $n' \in D(k)$. Let d_0 be the highest disc from $n'{\uparrow} \cup \{n\}$ that is bigger than n' and let d_1 be the first disc bigger than d_0 lying below d_0. Continuing in this manner we get the following chain of discs:

$$n' < d_0 < d_1 < \cdots < d_\nu = n\,;$$

it is possible that $\nu = 0$, that is, $n' < d_0 = n$. Then

$$\mathrm{sp}(\sigma) = \begin{cases} i, & \text{if } \nu \text{ is even}; \\ k, & \text{if } \nu \text{ is odd}. \end{cases}$$

For an example see the left picture in Figure 3.3. There $\nu = 2$ and thus peg i is special.

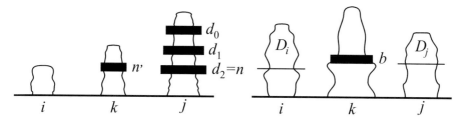

Figure 3.3: Special peg $i = \mathrm{sp}(\sigma)$ (left) and best buffer disc $b = \mathrm{bb}(k; D_i, D_j)$ (right)

A heuristic recipe in all P4 type tasks is to keep intermediate states before the moves of disc n as irregular as permitted; for a precise formulation of this practise, see [194, Lemmas 5 and 6]. Therefore, a typical sub-goal is to transfer a set of top discs $D_i \subseteq D(i)$ from peg i and a set of top discs $D_j \subseteq D(j)$ from peg j to peg k, leaving as many discs from $D(k)$ undisturbed as possible. Hence, the disc—if it exists—highest on peg k that is larger than all the discs in $D_i \cup D_j$ is called the *best buffer disc with respect to* k, D_i, D_j and denoted by $\mathrm{bb}(k; D_i, D_j)$. Buffer means that the disc hides away others below it, but that discs above it and from $D_i \cup D_j$ can be assembled on peg k without violating the divine rule. See the right-hand picture in Figure 3.3 for an example. If there is no such buffer disc, then let $\mathrm{bb}(k; D_i, D_j)$ be the bottom of peg k.

Now we are ready to present the solution (Algorithm 14) for the irregular to perfect problem as described in [243].

Algorithm 14 Algorithm irregular to perfect

Procedure $p3(n, \sigma, j)$
Parameter n: number of discs　　　$\{n \in \mathbb{N}_0\}$
Parameter σ: (ir)regular state　　　$\{\sigma \in \mathfrak{T}^n\}$
Parameter j: goal peg　　　$\{j \in T\}$
　　if $n \neq 0$ then
　　　　if disc n lies on peg j then
　　　　　　if disc n is the bottom disc **then**
　　　　　　　　recursively transfer the $n-1$ smallest discs to peg j
　　　　　　else　　　　　　　　　　　　　　　　　　　　{special case task}
　　　　　　　　$i \leftarrow \mathrm{sp}(\sigma)$
　　　　　　　　$k \leftarrow 3 - i - j$　　　　　　　　{the auxiliary peg different from i and j}
　　　　　　　　$b \leftarrow \mathrm{bb}(k; D(i), n\!\uparrow)$
　　　　　　　　recursively move all discs in $D(i) \cup n\!\uparrow$ to peg k leaving b fixed
　　　　　　　　move disc n from peg j to peg i
　　　　　　　　$b' \leftarrow \mathrm{bb}(k; \varnothing, D(j))$
　　　　　　　　recursively move all discs in $D(j)$ to peg k leaving b' fixed
　　　　　　　　move disc n from peg i to peg j
　　　　　　　　recursively transfer the $n-1$ smallest discs to peg j
　　　　　　end if
　　　　else　　　　　　　　　　　　　　　　　　{standard case task with $s_n \neq j$}
　　　　　　$i \leftarrow s_n$
　　　　　　$k \leftarrow 3 - i - j$　　　　　　　　{the auxiliary peg different from i and j}
　　　　　　$b \leftarrow \mathrm{bb}(k; n\!\uparrow, D(j))$
　　　　　　recursively move all discs in $D(j) \cup n\!\uparrow$ to peg k leaving b fixed
　　　　　　move disc n from peg i to peg j
　　　　　　recursively transfer the $n-1$ smallest discs to peg j
　　　　end if
　　end if

Algorithm 14 is illustrated in Figure 3.4. The left column schematically presents the solution for a special case task. The right column of the figure shows the steps of the algorithm in the standard case.

It is clear from this figure and the considerations in the previous section that Algorithm 14 solves a standard case P3-type task producing at most $2^n - 1$ disc moves and a special case P3-type task in at most $2^{n-2} + 2^n - 1$ moves. But Klein and Minsker claim even more [243, Theorem 3.4(iii)]:

Theorem 3.10. *Algorithm 14 returns an optimal sequence of moves for any task* $\sigma \to j^n$, $\sigma \in \mathfrak{T}^n$, $n \in \mathbb{N}_0$, $j \in T$.

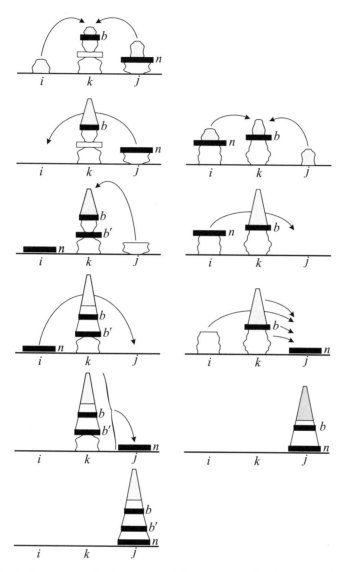

Figure 3.4: Performance of Algorithm 14 in a special (left) and a standard case (right)

To prove this, one has to show that no algorithm employing an intermediate peg different from the special peg for the first move of the largest disc in the special case or other buffer discs than the best buffers for the moves of the smaller discs can lead to a shorter solution. For details we refer to [243, p. 288–294]. In fact, with the extra observation that, similarly to the classical regular case (cf. Exercise 2.10 a)), the distances of σ to i^n and k^n are equal if and only if σ is semi-perfect on j [194, Lemma 8], one can even get uniqueness of the solution; cf. [194, Theorem 7] and [243, Corollary 3.5].

Theorem 3.11. *The task $\sigma \rightarrow j^n$, $\sigma \in \mathfrak{T}^n$, $n \in \mathbb{N}_0$, $j \in T$, has a unique optimal solution, except for the case of a semi-perfect but not perfect initial state on peg j, i.e. $s = j^n \neq \sigma$, where (obviously) two symmetric solutions just differ by exchanging the roles of the pegs different from j.*

The proof by induction on n can be found in [194, p. 317f]. The reader is invited to practise the execution of Algorithm 14 with the example of Exercise 3.4.

In Corollary 2.15 we saw that the average distance of a regular state from a fixed perfect one is $\frac{2}{3}(2^n - 1)$, and one may, of course, ask the same question for the average over all (irregular) states. Klein and Minsker [243, p. 294f] found a nice argument that this value will be of the order of 2^n as well when looked at for large n; see Exercise 3.5. Their question to find the exact value, however, seems quite ambitious indeed.

3.3   Exercises

3.1. Find an example of a P3 task in $\overrightarrow{H}_3^n$, $n \in \mathbb{N}_2$, whose solution needs $2^{n-2} + 2^n - 1$ moves.

3.2. Prove Proposition 3.6.

3.3. (Hinz [194, Example 3]) Consider the task $53||421 \rightarrow |1234|5 = 21^4$. Find out how many moves are necessary altogether, if disc 5 is to move once only, exactly twice, or three times, respectively.

3.4. Find out the length of the optimal solution for the task in $\overrightarrow{H}_3^{10}$ to get from the irregular state in Figure 3.1 to the perfect state on peg 2.

3.5. Show that the average length $\overline{d}_n$ of shortest solutions from $\sigma \in \mathfrak{T}^n$, $n \in \mathbb{N}_2$, to j^n, $j \in T$, in $\overrightarrow{H}_3^n$ satisfies

$$\frac{1}{24}2^n < \overline{d}_n < \frac{5}{4}2^n .$$

Chapter 4

Sierpiński Graphs

On several occasions in Chapter 2 we realized that a different labeling for the recursively obtained Hanoi graphs H_3^n (see Figure 2.12) would be desirable. We will realize this with the same recursive procedure, yielding graphs isomorphic to H_3^n, with the same vertex set, but with different edge sets.

We will denote the new graphs by S_3^n, and call them *Sierpiński graphs*. The reasons for such a choice of the name will be explained later, in Section 4.3. Moreover, we will show in the present chapter that the definition of graphs S_3^n can easily be generalized to a two-parametric family S_p^n, with *base* $p \in \mathbb{N}$ and *exponent* $n \in \mathbb{N}_0$, members of which will also be called Sierpiński graphs. The proofs of most of the properties of graphs S_p^n will not be much more difficult than the proofs in the special case $p = 3$. Later, in Chapter 5, we will introduce another two-parametric family of graphs (the Hanoi graphs H_p^n), but (alas!) it will turn out that to establish their properties for general p is much more difficult than to obtain them for $p = 3$.

4.1 Sierpiński Graphs S_3^n

We start, for $n = 0$, from the same one-vertex graph as in the recursive description of the graphs H_3^n, and then proceed as indicated in Figure 4.1.

Note that now all three subgraphs iS_3^n of S_3^{1+n} are obtained just by translation from one copy of the graph S_3^n and we have no need for any special positioning (compare the situation in Figure 2.13). Figure 4.2 shows the graphs emerging from the first three recursive steps of this construction (cf. Figure 2.11).

We observe for these four graphs, and then easily give an inductive proof for arbitrary n, that the vertex sets are just the same as for the corresponding Hanoi graphs, i.e. $V(S_3^n) = T^n = V(H_3^n)$. An $s \in T^n$ is again written as $s = s_n \ldots s_1$. The

© Springer International Publishing AG, part of Springer Nature 2018
A.M. Hinz et al., *The Tower of Hanoi – Myths and Maths*,
https://doi.org/10.1007/978-3-319-73779-9_5

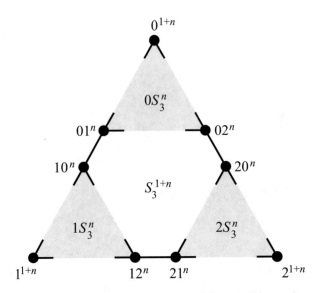

Figure 4.1: Recursive structure of Sierpiński graphs

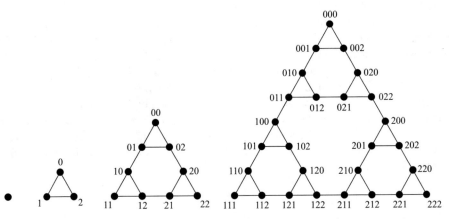

Figure 4.2: Sierpiński graphs S_3^0, S_3^1, S_3^2, and S_3^3

edge sets of the Sierpiński graphs S_3^n are given by

$$E(S_3^0) = \varnothing,$$
$$\forall n \in \mathbb{N}_0 : E(S_3^{1+n}) = \{\{ir, is\} \mid i \in T, \{r, s\} \in E(S_3^n)\}$$
$$\cup \{\{ij^n, ji^n\} \mid i, j \in T, i \neq j\}. \tag{4.1}$$

The construction from Figure 4.2 can actually be carried on as for the Sisyphean Hanoi graph to yield the *Sisyphean Sierpiński graph* S_3^∞ of Figure 4.3, where (infinitely many) leading 0s have been omitted from the labels.

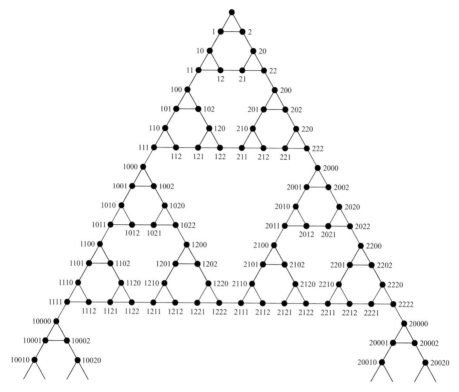

Figure 4.3: The Sisyphean Sierpiński graph S_3^∞

An isomorphism between S_3^∞ and AT mod 2 can now be defined much simpler than in the Hanoi case by

$$s \mapsto \left(\sum_{d=1}^{n} (s_d \neq 0) \cdot 2^{d-1}, \sum_{d=1}^{n} (s_d = 2) \cdot 2^{d-1} \right),$$

where again the vertex of AT mod 2 representing combinatorial number $\binom{k}{\ell}$ is given by the ordered pair (k, ℓ). The proof that this is in fact an isomorphism will be the same as before (cf. p. 154), but one needs a new distance formula corresponding to (2.8) which will be the base for a couple of simplifications. These can in turn be applied to Hanoi graphs, because restricting the isomorphisms to the first 2^n rows shows that H_3^n and S_3^n are isomorphic. A more straightforward proof of this fact can be obtained by comparing (4.1) with (2.12). The fact that we have the same recursive structure for both H_3^n and S_3^n enables us to construct an isomorphism directly.

It is obvious that any isomorphism from H_3^n onto S_3^n must map the set $\{0^n, 1^n, 2^n\}$ onto itself, since these vertices have degree two in both graphs, while all other vertices are of degree three in both graphs. Because of their special role and because of their special position in the graph S_3^n, the vertices i^n, $i \in T$, are called *extreme vertices* (of the graph S_3^n).

Note that there is precisely one isomorphism Φ from H_3^n to S_3^n mapping i^n to i^n for each $i \in T$. It can be given by

$$\forall s \in T^n \; \forall d \in [n]: \; \Phi(s)_d = s_n \vartriangle \cdots \vartriangle s_d, \qquad (4.2)$$

with the binary operation $\vartriangle$ on T as introduced in Section 2.2. We remind the reader that $\vartriangle$ is *not* associative, such that the evaluation of a chain as in (4.2) is to be understood as performed strictly from the right. That this is the desired isomorphism can easily be shown by induction on n.

The construction of Φ from (4.2) can already be found in [378, p. 98], where it was used to establish the existence of a drawing of H_3^n with straight lines of equal length, i.e. our standard representation. It is, however, not very efficient, because the components of $\Phi(s)$ have to be calculated, starting from $\Phi(s)_n = s_n$, without recourse to those already obtained; for instance, $\Phi(s)_{n-2} = s_n \vartriangle (s_{n-1} \vartriangle s_{n-2})$, which is, in general, not equal to $\Phi(s)_{n-1} \vartriangle s_{n-2} = (s_n \vartriangle s_{n-1}) \vartriangle s_{n-2}$. The task is facilitated by the following observation. For $i \in T$ let $\varphi_i \in \mathrm{Sym}(T)$ be the permutation which has the single fixed point i. Then $i \vartriangle j = \varphi_i(j)$ for each $j \in T$, and (4.2) reads

$$\forall s \in T^n \; \forall d \in [n]: \; \Phi(s)_d = \varphi_{s_n} \circ \cdots \circ \varphi_{s_{d+1}}(s_d). \qquad (4.3)$$

The composition $\circ$ on $\mathrm{Sym}(T)$ *is* associative, such that we can construct an algorithm for Φ by defining $\Phi_d = \varphi_{s_n} \circ \cdots \circ \varphi_{s_{d+1}} \in \mathrm{Sym}(T)$, whence in particular $\Phi_n = \mathrm{id}$. Given Φ_d, we write down $\Phi_d(s_d) = \Phi(s)_d$ and change, as long as $d > 1$, the permutation to $\Phi_{d-1} = \Phi_d \circ \varphi_{s_d}$. The algorithm can be realized with the automaton shown in Figure 4.4. For instance, if the current Φ_d is the upper left state of the automaton and we enter $s_d = 2$, then we obtain $\Phi_d(2) = 1$ for $\Phi(s)_d$ and move, if $d > 1$, to the lower right state $\Phi_{d-1} = \Phi_d \circ \phi_2$ in order to continue with the input of s_{d-1}. This procedure was for the first time explicitly described by Romik (who used the results of Hinz and Schief [214]) in [353, Theorem 2]. Romik conceives his H_3-to-S_3-automaton as a finite-state machine "translating" from Hanoi states to Sierpiński labelings. Recognizing H_3^n and S_3^n as different graphs, we prefer to use the language of isomorphisms of graphs.

The graphs H_3^n and S_3^n being isomorphic, uniqueness of the isomorphism Φ with the property $\forall i \in T: \; \Phi(i^n) = i^n$ follows from Lemma 2.8. Since Φ also transforms ij^{n-1} into ik^{n-1}, $\{i, j, k\} = T$, and vice versa, we have $\Phi^{-1} = \Phi$, and the H_3-to-S_3-automaton can be used as an S_3-to-H_3-automaton as well.

With the isomorphism Φ at hand and recalling the definition of $(s \vartriangle j)_d$ from p. 116, we have:

$$\forall j \in T \; \forall s \in T^n \; \forall d \in [n]: \; s_d = (s \vartriangle j)_d \Leftrightarrow \Phi(s)_d = j, \qquad (4.4)$$

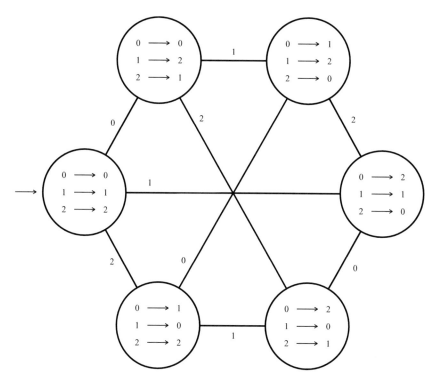

Figure 4.4: Automaton realizing the isomorphism between H_3^n and S_3^n

because $(s \vartriangle j)_d = \varphi_{s_{d+1}} \circ \cdots \circ \varphi_{s_n}(j)$ and $\Phi(s)_d = \varphi_{s_n} \circ \cdots \circ \varphi_{s_{d+1}}(s_d)$ by (4.3) and noting that $\varphi_k^{-1} = \varphi_k$. By virtue of (2.8), we get for $s \in T^n$:

$$d_S(\Phi(s), j^n) = d_H(s, j^n) = \sum_{d=1}^{n} (s_d \neq (s \vartriangle j)_d) \cdot 2^{d-1} = \sum_{d=1}^{n} (\Phi(s)_d \neq j) \cdot 2^{d-1},$$

where we write d_S and d_H for the distance in S_3^n and H_3^n, respectively. Hence,

$$\forall \sigma \in T^n \; \forall j \in T : \; d(\sigma, j^n) = \sum_{d=1}^{n} (\sigma_d \neq j) \cdot 2^{d-1}, \tag{4.5}$$

where d stands for d_S again.

An easy induction shows that the edge set of S_3^n can be written in the following way (cf. (2.11)):

$$E(S_3^n) = \left\{ \{\underline{s}ij^{d-1}, \underline{s}ji^{d-1}\} \mid i, j \in T, \; i \neq j, \; d \in [n], \; \underline{s} \in T^{n-d} \right\}. \tag{4.6}$$

Formula (4.6) gives rise to the interpretation of S_3^n as the state graph of a variant of the TH puzzle, namely the *Switching Tower of Hanoi* as described in the more general setting of graphs S_p^n in [235]. Vertices of S_3^n stand for legal distributions of n discs as before in H_3^n and the edge $\{\underline{s}ij^{d-1}, \underline{s}ji^{d-1}\}$ represents the switch of the single disc d with the (possibly empty) subtower consisting of all smaller discs between pegs i and j for any distribution $\underline{s}$ of discs larger than d.

The *P2 decision problem* for S_3^{1+n}, $n \in \mathbb{N}$, is to find out how many times the largest disc $n + 1$ moves in an optimal solution for the P2-type task *is* $\to$ *jt*, $s, t \in T^n$, $i, j \in T$, where we may assume that $i \neq j$, because otherwise the largest disc will not move at all and the task is reduced to one on a Sierpiński graph of lower exponent. We proceed by successively entering pairs (s_δ, t_δ), starting with $\delta = n$, into Romik's original automaton in Figure 2.28. Contrary to the procedure in Section 2.4.3, we do not have to change labels on the arcs of the automaton, as can be seen by looking at (4.4) or, in other words, comparing (2.8) with the distance formula (4.5) for Sierpiński graphs.

This leads to the simplified Algorithm 15.

Algorithm 15 P2 decision algorithm for S_3^{1+n}

Procedure p2S(n, s, t)
Parameter n: number of discs minus 1 $\quad \{n \in \mathbb{N}\}$
Parameter s: initial configuration $\quad \{s \in T^{1+n}\}$
Parameter t: goal configuration $\quad \{t \in T^{1+n}, t_{n+1} \neq s_{n+1}\}$
$\quad i \leftarrow s_{n+1}$, $j \leftarrow t_{n+1}$
$\quad$ start in state A of P2-automaton
$\quad \delta \leftarrow n$
$\quad$ **while** $\delta > 0$
$\quad\quad$ apply automaton to pair (s_δ, t_δ)
$\quad\quad\quad\quad$ {algorithm STOPs if automaton reaches terminating state D or E}
$\quad\quad \delta \leftarrow \delta - 1$
$\quad$ **end while**

In order to analyze the average running time in which the algorithm decides which of the alternatives is optimal in a P2 task, it is of utmost advantage to employ the theory of Markov chains. The reader who was previously not exposed to this concept may simply skip the next lines and move directly to Theorem 4.1, or look into some standard source for the basic theory of Markov chains, as for instance [400]. For a historical essay, see [184, Chapter 9].

Let the five states A, B, C, D, and E of Romik's automaton (Figure 2.28) be numbered 1, 2, 3, 4, and 5, respectively. Then we can consider the automaton as a Markov chain with states $1, 2, 3, 4, 5$, in which the process starts in state 1 and moves from one state to another with a certain probability. These probabilities are given by the *transition matrix* of the automaton

$$P = \begin{bmatrix} 2/9 & 1/9 & 0 & 6/9 & 0 \\ 2/9 & 3/9 & 2/9 & 1/9 & 1/9 \\ 0 & 1/9 & 2/9 & 0 & 6/9 \\ 0 & 0 & 0 & 1 & 0 \\ 0 & 0 & 0 & 0 & 1 \end{bmatrix},$$

that is, P_{uv} is the probability of a move from state u to state v. Note that $P_{44} = 1 = P_{55}$. Such states are called *absorbing*, and a Markov chain with at least one absorbing state is called an *absorbing Markov chain*. Note also that the transition matrix P is of the form

$$P = \begin{bmatrix} Q & R \\ 0 & I \end{bmatrix}.$$

We now collect facts about Markov chains needed here. The entry $(P^n)_{uv}$ is the probability that the Markov chain, starting in state u, will be in state v after n steps. In an absorbing Markov chain, $Q^n \to 0$ when $n \to \infty$ and $I - Q$ has an inverse, called the *fundamental matrix* (of the absorbing Markov chain),

$$N = (I - Q)^{-1} = \sum_{n=0}^{\infty} Q^n.$$

Now, N_{uv} is the expected number of times the chain is in state v provided that it has started in state u. In our case,

$$(I - Q)^{-1} = \begin{bmatrix} 7/9 & -1/9 & 0 \\ -2/9 & 2/3 & -2/9 \\ 0 & -1/9 & 7/9 \end{bmatrix}^{-1} = \begin{bmatrix} 180/133 & 9/38 & 9/133 \\ 9/19 & 63/38 & 9/19 \\ 9/133 & 9/38 & 180/133 \end{bmatrix}.$$

Since we start in state 1, the sum

$$\frac{180}{133} + \frac{9}{38} + \frac{9}{133} = \frac{63}{38}$$

is the expected number of times we will be in one of the states 1, 2, or 3. In conclusion, we have arrived at the following remarkable result due to Romik [353, Theorem 3]:

Theorem 4.1. *The average number of disc pairs checked by Romik's automaton is bounded above by and converges, as $n \to \infty$, to $\frac{63}{38}$.*

In other words, the decision problem whether the largest disc moves once or twice on an optimal path of a P2 task with the largest disc initially not in its goal position can be solved with the aid of Algorithm 15 by checking $\frac{101}{38}$, i.e. less than three pairs of discs on the average, namely the pair of largest discs to decide on which of the six versions of the automaton to employ and the expected $\frac{63}{38}$ pairs of smaller discs processed in Romik's automaton. Note that in any case at least two pairs of input data have to be processed by the algorithm.

Or do they? Well, looking at the automaton in Figure 2.28 again, we notice that the input of j in state A for the first component will lead to the terminating state D no matter what the second component is. Therefore, in that case, we only need half a pair of input, such that in A we just need to check $\frac{1}{3} \cdot \frac{1}{2} + \frac{2}{3} \cdot 1 = \frac{5}{6}$ pairs of tits or, in other words, save 3 out of 18 input data. The same applies in state C if the first input is k. So if we build in to the automaton these individual checks, we will need asymptotically, according to the above probabilistic analysis, only

$$\frac{5}{6} \frac{180}{133} + \frac{9}{38} + \frac{5}{6} \frac{9}{133} = \frac{27}{19}$$

pairs of input for the automaton, i.e. $\frac{46}{19}$ pairs in the algorithm which includes the very first one. All tasks of the form $ij * \ldots * \to j * * \ldots *$ need only $1\frac{1}{2}$ pairs of input. The graph being undirected, the same would, of course, apply to tasks $i ** \ldots * \to ji * \ldots *$, but we cannot use the trick twice because we have to decide to which of the two components of the input pairs we want to apply the extended automaton first.

In comparing Algorithms 13 and 15, one must not forget, however, that the latter is operating on S_3^{1+n}. So in order to solve the original P2 decision problem for Hanoi tasks with the latter algorithm, one has to make use of the isomorphism Φ described earlier, i.e. Romik's H_3-to-S_3-automaton. However, it is not necessary to convert both s and t completely (this would undo the advantage of checking fewer pairs of input), but only those pairs (s_d, t_d) which are taken in by the P2 decision automaton. Therefore one needs two identical copies of the H_3-to-S_3-automaton and replace, in every step of Algorithm 15, s_d by $\rho_s(s_d)$ and t_d by $\rho_t(t_d)$, respectively, where ρ_s and ρ_t mean one step into the corresponding H_3-to-S_3-automata. Examples for this approach can be found in Exercise 4.1.

Romik's automaton also leads to an alternative proof of the statement analogous to Proposition 2.40, namely

Proposition 4.2. *For every is $\in T^{1+n} \setminus \{0^{1+n}, 1^{1+n}, 2^{1+n}\}$, $n \in \mathbb{N}$, there is a $jt \in T^{1+n}$ such that there are two shortest paths between these vertices in S_3^{1+n}.*

Proof. Let $\{i, j, k\} = T$. We observe that in Romik's P2 decision automaton of Figure 2.28, for every $\ell \in T$ the input $(\ell, k \vartriangle (i \vartriangle \ell))$ in state B of the automaton will keep this state unchanged. Now every non-extreme vertex $is \in T^{1+n}$ is of the form $i^{1+n-d}k\bar{s}$ with $\bar{s} \in T^{d-1}$, $d \in [n]$. Define $t = k^{n-d+1}\bar{t}$ with $\forall \delta \in [d-1]: t_\delta = k \vartriangle (i \vartriangle s_\delta)$. Then, for the pair (is, jt), the first input (i, j) chooses the automaton as in the figure and the subsequent $n - d$ pairs (i, k) will keep it in state A. The next input pair is (k, k), such that we move to state B of the automaton. But then, according to the remark above and our definition of $\bar{t}$, the automaton will not leave B anymore.   □

The P2 decision automaton also makes it easy to see that, on the other hand, there *are* non-perfect TH states where one move of the largest disc is always sufficient to obtain an optimal path to another state; see Exercise 4.2.

Proposition 4.2 is Corollary 3.6 of [205]. In that paper, the considerations of Section 2.4.1 have been made for Sierpiński graphs, facilitated by the simpler formula for the distance from an extreme vertex. The differences are minor, however; just replace (2.29) with

$$\mathrm{d}_1(is, jt) \,\square\, \mathrm{d}_2(is, jt) \Leftrightarrow \mathrm{d}(s; j, k) + \mathrm{d}(t; i, k) \,\square\, 2^n \,,$$

where again $\square \in \{<, =, >\}$. The bijection from T^n to $\widetilde{T}^n$, $s \mapsto \widetilde{s}$ for (2.22) is now trivialized to $\widetilde{s}_d = (s_d \neq 0) - (s_d \neq 2)$ for $d \in [n]$ with $s_d = \widetilde{s}_d + 1$. By isomorphy, the definition of the functions z_n in (2.21) does not depend on whether one uses Hanoi or Sierpiński distance.

Let us mention that in [205, Theorem 3.2] a condition on the vertices u in the formula for $X(v)$ is missing; cf. the corresponding Theorem 2.42 for Hanoi graphs. Fortunately, this omission has no consequences for the rest of the paper.

4.2 Sierpiński Graphs S_p^n

Recalling that $T = [3]_0$ we can easily generalize the definition of Sierpiński graphs S_3^n by replacing the set T in (4.1) by $[p]_0$, where p is an arbitrary positive integer. The graphs obtained that way are denoted by S_p^n and also called *Sierpiński graphs*. That means that we define $V(S_p^n) = [p]_0^n$, $E(S_p^0) = \varnothing$, and then

$$\forall\, n \in \mathbb{N}_0 : E(S_p^{1+n}) = \big\{\{ir, is\}|\, i \in [p]_0,\, \{r, s\} \in E(S_p^n)\big\}$$
$$\cup \big\{\{ij^n, ji^n\}|\, i, j \in [p]_0,\, i \neq j\big\}\,. \tag{4.7}$$

Figure 4.5 shows the first four cases for $p = 4$.

It is easy to prove (Exercise 4.3) that the vertices $s = s_n \ldots s_1,\, s' = s'_n \ldots s'_1 \in [p]_0^n$ are adjacent in S_p^n if and only if there exists a $d \in [n]$ such that

$$(i)\ \forall\, k \in [n] \smallsetminus [d] : s_k = s'_k,$$
$$(ii)\ s_d \neq s'_d,$$
$$(iii)\ \forall\, k \in [d-1] : s_k = s'_d \wedge s'_k = s_d\,. \tag{4.8}$$

Note that if $d = 1$, then condition (iii) is void; so is (i) if $d = n$.

This was the original definition of Sierpiński graphs by Klavžar and Milutinović [235] and as in the case $p = 3$ it lends itself to the interpretation of S_p^n as the (general) Switching TH.

In a compact form, the edge sets can be described as (cf. (4.6))

$$E(S_p^n) = \big\{\{\underline{s}ij^{d-1}, \underline{s}ji^{d-1}\} \mid i, j \in [p]_0,\, i \neq j,\, d \in [n],\, \underline{s} \in [p]_0^{n-d}\big\}\,.$$

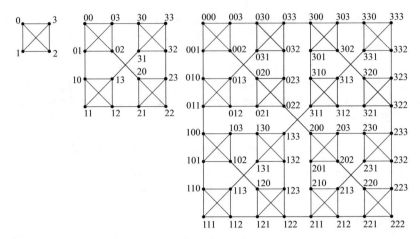

Figure 4.5: Sierpiński graphs S_4^0 to S_4^3

An immediate consequence is the size of S_p^n:

$$\|S_p^n\| = \binom{p}{2} \sum_{d=1}^{n} p^{n-d} = \frac{p}{2}\left(p^n - 1\right). \tag{4.9}$$

In particular, $\|S_p^1\| = \binom{p}{2}$; in other words, S_p^1 is isomorphic to the complete graph on p vertices.

For a fixed $\underline{s} \in [p]_0^{n-d}$, the subgraph of S_p^n induced by the set of vertices $\{\underline{s}\bar{s} \mid \bar{s} \in [p]_0^d\}$ is denoted by $\underline{s}S_p^d$ in analogy with the notation used in the recursive definition of Sierpiński graphs S_3^n as illustrated in Figure 4.1. The function mapping any $\underline{s}\bar{s}$ to $\bar{s}$ is an isomorphism between $\underline{s}S_p^d$ and S_p^d.

For $n \in \mathbb{N}$, each S_p^n contains p^{n-1} isomorphic copies of S_p^1, namely the subgraphs $\underline{s}S_p^1$ with $\underline{s} \in [p]_0^{n-1}$. They constitute p-*cliques*, i.e. cliques of order p in S_p^n. For $p \le 2$ these are the only maximal cliques (with respect to inclusion), such that the clique number is $\omega(S_p^n) = p$ in these cases. In fact, the formula carries over to general p.

Theorem 4.3. *Let* $p \in \mathbb{N}_3$, $n \in \mathbb{N}$. *The only maximal cliques in* S_p^n *are the* p-*cliques* $\underline{s}S_p^1$ *with* $\underline{s} \in [p]_0^{n-1}$ *and* 2-*cliques induced by edges not included in any of these. In particular,* $\omega(S_p^n) = p$.

Proof. Induction on n. The graph S_p^1 consists of just one p-clique. The edge $\{ij^n, ji^n\}$ induces a maximal 2-clique in S_p^{1+n}, because a larger subgraph including it would also contain, without loss of generality, a vertex $i\bar{s}$ with $\bar{s} \ne j^n$, but which can not be adjacent to ji^n. Each other maximal clique in S_p^{1+n} lies entirely in some subgraph iS_p^n and is therefore, by induction assumption, either of the form $i\underline{s}S_p^1$ with $\underline{s} \in [p]_0^{n-1}$ or it is induced by $\{iv, iw\}$, where $\{v, w\} \in E(S_p^n)$ is not contained in any p-clique of S_p^n and therefore neither is $\{v, w\}$ in any of the subgraphs $\underline{s}S_p^1$ of S_p^n. □

Remark 4.4. *For $p \in \mathbb{N}_3$, the p-cliques in S_p^n are induced by moves of disc $d = 1$ alone in the Switching TH, whereas the moves corresponding to maximal 2-cliques involve more than just the smallest disc. The number of these 2-cliques in S_p^n is therefore, by (4.9),*

$$\frac{p}{2}(p^n - 1) - p^{n-1}\frac{p(p-1)}{2} = \frac{p}{2}(p^{n-1} - 1).$$

4.2.1 Distance Properties

An easy induction argument based on the definition in (4.7) shows that S_p^n is connected; the canonical distance function will again be denoted by d. The following most fundamental result is a generalization of (4.5) to all Sierpiński graphs S_p^n and was given for the first time in [235, Lemma 4].

Theorem 4.5. *For any $j \in [p]_0$ and any vertex $s = s_n \ldots s_1$ of S_p^n,*

$$d(s, j^n) = \sum_{d=1}^{n}(s_d \neq j) \cdot 2^{d-1}, \qquad (4.10)$$

and there is exactly one shortest path between s and j^n.
In particular, for $i \in [p]_0 \smallsetminus \{j\}$,

$$d(i^n, j^n) = 2^n - 1.$$

Proof. By induction on n. The statement is trivial for $n = 0$. Let $n \in \mathbb{N}_0$ and $s = s_{n+1}\bar{s}$, $\bar{s} \in [p]_0^n$.

If $s_{n+1} = j$, then one can use the shortest path in S_p^n from $\bar{s}$ to j^n and add a j in front of each vertex. Hence

$$d(s, j^{1+n}) \leq \sum_{d=1}^{n}(s_d \neq j) \cdot 2^{d-1} = \sum_{d=1}^{n+1}(s_d \neq j) \cdot 2^{d-1} \ \left(< 2^{n+1}\right).$$

If $s_{n+1} \neq j$, we can compose a path from s to j^{1+n} by going from $s_{n+1}\bar{s}$ to $s_{n+1}j^n$ on a (shortest) path of length $\leq \sum_{d=1}^{n}(s_d \neq j) \cdot 2^{d-1}$, then moving to js_{n+1}^n on one extra edge and finally from here to j^{1+n} in another $2^n - 1$ steps, altogether

$$d(s, j^{1+n}) \leq \sum_{d=1}^{n+1}(s_d \neq j) \cdot 2^{d-1} \ \left(< 2^{n+1}\right).$$

To show that these are the unique shortest paths, respectively, we note that no optimal path from s to j^{1+n} can touch a subgraph kS_p^n for $s_{n+1} \neq k \neq j$. Consider any such path. Then it must contain (cf. Figure 4.6, where the corresponding edges are in red):

- the edge $\{ik^n, ki^n\}$ for some $i \neq k$, to enter kS_p^n;

- a path from ki^n to $k\ell^n$, $\ell \neq k$, inside kS_p^n;

- the edge $\{k\ell^n, \ell k^n\}$ to leave kS_p^n, so that $\ell \neq i$, because we are on a path;

- a path from some jh^n, $h \neq j$, to j^{1+n} to finish the path inside jS_p^n.

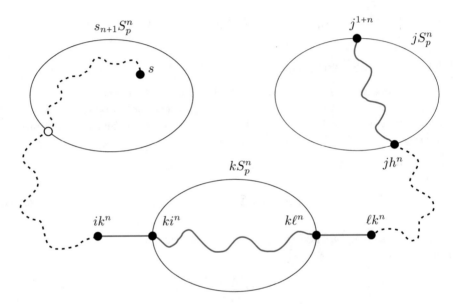

Figure 4.6: An s, j^{1+n}-path passing through subgraph kS_p^n

By induction assumption this would comprise at least $1+(2^n-1)+1+(2^n-1) = 2^{n+1}$ edges, such that the path cannot be optimal because we already found a strictly shorter one.   □

Theorem 4.5 has some interesting immediate consequences. For instance, if $j \in [p]_0$ does not appear as a **pit** in $s \in [p]_0^n$, then $d(s, j^n) = 2^n - 1$. Otherwise, the values $d(s, j^n) < 2^n - 1$ are strictly inceasing in the order in which the js appear (from left to right) in s; in particular, $d(s, s_n^n)$ is the unique minimum. Thus it is very easy to decide which extreme vertex different from 0^n is closest to an abandoned state $0\overline{s}$, namely j^n with the leftmost non-zero pit j in $\overline{s} \in [p]_0^{n-1}$; cf. Proposition 2.22 for the analogue question for Hanoi graphs.

Another consequence of Theorem 4.5 is the base for finding the best first move in optimal paths in Sierpiński graphs S_p^n (cf. [204, Lemma 1.2]).

Corollary 4.6. *Let $s \in [p]_0^n$ ($n \in \mathbb{N}$), $j \in [p]_0$, $s \neq j^n$.*
Then the optimal first move from s to j^n in S_p^n is from $\underline{s}s_\delta j^{\delta-1}$ to $\underline{s}js_\delta^{\delta-1}$, where $\delta \in [n]$ and $s_\delta \neq j$ (i.e. $\delta = \min\{d \in [n] \mid s_d \neq j\}$).

Proof. According to Theorem 4.5, the optimal path from s to j^n is unique and consequently so is its first edge which has to be of the form $\{s,t\}$ with $s = \underline{s}ik^{d-1}$ and $t = \underline{s}ki^{d-1}$ where $\{i,k\} \in \binom{[p]_0}{2}$, $d \in [n]$ and $\underline{s} \in [p]_0^{n-d}$. Moreover, $d(s,j^n) = d(t,j^n) + 1$, such that from (4.10) we obtain

$$(i \neq j) \cdot 2^{d-1} + (k \neq j)(2^{d-1} - 1) = (k \neq j) \cdot 2^{d-1} + (i \neq j)(2^{d-1} - 1) + 1,$$

which is equivalent to $(i \neq j) - (k \neq j) = 1$. This, in turn, can only be true if and only if $i \neq j = k$. So $i = s_\delta$ and $d = \delta$. $\qquad\square$

In the Switching TH interpretation this means that the best first move is to switch the smallest disc not on the goal peg with the tower of smaller ones on that peg.

The case $p = 1$ of Corollary 4.6 is, of course, void. For $p = 2$, the best first move to 1^n (0^n) corresponds to binary addition (subtraction) of 1 to (from) s. In particular, for $s = 0^n$, this describes the path graph S_2^n as shown for $n = 3$ in the lower part of Figure 1.4: obviously, S_2^n and the graph of the Chinese rings R^n are isomorphic, the isomorphism being given by the Gray/Gros code. As a consequence, the optimal path from 0^n to 1^n in *any* graph S_p^n, $p \in \mathbb{N}_2$, can be obtained by successive binary addition of 1 (cf. [235, p. 103]) since all states on it are "binary" because only pegs 0 and 1 are involved. More generally, a peg different from j and empty in s is not used during the execution of the optimal $s \to j^n$ solution.

Yet another application of Corollary 4.6 is the complete description of the shortest s,j^n-path until the goal is reached: if disc 1 is not on peg j, then move it there; otherwise make the only legal move involving also another disc. So moves of disc 1 alone are alternating with switching a larger disc with a non-trivial subtower, and the single discs are always moving to the goal peg j.

With the aid of formula (4.10) metric properties of CR and Hanoi graphs, as, e.g., Propositions 2.13 and 2.16, carry over to all Sierpiński graphs as noticed by Parisse in [332].

Corollary 4.7. *Let $p \in \mathbb{N}$ and $n \in \mathbb{N}_0$.*
For any $s \in [p]_0^n$,

$$\sum_{i=0}^{p-1} d(s,i^n) = (p-1)(2^n - 1).\tag{4.11}$$

Let j^n be a fixed extreme vertex. Then for $\mu \in [2^n]_0$:

$$|\{s \in [p]_0^n \mid d(s,j^n) = \mu\}| = (p-1)^{q(\mu)};\tag{4.12}$$

consequently,

$$\sum_{\mu=0}^{2^n-1}(p-1)^{q(\mu)} = p^n.$$

Proof. From (4.10) we get

$$
\sum_{i=0}^{p-1} \mathrm{d}(s, i^n) = \sum_{i=0}^{p-1} \sum_{d=1}^{n} (s_d \neq i) \cdot 2^{d-1}
$$
$$
= \sum_{d=1}^{n} \sum_{i=0}^{p-1} (s_d \neq i) \cdot 2^{d-1}
$$
$$
= \sum_{d=1}^{n} (p-1) \cdot 2^{d-1} = (p-1)(2^n - 1).
$$

For every non-zero bit of μ there are $p-1$ ways that the corresponding s_d in (4.10) is different from j, for a zero bit we necessarily have $s_d = j$. This proves (4.12).   □

Evidently, all results on eccentricities carry over from R^n and H_3^n to S_2^n and S_3^n, respectively. The methods developed for them can also be employed for general S_p^n. In particular, the eccentricity of an $s \in [p]_0^n$ is the maximum of its distances to extreme vertices j^n; see [332, Lemma 2.3]. Identity (4.10) tells us the following: if there is an empty peg i in state $s \in [p]_0^n$, i.e. $s([n]) \subseteq [p]_0 \smallsetminus \{i\}$, then $\varepsilon(s) = \mathrm{d}(s, i^n) = 2^n - 1$; otherwise we get (cf. [212, Equation (3.0)])

$$
\varepsilon(s) = 2^n - 2^m + \mathrm{d}(\overline{s}, s_m^{m-1}),
$$

where $\overline{s} := s_{m-1} \ldots s_1$ and $m \in [n-p+1]$ is the smallest disc which lies on the bottom of any of the pegs. This leads to (cf. [212, Corollary 3.5]; in that paper one can even find the standard deviation):

$$
\overline{\varepsilon}(S_p^n) = \left(1 - \binom{2p}{p-1}^{-1}\right) 2^n - \frac{p-1}{p} - \sum_{k=0}^{p-2} (-1)^{p-k} \frac{p-1-k}{2p-k} \binom{p}{k} \left(\frac{k}{p}\right)^n . \qquad (4.13)
$$

The respective formulas (1.4) and (2.28) are, of course, special cases of this. Note that for fixed p the asymptotic behavior of the average eccentricity is governed by the first term on the right-hand side of identity (4.13).

The task to find a shortest path between any two vertices of S_p^n can be reduced to putting together appropriate shortest paths from some vertices to extreme vertices (cf. [235, Theorem 5]). Let us first state an analogue to the boxer rule Lemma 2.32.

Lemma 4.8. *If on a geodesic of S_p^{1+n}, $n \in \mathbb{N}_0$, the largest disc is moved away from a peg in the Switching TH, it will not return to the same peg.*

Proof. Assume that disc $n+1$ leaves iS_p^n and eventually returns there on a geodesic. Then this geodesic contains a path $P = ij^n P' ik^n$, where P' is a ji^n, ki^n-path with $|\{i, j, k\}| = 3$, because two subgraphs iS_p^n and jS_p^n are linked by one edge only. (This already excludes $p = 2$; for $p = 1$ the statement is void anyway.) In turn, P'

must contain a $ji^n, j\ell^n$-path, $i \neq \ell \neq j$, such that $\|P'\| \geq 2^n - 1$ by Theorem 4.5. Hence $\|P\| > 2^n$, but the same theorem yields $\mathrm{d}(ij^n, ik^n) < 2^n$, such that P can not be part of a geodesic. $\qquad\square$

Theorem 4.9. *Let* $s = \underline{s}i\overline{s}, t = \underline{s}j\overline{t} \in [p]_0^{1+n}$ *with* $i \neq j$ *and* $\overline{s}, \overline{t} \in [p]_0^{d-1}$, $d \in [n+1]$. *Then*

$$\mathrm{d}(s,t) = \min\left\{\mathrm{d}(\overline{s}, j^{d-1}) + 1 + \mathrm{d}(\overline{t}, i^{d-1}),\right.$$
$$\left.\mathrm{d}(\overline{s}, k^{d-1}) + 1 + 2^{d-1} + \mathrm{d}(\overline{t}, k^{d-1}) \mid k \in [p]_0 \smallsetminus \{i,j\}\right\}.$$

Proof. By virtue of Lemma 4.8 we may assume that $d = n + 1$.

Any s, t-path in which disc $n + 1$ moves only once must lead from s to ij^n, then to ji^n and finally to t. By Theorem 4.5 and Lemma 4.8, there is a unique shortest one among these paths; its length is $\mathrm{d}(\overline{s}, j^n) + 1 + \mathrm{d}(\overline{t}, i^n) \leq 2^{n+1} - 1$.

Any s, t-path in which disc $n + 1$ is moving exactly twice must pass the vertices ik^n, ki^n, kj^n, and jk^n with $i \neq k \neq j$, such that, again by Theorem 4.5 and Lemma 4.8, the shortest such path is unique and has length $\mathrm{d}(\overline{s}, k^n) + 1 + 2^n + \mathrm{d}(\overline{t}, k^n)$.

A path including more than two moves of disc $n + 1$ has to contain, by Lemma 4.8, the passage through two subgraphs kS_p^n, ℓS_p^n, $|\{i,j,k,\ell\}| = 4$, costing at least $2^{n+1} + 1$ moves, such that it can not be minimal. $\qquad\square$

An immediate consequence is the diameter of Sierpiński graphs.

Corollary 4.10. *For any* $p \in \mathbb{N}_2$ *and any* $n \in \mathbb{N}_0$, $\mathrm{diam}(S_p^n) = 2^n - 1$.

From the computational point of view, Theorem 4.9 can be used to calculate $\mathrm{d}(s,t)$ in $O(pn)$ time. This can be improved by noting (again using Theorem 4.5) that $\mathrm{d}(i\overline{s}, j\overline{t})$, where without loss of generality $i \neq j$, is obtained from Theorem 4.9 as the minimum of

$$\mathrm{d}(\overline{s}, j^{n-1}) + 1 + \mathrm{d}(\overline{t}, i^{n-1})$$

and

$$\mathrm{d}(\overline{s}, k^{n-1}) + 1 + 2^{n-1} + \mathrm{d}(\overline{t}, k^{n-1}), \; k \in \{s_{n-1}, t_{n-1}\} \smallsetminus \{i,j\}.$$

Therefore:

Theorem 4.11. *The distance between any two vertices of* S_p^n *can be computed in* $O(n)$ *time.*

With some further effort (see [235, Theorem 6]; cf. [204, Corollary 1.1]) it can also be proved that there are at most two shortest paths between any two vertices is and jt of S_p^{1+n}, namely either the direct path with one move of the largest disc $n+1$ or a path where disc $n+1$ passes a uniquely determined peg $k \in [p]_0 \smallsetminus \{i,j\}$. The decision which of the alternatives is optimal can be made by employing an algorithm similar to the one of Romik for $p = 3$; see the paper [204] of Hinz and

C. Holz auf der Heide. In contrast to the $p = 3$ case, where $k = 3 - i - j$, the value of k has to be fixed too for $p \geq 4$ if the largest disc moves twice. As soon as the "charming" algorithm (Bing Xue in MathSciNet®), which as Romik's is based on an automaton shown in Figure 4.7 (cf. [207, Figure 10]), has provided this information, it is easy to construct the shortest paths, because they are made up from optimal s, j^n-paths.

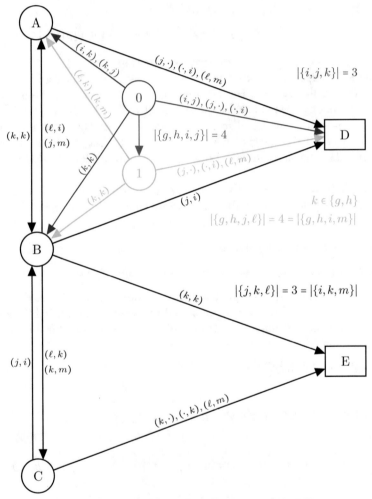

Figure 4.7: The P2 decision automaton for S_p^n

The paper [204] solves one of the open problems from Chapter 9 in the first edition of this book. It also contains a complexity analysis, based on the Markov chain method as before, which reveals [204, Theorem 3.1] that on the average asymptotically 3 pairs of **qits** (**qu**aternary dig**its**) have to be employed in the decision algorithm for $p = 4$; compare this with the $\frac{101}{38}$ pairs for $p = 3$.

Among the metric graph parameters which could be determined for Sierpiński graphs we mention the *metric dimension* μ, i.e. the size of a smallest **resolving set**. Klavžar and S. S. Zemljič found out in [242, Corollary 6] that any $p - 1$ out of p extreme vertices, $p \in \mathbb{N}_2$, form(s) a minimal resolving set, i.e. $\mu(S_p^{1+n}) = p - 1$, $n \in \mathbb{N}_0$, and that it is in fact a general feature of minimal resolving sets for S_p^{1+n} to be distributed among $p - 1$ of the subgraphs jS_p^n, $j \in [p]_0$. The metric dimension of Sierpiński graphs was independently obtained by A. Parreau in [334, Théorème 3.6].

Due to the similarity of the metric structure of Sierpiński graphs S_p^n, $p \geq 4$, to the cases $p = 2, 3$, we can even extend the methods leading to the average distances like in (1.5), Proposition 2.46, and Theorem 2.47 to general S_p^n. Wiesenberger showed [434, Satz 3.1.10] that the average distance to a fixed extreme vertex is $\frac{p-1}{p}(2^n - 1)$ and determined the numbers of tasks whose optimal solutions need one or two moves of the largest disc or with a draw [434, Satz 3.1.8]. Based on this, he obtained the ultimate result for the average distance [434, Satz 3.1.11] which we only present here to impress the reader!

Proposition 4.12. *If we put, for* $p \in \mathbb{N}$,

$$\alpha_p = p^4 - 12p^3 + 56p^2 - 104p + 68, \quad \lambda_{p,\pm} = \frac{1}{2}p^2 - p + 1 \pm \frac{1}{2}\alpha_p^{\frac{1}{2}},$$

$$\gamma_{p,\pm} = (p^2 + 3p - 2) \mp (p^4 + p^3 - 30p^2 + 58p - 36)\alpha_p^{-\frac{1}{2}},$$

then for all $n \in \mathbb{N}_0$:

$$\overline{d}(S_p^n) = \frac{(p - 1)\left(2p^4 + 6p^3 - 17p^2 + 26p - 16\right)}{p(2p - 1)\left(p^3 + 4p^2 - 4p + 8\right)} 2^n - \frac{p - 2}{p} + \frac{p^2 + 3p - 6}{(2p - 1)\left(p^2 - 7p + 8\right)} p^{-n}$$
$$- \frac{p(p - 1)\gamma_{p,+}}{2\left(p^2 - 7p + 8\right)\left(p^3 + 4p^2 - 4p + 8\right)} \left(\frac{\lambda_{p,+}}{p^2}\right)^n$$
$$- \frac{p(p - 1)\gamma_{p,-}}{2\left(p^2 - 7p + 8\right)\left(p^3 + 4p^2 - 4p + 8\right)} \left(\frac{\lambda_{p,-}}{p^2}\right)^n.$$

At the end of Section 1.1 we mentioned random walks on the CR graph $R^n \cong S_2^n$ and in Section 2.2.2 we considered random walks on the state graph of the TH with three pegs, i.e. on $H_3^n \cong S_3^n$. One of the problems addressed was the expected number of moves in the classical task to get from one perfect state (or extreme vertex) to another. Wiesenberger[49] generalized these results by the following formula for expected distances in S_p^n, $p \in \mathbb{N}_2$, [434, Folgerung 4.3.1]:

$$d_e(0^n, 1^n) = \frac{p}{2}(p^n - 1)\left(\left(\frac{p + 2}{p}\right)^n - 1\right) = \|S_p^n\|\left(\left(\frac{p + 2}{p}\right)^n - 1\right).$$

---

[49] with some input from T. Berger

4.2.2   Other Properties

The previous subsection demonstrates that the distance function is well-understood in Sierpiński graphs. Among the topological questions let us first mention eulericity and hamiltonicity which are dealt with in Exercises 4.4 and 4.5, respectively. We next present some selected other appealing properties of Sierpiński graphs.

Symmetries

The automorphism group of S_3^n is isomorphic to $\mathrm{Sym}(T)$; see Theorem 2.29. As noticed by Klavžar and B. Mohar for [241, Lemma 2.2], the proof of Theorem 2.29 extends to S_p^n for every $p \in \mathbb{N}_3$. One first observes that any permutation of extreme vertices of S_p^n leads to an automorphism of S_p^n. Moreover, these are the only symmetries, which follows by the fact, immediate from (4.10), that any vertex $s \in [p]_0^n$ is uniquely determined by the values $\mathrm{d}(s, k^n)$, $k \in [p]_0$. Theorem 2.29 thus generalizes as follows:

Theorem 4.13. *For any* $n, p \in \mathbb{N}$, $\mathrm{Aut}(S_p^n) \cong \mathrm{Sym}([p]_0)$.

Domination type invariants

Recall from Exercise 1.4 that the state graph of the Chinese rings R^n contains precisely two perfect codes if n is odd, and precisely one if n is even. This result was extended to Hanoi graph H_3^n in Section 2.3. Now, R^n is isomorphic to S_2^n and H_3^n is isomorphic to S_3^n, hence the following result due to Klavžar, Milutinović, and C. Petr [239, Theorem 3.6] complements these developments.

Theorem 4.14. *For any* $p, n \in \mathbb{N}$, *the graph* S_p^n *has a unique perfect code, if* n *is even, and there are exactly* p *perfect codes, if* n *is odd. In the latter case each perfect code is characterized by the only extreme vertex it contains.*

Figure 4.8 shows perfect codes in S_4^n for $n = 0, 1, 2, 3$. The reader is invited to compare these codes with those of S_3^n given in the lower part of Figure 2.16.

From Corollary 2.28 we know that if a graph has a perfect code then its domination number and the size of the code are equal. Hence Theorem 4.14 also implies (cf. Exercise 4.6) that for any $p, n \in \mathbb{N}$,

$$\gamma(S_p^n) = \frac{p^n + p^{(n \text{ even})}}{p + 1}. \tag{4.14}$$

For a proof of Theorem 4.14 shorter than the original one (by considering more general almost perfect codes) see S. Gravier, Klavžar, and M. Mollard [164, Section 2]. In this paper perfect codes were then applied to solve a labeling problem for Sierpiński graphs (determining an optimal $L(2, 1)$-labeling), a problem that has its origins in the theory of frequency assignment. Another invariant of Sierpiński

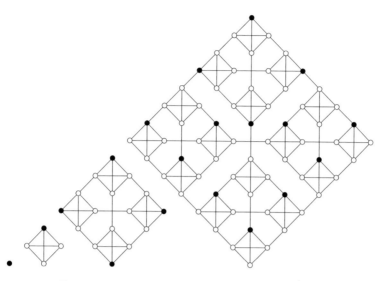

Figure 4.8: Perfect codes in S_4^n for $n = 0, 1, 2, 3$

graphs, the hub number, which plays a role in allocation network theory and is also closely related to the domination number, was determined by C.-H. Lin, J.-J. Liu, Y.-L. Wang, and W. C.-K. Yen [267, Theorem 9].

Planarity

We already know that all S_3^n are planar graphs. The same clearly holds for the graphs S_1^n and S_2^n. The only other planar Sierpiński graphs are $S_p^0 \cong K_1$, $S_4^1 \cong K_4$, and S_4^2; for the planarity of the latter and the fact that no other Sierpiński graph is planar, see Exercise 4.7.

The crossing number of Sierpiński graphs was studied in [241]. It was proved [241, Proposition 3.2] that for $n \in \mathbb{N}_3$:

$$\frac{3}{16}4^n \le \operatorname{cr}(S_4^n) \le \frac{1}{3}4^n - \frac{4}{3}(3n - 2).$$

In particular, $\operatorname{cr}(S_4^3) = 12$; see Figure 4.9 for a drawing with 12 crossings.

It is obvious from the figure that each of the 6 connecting edges $\{ijj, jii\}$ contributes 2 crossings and that these could be avoided by gluing a "handle" to the plane (or sphere) for each such edge and leading the latter over the surface of that handle. We therefore have an information on the genus, namely $g(S_4^3) \le 6$ [371, Theorem 4.10]; a better estimate is not known.

For two regularizations of Sierpiński graphs the crossing numbers were expressed in [241, Theorem 4.1] in terms of those of complete graphs, thus establishing the first non-trivial families of graphs of "fractal" type whose crossing number

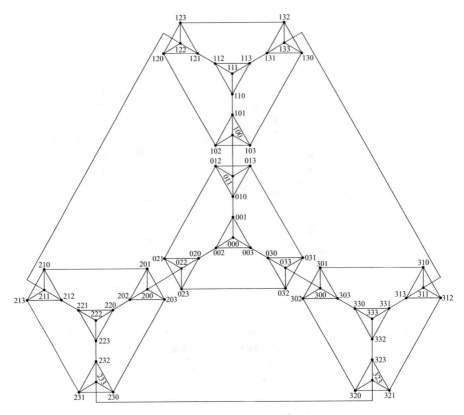

Figure 4.9: Drawing in the plane of S_4^3 with 12 crossings

is known, provided that Guy's conjecture or any other statement about the values of $\mathrm{cr}(K_p)$ is confirmed.

Based on the same kind of arguments, T. Köhler showed [249, Satz 3.11] that

$$\forall\, p \in \mathbb{N}: \ \mathrm{cr}(S_p^2) = p \cdot \mathrm{cr}(K_{p+1}^-) + \mathrm{cr}(K_p), \tag{4.15}$$

where for $q \geq 2$, the graph K_q^- is obtained from K_q by deleting one edge. In order to get an explicit value for the right-hand side of (4.15), Köhler also proved [249, Lemma 3.22] that

$$\mathrm{cr}(K_q^-) \leq \frac{1}{4} \left\lfloor \frac{q}{2} \right\rfloor \left\lfloor \frac{q-1}{2} \right\rfloor \left\lfloor \frac{q-2}{2} \right\rfloor \left\lfloor \frac{q-3}{2} \right\rfloor - \frac{1}{2} \left\lfloor \frac{q-1}{2} \right\rfloor \left\lfloor \frac{q-3}{2} \right\rfloor \tag{4.16}$$
$$= \frac{1}{4} \left\lfloor \frac{q+2}{2} \right\rfloor \left\lfloor \frac{q-1}{2} \right\rfloor \left\lfloor \frac{q-3}{2} \right\rfloor \left\lfloor \frac{q-4}{2} \right\rfloor.$$

For $q \leq 8$, the inequality in (4.16) was even shown to be an identity, such that the crossing numbers for S_p^2 are established for $p \leq 7$. One is, of course, tempted to conjecture that equality holds for all q in (4.16).

Connectivity

The connectivity $\kappa\left(S_3^n\right) = 2$ of S_3^n was determined in Proposition 2.25. An extension of this result to all Sierpiński graphs is given in Exercise 4.8.

Colorings

Recall from Proposition 2.24 that $\chi(S_3^n) = 3$ holds for any $n \in \mathbb{N}$. Parisse [332, p. 147] noticed that labeling each vertex $s = s_n \ldots s_1$ of S_p^n with s_1 yields a p-coloring of S_p^n. Since S_p^n contains $S_p^1 \cong K_p$ as a subgraph, $\chi(S_p^n) = p$ holds for any $p, n \in \mathbb{N}$. Using this fact it is also not difficult to determine the **independence number** of Sierpiński graphs, denoted by $\alpha\left(S_p^n\right)$, see Exercise 4.9.

By Proposition 2.24, $\chi'(S_3^n) = 3$ holds for any $n \in \mathbb{N}$. It was further shown that an edge-coloring of S_3^n using three colors is unique [234, Corollary 3.3]. The edge chromatic number of all Sierpiński graphs was determined by M. Jakovac and Klavžar, [222, Theorem 4.1]: $\chi'(S_p^n) = p$ (for $p, n \in \mathbb{N}_2$). To construct a p-edge-coloring is easy when p is even: color the p-cliques using $p-1$ colors and use the last color on all the other edges. To obtain such a coloring for odd p is more subtle, see [211] and [222] for two different constructions.

Total colorings of Sierpiński graphs were first studied in [222], but the complete answer was obtained by Hinz and Parisse [211, Theorem 4]: $\chi''(S_p^n) = p+1$ (for $p, n \in \mathbb{N}_2$). Since $\Delta(S_p^n) = p$, it is clear that $\chi''(S_p^n) \geq p+1$. See [211, Lemmas 4' and 5'] for a thoughtful construction that proves the reverse inequality.

Additional properties

Numerous additional properties of Sierpiński graphs were investigated, too many to be listed here. The interested reader may find them in the comprehensive survey [207]. This article also provides a strict distinction between Sierpiński graphs and *Sierpiński triangle graphs*, which are often mistaken for them.[50]

For two additional aspects of S_3^n that appeared after the survey see [79, 294]. In the first of these papers the asymptotic number of independent sets in S_3^n is established. More precisely, if $\sigma(G)$ denotes the number of independent sets of a graph G, then it is proved that

$$\lim_{n \to \infty} \frac{\sigma(S_3^n)}{|S_3^n|} = 0.42433435855938823\ldots$$

The second paper deals with several degree-based invariants of S_3^n and related families of graphs, for the latter families their metric dimension is also addressed.

---

[50]Roughly speaking the latter correspond to the drawings of "graphs" (without vertices/dots) in the bottom row of Figure 0.19.

4.2.3   Sierpiński Graphs as Interconnection Networks

It should be clear by now that Sierpiński graphs play an important role in the theory of the Tower of Hanoi. This was the main motivation for their extensive exploration. On the other hand, in parallel and completely independent of the above described developments, these graphs were proposed and studied in computer science as a model for interconnection networks. We now briefly describe this point of view.

In order to extend the range of applications of supercomputers, G. Della Vecchia and C. Sanges [99] proposed a class of recursive scalable networks named *WK-recursive networks.* Their goal was to construct networks that would have a high degree of regularity and scalability, keep internode distances (very) small, admit self-routing techniques, show a high degree of local density, and more. With these requirements in mind they proposed the networks $WK(p,n)$,[51] for which the removal of the so-called open edges from each of the extreme vertices leads to graphs isomorphic to S_p^n.

The open edges of $WK(p,n)$ are aimed to be used for further expansions. Considering these links as edges at extreme vertices, $WK(p,n)$ is then a p-regular graph. (For two other ways how to regularize Sierpiński graphs see [241].) In the seminal paper [99], the authors not only introduced WK-recursive networks but also discussed an actual VLSI (very-large-scale integration) implementation of these networks and proposed a message exchange algorithm based on a self-routing technique. G.-H. Chen and D.-R. Duh followed with the paper [77] in which they investigated several structural properties of WK-networks. They determined their diameter [77, Theorem 2.1] (our Corollary 4.10), their connectivity [77, Theorem 2.3] (Exercise 4.8), and found out that they are hamiltonian [77, Theorem 2.2] (Exercise 4.5). They also addressed the routing problem (to transmit a message from a node to another node) and the broadcasting problem (to transmit a message from a node to all other nodes) in WK-networks by studying (shortest) paths in $WK(p,n)$. J.-F. Fang, W.-Y. Liang, H.-R. Chen, and K.-L. Ng [145] presented a simple broadcasting algorithm that, starting from an arbitrary node of the $WK(p,n)$ network, transmits a message to each node of the network exactly once in $2^n - 1$ steps.

As Exercise 4.5 asserts, Sierpiński graphs are hamiltonian. For communication networks, further traversability properties are important. We say that a graph is *pancyclic* if it contains cycles of all lengths between 3 and its order. M. Hoseiny Farahabady, N. Imani, and H. Sarbazi-Azad showed in [218] that the networks $WK(4,n)$ are almost pancyclic—they contain cycles of all lengths but of length 5. Another important issue in communication networks is how they behave under faulty nodes. In this direction J.-S. Fu [155] proved that in the network which remains after removing $p - 4$ arbitrary nodes from $WK(p,n)$, any two nodes are connected by a hamiltonian path.

Yet another name was given to Sierpiński graphs in the computer science

---

[51]The notation is not standardized; $WK_{p,n}$ and $K(p,n)$ are also used for the same networks.

community—*iterated complete graphs*. They have been investigated since 2001 in the frame of the *Research Experiences for Undergraduates Program in Mathematics* at Oregon State University under the supervision of Cull. Numerous related manuscripts were written, but as far as we know, none was eventually published. The manuscripts mainly consider problems on iterated complete graphs related to codes and variants of the TH derived from these graphs. Unfortunately, the authors of these investigations seemed not to be aware that Sierpiński graphs and WK-networks had been introduced earlier.

4.3 Connections to Topology: Sierpiński Curve and Lipscomb Space

We encountered the Sierpiński triangle in Chapters 0 and 2. In this section we take a more formal topological view to this object and along the way explain the original motivation for the introduction of Sierpiński graphs.

4.3.1 Sierpiński Spaces

For $p \in \mathbb{N}$, the $(p-1)$-dimensional *Sierpiński space* Σ_p can formally be described as a subset of $\mathbb{R}^{p-1}$ as follows.

Let a_i, $i \in [p]_0$, be the vertices of a regular $(p-1)$-simplex in $\mathbb{R}^{p-1}$ and define the contracting similarity mappings φ_i on $\mathbb{R}^{p-1}$ by $x \mapsto (x + a_i)/2$. Then there is a unique compact non-empty set $\Sigma_p \subseteq \mathbb{R}^{p-1}$ with

$$\Sigma_p = \bigcup_{i=0}^{p-1} \varphi_i(\Sigma_p).$$

Σ_p is the *self-similar set* corresponding to the *iterated function system* $\{\varphi_i \mid i \in [p]_0\}$. It fulfills the *open set condition* and therefore its *Hausdorff dimension* is $\alpha = \mathrm{lb}(p)$; cf. [34, Section 2]. Using the α-dimensional *Hausdorff measure*, C. Bandt and T. Kuschel found in [34, Section 4] that the average distance of Σ_p is

$$\frac{(p-1)\left(2p^4 + 6p^3 - 17p^2 + 26p - 16\right)}{p(2p-1)\left(p^3 + 4p^2 - 4p + 8\right)},$$

assuming that the underlying regular simplex has side length 1. This is in perfect accordance with Wiesenberger's formula in Proposition 4.12.

Σ_1 is a point with dimension and average distance 0. The set Σ_2 is a unit line with dimension 1 and average distance $1/3$. For $p = 3$ we get the Sierpiński triangle Σ_3 with Hausdorff dimension $\mathrm{lb}(3)$ and the famous average distance of $466/885$. The set Σ_4 has Hausdorff dimension 2 and an average distance of $89/140$.

4.3.2 Sierpiński Triangle

Here we want to compare the two constructions of the Sierpiński triangle ST as shown in the two bottom lines of Figure 0.19. Let us denote the corresponding sequences of point sets in $\mathbb{R}^2$ by ST_n (with filled triangles) and st_n (triangle lines only), respectively; $n \in \mathbb{N}_0$. Setting $ST := \bigcap_{n \in \mathbb{N}_0} ST_n$ and $st := \bigcup_{n \in \mathbb{N}_0} st_n$, we have $ST = \overline{st}$. It is clear that st is dense in ST, because every $x \in ST_n$ lies close to a point in st_n. In order to show that taking the closure is necessary, we characterize the points of ST as follows. (Recall that $T = \{0, 1, 2\}$.)

Lemma 4.15. *There is a surjective mapping from $T^{\mathbb{N}}$ onto ST.*

Proof. Let $s \in T^{\mathbb{N}}$ and define $x \in ST$ as follows: x lies in the upper subtriangle of ST_1 (the subdivision of ST_0), if $s_1 = 0$; x lies in the lower left subtriangle of ST_1, if $s_1 = 1$; and in the lower right one, if $s_1 = 2$. This procedure is repeated recursively for each stage of subdivision of ST_n. The point x is uniquely determined as the limit in a nested sequence of compacta with diameters going to 0. By the definition of ST it is clear that every element of ST can be reached this way. □

The mapping constructed in the preceding proof is *not* injective. But it is easy to see that the preimage of any $x \in ST$ consists of at most two elements. If there are two (this applies exactly to those points where two subtriangles meet), then they have the form $\underline{s}\alpha\beta\dots$ and $\underline{s}\beta\alpha\dots$ with $\alpha, \beta \in T$ and $\underline{s} \in T^n$ for some $n \in \mathbb{N}_0$.

Those points in st which are not intersection points of lines (or triangles for that matter) can only sit on a line with a specific orientation, either horizontally (or opposite to 0), or inclined from 0 to 2 (opposite to 1) or to 1 (opposite to 2). But then their corresponding (unique) sequences cannot contain infinitely many 0s, 1s or 2s, respectively, because this would bring them off the line. This means that all those $x \in ST$ whose coordinate sequences contain infinitely many 0s, 1s *and* 2s make up the set $ST \smallsetminus st$.[52]

The curves approximating ST in the top line of Figure 0.19 converge to the (continuous!) *Sierpiński curve* whose image is closed and therefore equal to the point set ST, because all points of the approximating curves lie in st. However, the points on an individual curve of that sequence are even sparser than those on the corresponding triangle lines, their total length making up only 1/3. This is even true for every single line. To see this, consider the number y_n of intervals (of length 2^{-n}) lying on the base line of the nth curve and the number x_n of intervals on one of its flanks. Then $x_{n+1} = x_n + y_n$ and $y_{n+1} = 2x_n$ for $n \in \mathbb{N}_0$, such that $x_{n+2} = x_{n+1} + 2x_n$. Since $x_0 = 0$ and $x_1 = 1$, the sequence x is equal to the Jacobsthal sequence J (cf. p. 74). As $J_n/2^n \to 1/3$, only 1/3 of the points on the triangle lines are matched by the curves asymptotically.

---

[52]Hint from Uroš Milutinović (Maribor).

Cantor sets

The Sierpiński triangle has the following set theoretical property:

Proposition 4.16. ST *is a* Cantor set, *i.e. the set (as a subset of* $\mathbb{R}^2$*) is*

1. compact *(closed* $(ST' \subseteq ST)$ *and bounded* $(ST \subseteq \overline{B}_R(0)$ *for some* $R > 0))$,

2. nowhere dense $(\overset{\circ}{\overline{ST}} = \varnothing)$,

3. perfect $(ST' = ST)$;

here ST' *is the* derived set *of* ST, *i.e. the set of all its* limit points *(*$x \in \mathbb{R}^2$ *such that* $\forall\, \varepsilon > 0\; \exists\, x \neq y \in ST : \|x - y\| < \varepsilon$*).*

Proof. Statement 1 follows from the definition of ST. In particular, $\overline{ST} = ST$ and every point of ST can be approached by points outside ST (from the middle triangles left out in the construction procedure); this proves statement 2. By the same argument $ST \subseteq ST'$, because every point of ST can be approached by points from subtriangles *not* left out in the construction procedure. So from statement 1 we get $ST = ST'$, i.e. statement 3.   □

The prototype of a Cantor set is *the Cantor set* CS, defined in a similar way as the Sierpiński triangle: start with $CS_0 = [0, 1] \subseteq \mathbb{R}$, take out the central open third, i.e. $CS_1 = [0, \frac{1}{3}] \cup [\frac{2}{3}, 1]$ and keep on deleting middle third triangles to obtain the sequence of sets CS_n; then $CS := \bigcap_{n \in \mathbb{N}_0} CS_n$. Now we have

Lemma 4.17. *There is a bijective mapping from* $B^{\mathbb{N}}$ *to* CS.

Proof. Let $x \in CS$. Then for every $n \in \mathbb{N}$ the point x lies either in the left third of its interval of length 3^{1-n} in CS_{n-1} or in the right one; we put $s_n = 0$ or $s_n = 1$, respectively.   □

The union cs of the sets cs_n of endpoints of the intervals occuring in CS_n corresponds to the set of elements of $B^{\mathbb{N}}$ which are eventually constant. The set cs is dense in CS and countable. However, CS is uncountable as can be seen by Cantor's second diagonal method: assume that $\beta_n = \beta_{n,1}, \beta_{n,2}, \ldots$ $(n \in \mathbb{N})$ forms a list of elements of $B^{\mathbb{N}}$ and set $\beta_{0,n} = 1 - \beta_{n,n}$; then β_0 can not be in that list. The properties of a general Cantor set are also clear.

We have

$$cs_n = \left\{ 2 \sum_{k=1}^{n} \beta_k \cdot 3^{-k} \mid \beta \in B^n \right\} \cup \left\{ 2 \sum_{k=1}^{n} \beta_k \cdot 3^{-k} + 3^{-n} \mid \beta \in B^n \right\}.$$

Since $3^{-n} = 2 \sum_{k=n+1}^{\infty} 3^{-k}$ (cf. Lemma 2.18), we see that CS is the set of real numbers in $[0, 1]$ whose ternary representations lack the digit 1.

Connections to Sierpiński and Hanoi graphs

The endpoints of the intervals composing CS_n have coordinates of the form $\underline{\beta}, \beta_{n+1}, \dots$ with $\underline{\beta} \in B^n$ and $\beta_{n+1} \in B$. We associate the finite string $\underline{\beta}\beta_{n+1} \in B^{n+1}$ with this sequence of coordinates. Then CS_n can be viewed as a drawing of the Hanoi graph H_2^{n+1} with cn_n corresponding to the set of vertices B^{n+1} and the intervals joining the endpoints in CS_n representing the edges of H_2^{n+1}.

Similarly, let $s_1, \dots, s_m, s_{m+1} \dots$ with $m \in [n]$, and $s_m \neq s_{m+1}$ be the coordinates of an *interior* corner of one of the filled triangles in ST_n. Then we associate the finite string $s_1, \dots, s_{m-1}, 3 - s_m - s_{m+1} \in T^m$ with it. The *boundary* corners $k, k, \dots$ $(k \in T)$ are mapped to the empty strings $\hat{k}$, respectively. Then we can define the *Sierpiński triangle graph* $\widehat{S}^n$ on the vertex set $V(\widehat{S}^n) = \{\hat{0}, \hat{1}, \hat{2}\} \cup \bigcup_{m=1}^{n} T^m$ with two vertices being joined by an edge, if the corresponding points form an edge of a filled triangle in ST_n; cf. [207, Section 0.2.1].

Another way to associate a graph with ST_n is to take its set of filled triangles as the vertex set, the vertices being labelled as in the construction of the proof of Lemma 4.15, and joining two vertices by an edge if the corresponding triangles have a point in common (cf. [214, Definition 6]). This leads to Sierpiński graph S_3^n which is isomorphic to the Hanoi graph H_3^n. This can also be viewed in the following way: if we replace the vertices in the canonical drawing of H_3^n (or equivalently S_3^n) by filled equilateral triangles of side length 3^{-n}, we get ST_n.

4.3.3 Sierpiński Curve

The Sierpiński triangle is also called *triangular Sierpiński curve* and can formally be described as a subset of $\mathbb{R}^2$ as follows. For a triangle Σ in the plane $\mathbb{R}^2$ with vertices e^0, e^1, e^2, let $\varphi_1, \varphi_2, \varphi_3 : \Sigma \longrightarrow \Sigma$ be the *homotheties* (central similarity transformations) with coefficients $1/2$ and centers e^0, e^1, e^2, respectively. It is obvious that the set obtained from Σ by m removals of the open middle triangles (cf. the center row in Figure 0.19) may be described as

$$\Sigma_m = \bigcup_{(\lambda_1, \dots, \lambda_m) \in \Lambda^m} \varphi_{\lambda_1} \circ \cdots \circ \varphi_{\lambda_m} \Sigma, \tag{4.17}$$

where $\Lambda = \{0, 1, 2\} = T$. We denote this set by Λ, since we will show later that any set may be used in an analogous construction.

After that, the triangular Sierpiński curve $\Sigma(3)$ is obtained as the intersection of all these unions:

$$\Sigma(3) = \bigcap_{m \in \mathbb{N}} \Sigma_m.$$

In the theory of fractals, the set $\{\varphi_0, \varphi_1, \varphi_2\}$ is called an *iterated function system*, and $\Sigma(3)$ is its attractor (cf. [396, Section 4]). Moreover, the mapping $X \mapsto \varphi_0 X \cup \varphi_1 X \cup \varphi_2 X$ (defined on the subsets X of Σ) is called the *Hutchinson operator*, and $\Sigma(3)$ is the unique compact fixed point of the Hutchinson operator.

Figure 4.10 shows what happens to Σ after three applications of the Hutchinson operator, i.e. the sets Σ_1, Σ_2, and Σ_3. To make the figure simpler and more transparent, $\varphi_i \Sigma$ is shortened to i, $\varphi_i \circ \varphi_j \Sigma$ to ij, and $\varphi_i \circ \varphi_j \circ \varphi_k \Sigma$ to ijk. We interpret this as graphs in which triangles are vertices, pairs of which are joined by an edge precisely when they have a point in common, i.e. when they are not disjoint (cf. [214, Definition 6]). We then get the Sierpiński graphs S_3^0, S_3^1, S_3^2, and S_3^3, see Figure 4.11.

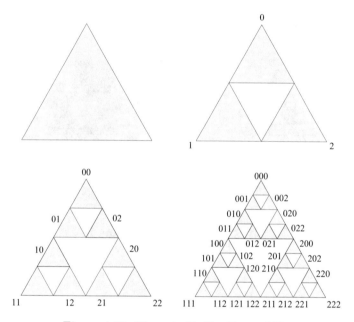

Figure 4.10: The sets Σ, Σ_1, Σ_2, and Σ_3

Now let $e^0 = (1,0,0)$, $e^1 = (0,1,0)$, $e^2 = (0,0,1)$ be points of $\mathbb{R}^3$ and let Σ be the convex hull of these three points (i.e. the standard 2-simplex), cf. Figure 4.12. This choice of the triangle vertices enables us to give a simple characterization of the points belonging to $\Sigma(3)$:

Theorem 4.18. *For any $x = (x_0, x_1, x_2) \in \mathbb{R}^3$, $x \in \Sigma(3)$ if and only if there is a sequence $(\lambda_k)_{k \in \mathbb{N}} \in T^{\mathbb{N}}$ such that for every $i \in T$,*

$$x_i = \sum_{k=1}^{\infty} (\lambda_k = i) \cdot 2^{-k} . \tag{4.18}$$

For a proof and a generalization, see [310, p. 347f]. Note that if $x = (x_0, x_1, x_2)$ satisfies (4.18), then one easily sees that $x_0 + x_1 + x_2 = 1$, whence $x \in \Sigma$.

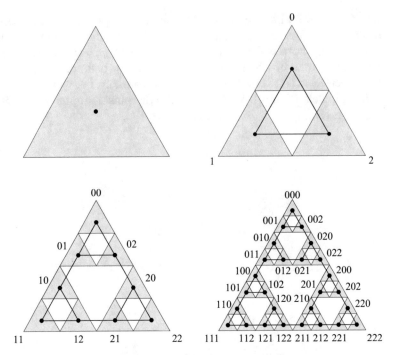

Figure 4.11: S_3^0, S_3^1, S_3^2, and S_3^3 obtained from Σ, Σ_1, Σ_2, and Σ_3

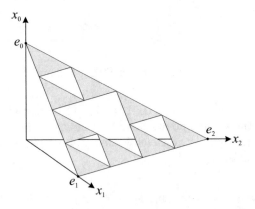

Figure 4.12: The set Σ_2 as obtained from the standard 2-simplex Σ

Hanoi attractors

In [13] Patricia Alonso Ruiz employed the same method based on iterated function systems as described above to introduce an interpolating family of *Hanoi attrac-*

tors[53] $\Sigma^\tau \subseteq \mathbb{R}^2$ for the real parameter $\tau \in {]}0,1{[}$. Roughly speaking this family leads from a rescaled Sisyphean Sierpiński graph ("$\tau = 1$") to the Sierpiński curve ("$\tau = 0$") by modifying the lengths of the critical edges linking the subtriangles by a factor $\tau/3$; see Figure 4.13, where the respective second approximation to Σ^τ is shown for $\tau \in \{1, 2/3, 1/3, 0\}$.

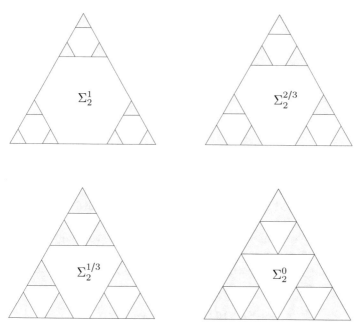

Figure 4.13: Approximations leading to Hanoi attractors Σ^τ

Note that, in contrast to $\Sigma = \Sigma^0$ the sets Σ^τ are not self-similar for $\tau \in {]}0,1{[}$, a fact which makes their analysis more demanding. Nevertheless, it was shown in [13, Theorem 2.2.1] that the Σ^τ converge, as $\tau \to 0$, to Σ^0 in the *Hausdorff space*, i.e. the set of non-empty compact subsets of $\mathbb{R}^2$ together with the distance function[54]

$$d_H(X,Y) = \max\left\{\max\{d(x,Y) \mid x \in X\}, \max\{d(X,y) \mid y \in Y\}\right\},$$

where $d(x,Y) = \min\{d(x,y) \mid y \in Y\}$ and $d(X,y) = \min\{d(x,y) \mid x \in X\}$ with the euclidean distance d on $\mathbb{R}^2$. Moreover, [13, Theorem 2.2.7] says that for the

---

[53] A better expression would have been "Sierpiński attractors", but again the name "Hanoi" seems to be more attractive!

[54] If you feel uncomfortable with this definition, think of the following scenario. Assume that for people in (continental) France the only way to get a good beer is to travel to (continental) Germany and that for people in Germany the only way to get a good vine is to travel to France. Then the Hausdorff distance between France and Germany is the worst case of the travel distance needed in order to get a good beer starting from France or to get a good vine starting from Germany.

Hausdorff dimension we have with $\sigma := 6/(3-\tau)$ $(\in]2,3[)$ (the side length of the subtriangles of order $n \in \mathbb{N}_0$ is σ^{-n}) that

$$\dim_H(\Sigma^\tau) = \mathrm{lt}(\sigma)^{-1} \to \dim_H(\Sigma), \text{ as } \tau \to 0.$$

The goal of [13] was to analyze diffusion processes on fractals by studying characteristic features such as the *spectral dimension*. Energy forms were defined on Σ ([13, Chapter 3]) and Σ^τ, $\tau \in]0,1[$ ([13, Chapter 4]) leading to *Laplacian operators* with spectral dimension $2/\mathrm{lt}(5)$, independent of τ ([13, Theorem 3.4.2 and Corollary 4.4.20]. This lead Alonso-Ruiz to summarize that you can see the difference between the Hanoi attractors, but not hear it! They were used as examples for applying a fractal quantum graph[55] approach to the existence of resistance forms in [14].

Lipscomb space

The definition of Sierpiński graphs in [235] was motivated by a different (but related) topological construction. In [271, 272] S. L. Lipscomb introduced the space $\mathcal{J}(\tau)$, now called the *Lipscomb space*, in the following way. Let Λ be any set of cardinality τ. Then the set $\Lambda^{\mathbb{N}}$ of all sequences in Λ carries the product topology, where Λ itself is equipped with the discrete topology. Further, $\mathcal{J}(\tau)$ is defined as the quotient topology space $\Lambda^{\mathbb{N}}/\approx$, where the equivalence $\approx$ is defined as follows. For any pair of distinct sequences λ and μ, let $\lambda \approx \mu$ if and only if there exists a $d \in \mathbb{N}$ such that

(i) $\forall k \in [d-1] : \lambda_k = \mu_k,$

(ii) $\lambda_d \neq \mu_d,$

(iii) $\forall k \in \mathbb{N} \smallsetminus [d] : \lambda_k = \mu_d \wedge \mu_k = \lambda_d.$ (4.19)

The corresponding equivalence classes either contain two elements (non-constant, eventually constant sequences) or one element (any other sequence, including constant ones). The first type plays the role of "rationals", the second of "irrationals" in the quotient space.

The Sierpiński graphs S_p^n were introduced in [235] after Λ was chosen to be $[p]_0$ and (4.19), employed in the definition of the equivalence $\approx$ on $\Lambda^{\mathbb{N}}$, was modified to (4.8) and used as the definition of adjacency.

Lipscomb proved that $\mathcal{J}(\tau)$ is a one-dimensional metrizable space of weight τ, and that the topological product of $n+1$ copies of it contains a subspace, which is a universal n-dimensional metrizable space of weight τ, meaning that it is itself an n-dimensional metrizable space of weight τ, and that every n-dimensional metrizable space of weight τ topologically embeds into it.

Since this is quite far away from our topic, we will not give any details here. The interested reader may consult Lipscomb's book [273], written in an extremely

---

[55]Quantum graphs are currently very fashionable in physics, but, of course, are not graphs.

reader-friendly style. It offers a wealth of material on topological and fractal-theoretic aspects and includes detailed historical and motivational background. For a more compact presentation see [274].

In [310] the *generalized Sierpiński curve* $\Sigma(\tau)$ was introduced in the same manner as $\Sigma(3)$, after replacing the plane and the triangle by an appropriate Hilbert space and by the closed convex hull of a subset containing τ independent vectors, respectively. More specifically, for infinite τ, the Hilbert space $\ell_2(\tau) = \{x = (x_\lambda) \in \mathbb{R}^\Lambda \mid \sum_{\lambda \in \Lambda} x_\lambda^2 < \infty\}$ and $\Sigma = \{(x_\lambda) \in \ell_2(\tau) \mid \sum_{\lambda \in \Lambda} x_\lambda \leq 1 \wedge \forall \lambda \in \Lambda : 0 \leq x_\lambda \leq 1\}$ were employed. Using homotheties $\varphi_\lambda : \Sigma \longrightarrow \Sigma$ with coefficients $1/2$, defined by

$$(\varphi_\lambda(x))_\mu = \tfrac{1}{2}\left(x_\mu + (\mu = \lambda)\right),$$

the generalized Sierpiński curve was obtained as

$$\Sigma(\tau) = \bigcap_{m \in \mathbb{N}} \Sigma_m,$$

with Σ_m as in Equation (4.17). In [310] it was proved that the Lipscomb space $\mathcal{J}(\tau)$ is homeomorphic to the generalized Sierpiński curve $\Sigma(\tau)$. Hence, the Sierpiński graphs are related to the Lipscomb space not only through the adjacency-equivalence link, but also through the (generalized) Sierpiński curve and the Lipscomb space being topologically isomorphic.

To conclude the chapter we mention a recent related topological investigation [148] by H. Fischer and A. Zastrow, in which the Tower of Hanoi plays a significant role. In the paper a construction of a generalized Cayley graph for the fundamental group of the Menger universal curve is given. This is done in terms of word sequences which is in turn based on a representation of the Menger universal curve as the limit of an inverse sequence of finite 4-regular graphs X_n. Now, these graphs are selected in such a way that X_n can be represented as a certain variant of the Tower of Hanoi. In this variant, discs are white on one side and black on the other and can be turned over during a move. In the state graph that suits the purpose, the vertices correspond to moments when discs are in transition, while the edges correspond to the situation when discs are on pegs. Moreover, the states are restricted to those of the classical TH task, where a certain backtracking is also allowed. We do not want to go into additional details here, the interested reader can find them in [148, Sec. 5].

4.4 Exercises

4.1. Let the start distribution in the TH with 4 discs be given by 0220, and consider five different goal distributions to be reached directly from it:
a) 2022, **b)** 2210, **c)** 2211, **d)** 2112, and **e)** 2001.
Decide about the number of moves of disc 4 in the corresponding optimal paths.

4.2. Show that all tasks $i^{1+n-d} k j^{d-1} \to t$ with $n \in \mathbb{N}$, $d \in [n]$, $\{i, j, k\} = T$, and $t \in T^{1+n}$ of the Switching TH can be solved optimally with at most one move of the largest disc.

4.3. Prove that $s, s' \in S_p^n$ are adjacent if and only if (4.8) holds.

4.4. Determine which Sierpiński graphs are eulerian and which are semi-eulerian.

4.5. Show that S_p^n is hamiltonian for any $p \in \mathbb{N}_3$ and $n \in \mathbb{N}$.

4.6. Determine the domination number of Sierpiński graphs, that is, prove (4.14).

4.7. Show that S_4^2 is planar, but S_4^3 is not.

4.8. Show that $\kappa\left(S_p^n\right) = p - 1$ for any $p, n \in \mathbb{N}$.

4.9. Show that for any $p, n \in \mathbb{N}$, $\alpha\left(S_p^n\right) = p^{n-1}$. (The case $p = 3$ was first observed by A. N. Nurlybaev in [324, Теорема 3].)

Chapter 5

The Tower of Hanoi with More Pegs

In Chapter 2 we have described the fundamental object of the book—the classical TH with three pegs. We have revealed its secrets and hopefully convinced the reader that it contains exciting problems with not too difficult solutions. Now, a well-known mathematical metatheorem asserts that it is easy to generalize. In this chapter we are going to study the most natural and obvious generalization of the TH with three pegs, namely the TH with more than three pegs. The main message of the chapter is that in this particular case the abovementioned metatheorem is as wrong as it can possibly be. In order to avoid confusion with the introduction of a second parameter besides the number n of discs, namely the number p of pegs, we will focus on $p = 4$ in the first part of this chapter. Later we will extend our discussion to the general case.

5.1 The Reve's Puzzle and the Frame-Stewart Conjecture

As already said in the introduction, the first explicit extension of the classical puzzle to four pegs was made popular in the book [116] back in 1907 by Dudeney. It is now time to take a closer look at this puzzle, known as *The Reve's puzzle*.

The terms *regular state* and *perfect state* will have the same meaning as in Chapter 2. The discs are labeled from 1 to n in increasing order of diameter as before, and a regular state is denoted by $s = s_n \dots s_1$ where now the position s_d of disc $d \in [n]$ is an element of $Q = \{0, 1, 2, 3\}$ (Q for *quaternary*), the pegs being called $0, 1, 2, 3$. We concentrate on the classical problem P0 to get from a perfect state to another perfect state, from peg 0 to peg 2, say, as before. Then peg 3 is the extra peg, the *Devil's peg*. It is obvious that we do not need more moves than with three pegs: just ignore the Devil. However, it is also immediately clear that, for $n > 2$, we need *strictly* fewer moves than for three pegs: move disc 1 to peg 3,

© Springer International Publishing AG, part of Springer Nature 2018
A.M. Hinz et al., *The Tower of Hanoi – Myths and Maths*,
https://doi.org/10.1007/978-3-319-73779-9_6

discs 2 to n to peg 2 avoiding peg 3 and finally disc 1 to peg 2. This will take $1 + (2^{n-1} - 1) + 1 = 2^{n-1} + 1 < 2^n - 1$ moves altogether. But why not place a whole subtower of, say, m smallest discs on the Devil's peg, then transfer discs $m + 1$ to n to the goal, avoiding peg 3, and bring the smaller discs to the goal thereafter? This would take $2f(m) + 2^{n-m} - 1$ moves, if we already know how to transfer m discs using all four pegs in $f(m)$ moves. We then can still optimize with respect to the parameter m and define the so-called *Frame-Stewart numbers* recursively:

Definition 5.1. *For all* $n \in \mathbb{N}_0$: $FS_3^n = 2^n - 1 = M_n$.

$$
\begin{aligned}
FS_4^0 &= 0, \\
\forall n \in \mathbb{N}: \quad FS_4^n &= \min\{2FS_4^m + FS_3^{n-m} \mid m \in [n]_0\}.
\end{aligned}
$$

Obviously (put $m = 0$), $FS_4^n \le FS_3^n$, with $FS_4^n = FS_3^n$, if and only if $n \le 2$, and $FS_4^3 = 5 < 7 = FS_3^3$ with m chosen to be 1.

The name of these numbers derives from the solutions given by Frame [150] and Stewart [391] to a problem posed by the latter [390], namely to find an optimal solution for the general p-peg case, to which we will return later in Section 5.4.

In fact, Frame's and Stewart's solutions formally differ from each other. While Stewart's approach leads to the above Definition 5.1, Frame introduced a recursive solution based on the idea of looking at the half-way situation, i.e. the distribution of n discs just before the (only) move of the largest disc $n + 1$. We will analyze this approach later for any number of pegs. For up to four pegs it gives the following sequences $(F_3^n)_{n \in \mathbb{N}_0}$ and $(F_4^n)_{n \in \mathbb{N}_0}$:

Definition 5.2. *For all* $n \in \mathbb{N}_0$: $F_3^n = 2^n - 1 = M_n$.

$$
\begin{aligned}
F_4^0 &= 0, \\
\forall n \in \mathbb{N}_0: \quad \overline{F}_4^n &= \min\{F_4^m + F_3^{n-m} \mid m \in [n+1]_0, \ 2m \ge n\}, \\
F_4^{n+1} &= 2\overline{F}_4^n + 1.
\end{aligned}
$$

The numbers $\overline{F}_4^n$ characterize the half-way solution and will be called *Frame numbers*. On the other hand, there is no need to distinguish the sequences F and FS. $F_3 = FS_3$ is obvious. Noting that

$$
\forall n \in \mathbb{N}: F_4^n = \min\{2F_4^m + F_3^{n-m} \mid m \in [n]_0, \ 2m \ge n - 1\},
$$

we immediately see that $FS_4^n \le F_4^n$. To get rid of the restriction on m in Frame's half-way approach, namely that the number m of smaller discs transferred using 4 pegs is not smaller than the number $n - m$ of larger ones restricted to 3 pegs, we observe the following: from Definition 5.1 we get (put $n = k + 1$ and $m = k$):

$$
FS_4^{k+1} \le 2FS_4^k + FS_3^1 \le FS_4^k + FS_3^k + FS_3^1 = FS_4^k + 2^k,
$$

such that

$$
\forall k \in \mathbb{N}_0: FS_4^{k+1} - FS_4^k \le 2^k.
$$

Summing this from m to $\ell - 1 \geq m$, we obtain

$$FS_4^\ell - FS_4^m = \sum_{k=m}^{\ell-1}(FS_4^{k+1} - FS_4^k) \leq \sum_{k=m}^{\ell-1} 2^k = 2^\ell - 2^m = FS_3^\ell - FS_3^m,$$

whence

$$\forall m \in [\ell]_0 : \ FS_4^\ell + FS_3^m \leq FS_4^m + FS_3^\ell. \tag{5.1}$$

We arrive at the following proposition:

Proposition 5.3. $\forall n \in \mathbb{N}_0 : \ F_4^n = FS_4^n$.

Proof. We proceed by induction. The case $n = 0$ is immediate from the definitions. So let $n \in \mathbb{N}_0$ and assume that $\forall k \in [n+1]_0 : \ F_4^k = FS_4^k$. Then

$$
\begin{aligned}
F_4^{n+1} &= 2\overline{F}_4^n + 1 \\
&= 2\min\{F_4^m + F_3^{n-m} \mid m \in [n+1]_0, \ 2m \geq n\} + 1 \\
&= 2\min\{FS_4^m + FS_3^{n-m} \mid m \in [n+1]_0, \ 2m \geq n\} + 1 \\
&= 2\min\{FS_4^m + FS_3^{n-m} \mid m \in [n+1]_0\} + 1 \\
&= \min\{2FS_4^m + FS_3^{n+1-m} \mid m \in [n+1]_0\} \\
&= FS_4^{n+1};
\end{aligned}
$$

here we have made use of the induction assumption and (5.1), where we put $\ell = n - m$ to get

$$\forall m \in [n+1]_0, \ 2m < n : \ FS_4^{n-m} + FS_3^m \leq FS_4^m + FS_3^{n-m},$$

such that for these cases we can just switch m and $n - m$. □

On the other hand, Dunkel, the editor of the problem section of the *American Mathematical Monthly*, immediately pointed out in [121] that a proof of minimality was lacking for the presumed minimal solutions of Frame and Stewart. Writing $\mathrm{d}(s,t)$ again for the minimal number of moves to get from state s to state t in Q^n, the following had been open since 1941 for more than seven decades:

Frame-Stewart Conjecture (FSC). $\forall n \in \mathbb{N}_0 : \ \mathrm{d}(0^n, 2^n) = FS_4^n$.

The conjecture can easily be verified for the first few cases by hand. It has been tested by R. E. Korf and A. Felner in [253] to be true up to and including $n = 30$ using a refined search algorithm on a computer, see also [252] for more details. We will discuss these numerical experiments in Section 5.3.

A recursive algorithm realizing the solutions given by Frame and Stewart (cf. also [116, p. 131f]) can be based on *partition numbers* for an n-disc tower, that is on numbers m_n for which the minimum in Definition 5.1 is achieved. Note that for given $n \in \mathbb{N}$ partition numbers need not be unique: $m_4 = 1$ and $m_4 = 2$ lead to

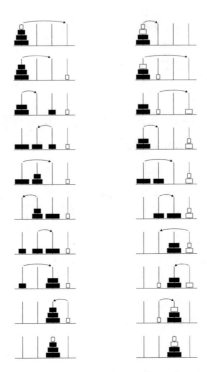

Figure 5.1: Two optimal solutions for 4 discs

essentially different solutions both of length $FS_4^4 = 9$; see Figure 5.1. Of course, solutions which differ by an exchange of the roles of the two auxiliary pegs 1 and 3 only are not considered essentially different.

In preparation for the *recursive Frame-Stewart algorithm* we need a revised version of Algorithm 4, namely Algorithm 16. This algorithm will transfer a *subtower* consisting of n consecutive discs starting with d from peg i to peg j using the auxiliary peg k in the optimal number of $2^n - 1$ moves, where now the three different pegs are labeled with arbitrary natural numbers.

Now Algorithm 17 will realize a solution of length FS_4^n to transfer $n \in \mathbb{N}_0$ discs from peg i to peg j via auxiliary pegs k and l as long as i, j, k, l are mutually different.

For the example with $n = 10$ discs, see Figure 5.2, where the number of individual disc moves is shown above the arrows.

It is easy to derive an algorithm assigning a(ll) partition number(s) and the value of FS_4^n to a given n directly from Definition 5.1 (see Table 5.1).

Algorithm 16 Revised recursive algorithm for 3 pegs

Procedure FS3(n, d, i, j, k)

Parameter n: number of discs to be transferred $\{n \in \mathbb{N}_0\}$

Parameter d: smallest disc to be transferred $\{d \in \mathbb{N}\}$

Parameter i: source peg $\{i \in \mathbb{N}_0\}$

Parameter j: goal peg $\{j \in \mathbb{N}_0,\ j \neq i\}$

Parameter k: auxiliary peg $\{k \in \mathbb{N}_0,\ i \neq k \neq j\}$

    if $n \neq 0$ **then**

        FS3$(n-1, d, i, k, j)$ {transfers $n-1$ smallest discs to auxiliary peg}

        move disc $d + n - 1$ from i to j {moves largest disc to goal peg}

        FS3$(n-1, d, k, j, i)$ {transfers $n-1$ smallest discs to goal peg}

    end if

Algorithm 17 Recursive Frame-Stewart algorithm for 4 pegs

Procedure FS4(n, i, j, k, l)

Parameter n: number of discs $\{n \in \mathbb{N}_0\}$

Parameter i: source peg $\{i \in \mathbb{N}_0\}$

Parameter j: goal peg $\{j \in \mathbb{N}_0 \setminus \{i\}\}$

Parameter k: the Devil's peg $\{k \in \mathbb{N}_0 \setminus \{i, j\}\}$

Parameter l: the other auxiliary peg $\{l \in \mathbb{N}_0 \setminus \{i, j, k\}\}$

    if $n \neq 0$ **then**

        $m \leftarrow$ partition number for n {partition number assigned}

        FS4(m, i, k, j, l)

        FS3$(n - m, m + 1, i, j, l)$

        FS4(m, k, j, i, l)

    end if

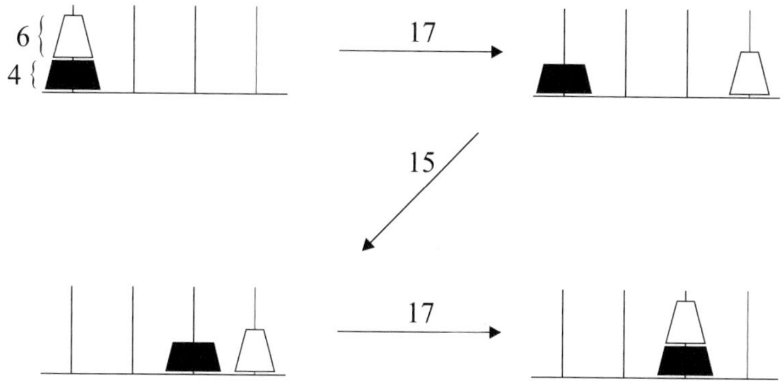

Figure 5.2: Recursive solution for 10 discs on 4 pegs. Only subtower moves are shown.

ν	Δ_ν	x	n	FS_3^n	m	FS_4^n	$FS_4^n - FS_4^{n-1}$	$\overline{FS}_4^n$
0	0	0	0	0	0	0		0
1	1	0	1	1	0	1	1	1
1	1	1	2	3	1,0	3	2	2
2	3	0	3	7	1	5	2	4
2	3	1	4	15	2,1	9	4	6
2	3	2	5	31	3,2	13	4	8
3	6	0	6	63	3	17	4	12
3	6	1	7	127	4,3	25	8	16
3	6	2	8	255	5,4	33	8	20
3	6	3	9	511	6,5	41	8	24
4	10	0	10	1023	6	49	8	32
4	10	1	11	2047	7,6	65	16	40

Table 5.1: Frame-Stewart numbers (Δ_ν triangular number, x excess, m partition number(s))

5.2 Frame-Stewart Numbers

To facilitate their construction, it would be desirable to obtain a closed formula for Frame-Stewart numbers and their partitions (cf. [355], [193, Theorem 1], and [296, Theorem 3.2]). As had already been observed by Dudeney, a look at the first few cases of n in Table 5.1 reveals the special status of the sequence $(\Delta_\nu)_{\nu \in \mathbb{N}_0}$ of triangular numbers; recall from Chapter 0 that each Δ_ν is given by

$$\Delta_\nu = \frac{\nu(\nu+1)}{2}$$

and that every $n \in \mathbb{N}_0$ can uniquely be written as $n = \Delta_\nu + x$, with index $\nu = \left\lfloor \dfrac{\sqrt{8n+1}-1}{2} \right\rfloor \in \mathbb{N}_0$ and excess $x \in [\nu+1]_0$. We arrive at the following comprehensive result.

Theorem 5.4. *For every $\nu \in \mathbb{N}_0$ and $x \in [\nu+1]_0$:*

$$FS_4^{\Delta_\nu + x} = (\nu - 1 + x)2^\nu + 1.$$

Moreover, for $\nu \in \mathbb{N}$, $\Delta_{\nu-1} + x$ is a partition number for $\Delta_\nu + x$. It is the only one if $x = 0$, and there is precisely one further partition number, namely $\Delta_{\nu-1} + x - 1$, otherwise.

Proof. We will proceed by induction on $n = \Delta_\nu + x$. For $n = 0$ we have $\nu = 0 = x$ and $FS_4^0 = 0 = (0 - 1 + 0)2^0 + 1$. Similarly, the case $\nu = 1$ with excesses $x = 0, 1$, that is $n = 1, 2$, can easily be checked (cf. Table 5.1). So let $n > 2$ and assume that the theorem is true for all $m = \Delta_\mu + y < n$. We have to show that the function f on

$[n]_0$ given by $f(m) = 2FS_4^m + FS_3^{n-m}$ attains its minimum at $m_n := \Delta_{\nu-1} + x$, that it has no other minimum point for $x = 0$ and precisely one more, namely $m_n - 1$, otherwise, and that $f(m_n) = (\nu - 1 + x)2^{\nu} + 1$; cf. Figure 5.3.

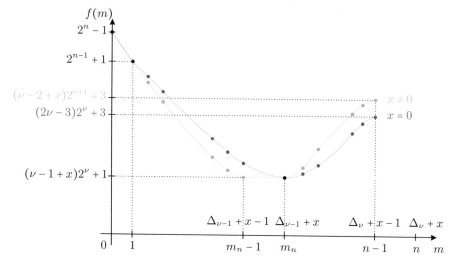

Figure 5.3: Induction step for Theorem 5.4

We begin with the latter. By induction assumption we have, making use of $n - m_n = \Delta_{\nu} - \Delta_{\nu-1} = \nu$,

$$f(m_n) = f(\Delta_{\nu-1} + x) = 2\left((\nu - 1 - 1 + x)2^{\nu-1} + 1\right) + 2^{\nu} - 1 = (\nu - 1 + x)2^{\nu} + 1,$$

for $x \in [\nu]_0$ and

$$f(m_n) = f(\Delta_{\nu}) = 2\left((\nu - 1)2^{\nu} + 1\right) + 2^{\nu} - 1 = (2\nu - 1)2^{\nu} + 1 = (\nu - 1 + x)2^{\nu} + 1,$$

if $x = \nu$.

In order to locate the point(s) of minimum of f, we define the function g on $[n - 1]$ by $g(m) = f(m) - f(m - 1)$; it then suffices to show that

$$g(m) \begin{cases} < 0, & m \le m_n; \\ > 0, & m > m_n; \end{cases}$$

if $x = 0$ (red case in Figure 5.3) and

$$g(m) \begin{cases} < 0, & m < m_n; \\ = 0, & m = m_n; \\ > 0, & m > m_n; \end{cases}$$

if $x \ne 0$ (green case in Figure 5.3). This can be reduced to a bookkeeping task observing that (recall that $m = \Delta_{\mu} + y$),

$$g(m) = 2\left(FS_4^m - FS_4^{m-1}\right) - 2^{n-m} = 2^{\mu + (y \ne 0)} - 2^{n-m},$$

such that only the exponents have to be compared in the different cases above. This is left to the reader as Exercise 5.1.   □

Motivated by the definition of Frame-Stewart numbers, A. Matsuura [305] studied recurrence relations of the following form. Let $p, q \in \mathbb{N}$ be given, then set $T(0, p, q) = 0$ and for $n \in \mathbb{N}$ let

$$T(n, p, q) = \min\{pT(m, p, q) + qFS_3^{n-m} \mid m \in [n]_0\}.$$

In this notation, $FS_4^n = T(n, 2, 1)$.

It is not difficult to see (by induction on n) that $T(n, p, q) = qT(n, p, 1)$. To present the main result of Matsuura, the following sequences are crucial. Let $s^{(1)}$ denote the constant sequence of 1s on $\mathbb{N}$ and for $\ell \in \mathbb{N}_2$, let $s^{(\ell)} = (s_i^{(\ell)})_{i \in \mathbb{N}}$ be the non-decreasing sequence of numbers of the form $2^j \ell^k$, $j, k \in \mathbb{N}_0$. For instance,

$$s^{(2)} = (1, 2, 2, 4, 4, 4, 8, 8, 8, 8, 16, 16, 16, 16, 16, 32, \ldots)$$

and

$$s^{(3)} = (1, 2, 3, 4, 6, 8, 9, 12, 16, 18, 24, 27, 32, 36, 48, 54, \ldots).$$

The numbers $2^j 3^k$ are known as the 3-*smooth numbers* and the sequence $s^{(3)}$ as the 3-*smooth sequence*. We will meet them again in Chapter 8 where the numbers $T(n, 3, 2)$ will find an attractive interpretation.

Theorem 5.5. (Matsuura [305, Corollary 2.1]) *For any integers $p, q \in \mathbb{N}$ and every $n \in \mathbb{N}_0$,*

$$T(n, p, q) = q \sum_{i=1}^{n} s_i^{(p)}.$$

Corollary 5.6. *For any $n \in \mathbb{N}_0$,*

$$FS_4^n = \sum_{i=1}^{n} s_i^{(2)}.$$

Theorem 5.5 was further generalized by J. Chappelon and Matsuura in [76].

An iterative algorithm for the Frame-Stewart solution can be obtained from the following interesting observation (cf. [189, p. 30f]). We divide the n-tower into N *subtowers* of consecutive discs and regard each subtower as a *superdisc* $D \in [N]$ containing n_D individual discs somehow glued together. The superdiscs then move from peg 0 to peg 2 using only peg 1 as an auxiliary peg as in the classical 3-peg version. The moves of these superdiscs in turn are viewed as transfers of towers with relabeled discs, these transfers being legal by using the Devil's peg 3 which is empty before each move of a superdisc. See Figure 5.4 for the example of $n = 10$ discs with 4 subtowers=superdiscs of size $1, 2, 3, 4$, respectively; the number of individual disc moves during each transfer of a superdisc is given on the arrows. (Note that the sequence of moves is essentially the same as for the recursive Algorithm 17 in Figure 5.2, just the roles of pegs 1 and 3 have been switched depending on the different parts the Devil's peg plays in the two algorithms. Note also that the superdiscs consistently follow the moves of the solution of the TH with four discs presented in Figure 2.2.)

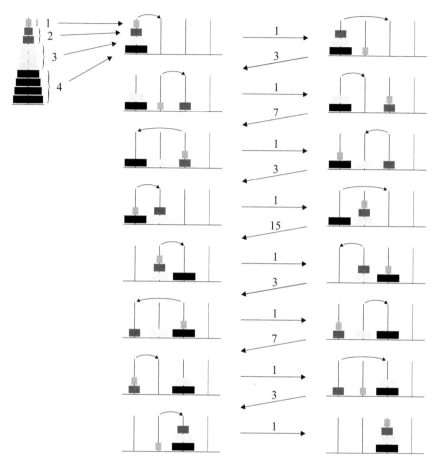

Figure 5.4: Iterative solution for 10 discs on 4 pegs. Only subtower moves are shown.

With the notation $d_D = \sum_{\delta=1}^{D} n_\delta$ we see that superdisc D contains the discs from $d_{D-1} + 1$ to d_D, and we arrive at Algorithm 18 which is designed to transfer n discs from peg 0 to peg 2 making use of auxiliary pegs 1 and 3 in FS_4^n moves.

According to Proposition 2.4, Algorithm 18 needs

$$\sum_{D=1}^{N} 2^{N-D} FS_3^{n_D} = \sum_{k=0}^{N-1} 2^k FS_3^{n_{N-k}} = \sum_{k=0}^{N-1} 2^k FS_3^{d_{N-k}-d_{N-k-1}}$$

Algorithm 18 Iterative Frame-Stewart algorithm for 4 pegs

Procedure FS4i(n)

Parameter $n = n_1 + \cdots + n_N$: number of discs $\{n \in \mathbb{N}_0\}$

   $d_0 = 0$

   for $\delta = 1$ **to** N

      $d_\delta = d_{\delta-1} + n_\delta$

   end for

   for $l = 1$ **to** $2^N - 1$ {moves of superdiscs}

      p0m$(N, 0, 2, l)$, $D \leftarrow d$ {superdisc D to be moved from i_l to j_l in move l}

      FS3$(n_D, d_{D-1} + 1, i_l, j_l, 3)$ {transfer of superdisc in move l}

   end for

---

moves, and it only remains to choose a partition $n = n_1 + \cdots + n_N$ in Algorithm 18 such that this number of moves is equal to FS_4^n. But this is the case if and only if d_{N-k-1} is a partition number for d_{N-k}, as an N-fold iteration of the defining formula for FS_4^n shows. If for n as in Theorem 5.4 we always choose the smallest partition number, we get $N = \nu$ and $n_D = D + (D > \nu - x)$; cf. [295]. Note that although, in contrast to the other superdiscs, $D = 1$ has 4 pegs to move on, this is not a real advantage, because it contains at most 2 discs; cf. also Table 5.1 and Figure 5.1.

Remark. Instead of the recursive Algorithm 16 we could have used any other algorithm for the solution of the classical 3-peg problem, provided such an algorithm is adapted to the range of discs to be moved. In that way, Algorithm 18 can be modified to a purely non-recursive one.

Every disc moves a power of 2 times in the Frame-Stewart algorithms. Moreover, the largest disc moves exactly once; this is obvious for the recursive algorithm and follows for the iterative one from the fact that superdisc N, which contains disc n, moves exactly once using three pegs only. (Recall from Proposition 2.2 that in the optimal solution for the classical task for *three* pegs, the largest disc moves exactly once.) So for $n \in \mathbb{N}_0$, the numbers

$$\overline{FS}_4^n := \frac{1}{2}\left(FS_4^{1+n} - 1\right)$$

(cf. Table 5.1) give the number of moves made by the algorithm to distribute n discs from peg 0 to two other pegs 1 and 3, say (to allow disc $n + 1$ to move from 0 to 2 in one step). This is the original approach taken by Frame and according to Definition 5.2 and Proposition 5.3 we know that $\overline{FS}_4 = \overline{F}_4$, the sequence of Frame numbers. In fact, in order to prove the Frame-Stewart conjecture it would suffice to show that this procedure is optimal, because we can prove the following (cf. [56, p. 119]):

Proposition 5.7. *In an optimal solution for The Reve's puzzle, the largest disc moves exactly once.*

Proof. Let μ_n be the number of moves necessary and sufficient to move $n \in \mathbb{N}_0$ discs from one peg to two other ones. Then an $(n + 1)$-tower can be transferred from one peg to another one in at most $2\mu_n + 1$ moves.

Any solution of this problem needs at least μ_n moves before the first move of disc $n + 1$ and at least μ_n moves after the last one. Therefore, it would need at least $2\mu_n + 2$ moves if disc $n + 1$ were to move more than once, such that this cannot be optimal. □

By virtue of Theorem 5.4, we get

Corollary 5.8. $\forall \nu \in \mathbb{N}_0 \ \forall x \in [\nu + 1]_0 : \overline{FS}_4^{\Delta_\nu + x} = (\nu + x)2^{\nu - 1}.$

An equivalent formulation of the Frame-Stewart conjecture is therefore:

Frame-Stewart Conjecture*. *The task to move* $n = \Delta_\nu + x$ *discs,* $\nu \in \mathbb{N}_0$, $x \in [\nu + 1]_0$, *from a single peg to two others can not be achieved in less than* $(\nu + x)2^{\nu - 1}$ *moves.*

It is this form of the Frame-Stewart conjecture that Korf and Felner checked up to $n = 29$ to arrive at the result for $n = 30$ mentioned in Section 5.1 in connection with the original FSC.

The FSC* can be proved without recourse to numerical computations for small numbers of discs only; cf. Exercise 5.2. So where is the problem for a proof of the general case? Well, in addition to the assumption, proved to be justified in Proposition 5.7, that the largest disc $n + 1$ moves directly from peg 0 to peg 2, Frame's approach uses no less than three other assumptions! Realizing that minimal solutions might not be unique, Frame coined the term *most economical* for any solution requiring the least numbers of moves. We now put Frame's original words from [150, p. 216] into frames. First, he writes for the half-way situation of his proposed most economical solution about the n smaller discs (which he calls "washers") that[56]

> ... we may assume that the $n - m$ largest of these washers are on the first peg ... and the m smallest ones on the last.

This amounts to the assumption that among the optimal solutions, there is always what we will call a *subtower solution*, i.e. half-way through the solution the two auxiliary pegs each contain only consecutive discs. There are optimal non-subtower solutions, but computational evidence suggests that they are restricted to $n = 3$ (e.g., placing discs 3 and 1 on one and disc 2 on the other auxiliary peg in altogether 4 moves) and $n = 4$, but in each of these cases there are also subtower solutions. Nevertheless, Frame's first assumption has neither been proved nor refuted yet. One has to be careful, however: there are examples for *non-optimal* partitions of the

---

[56]The quotes in frames are original except for an adaptation to our notation.

n-tower, where a non-subtower solution may be strictly shorter than a subtower solution; see Figure 5.5, where $n = 8$ discs are divided into two sets of 4 discs each. After 21 moves, a half-way situation is reached where disc $n + 1 = 9$ could move. Note that a subtower solution for the partition number 4 would take 24 moves. (We cannot prove this fact right now, but the reader will be able to do so later on; see p. 233.) This can not be an optimal partition number though, as can be seen from the penultimate state in Figure 5.5 which has partition number 5 and can be reached in 20 moves; cf. also Table 5.1. Note further that the partition numbers in that table refer to subtower solutions; optimal partition numbers for non-subtower solutions could, in principle, be different.

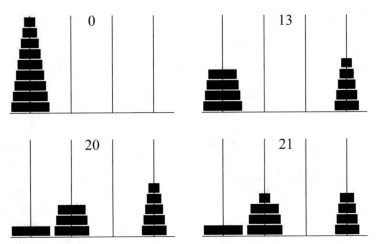

Figure 5.5: Four stages of a non-subtower solution; states after 0, 13, 20, and 21 moves, respectively

Although the example shows that it is not feasible to convert a non-subtower solution into a subtower solution of the same, or even smaller length in general, Frame emphasizes his subtower assumption by saying that

> ... if the smallest washer is to cover $m-1$ others at this stage, it is a most economical method to have these be the smallest washers, so that these in turn do not block other pegs.

Frame continues with another assumption, namely

> It is also a most economical method to have the larger pile contain the smaller washers, since the latter have access to more pegs at the time of their transfer.

We have seen in Proposition 5.3 that this premise, namely $m \geq n - m$, is no real restriction, provided we understand the previous assumption to comprise the last and probably most restrictive assumption of Frame, namely that the subtowers may be moved one after the other, the smaller m discs using 4 pegs, the others only the 3 remaining after disc 1 has reached its terminal position on the Devil's peg. This is not said explicitly, but only, in reversed order, for the second part of the solution when following the half-way situation

> ... we move the largest washer to its destination, then move the $n - m$ next largest washers onto it using one auxiliary peg, ... and finally move the m smallest washers using two auxiliary pegs.

So there are two unproven assumptions in Frame's solution, and they are also implicit in Stewart's approach, namely,

1. Among the optimal solutions for the half-way problem there is always a subtower solution.

2. An optimal subtower solution can be obtained by first transferring the subtower containing the smaller discs using four pegs and then moving the other subtower on three pegs.

This was precisely Dunkel's objection in [121], who added

> It would be desirable to have a brief and rigorous proof of these lemmas.

This wish has not been fulfilled until today! Dunkel pointed out that it would be sufficient to prove the following:

> If the first m washers from the top of the initial peg are placed on a single auxiliary peg, say peg 3; the next $n - m$ on peg 1; and ... the largest washer is placed with one move on the final peg 2 where it is alone; then, for a suitable value of m, this plan for the removal of all the washers from the initial peg requires as small a number of moves as any other.

This angle to look at the Frame-Stewart conjecture can be pronounced in a more concise fashion by starting from an infinite tower of discs $d \in \mathbb{N}$ on peg 0. Then

Frame-Stewart Conjecture ** (Dunkel's Lemma).** *After* $(\nu + x)2^{\nu-1}$ *moves,* $\nu \in \mathbb{N}_0$, $x \in [\nu + 1]_0$, *at most* $\Delta_\nu + x$ *discs have left peg* 0.

In view of this approach it would be advisable to consider problems of type P1 also for the four-peg version of the TH. The argument in the proof of Proposition 5.7 shows that the largest disc moves only once in this case too.

An interesting observation about Frame numbers is that they contain a subsequence consisting of powers of 2. The first few instances in Table 5.1 are at $n = 1, 2, 3, 5, 7, 10$. To obtain a complete listing, we need a little lemma.

Lemma 5.9. $\forall\, m \in \mathbb{N}_0\; \exists_1\, \mu \in \mathbb{N}_0,\; \rho \in [2^\mu + 1]_0 : m = \mu + 2^\mu - 1 + \rho$.

The proof follows easily from the fact that the sequence on $\mathbb{N}_0$, given by $\delta_\mu = \mu + 2^\mu - 1$ (A052944), satisfies the recurrence

$$\delta_0 = 0,\; \delta_{\mu+1} = \delta_\mu + 2^\mu + 1,$$

and is therefore obviously strictly monotone increasing, the ρ then covering all natural numbers in between.

We now define the sequence a on $\mathbb{N}_0$ by

$$a_m = \Delta_{2^\mu - 1 + \rho} + 2^\mu - \rho + 1 - 0^\rho,$$

where m is represented according to Lemma 5.9. (We remind the reader that $0^k = (k = 0)$ by convention.) This sequence fulfills the recurrence

$$a_0 = 1,\; \forall\, m \in \mathbb{N}_0 :\; a_{m+1} = a_m + 2^\mu - 1 + \rho + 0^\rho.$$

We now get the following remarkable result.

Proposition 5.10. $\forall\, m \in \mathbb{N}_0 :\; \overline{FS}_4^{a_m} = 2^m$.

Proof. In order to make use of Corollary 5.8, we have to represent a_m as $\Delta_\nu + x$ with $x \in [\nu + 1]_0$ as before. If $\rho = 0$, then $\nu = 2^\mu$ and $x = 0$, such that

$$\overline{FS}_4^{a_m} = FS_4^{\Delta_{2^\mu}} = 2^\mu 2^{2^\mu - 1} = 2^m.$$

For $\rho \neq 0$, we have $\nu = 2^\mu - 1 + \rho$ and $x = 2^\mu - \rho + 1$, such that

$$\overline{FS}_4^{a_m} = FS_4^{\Delta_{2^\mu - 1 + \rho} + 2^\mu - \rho + 1} = \left((2^\mu - 1 + \rho) + (2^\mu - \rho + 1)\right) 2^{(2^\mu - 1 + \rho) - 1} = 2^{\mu + 2^\mu - 1 + \rho} = 2^m.$$

$\square$

Remark 5.11. *The sequence a also fulfills the recurrence*

$$a_0 = 1,\, a_{m+1} = a_m + \left\lfloor \frac{\sqrt{8a_m + 1} - 1}{2} \right\rfloor,$$

as can be seen by looking at the proof of Proposition 5.10 and the formula for ν on p. 51. This might be used to calculate members of the sequence more efficiently; it starts $1, 2, 3, 5, 7, 10, 14, 18, 23, 29, 36, 44, 52, \ldots$.

Since $\overline{FS}_4^n$ is strictly increasing and the statement in Proposition 5.10 is exhaustive in m, the sequence a covers all n for which $\overline{FS}_4^n$ is a power of 2. The sequence a' given by $a'_m = a_{m-1} + 1$, $m \in \mathbb{N}$, characterizes those disc numbers where the FS numbers are of the form $2^m + 1$.

5.3   Numerical Evidence for The Reve's Puzzle

As long as one had been unable to prove or disprove the Frame-Stewart conjecture (FSC), one was tempted to try it with the aid of a computer. A lot of work has indeed been done in this direction. Along the way, computations revealed some completely unexpected facts about The Reve's puzzle, thus giving further evidence for its intrinsic hardness.

The first significant computer experiment was published in 1999 by J.-P. Bode and Hinz [56] who confirmed the FSC for up to and including 17 discs. They employed a breadth-first search (BFS). Recall from the solution to Exercise 0.8 that in an ordinary breadth-first search algorithm the root vertex is first put into level 0, all its neighbors into level 1, then all unvisited neighbors of vertices in level 1 are placed into level 2, and so forth. A vertex is marked as visited immediately after it is put into its level to prevent repeated visits. This procedure ends when all vertices (of the component) have been included into the resulting tree structure. In our case, it suffices to store just three levels at a time and to stop as soon as the goal, namely one of the other perfect states (or rather a half-way state), has been reached. The level number when this happens is just the distance between root and goal (one half of the distance minus 1). Another reduction takes into account the symmetries of the problem by limiting the set of vertices on a level to non-equivalent representatives.

This method of a *frontier search* combined with a *delayed duplicate detection (DDD)* was also employed by Korf [251] who in 2004 extended the previous result by confirming the conjecture for up to and including 24 discs. This was topped by S. Strohhäcker [411] to 25 in 2005. Finally, as already mentioned, in 2007 Korf and Felner demonstrated the truth of the FSC for up to and including 30 discs. They employed heuristics based on a complete search for the 22-disc problem. The computation for 30 discs took over 17 days to run and required a maximum of 398 gigabytes of disk[57] storage. The limiting resource was CPU time. The computation was also executed to verify FSC for 31 discs. It ran more than three months and used two terabytes of disk storage. Unfortunately, an unrecoverable disk error occurred at depth 419. An analysis of this error showed that there is a one in 191 million probability that the confirming 31-disc calculation is incorrect.

The tree built up from the perfect state is, of course, an analogue to the tree with vertex set T^n and root 0^n we introduced for the case of three pegs in Section 2.3; we just have to replace $T = [3]_0$ by $Q = [4]_0$ and to represent states by elements in Q^n as before. The resulting tree for four pegs is a subgraph of what we will call H_4^n, namely the *Hanoi graph* with vertex set Q^n and where edges refer

---

[57]It was Korf in [252, footnote 2] who suggested to distinguish the spellings of a magnetic disk and a TH disc.

to legal moves of discs. More formally (cf. (2.11)),

$$V(H_4^n) = Q^n,$$
$$E(H_4^n) = \left\{ \{\underline{s}i\overline{s}, \underline{s}j\overline{s}\} \mid i, j \in Q,\ i \neq j,\ d \in [n],\ \underline{s} \in Q^{n-d},\ \overline{s} \in (Q \setminus \{i, j\})^{d-1} \right\}.$$

For an element of $E(H_4^n)$, disc d moves from peg i to peg j with the $d-1$ smaller discs not lying on either of these pegs and the $n-d$ larger discs in arbitrary position. Just like before, the graphs H_4^n can be constructed recursively as shown in Figure 5.6.

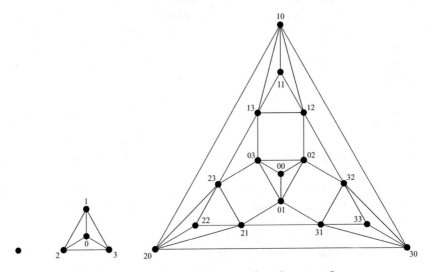

Figure 5.6: Hanoi graphs H_4^0, H_4^1, and H_4^2

Formally, the recursive definition of the edge sets is (cf. Equation (2.12))

$$E(H_4^0) = \varnothing,$$
$$\forall\, n \in \mathbb{N}_0 : E(H_4^{1+n}) = \left\{ \{ir, is\} \mid i \in Q,\ \{r, s\} \in E(H_4^n) \right\}$$
$$\cup \left\{ \{is, js\} \mid i, j \in Q,\ i \neq j,\ s \in (Q \setminus \{i, j\})^n \right\}. \qquad (5.2)$$

As before, connectivity is obvious from (5.2), and we will again denote the distance between two vertices $s, t \in Q^n$ by $\mathrm{d}(s, t)$. In contrast to all Sierpiński graphs, two subgraphs iH_4^n and jH_4^n of H_4^{1+n}, $n \in \mathbb{N}$, are joined by more than one edge, namely by 2^n edges. This fact makes it easier to find paths between two vertices, but much harder to find a shortest one. Unlike the invariant for Sierpiński graphs from (4.11) in Corollary 4.7, the sum of distances from some state s to all perfect states is not a constant anymore: in H_4^2 this sum is 9 for perfect states, but 8 for all other vertices, as can easily be seen from Figure 5.6.

In Figure 5.7 we try to give an impression of the distance distribution (cf. [67, Section 9.3]) on Hanoi graphs as compared with Sierpiński graphs. In the figure, the distance matrices are shown for $p = 4$ and $n = 5$ with the vertices arranged in lexicographic order. The distance 0 is represented by a black square, the distance equal to the diameter by a white box with the intermediate distances following a corresponding gray scale. Such representations are known as *heat maps*. We can deduce the higher complexity of the Hanoi case.

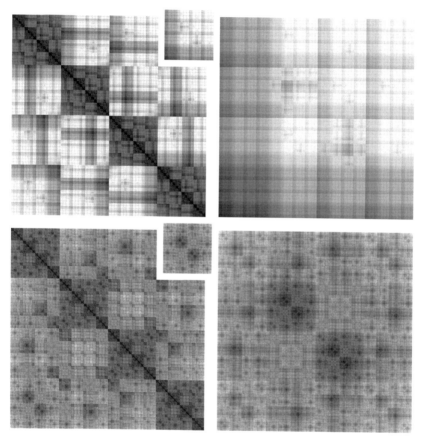

Figure 5.7: Heat maps for distances in S_4^5 (top left), H_4^5 (bottom left) and enlarged submatrices of pairs of vertices $(0Q^4, 3Q^4)$

Along with his numerical attempts to verify the FSC, Korf [251] computed the eccentricities of perfect states in H_4^n for $n \leq 20$, while Korf and Felner [253] extended this to $n = 22$; see also [252]. The computation for 22 discs took 9 days and 18 hours to run. The computer had dual two gigahertz processors, two gigabytes of internal memory, and the problem required over 2.2 terabytes of external storage.

We computed the eccentricities of perfect states in the graphs H_4^{23} to H_4^{25} in the frame of our project pr87mo *Eccentricities in Hanoi Graphs* at SuperMUC (Bayerische Akademie der Wissenschaften). We implemented a delayed duplicate detection (DDD) BFS algorithm. This approach is practical when levels are so large that they cannot fit entirely into the internal memory (RAM). Also there is no need for a special data structure in order to document which vertices have already been visited. The DDD-BFS algorithm is suitable for massive paralleliza- tion. For generating level nodes we used a high-performance parallel file system. Level 503 in the tree rooted in a perfect state in H_4^{25} is the largest and contains $12\,764\,800\,322\,652$ states. The main limiting resources with the implementation were the speed of reading/writing from/to the file system, querying meta-data servers, and creating/deleting numerous files.

The results are collected in Table 5.2 where *ex* stands for the *excess function* defined by $ex(n) = \varepsilon(0^n) - d(0^n, 3^n)$.

n	1	2	3	4	5	6	7	8	9	10	11
$\varepsilon(0^n)$	1	3	5	9	13	17	25	33	41	49	65
$ex(n)$	0	0	0	0	0	0	0	0	0	0	0

n	12	13	14	15	16	17	18	19	20	21	22
$\varepsilon(0^n)$	81	97	113	**130**	161	193	225	257	**294**	**341**	**394**
$ex(n)$	0	0	0	**1**	0	0	0	0	**5**	**20**	**9**

n	23	24	25
$\varepsilon(0^n)$	**453**	**516**	**585**
$ex(n)$	**4**	**3**	**8**

Table 5.2: Eccentricities of perfect states and their excess in H_4^n

Table 5.2 reveals a very surprising anomaly: for $n = 15$ and $n = 20$ to 25, there exist regular states $s \in Q^n$ such that $d(0^n, s) > d(0^n, 3^n)$. We call this feature of H_4^n the *Korf phenomenon*. It is first of all amazing that there are such integers n, and, moreover, that the first one occurs as late as $n = 15$, and that there is a gap before $n = 20$. No explanation for this surprising phenomenon is known, but Korf and Felner proposed:

Conjecture 5.12. *For any $n \geq 20$, $ex(n) > 0$.*

Note that the eccentricity of a perfect state of H_4^n is a lower bound for the diameter of H_4^n. Hence Conjecture 5.12 could be weakened by asserting that $EX(n) > 0$ holds for any $n \geq 20$, where the function EX is defined by $EX(n) = \mathrm{diam}(H_4^n) - d(0^n, 3^n)$, $n \in \mathbb{N}$. In the examples which have been calculated, the excess function *ex* is not monotonic; see Figure 5.8. This is a bit disturbing. It may turn out that EX behaves more regularly, so we are inclined to propose:

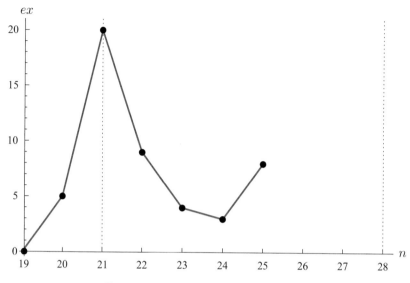

Figure 5.8: Shape of the function ex

Conjecture 5.13. *The function EX is eventually (strictly) monotone increasing.*

In [308] it was shown that $EX(n) = 0$ for $n \leq 13$. Later we were able to extend this to $n = 14$. Moreover, $EX(n) \geq ex(n) > 0$ for $n = 15, 20, 21, 22, 23, 24, 25$. In the range $n = 15, 16, 17$ Korf confirmed $EX(n) = ex(n)$ computationally.[58] If Conjecture 5.13 proves to be true, this together with the fact that ex is not monotonic will show that perfect states are not special anymore. On the other hand, no example is known for $EX(n) > ex(n)$, i.e. perfect states *not* lying in the **periphery**, which would be necessary, at least for some n, for Conjecture 5.13 to hold. This means that in order to decide about this question, one has to look at tasks of type P2.

In Table 5.3 we have collected computational results of Strohhäcker [411, Table 6] and M. Meier [308, p. 29–31] on the eccentricity of the states of H_4^n, $n \in [8]$. For instance, $\mathrm{rad}(H_4^6) = 13$, $\mathrm{diam}(H_4^6) = 17$, and there are 1776 states in H_4^6 with eccentricity 14. Moreover, among these there are 74 pairwise non-equivalent states if we consider two states to be *equivalent* if one state emanates from the other by a permutation of the pegs:

$$s \approx s' \Leftrightarrow \exists \sigma \in \mathrm{Sym}(Q) \; \forall d \in [n] : s'_d = \sigma(s_d).$$

We will address the problem of symmetries of H_4^n and how to compute the number of equivalence classes, which is $|[Q^n]| = \frac{1}{6}(4^{n-1} + 3 \cdot 2^{n-1} + 2)$ (cf A007581), and their sizes as well as identifying representatives in Section 5.6.

---

[58] R. E. Korf, private communication, 2016

n	$\mathrm{rad}(H_4^n), \mathrm{rad}(H_4^n)+1, \ldots, \mathrm{diam}(H_4^n)$ number of states with given eccentricity number of non-equivalent states with given eccentricity										diam − rad + 1 $\lvert Q^n \rvert = 4^n$ $\lvert Q^n / \approx \rvert$
1	1										1
	4										4
	1										1
2	3										1
	16										16
	2										2
3	4	5									2
	24	40									64
	1	4									5
4	7	8	9								3
	144	96	16								256
	6	7	2								15
5	10	11	12	13							4
	528	360	120	16							1024
	22	18	9	2							51
6	13	14	15	16	17						5
	168	1776	1644	492	16						4096
	7	74	74	30	2						187
7	18	19	20	21	22	23	24	25			8
	624	3840	6600	3300	1740	216	48	16			16384
	26	160	275	148	89	12	3	2			715
8	24	25	26	27	28	29	30	31	32	33	10
	3024	12648	19464	16968	9312	3120	720	216	48	16	65536
	126	527	811	716	407	148	43	12	3	2	2795

Table 5.3: Eccentricities in H_4^n

In the frame of our project at SuperMUC we meanwhile computed all eccentricities in all graphs H_4^n up to and including $n = 15$. Figures 5.9 and 5.10 show an attempt to locate center and periphery vertices in H_4^{15}, i.e. those with minimal or maximal eccentricity, respectively. Eccentricities are represented in a matrix-layout ($16\,384 \times 16\,384$), the extreme values marked as 64×64 patches.

Let us now look at Table 5.4, where eccentricity results for H_4^n are summarized for $n \in [15]$ and where

$$C(G) = \{u \in V(G) \mid \varepsilon(u) = \mathrm{rad}(G)\}$$

denotes the center of graph G and

$$P(G) = \{u \in V(G) \mid \varepsilon(u) = \mathrm{diam}(G)\}$$

is its periphery. With the exception of the last column of the table, which contains the average eccentricity, the data in the first eight rows are already included in

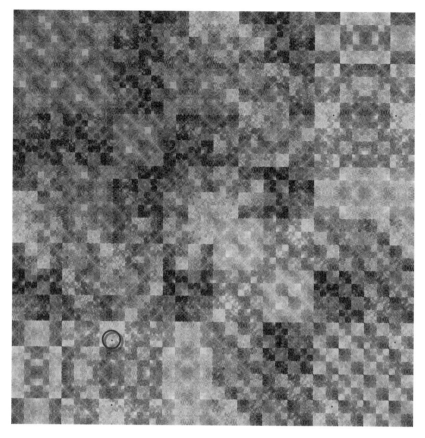

Figure 5.9: All eccentricities in H_4^{15} with blue patches marking areas with center vertices and red patches indicating vertices in the periphery

Table 5.3. However, adding the next rows uncovered, for $n = 10$, another surprise which we called *Peripheral phenomenon*.

For $n = 1$ and $n = 2$, all the states are peripheral (and central); the case $n = 3$ is sporadic because $\mathrm{diam}(H_4^3) - \mathrm{rad}(H_4^3) = 1$. After these small cases, $P(H_4^n)$, $4 \le n \le 9$, consists of the states of the form ij^{n-1}, $i, j \in Q$, whence $|P(H_4^n)| = 16$. But for $n = 10$, our computations revealed that the states $iiiji^6$ and $jiiji^6$, $i \ne j$, also belong to the periphery of H_4^{10}. Hence $|P(H_4^{10})| = 40$. However, $P(H_4^n)$ for $11 \le n \le 14$ have again only 16 elements each, so we are justified to speak about the Peripheral phenomenon, which we still cannot explain. In particular, the case $n = 15$ stands out with its small center and large periphery. The statement $EX(15) = 1 = ex(15)$ has been confirmed though. The calculations for the corresponding values for $n \ge 16$ have not been embarked yet.

In his determination of the diameters of H_4^{16} and H_4^{17}, Korf employed an

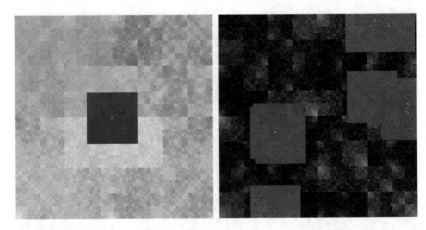

Figure 5.10: Zoom into areas with central (left) and peripheral (right) vertices marked with light green dots

n	rad	diam	diam − rad + 1	$|C|$	$|C/\approx|$	$|P|$	$|P/\approx|$	$\bar{\varepsilon}$
1	1	1	1	4	1	4	1	1.0000
2	3	3	1	16	2	16	2	3.0000
3	4	5	2	24	1	40	4	4.6250
4	7	9	3	144	6	16	2	7.5000
5	10	13	4	528	22	16	2	10.6328
6	13	17	5	168	7	16	2	14.6123
7	18	25	8	624	26	16	2	20.1594
8	24	33	10	3024	126	16	2	26.4672
9	30	41	12	1416	59	16	2	33.7114
10	37	49	13	216	9	40	4	42.5358
11	47	65	19	1056	44	16	2	53.8479
12	58	81	24	888	37	16	2	66.6067
13	71	97	27	2208	92	16	2	80.7622
14	85	113	29	288	12	16	2	96.9233
15	101	130	30	48	2	2356	99	115.8447

Table 5.4: Radius, diameter, center, periphery, and average eccentricity of H_4^n

approach without calculating *all* eccentricities, based on an algorithm anticipated in [415]. A flavor of this method is given in Exercise 5.3.

We will report on further numerical experiments, e.g., in search for non-subtower solutions, after the theoretical discussion of the TH with more than four pegs.

5.4 Even More Pegs

Throughout we will use the variables q, ν, $x \in \mathbb{N}_0$ and $h \in \mathbb{N}$.

Recall from Section 5.1 that Frame and Stewart proposed formally different strategies to solve The Reve's puzzle. In Proposition 5.3 we proved, however, that both strategies lead to the same number of moves—the Frame-Stewart numbers. In fact, they proposed different approaches for any number $p \geq 4$ of pegs as follows.

In Frame's strategy, the upper n discs of the initial perfect $(n+1)$-tower are collected into $p-2$ subtowers, considered as superdiscs $D_1, D_2, \ldots, D_{p-2}$, consisting of $n_1, n_2, \ldots, n_{p-2}$ consecutive discs, respectively, such that $\sum_{i=1}^{p-2} n_i = n$ and $n_1 \geq n_2 \geq \cdots \geq n_{p-2}$. Then D_1 is transferred to an auxiliary peg using all pegs, D_2 to another auxiliary peg using $p-1$ pegs and so on until finally D_{p-2} is transferred to the remaining auxiliary peg using 3 pegs. Then the move of disc $n+1$ to the goal peg is made, after which the superdiscs are transferred to the goal peg in reverse order. The strategy is schematically shown in Figure 5.11 for the case $p = 5$.

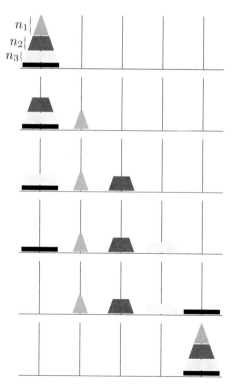

Figure 5.11: Frame's strategy on five pegs

This approach leads to the following numbers. Let $p \in \mathbb{N}_3$. Then set $F_p^0 = 0$

and for any $n \in \mathbb{N}_0$ let

$$\overline{F}_p^n = \min \left\{ F_p^{n_1} + \cdots + F_3^{n_{p-2}} \mid \sum_{i=1}^{p-2} n_i = n, \ 0 \le n_{p-2} \le \cdots \le n_1 \right\}, \ F_p^{n+1} = 2\overline{F}_p^n + 1 \,.$$

Note that this definition is compatible with Definition 5.2; in particular, we will call the $\overline{F}_p^n$ *Frame numbers* again. The condition $n_1 \ge n_2 \ge \cdots \ge n_{p-2}$ is called the *monotonicity condition* and was assumed by Frame to be self-evident. However, it seems natural to define numbers $\hat{F}_p^n$ in the same way as F_p^n, except that the monotonicity condition is not required:

$$\hat{\overline{F}}_p^n = \min \left\{ \hat{F}_p^{n_1} + \cdots + \hat{F}_3^{n_{p-2}} \mid \sum_{i=1}^{p-2} n_i = n, \ n_i \in \mathbb{N}_0 \right\}, \ \hat{F}_p^{n+1} = 2\hat{\overline{F}}_p^n + 1 \,.$$

Stewart's strategy is conceptually a bit simpler: split the starting perfect n-tower into subtowers $\overline{D}$ and $\underline{D}$, consisting of the $m \in [n]_0$ upper discs and the $n-m$ lower ones, respectively. Then transfer the top m discs to an auxiliary peg using all pegs, transfer $\underline{D}$ to the goal peg using $p-1$ pegs, and complete the task by transferring $\overline{D}$ to the goal peg. Now we are led to the following numbers FS_p^n, which will again be called *Frame-Stewart numbers*. As already defined, $FS_3^n = 2^n - 1$ for any $n \in \mathbb{N}_0$. Let $p \in \mathbb{N}_4$. Then $FS_p^0 = 0$ and for any $n \in \mathbb{N}$ let:

$$FS_p^n = \min \left\{ 2FS_p^m + FS_{p-1}^{n-m} \mid m \in [n]_0 \right\} \,. \tag{5.3}$$

Again this is consistent with Definition 5.1. One might be tempted to introduce numbers defined as FS_p^n, but with the additional monotonicity condition $m \ge n-m$. However, this would lead to larger values which is not what we want to have. For instance, the minimum of $2FS_5^m + FS_4^{9-m}$ is assumed for $m = 3$ and $m = 4$, but not for $m \ge 5$; cf. [238, p. 146].

On the other hand, we can extend the number of configurations involved by considering *all* partitions into superdiscs; cf. Exercise 5.4. More precisely, let $\hat{A}_p^0 = 0$ for every $p \in \mathbb{N}_2$ and assume the convention that $\hat{A}_2^1 = 1$ and $\hat{A}_2^n = \infty$ for $n \in \mathbb{N}_2$. (This corresponds to a TH with only two pegs, where obviously only at most 1 disc can be moved.) Then, for $p \in \mathbb{N}_3$ and $n \in \mathbb{N}$, set

$$\hat{A}_p^n = \min \bigcup_{k=2}^{p-1} \left\{ 2(\hat{A}_p^{n_1} + \cdots + \hat{A}_{k+1}^{n_{p-k}}) + \hat{A}_k^{n_{p-k+1}} \mid \sum_{i=1}^{p-k+1} n_i = n, \ n_i \in \mathbb{N}_0, \ n_{p-k+1} \ne 0 \right\} \,.$$

Moreover, let A_p^n be defined as $\hat{A}_p^n$ with the addition that the monotonicity condition is required in all the partitions. Then the main result of Klavžar, Milutinović, and Petr [238, Theorem 6.5] reads as follows:

Theorem 5.14. *For any $p \in \mathbb{N}_3$ and every $n \in \mathbb{N}_0$,*

$$FS_p^n = F_p^n = \hat{F}_p^n = A_p^n = \hat{A}_p^n \,.$$

The proof of this theorem is rather lengthy and technical, the interested reader can verify it in [238]. It seems that the technical difficulties are intrinsic as can be seen already in the proof of Proposition 5.3 which is a rather special case of Theorem 5.14.

Therefore it comes as no surprise that the analogue of FSC (cf. p. 209) for more than four pegs will be an even bigger challenge. We cannot resist but to quote Knuth (see [293]), whose opinion is based on "a solid week of working [he spent] on it pretty hard" (cf. [244, p. 321]): "I doubt if anyone will ever resolve the conjecture; it is truly difficult."

Strong Frame-Stewart Conjecture (SFSC).

$$\forall\, p \in \mathbb{N}_3 \;\, \forall\, n \in \mathbb{N}_0 : \; \mathrm{d}\left(0^n, (p-1)^n\right) = FS_p^n \,.$$

Recall from Theorem 5.4 that for $p = 4$, the Frame-Stewart numbers can be written explicitly as

$$FS_4^{\Delta_\nu + x} = (\nu - 1 + x)2^\nu + 1\,,$$

where $\Delta_\nu = \frac{\nu(\nu+1)}{2}$ and $x \in [\nu + 1]_0$. Lu in [278, Theorem 3] and also Klavžar and Milutinović in [236, Theorem 3.1] extended this formula to any number of pegs. Setting, for $q \in \mathbb{N}_0$,

$$\Delta_{q,\nu} = \binom{q + \nu - 1}{q} \tag{5.4}$$

one gets the following.

Theorem 5.15. *For all $p \in \mathbb{N}_3$, $\nu \in \mathbb{N}_0$, and $0 \le x \le \Delta_{p-3,\nu+1}$,*

$$FS_p^{\Delta_{p-2,\nu} + x} = (P_{p-3}(\nu) + x)2^\nu + (-1)^p\,, \tag{5.5}$$

where P_q is the polynomial of degree $q \in \mathbb{N}_0$ defined by

$$P_q(\nu) = (-1)^q \left(1 + \sum_{k=1}^{q}(-1)^k \Delta_{k,\nu}\right). \tag{5.6}$$

Note that $\Delta_{2,\nu} = \Delta_\nu$. In addition, the reader is invited to verify that Theorem 5.15 reduces to the formula in Theorem 5.4 in the case $p = 4$ and that for $p = 3$ it gives the number of moves of the 3-pegs problem. The case $p = 5$, where $\Delta_{3,\nu}$ are the tetrahedral numbers, was considered by A. Brousseau in [66] who also conjectured the general formula writing the polynomial as $\widetilde{P}_q(\nu) = \sum_{k=0}^{q}(-2)^k \binom{q+\nu}{q-k}$. The reader should prove in Exercise 5.5 that indeed $\widetilde{P}_q = P_q$. We will come back to the *hypertetrahedral numbers* $\Delta_{q,\nu}$ and provide a proof for Theorem 5.15 in Section 5.5.3 below.

Cull and E. F. Ecklund noticed in [87, Theorem 2] that for $n \geq p-1$ the partition according to Stewart is unique if and only if $n = \Delta_{p-2,\nu}$ for some $\nu \geq 2$. The general question of partition numbers has been addressed in [236, Theorem 2.7].

An iterative algorithm realizing FS_p can be constructed from one for FS_{p-1} by replacing 4 and 3 in Algorithm 18 by p and $p-1$, respectively and employing the same argument as there to choose appropriate partition numbers. We arrive at another representation for the FS numbers, namely (cf. [297, Theorem 2.2])

$$FS_p^n = \sum_{D=1}^{N} 2^{N-D} FS_{p-1}^{n_D}, \tag{5.7}$$

where again N is the number of superdiscs and n_D is the number of individual discs composing superdisc $D \in [N]$. The formula in (5.7) actually applies to $p = 3$ as well, where the notions of superdiscs and discs coincide and putting $FS_2^1 = 1$, because on two pegs only one disc can be moved; cf. [297, Remark 2.1].

One would, of course, like to attempt a proof of the SFSC on the lines of Frame's and Stewart's approaches. In fact, the argument of Proposition 5.7 carries over to $p > 4$, such that the largest disc will move exactly once in an optimal solution. As before, the problem for $n + 1$ discs can thus be reduced to find an optimal half-path solution, i.e. the task to distribute n discs from a tower on one peg onto $p-2$ other pegs in a minimum number of moves. Again, the largest disc n will only move once (and once more after the move of disc $n + 1$). To see this, we first note that the proof of the boxer rule Lemma 2.32 (cf. also Lemma 4.8) carries over verbally to $p > 3$ pegs, such that we have

Lemma 5.16. *If on a geodesic of H_p^{1+n}, $n \in \mathbb{N}_0$, the largest disc is moved away from a peg, it will not return to the same peg.*

On the other hand, for a P1-type task, the proof of Theorem 2.7 can *not* be adapted easily to the p-pegs situation because it relies intimately on the knowledge of the (length of a) solution for the P0 problem. We therefore have to resort to an ingenious observation of S. Aumann [23, Lemma 5.19]:

Lemma 5.17. *Let P be a geodesic in H_p^{1+n}, $p \in \mathbb{N}_3$, $n \in \mathbb{N}_0$, containing a vertex t associated with a state in which pegs i and j, $i \neq j$, are empty, i.e. $t([1+n]) \subseteq [p]_0 \setminus \{i,j\}$. Then no edge between iH_p^n and jH_p^n is contained in P.*

Proof. Suppose that such an edge exists in P. Then we may assume that it corresponds to the first move on a shortest path from is to t with some $s \in ([p]_0 \setminus \{i,j\})^n$. Consider the subpath from js to t and switch the roles of i and j in it (cf. [359, p. 17]). This results in a path from is to t which is shorter by one edge than the original one, leading to a contradiction. □

We can now prove the following (cf. [23, Theorem 5.20]).

Theorem 5.18. *Let* $s, t \in [p]_0^{1+n}$, $p \in \mathbb{N}_3$, $n \in \mathbb{N}_0$. *If either* s *or* t *is perfect, then in an optimal solution to get from* s *to* t, *the largest disc* $n + 1$ *moves precisely once, if* $s_{n+1} \neq t_{n+1}$, *and not at all otherwise.*

Proof. We may assume that $t = 0^{1+n}$. If $s_{n+1} = 0$, then, by Lemma 5.16, disc $n + 1$ does not move in an optimal solution. Otherwise, according to Lemma 5.17, there is no move of disc $n + 1$ between i and j if $|\{0, i, j\}| = 3$. The only possible moves of $n + 1$ are therefore to or from peg 0, such that by Lemma 5.16 it can only move once, namely from s_{n+1} to 0. □

This theorem now allows us to come back to the question from p. 218; see Exercise 5.6. As a direct consequence of Theorem 5.18 we get the following formula for the distance between perfect states.

Corollary 5.19. *Let* $p \in \mathbb{N}_3$ *and* $n \in \mathbb{N}_0$. *Then*

$$d(0^{1+n}, (p-1)^{1+n}) = 1 + 2\min\{d(0^n, s) \mid s \in [p-2]^n\}.$$

Apart from $n = 0$ and $n = 1$ (case $\nu = 0$ of Theorem 5.15), there are, of course, a couple of other trivial cases, where the SFSC holds, namely ($\nu = 1$) when $2 \le n \le p - 1$: to reach the half-way distribution, each of the smaller $n - 1$ discs has to move at least once, so $d(0^n, (p-1)^n) \ge 2(n-1) + 1 = 2n - 1 = FS_p^n$. For $\nu = 2$ in Theorem 5.15 we get:

Proposition 5.20. *For* $p \in \mathbb{N}_3$ *and* $n \in \mathbb{N}_p$,

$$d(0^n, (p-1)^n) \ge 4n - 2p + 1;$$

in particular, the SFSC holds for $n \le \binom{p}{2}$.

Proof. Those among the $n - 1$ smaller discs which move only once to reach half-way must then lie on the bottom of their peg, since the first move of a disc from a perfect state always goes there. Because of availability of pegs, this is possible for at most $m \le p - 2$ of these discs. The other $n - 1 - m$ have to move at least twice to half-way. We therefore have

$$d(0^n, (p-1)^n) \ge 1 + 2m + 4(n - 1 - m) = 4n - 3 - 2m \ge 4n - 2p + 1.$$

For $p \le n \le \binom{p}{2}$, this lower bound is precisely the value of FS_p^n. □

This proposition leads to a useful representation of the Frame numbers; see Exercise 5.7.

M. Szegedy [414] used a similar approach to prove a non-trivial lower bound on $d(0^n, (p-1)^n)$. Instead of attacking the problem directly, he suggested to consider the function g defined as follows. For a regular state s, let $\bar{g}(s)$ be the minimum number of moves needed to move each disc at least once. Then set

$$g(p, n) = \min\{\bar{g}(s) \mid s \in [p]_0^n\}. \tag{5.8}$$

Since clearly $d\left(0^n, (p-1)^n\right) \geq g(p,n)$, Szegedy suggested to bound $g(p,n)$ from below. In the following we describe how this can be achieved.

Consider the P0 problem on $p \in \mathbb{N}_4$ pegs with n discs and let m be an integer such that $0 < m < \frac{n}{2p}$. Let S be the set of the smallest $n - 2pm$ discs and let s be an arbitrary regular state. By the Pigeonhole principle there exists a peg, say peg 0, holding at least $2m$ discs among the largest $2pm$ discs. Let L_1 be the set of the m largest discs on peg 0 and let L_2 be the set of the next m largest discs on the same peg. The sets of discs S, L_1, and L_2 are pairwise disjoint. Note also that every disc from L_1 is larger than any disc from L_2, and any disc from L_2 is larger than any disc from S.

Let $\mathcal{M}$ be a sequence of moves starting in s and moving each disc at least once. Split $\mathcal{M}$ into $\mathcal{M}_1$ and $\mathcal{M}_2$, where $\mathcal{M}_1$ are the starting moves up to and excluding the first move of the smallest disc from L_1, while $\mathcal{M}_2$ contains the first move of that disc as well as all the remaining moves. Clearly, during the sequence $\mathcal{M}_1$ each disc of L_2 is moved (at least once), and during $\mathcal{M}_2$ each disc of L_1 is moved. Suppose now that during $\mathcal{M}_1$, at least one of the discs from S is not moved. Then the peg of this idle disc cannot be used for the moves of the discs from L_2. In other words, if some disc from S is not moved during $\mathcal{M}_1$, only $p-1$ pegs are used to move discs from L_2. It follows that the sequence $\mathcal{M}_1$ has at least

$$\min\{g(p, n-2pm), g(p-1, m)\}$$

terms. Similarly, during $\mathcal{M}_2$ either all the discs from S were moved at least once or the discs from L_1 were moved using at most $p-1$ pegs. Therefore,

$$g(p,n) \geq 2\,\min\{g(p, n-2pm), g(p-1, m)\}\,.$$

After some technical calculation for which we refer to [414], this inequality leads to the following:

Theorem 5.21. (Szegedy [414, Theorem 1]) *For any $p \in \mathbb{N}_3$,*

$$d\left(0^n, (p-1)^n\right) \geq 2^{(1+o(1))C_p n^{1/(p-2)}}\,,$$

where

$$C_p = \frac{1}{2}\left(\frac{12}{p(p-1)}\right)^{1/(p-2)}.$$

Theorem 5.21 needs several comments. First, it uses the little-o notation which in the particular case o(1) means that the term tends to 0 when n tends to infinity. Observe also that we have developed the theorem for $p \geq 4$, but it is stated for $p \geq 3$. This is justified in Exercise 5.8.

Xiao Chen and J. Shen [80] adopted Szegedy's strategy and proved that

$$d\left(0^n, (p-1)^n\right) \geq 2^{(1+o(1))(n(p-2)!)^{1/(p-2)}}\,,$$

in other words, they improved the constant C_p from Theorem 5.21 to $(p-2)!^{1/(p-2)}$.

In 2016, Codruţ Grosu gave the lower bound a more explicit form. Making use of Bousch's ideas from [62] (see the next section) his statement [169, Theorem 3] is that

$$d\left(0^n, (p-1)^n\right) \geq \frac{\nu + \mu}{4^{p-2}} 2^{\nu}$$

for $p \in \mathbb{N}_4$ and $n \in \mathbb{N}$ represented as $\Delta_{p-2,\nu} + \Delta_{p-3,\mu} + y$ with $\mu, \nu \in \mathbb{N}_0$, $\mu \leq \nu$, and $y \in [\Delta_{p-4,\mu+1}]$. Compare this with (5.5).

Coming back to the Frame-Stewart approach, Theorem 5.14 implies, in particular, that Frame's monotonicity condition constitutes no restriction. However, we are facing the same confinements as for the case $p = 4$ of The Reve's puzzle, namely the two assumptions of a subtower solution and the sequential transfer of these subtowers/superdiscs; cf. the discussion (starting on p. 217) of the original statements of Frame and Dunkel, who actually formulated them referring to the general case $p \in \mathbb{N}_3$. We will analyze these assumptions further with a look at numerical experiments in Section 5.7. As a preparation we come back to the representation of the TH by graphs in Section 5.6. But now we turn to the solution of The Reve's Puzzle.

5.5 Bousch's Solution of The Reve's Puzzle

73 years after Dunkel asked for a "brief and rigorous proof" for his lemma which would lead to the SFSC being confirmed, Thierry Bousch came up with a solution for the $p = 4$ case, i.e. The Reve's Puzzle. His rigorous proof, which is the topic of the present section, is *not* brief though! It results from two ingenious ideas of Bousch, namely to look back more carefully to the indications contained in the works of Dudeney and to introduce the function Ψ, a prototype of which we already encountered in Exercise 2.14. We therefore begin this section by collecting some material developed from the scattered explanations of Dudeney. Throughout this section we will use the variables q, ν, $x \in \mathbb{N}_0$ and $h \in \mathbb{N}$.

5.5.1 Some Two-Dimensional Arrays

Recall from (5.4) that for every *dimension* q the hypertetrahedral sequence Δ_q is defined by

$$\Delta_{q,\nu} = \binom{q + \nu - 1}{q}.$$

Viewed as a two-dimensional array, this is built up just like Pascal's Arithmetical triangle for $\binom{q+\nu}{q}$, but with the 1s in the 0th column replaced by 0s,[59] i.e. $\Delta_{q,0} = 0$.

---

[59]Pascal labelled rows and columns starting from 1.

For instance,

$$
\begin{aligned}
\Delta_{0,\nu} &= (\nu \in \mathbb{N}) & &\text{(characteristic function of } \mathbb{N}), \\
\Delta_{1,\nu} &= \nu & &\text{(natural numbers)}, \\
\Delta_{2,\nu} &= \Delta_\nu & &\text{(triangular numbers)}, \\
\Delta_{3,\nu} &= T_\nu & &\text{(tetrahedral numbers)}.
\end{aligned}
$$

In general,

$$\Delta_{q+1,\nu+1} = \Delta_{q+1,\nu} + \Delta_{q,\nu+1}. \tag{5.9}$$

Then $\overline{\Delta_{q+1}} = \Delta_q$, $\Sigma(\Delta_q) = \Delta_{q+1}$, and $\Delta_{q+1,\nu} \geq \Delta_{q,\nu}$ (cf. Table 5.5).

$q \setminus \nu$	0	1	2	3	4	5	6	7	8	9	10
0	0	1	1	1	1	1	1	1	1	1	1
1	0	1	2	3	4	5	6	7	8	9	10
2	0	1	3	6	10	15	21	28	36	45	55
3	0	1	4	10	20	35	56	84	120	165	220
4	0	1	5	15	35	70	126	210	330	495	715
5	0	1	6	21	56	126	252	462	792	1287	2002
6	0	1	7	28	84	210	462	924	1716	3003	5005
7	0	1	8	36	120	330	792	1716	3432	6435	11440
8	0	1	9	45	165	495	1287	3003	6435	12870	24310
9	0	1	10	55	220	715	2002	5005	11440	24310	48620

Table 5.5: The hypertetrahedral array $\Delta_{q,\nu}$

Next we define for each q the sequence P_q by

$$P_{q,\nu} = (-1)^q \left(1 + \sum_{k=1}^{q} (-1)^k \Delta_{k,\nu} \right).$$

Note that this is just to reinterpret the polynomial P_q in ν from (5.6) as an integer sequence. We immediately see that $P_{q,0} = (-1)^q$, $P_{0,\nu} = 1$ and from

$$P_{h,\nu} + P_{h-1,\nu} = \Delta_{h,\nu} \tag{5.10}$$

that

$$P_{1,\nu} = \nu - 1, \ P_{2,\nu} = \Delta_{\nu-1} + 1, \ P_{3,\nu} = T_{\nu-1} + \nu - 1.^{60}$$

More advanced properties are

$$P_{q+1,\nu+1} = P_{q+1,\nu} + P_{q,\nu+1} \tag{5.11}$$

---

[60] For $n \in \mathbb{N}_0$ the sequences $P_{2,n+1}$ and $P_{3,n+1}$ are A000124 and A003600, respectively.

$q \setminus \nu$	0	1	2	3	4	5	6	7	8	9	10
0	1	1	1	1	1	1	1	1	1	1	1
1	-1	0	1	2	3	4	5	6	7	8	9
2	1	1	2	4	7	11	16	22	29	37	46
3	-1	0	2	6	13	24	40	62	91	128	174
4	1	1	3	9	22	46	86	148	239	367	541
5	-1	0	3	12	34	80	166	314	553	920	1461
6	1	1	4	16	50	130	296	610	1163	2083	3544
7	-1	0	4	20	70	200	496	1106	2269	4352	7896
8	1	1	5	25	95	295	791	1897	4166	8518	16414
9	-1	0	5	30	125	420	1211	3108	7274	15792	32206

Table 5.6: The P-array $P_{q,\nu}$

and (cf. the solution of Exercise 5.5)

$$2P_{q,\nu+1} = P_{q,\nu} + \Delta_{q,\nu+1}. \tag{5.12}$$

Formula (5.11) shows that the construction of the array $P_{q,\nu}$ is as the previous one for $\Delta_{q,\nu}$, but now the 0th column contains the alternating sign sequence $(-1)^q$ (cf. Table 5.6).

Proof of (5.11).

$$
\begin{aligned}
P_{q+1,\nu+1} - P_{q+1,\nu} &= (-1)^{q+1}\left(\sum_{k=1}^{q+1}(-1)^k \Delta_{k,\nu+1} - \sum_{k=1}^{q+1}(-1)^k \Delta_{k,\nu}\right) \\
&= (-1)^{q+1}\sum_{k=1}^{q+1}(-1)^k \Delta_{k-1,\nu+1} \\
&= (-1)^q\left(1 + \sum_{k=1}^{q}(-1)^k \Delta_{k,\nu+1}\right) \\
&= P_{q,\nu+1}.
\end{aligned}
$$

$\square$

Proof of (5.12).

$$
\begin{aligned}
(-1)^q \cdot 2P_{q,\nu+1} &= 2 + \sum_{k=1}^{q}(-1)^k \Delta_{k,\nu+1} + \sum_{k=1}^{q}(-1)^k \Delta_{k,\nu+1} \\
&= 2 + \sum_{k=1}^{q}(-1)^k \Delta_{k,\nu+1} + \sum_{k=1}^{q}(-1)^k \Delta_{k,\nu} + \sum_{k=1}^{q}(-1)^k \Delta_{k-1,\nu+1} \\
&= 1 + \sum_{k=1}^{q}(-1)^k \Delta_{k,\nu} + (-1)^q \Delta_{q,\nu+1} \\
&= (-1)^q \left(P_{q,\nu} + \Delta_{q,\nu+1}\right).
\end{aligned}
$$

$\square$

We also have

$$\sum_{k=0}^{q}(-2)^{k}\binom{q+\nu}{q-k} = P_{q,\nu} = 2^{-\nu}\left((-1)^{q} + \sum_{k=0}^{\nu-1}2^{k}\Delta_{q,k+1}\right).$$

This is so since the left and right terms are $(-1)^{q}$ for $\nu = 0$ and fulfil the recurrence relation in (5.12). For the left identity (anticipated by Brousseau [66, p. 176]) see Theorem 5.15 and the discussion following it, the right one also follows from (5.12) and Lemma 2.18.

5.5.2 Dudeney's Array

In the tradition of Édouard Lucas, Henry Ernest Dudeney published mathematical puzzles in English magazines starting in the 1890s. Since these sources are difficult to access,[61] we will present a detailed analysis here. Without mentioning his French antetype Dudeney presented *The Tower of Bramah* in 1896 in [111] under the pen name "Sphinx", thereby citing Ball's famous description (cf. page 1) almost verbally, even with the strange spelling of the Hindu God of Creation.[62] Dudeney posed the usual questions for the length of a minimal solution for the TH with three pegs and 8 or 64 discs. For the first time he also asked for the best first move. His answers contain a clerical error (225 moves for 8 discs instead of 255) and no proofs.

Only three months later and again in *The Weekly Dispatch*, Dudeney, or rather "Sphinx Junior", came up with the challenge to solve the problem in the presence of four pegs (or pins) with 10 discs [112]. The accompanying illustration shows the four pins in a row and ten discs of equal diameter, but numbered from 1 to 10, where the values replace the diameters of the original puzzle. So the *teminus ante quem* for the TH with four pegs is 1896–11–15.

In the solution, Dudeney claims two weeks later that "[t]he feat can be performed in forty-nine moves, which is the correct answer." Just like Fermat he denounces the lack of space for not going "fully into the subject", but reveals "the Junior's method of working the puzzle", namely to make successive piles of 6 (1 to 6) and 3 (7 to 9) discs, then removing disc 10 taking 17, 7, and 1 moves, respectively. Finally, the two piles are replaced on disc 10 in inverse order, so that in total 49 moves are performed, "the fewest possible". Again, there is no argument given for the latter claim. It would take more than a hundred years before the claim was eventually proved in [56]! But the stage was set for what would later become known as the *Frame-Stewart conjecture*, and *Frame's algorithm* was anticipated.

In *The London Magazine* of 1902 Dudeney posed the problem to find a shortest solution for the TH with four pegs again; see [113, p. 367f]. He does not mention the TH, though, but dresses his question into the fantastic story of *The Reve's*

---

[61] Most of the documents we obtained from P. K. Stockmeyer.
[62] Only in 1908, Dudeney would mention the name of the inventor in [117, p. 784].

puzzle,[63] in fact the first of his collection of *Canterbury puzzles*. Here the pegs are replaced by stools and the discs by "eight cheeses of graduating sizes". The text was then verbally adopted for the book [116, p. 1f] (with the inscrutable exception of "treat to" being replaced by "give"). The challenge is to solve the problem "in the fewest possible moves, first with 8, then with 10, and afterwards with 21 cheeses".

The solution given for $n = 8$ in [113, p. 480] (which has *not* been taken over for [116]) is an example of what has later been called *Frame's algorithm* (see [150]; cf. also above, Section 5.2).[64] Dudeney mentions both possible partitions of the topmost 7 cheeses, namely $4|3$ and $5|2$, leading to the conclusion that "[t]he least number of moves in which the cheeses can be so removed is thirty and three." However, no attempt has been made to *prove* minimality! For the remaining cases $n = 10$ and $n = 21$ again the claim of minimality of 49 and 321 moves, respectively, is made.[65] It is also noted that the partitions are unique, respectively.

The Reve's puzzle [sic!] was taken up by Dudeney back in the *Dispatch* as puzzle no. 447 in [114], citing several passages from [113]. Strangely enough, it is claimed that 6 cheeses "may all be removed . . . in sixty-three moves."; this is in fact the value if only three stools are used, i.e. $d_3(0^6, 1^6) = 63$, for four stools $d_4(0^6, 1^6) = 17$ moves are sufficient (and necessary). The "half-guinea prize"[66] was then offered for finding the "fewest moves in which *thirty-six* cheeses may be so removed", i.e. $d_4(0^{36}, 1^{36})$.

Dudeney was quite surprised that he received the correct solution[67] within the time limit by one competitor only. This may have been the reason why he explains his own approach in great detail. Assuming that the largest cheese is moved only once,[68] he divides the solution into the three steps to form two piles with all cheeses except the largest, to move the latter and finally to reunite the others on top of it. He also observes that the first of these steps is the crucial one, the last one being just its reversal. He then presents "the curious point of the thing", namely triangular numbers, for which he claims that the formation of the two piles is unique, whereas there are two different ways otherwise. He characterizes triangular numbers as those n for which $8n + 1$ is a square, i.e. $n = \frac{\nu(\nu+1)}{2}$; the partition of the $n - 1$ smaller discs is then given by $\frac{\nu(\nu-1)}{2}|\nu - 1$. He finds this fact "[r]ather peculiar". For $n = 36$, i.e. $\nu = 8$, one now has to pile up the $\Delta_7 = 28$ smallest cheeses to one auxiliary stool in 769 moves, then the next 7 cheeses to the other auxiliary stool in 127 moves, to accomplish the first step in altogether 896, so that the whole solution takes $2 \cdot 896 + 1 = 1793$ moves.

---

[63] Alluding to the *Canterbury tales* of Geoffrey Chaucer; Dudeney here employs the spelling of "Reve" from Chaucer's time for historical consistency, but later he also used "Reeve".

[64] The respective partial move numbers contain a little clerical error.

[65] The minimality of the first two values, i.e. $d_4(0^8, 1^8)$ and $d_4(0^{10}, 1^{10})$, where d_p denotes the distance in graphs H_p^n, was shown in [56] and of the third one, $d_4(0^{21}, 1^{21})$, in [251].

[66] This was the weekly prize awarded by the journal. A half-guinea was 10s6d or 10/6 as it can be seen on The Hatter's hat in Lewis Carroll's *Alice's Adventures in Wonderland*.

[67] Again up to minimality which has been open until quite recently!

[68] This assumption has been justified for the first time in [56, p. 119]; cf. Proposition 5.7.

The most remarkable feature of Dudeney's exposition is a small table of three rows and eight columns. The first row contains the powers of two 2^ν for $\nu \in [8]$, the second row are these numbers reduced by 1, i.e. the *Mersenne numbers* M_ν, and for the sequence c of numbers in the third row Dudeney gives the formative rule $1 = c_1$, $2c_\nu + M_{\nu+1} = c_{\nu+1}$. He then deduces that $c_8 = 1793$ is the number of moves for the eighth triangular number 36. This, in fact, anticipates *Stewart's algorithm* (see [391]; cf. also Definition 5.1), provided that the number of cheeses is a triangular number.

The sparse reaction to his puzzle no. 447 might also have caused Dudeney's "intention to advance the subject one stage further." This he did in puzzle no. 494 in [115] by proposing "the case where there is one stool more and one cheese less", i.e. five stools and 35 cheeses. Two weeks later he revealed the solution he had in mind, namely 351 moves, but postponed the explanation to the next issue of the journal. This time ten readers had submitted the desired number.

Dudeney's justification starts with a table for $\Delta_{h,\nu}$, with $h \in [3]$ representing the number of auxiliary stools and $\nu \in [7]$, but he stresses that this table could be extended "to any required length". It shows the figurate numbers of cheeses he considers for three ($h = 1$), four ($h = 2$), and five ($h = 3$) stools, respectively. For the construction of this table Dudeney says that the sequence in the second row is made up from the partial sums of the first and similarly for the third row of tetrahedral numbers, which he calls "pyramidal". The first two rows of a second table repeat the last two rows of the table in reference [114], i.e. M_ν and c_ν, shortened to $\nu \in [7]$. Again the third row of this new table is obtained from the second as the latter is from the first. The claim is now that the entries in this table yield the optimal number of moves for the respective number of stools and cheeses from the first table. Since 35 is the fifth number in the third row of the first table, i.e. the fifth tetrahedral number, the previously announced solution 351 can be found at the fifth position of the third row in the second table.

We generalize these ideas and define the array a by (cf. Table 5.7)

$$a_{q,0} = 0, \quad a_{0,\nu+1} = 1, \quad a_{q+1,\nu+1} = 2a_{q+1,\nu} + a_{q,\nu+1}. \tag{5.13}$$

Dudeney's claim is now that given the number $p = h + 2$ of stools and a figurate number $n = \Delta_{h,\nu}$ of cheeses, we can deduce the corresponding value for ν from Table 5.5 and with this ν read the number of moves $a_{h,\nu}$ for his algorithm from Table 5.7. Dudeney also observes that $\Delta_{h,\nu+1}$ cheeses on $h + 2$ stools have to be split successively into piles of $\Delta_{h,\nu}$, $\Delta_{h-1,\nu}$, ..., $\Delta_{1,\nu}$, and 1 discs from small to the largest. So it can be said that Dudeney has provided an argument for the statement

$$d_{h+2}\left(0^{\Delta_{h,\nu}}, 1^{\Delta_{h,\nu}}\right) \le a_{h,\nu}.$$

The assumption that Dudeney was aware of the general method, at least for hypertetrahedral disc numbers, is supported by references in another London monthly, namely *The Strand Magazine*. In fact, in [117, p. 784] he asserts that he has "given elsewhere the general solution for any number of needles". In [118]

$q \setminus \nu$	0	1	2	3	4	5	6	7	8	sequence
0	0	1	1	1	1	1	1	1	1	
1	0	1	3	7	15	31	63	127	255	Mersenne
2	0	1	5	17	49	129	321	769	1793	Dudeney
3	0	1	7	31	111	351	1023	2815	7423	
4	0	1	9	49	209	769	2561	7937	23297	
5	0	1	11	71	351	1471	5503	18943	61183	
6	0	1	13	97	545	2561	10625	40193	141569	
7	0	1	15	127	799	4159	18943	78079	297727	

Table 5.7: Dudeney's array $a_{q,\nu}$

and [120] we meet problems equivalent to finding $d_5(0^{10}, 1^{10})$, $d_5(0^{20}, 1^{20})$ and $d_6(0^{15}, 1^{15})$, respectively. The solutions give the "right" numbers of moves, namely $a_{3,3}$, $a_{3,4}$ and $a_{4,3}$ in our notation, together with the corresponding (unique) decompositions according to "Junior's method", but no proof of minimality. In fact, Proposition 5.20 shows that the first and the last value are indeed optimal; however, the question has only been settled recently for the second one; see [213, Proposition 1]!

In the last column of [115], Dudeney also deals with the problem of how to proceed if the number of cheeses is *not* tetrahedral in the case of five stools. He presents a "little subsidiary table" of three rows and six columns. The first row contains $\Delta_{3,\nu}$ for $\nu \in [6]$, the second has the powers of two $2^{\nu-1}$, and for the third row s he gives the rule $0 = s_1$, $\Delta_{3,\nu} \cdot 2^{\nu-1} + s_\nu = s_{\nu+1}$. He then presents, with two examples, his recipe to get the (presumed) optimal number of moves $\widetilde{d}_5(0^n, 1^n)$ for n cheeses, which amounts to

$$\mu := \min\{\nu \in \mathbb{N} \mid \Delta_{3,\nu} > n\}, \quad \widetilde{d}_5(0^n, 1^n) = n2^{\mu-1} - s_\mu. \tag{5.14}$$

Although Dudeney mentions that for non-tetrahedral cheese numbers there "will always be more than one way in which the cheeses may be piled", he does *not* say how.

This question, also for the case of non-triangular cheese numbers with four stools, is left to "the reader to work out for himself" in the solutions section of [116, p. 131f], which repeats the first two tables from the previous discussion, but not the subsidiary table and its interpretation. Maybe the author had meanwhile realized that $s_\mu = a_{4,\mu-1}$, as we will see below, such that all information is already contained in the two tables for hypertetrahedral and Dudeney numbers. Even more remarkable is the fact that in his final solution Dudeney is *not* addressing the question of optimality at all, but writes that his "8 cheeses can be removed in 33 moves; 10 cheeses in 49 moves; and 21 cheeses in 321 moves."

We will now analyze properties of the Dudeney array. For instance,

$$a_{h,\nu} = P_{h-1,\nu} \cdot 2^\nu + (-1)^h; \tag{5.15}$$

in particular (cf. (5.10)):

$$a_{q+1,\nu} + a_{q,\nu} = 2^{\nu} \Delta_{q,\nu} \,. \tag{5.16}$$

and

$$a_{1,\nu} = 2^{\nu} - 1, \ a_{2,\nu} = (\nu-1)2^{\nu} + 1, \ a_{3,\nu} = (\Delta_{\nu-1}+1)2^{\nu} - 1, \ a_{4,\nu} = (T_{\nu-1}+\nu-1)2^{\nu} + 1.^{[69]}$$

Note that a_2 is Dudeney's original sequence c; it will be called the *Dudeney sequence*.

Proof of (5.15). Let the right-hand term in (5.15) be called $\widetilde{a}_{h,\nu}$. Then $\widetilde{a}_{h,0} = 0$, $\widetilde{a}_{1,\nu} = 2^{\nu} - 1$, and with the aid of (5.11) we get

$$
\begin{aligned}
\widetilde{a}_{h+1,\nu+1} &= P_{h,\nu+1} \cdot 2^{\nu+1} + (-1)^{h+1} \\
&= P_{h,\nu} \cdot 2^{\nu+1} + 2(-1)^{h+1} + P_{h-1,\nu+1} \cdot 2^{\nu+1} + (-1)^{h} \\
&= 2\widetilde{a}_{h+1,\nu} + \widetilde{a}_{h,\nu+1} \,.
\end{aligned}
$$

So $\widetilde{a}$ fulfills the recurrence (5.13) starting at $q = h = 1$. □

Together with (5.12) we get

$$
\begin{aligned}
a_{h,\nu+1} - a_{h,\nu} &= P_{h-1,\nu+1} \cdot 2^{\nu+1} - P_{h-1,\nu} \cdot 2^{\nu} \\
&= 2^{\nu}(2P_{h-1,\nu+1} - P_{h-1,\nu}) \\
&= 2^{\nu} \Delta_{h-1,\nu+1} \tag{5.17}
\end{aligned}
$$

and consequently

$$a_{h,\nu} = \sum_{k=0}^{\nu-1} \Delta_{h-1,k+1} \cdot 2^{k} \,. \tag{5.18}$$

With the aid of (5.17) (put $h = 4$ and $\nu = \mu - 1$) we can now perform the induction step in the proof of the above mentioned identity $s_{\mu} = a_{4,\mu-1}$ for Dudeney's subsidiary sequence s.

Before we turn to the relation between the array a and the Frame-Stewart numbers of Section 5.4, we note an interesting alternative representation (cf. Corollary 5.6, [62, Lemme 2.1] for $h = 2$ and [236, Theorem 4.6] for the general case). We will make use of the *hypertetrahedral root* $\nabla_{h,k}$ of $k \in \mathbb{N}_0$, namely

$$\nabla_{h,k} = \max\{\mu \in \mathbb{N}_0 \mid \Delta_{h,\mu} \le k\} \,;$$

for instance, $\nabla_{1,k} = k$ (*trivial root*) and $\nabla_{2,k} = \left\lfloor \frac{\sqrt{8k+1}-1}{2} \right\rfloor$ (*triangular root*). (There seems to be no simple closed expression for the *tetrahedral root* ∇_3 though; it behaves asymptotically like $\sqrt[3]{6k}$. In general, $\nabla_{h,k} \sim \sqrt[h]{h!k}$.) From $\Delta_{h,\mu} \le \Delta_{h+1,\mu}$ (cf. (5.9)) we get

$$\nabla_{h+1,k} \le \nabla_{h,k} \le k \,.$$

---

[69] For $n \in \mathbb{N}_0$ the sequences $a_{2,n}$, $a_{3,n+1}$ and $a_{4,n+1}$ are A000337, A055580 and A027608, respectively.

We also have

$$\forall\, x \in [\Delta_{h-1,\nu+1}]_0 : \nabla_{h,\Delta_{h,\nu}+x} = \nu\,;\tag{5.19}$$

in particular, $\nabla_{h,\Delta_{h,\nu}} = \nu$, but $\mu - \Delta_{h,\nabla_{h,\mu}} \in [\Delta_{h-1,\nabla_{h,\mu}+1}]_0$. Moreover,

$$\nabla_{h,k} - \nabla_{h,k-1} = \left(k \in \{\Delta_{h,\nu} \mid \nu \in \mathbb{N}\}\right)\,;\tag{5.20}$$

this also shows that the sequence ∇_h is (not strictly) increasing. It is interesting to note that we have what Bousch calls a *Galois correspondence* in [62, p. 897]:

$$\Delta_{h,\mu} \le \nu \Leftrightarrow \mu \le \nabla_{h,\nu} \ (\text{or } \nu < \Delta_{h,\mu} \Leftrightarrow \nabla_{h,\nu} < \mu)\,,$$

i.e. Δ_h and ∇_h engender a *Galois connection* in $\mathbb{N}_0$.

For later purpose we consider the following statement.

Lemma 5.22. *For all $h \in \mathbb{N}_2$ we have $\Delta_{h,\nabla_{h-1,n}+1} - \Delta_{h,\nabla_{h-1,n+1}-1} \ge n+1$ with equality if and only if $n + 1 = \Delta_{h-1,\nabla_{h-1,n}+1}$.*

Proof. Let $n = \Delta_{h-1,\nu} + x$ with $x \in [\Delta_{h-2,\nu+1}]_0$ (cf. (5.19)). We may assume that $n \ne 0$, i.e. $\nu \in \mathbb{N}$. Then $\nu = \nabla_{h-1,n} = \nabla_{h-1,n+1} - (n + 1 = \Delta_{h-1,\nu+1})$ (cf. (5.20)) and

$$
\begin{aligned}
\Delta_{h,\nabla_{h-1,n}+1} - \Delta_{h,\nabla_{h-1,n+1}-1}
&= \Delta_{h,\nu+1} - \Delta_{h,\nu+(n+1=\Delta_{h-1,\nu+1})-1} \\
&= \Delta_{h,\nu+1} - \Delta_{h,\nu} + (n + 1 \ne \Delta_{h-1,\nu+1})\left(\Delta_{h,\nu} - \Delta_{h,\nu-1}\right) \\
&= \Delta_{h-1,\nu+1} + (n + 1 \ne \Delta_{h-1,\nu+1})\Delta_{h-1,\nu} \\
&= \Delta_{h-1,\nu+1} + (n + 1 \ne \Delta_{h-1,\nu+1})(n - x)
\end{aligned}
$$

The right hand side is equal to $n + 1$, if $n + 1 = \Delta_{h-1,\nabla_{h-1,n}+1}$ and greater than or equal to $\Delta_{h-1,\nu+1} + n + 2 - \Delta_{h-2,\nu+1} = n + 2 + \Delta_{h-1,\nu}$ otherwise. □

The statement on a now reads as follows.

Proposition 5.23.

$$a_{h,\nu} = \sum_{k=0}^{\Delta_{h,\nu}-1} 2^{\nabla_{h,k}}\,.\tag{5.21}$$

The proof is by induction on ν. The case $\nu = 0$ is trivial. Making use of the induction assumption, (5.19), (5.15), and (5.12), we get:

$$
\begin{aligned}
\sum_{k=0}^{\Delta_{h,\nu+1}-1} 2^{\nabla_{h,k}}
&= \sum_{k=0}^{\Delta_{h,\nu}-1} 2^{\nabla_{h,k}} + \sum_{k=\Delta_{h,\nu}}^{\Delta_{h,\nu+1}-1} 2^{\nabla_{h,k}} \\
&= a_{h,\nu} + \Delta_{h-1,\nu+1}\cdot 2^{\nu} \\
&= P_{h-1,\nu}\cdot 2^{\nu} + (-1)^{h} + \Delta_{h-1,\nu+1}\cdot 2^{\nu} \\
&= (P_{h-1,\nu} + \Delta_{h-1,\nu+1})2^{\nu} + (-1)^{h} \\
&= P_{h-1,\nu+1}\cdot 2^{\nu+1} + (-1)^{h} \\
&= a_{h,\nu+1}\,. \qquad\qquad\qquad\qquad\qquad\quad □
\end{aligned}
$$

Remark. Formula (5.21) also makes sense for $h = 0$, if we interpret $\nabla_{0,0}$ as 0.

We define the right-hand side of (5.21) as $\Phi_h(\Delta_{h,\nu})$ and by recourse to (5.19) we can extend this concept to all n as

$$\forall\, n \in \mathbb{N}_0 : \quad \Phi_h(n) = \sum_{k=0}^{n-1} 2^{\nabla_{h,k}} \tag{5.22}$$

by putting $n = \Delta_{h,\nu} + x$ with $x \in [\Delta_{h-1,\nu+1}]_0$. Note that $\Phi_1(n) = M_n$. With (5.18) and (5.15) we have (cf. [405, Lemma 1 and Theorem 2]):

Lemma 5.24. *For all $x \le \Delta_{h-1,\nu+1}$:*

$$\Phi_h(\Delta_{h,\nu} + x) = a_{h,\nu} + x \cdot 2^\nu = \sum_{k=0}^{\nu-1} \Delta_{h-1,k+1} \cdot 2^k + x \cdot 2^\nu = (P_{h-1,\nu} + x) \cdot 2^\nu + (-1)^h \, .$$

The most important feature of the sequences Φ_h is an inequality:

Lemma 5.25. *For all $m, \ell \in \mathbb{N}_0$: $\Phi_h(m + \ell) \le 2\Phi_h(m) + \Phi_{h-1}(\ell)$.*
Here equality holds if either $\ell = \Delta_{h-1,\nu}$ and $\Delta_{h,\nu} \le m+\ell \le \Delta_{h,\nu}+\Delta_{h-1,\nu}$ or $m = \Delta_{h,\nu}$ and $\Delta_{h,\nu} + \Delta_{h-1,\nu} \le m + \ell \le \Delta_{h,\nu+1}$ for some ν.

Proof. The case $h = 1$ is only stated for formal completeness; here $\Phi_0(\ell)$ has to be interpreted as ℓ, if $\ell \in B$, and as ∞ otherwise; this is formally compatible with (5.22), if we interpret $\nabla_{0,k}$ as ∞ for $k \in \mathbb{N}$, i.e. $\nabla_{0,k} = -\log(1 - \Delta_{0,k})$ for $k \in \mathbb{N}_0$. We now suppose that $h \in \mathbb{N}_2$.

We first prove the inequality.[70] We fix h and proceed by induction on m. For $m = 0$ we have to show that

$$\forall\, \ell \in \mathbb{N}_0 : \quad \Phi_h(\ell) \le \Phi_{h-1}(\ell) \, ,$$

but this follows from the definition of the Φs since $\nabla_{h,k} \le \nabla_{h-1,k}$.

Now let $m \in \mathbb{N}_0$ and assume that

$$\forall\, \ell \in \mathbb{N}_0 : \quad \Phi_h(m + \ell) \le 2\Phi_h(m) + \Phi_{h-1}(\ell) \, . \tag{5.23}$$

Then clearly $0 \le \Phi_h(m + 1 + 0) \le 2\Phi_h(m + 1) + \Phi_{h-1}(0)$ and we assume that for $\ell \in \mathbb{N}_0$,

$$\Phi_h(m + 1 + \ell) \le 2\Phi_h(m + 1) + \Phi_{h-1}(\ell) \, . \tag{5.24}$$

Now, making use of (5.23),

$$
\begin{aligned}
\Phi_h(m + 1 + \ell + 1) &= \Phi_h(m + \ell + 2) \\
&\le 2\Phi_h(m) + \Phi_{h-1}(\ell + 2) \\
&= 2\Phi_h(m) + \Phi_{h-1}(\ell + 1) + 2^{\nabla_{h-1,\ell+1}} \\
&= 2\left(\Phi_h(m) + 2^{\nabla_{h-1,\ell+1}-1}\right) + \Phi_{h-1}(\ell + 1) \\
&\le 2\left(\Phi_h(m) + 2^{\nabla_{h,m}}\right) + \Phi_{h-1}(\ell + 1) \\
&= 2\Phi_h(m + 1) + \Phi_{h-1}(\ell + 1) \, ,
\end{aligned}
$$

---

[70] Following Pascal Stucky; private communication, 2015.

where the last inequality is only true if $\nabla_{h-1,\ell+1} - 1 \le \nabla_{h,m}$. If this is not the case, then $m < \Delta_{h,\nabla_{h-1,\ell+1}-1} = \Delta_{h,\nabla_{h-1,\ell}+1} - \ell - 1 - \left(\Delta_{h,\nabla_{h-1,\ell}+1} - \Delta_{h,\nabla_{h-1,\ell+1}-1} - \ell - 1\right)$. From Lemma 5.22 we obtain $m + \ell + 1 < \Delta_{h,\nabla_{h-1,\ell}+1}$ and therefore $\Delta_{h,\nabla_{h,m+\ell+1}} \le m + \ell + 1 < \Delta_{h,\nabla_{h-1,\ell}+1}$, whence $\nabla_{h,m+\ell+1} < \nabla_{h-1,\ell} + 1$, i.e. $\nabla_{h,m+\ell+1} \le \nabla_{h-1,\ell}$. With (5.24) we arrive at

$$
\begin{aligned}
\Phi_h(m+1+\ell+1) &= \Phi_h(m+1+\ell) + 2^{\nabla_{h,m+1+\ell}} \\
&\le 2\Phi_h(m+1) + \Phi_{h-1}(\ell) + 2^{\nabla_{h-1,\ell}} \\
&= 2\Phi_h(m+1) + \Phi_{h-1}(\ell+1).
\end{aligned}
$$

We note that the above argument is simpler for the case $h = 2$ where Lemma 5.22 is obvious.

Now let us turn to the cases of equality. We begin with $\ell = \Delta_{h-1,\nu}$ and $\Delta_{h,\nu} \le m + \ell \le \Delta_{h,\nu} + \Delta_{h-1,\nu}$; then, by Lemma 5.24 and (5.13),

$$
\begin{aligned}
\Phi_h(m+\ell) &= a_{h,\nu} + (m + \ell - \Delta_{h,\nu})2^{\nu} \\
&= a_{h,\nu} + (m - \Delta_{h,\nu-1})2^{\nu} \\
&= 2a_{h,\nu-1} + a_{h-1,\nu} + (m - \Delta_{h,\nu-1})2^{\nu} \\
&= 2\left(a_{h,\nu-1} + (m - \Delta_{h,\nu-1})2^{\nu-1}\right) + a_{h-1,\nu} \\
&= 2\Phi_h(m) + \Phi_{h-1}(\ell),
\end{aligned}
$$

where the last equality comes from Lemma 5.24 again, because $\Delta_{h,\nu-1} \le m \le \Delta_{h,\nu}$.

Now let $m = \Delta_{h,\nu}$ and $\Delta_{h,\nu} + \Delta_{h-1,\nu} \le m + \ell \le \Delta_{h,\nu+1}$; then, again with Lemma 5.24,

$$
\begin{aligned}
\Phi_h(m+\ell) &= a_{h,\nu} + (m + \ell - \Delta_{h,\nu})2^{\nu} \\
&= 2\Phi_h(m) + \ell \cdot 2^{\nu} - a_{h,\nu} \\
&= 2\Phi_h(m) + \Phi_{h-1}(\ell) - (a_{h,\nu} + a_{h-1,\nu}) + \Delta_{h-1,\nu} \cdot 2^{\nu} \\
&= 2\Phi_h(m) + \Phi_{h-1}(\ell),
\end{aligned}
$$

where in the last line (5.16) has been employed and the penultimate equality follows from Lemma 5.24 since $\Delta_{h-1,\nu} \le \ell \le \Delta_{h-1,\nu+1}$. ◻

Remark. When we fix $n := m + \ell \in \mathbb{N}_0$, the values of m and $\ell = n - m$ where equality holds are, in general, not the only ones, just those with maximal m. (Note that except for $n = 0$ we always have $\ell \in \mathbb{N}$.) The minimal m is achieved if either $\Delta_{h,\nu} \le n \le \Delta_{h,\nu+1} - \Delta_{h-1,\nu}$ and $m = \Delta_{h,\nu-1}$ or $\Delta_{h,\nu+1} - \Delta_{h-1,\nu} \le n \le \Delta_{h,\nu+1}$ and $\ell = \Delta_{h-1,\nu+1}$. All intermediate values of m are also admissible, as can be seen with similar arguments as in the proof of Lemma 5.25. A proof to show that this covers *all* cases of equality can actually be employed as an alternative to verify the inequality in that lemma; see [405, Theorem 1].

5.5.3   Frame-Stewart Numbers Revisited

So far everything could be formulated and proved without reference to the TH or to the Frame-Stewart numbers. Let us recall their definition in (5.3):

Definition 5.26. *For all $n \in \mathbb{N}_0$: $FS_3^n = 2^n - 1 = M_n$. Let $p \in \mathbb{N}_3$; then*

$$
\begin{aligned}
FS_{p+1}^0 &= 0, \\
\forall n \in \mathbb{N}: \quad FS_{p+1}^n &= \min\{2FS_{p+1}^m + FS_p^{n-m} \mid m \in [n]_0\}, \\
\forall n \in \mathbb{N}_0: \quad \overline{FS}_p^n &= \tfrac{1}{2}(FS_p^{n+1} - 1).
\end{aligned}
$$

There have been many attempts at deducing explicit formulas from the recursive Definition 5.26. If we interpret $h := p-2$ as the number of auxiliary pegs in the TH, i.e. those which are neither the starting nor the goal peg, the case $h = 1$ is classical. For $h = 2$, see [193, Theorem 1] and, more comprehensive, Theorem 5.4; cf. also [62, (1.2)].[71] The general case is Theorem 5.15 and goes back to [278, Theorem 3] and [236, Theorem 3.1]. All these results are, via Lemma 5.24, contained in the following theorem (cf. [405, Corollary 1]).

Theorem 5.27. $FS_{h+2}^n = \Phi_h(n)$, $\overline{FS}_{h+2}^n = \tfrac{1}{2}(\Phi_h(n+1) - 1)$.

Proof. We only have to show that Φ_h fulfills the recurrence for FS_{h+2}. We already saw that $\Phi_1 = M$, and $\Phi_h(0) = 0$ for $h \in \mathbb{N}_2$ is clear. Moreover, since every $n \in \mathbb{N}$ lies in some $[\Delta_{h,\nu+1}]_0 \setminus [\Delta_{h,\nu}]_0$ with $\nu \in \mathbb{N}$, Lemma 5.25 (together with the remark following it) guarantees that

$$
\forall h \in \mathbb{N}_2: \quad \Phi_h(n) = \min\{2\Phi_h(m) + \Phi_{h-1}(n-m) \mid m \in [n]_0\}.
$$

This completes the proof.   □

Remark. The statement in Theorem 5.27 also makes (some) sense for $h = 0$; cf. the first sentence in the proof of Lemma 5.25: the TH with only two pegs can be solved only for $n \in B$ making n moves. In that spirit Definition 5.26 could start with $p = 2$.

From Theorem 5.27, Lemma 5.24, (5.10) and (5.16) we get, if we put $n = \Delta_{h,\nabla_{h,n}} + x$:

Corollary 5.28. $\forall n \in \mathbb{N}_0$: $FS_{h+2}^n = (n - P_{h,\nabla_{h,n}})2^{\nabla_{h,n}} + (-1)^h = n \cdot 2^{\nabla_{h,n}} - a_{h+1,\nabla_{h,n}}$.

For $h = 3$, i.e. $p = 5$, this corresponds to Dudeney's formula (5.14). The first few Frame-Stewart numbers are presented in Table 5.8.

By definition in (5.22), $\Phi_h(n)$ is a sum of n powers of 2 which are, however, not all different. Lemma 5.24 shows that if $n = \Delta_{h,\nu} + x$ with $x \le \Delta_{h-1,\nu+1}$ all powers with exponent $k \in [\nu]_0$ occur with multiplicity $\Delta_{h-1,k+1}$ and $k = \nu$ with multiplicity x, respectively. In [405, Corollary 2] Stockmeyer observes that each of the n powers of 2 in $\Phi_h(n)$, and therefore in FS_{h+2}^n, corresponds to the number of moves of an individual disc in the algorithm of Stewart to solve the $(h+2)$-peg, n-disc TH task to get from 0^n to 1^n, say. To be specific in that algorithm we always choose the partition number m according to the values in Lemma 5.25; the fact itself does not depend on the choice of m.

---

[71] On page 896 the author says that it can be shown by simple calculations; on page 897, however, he admits that it is less easy to verify!

$h \setminus n$	0	1	2	3	4	5	6	7	8	9	10	11	12
1	0	1	3	7	15	31	63	127	255	511	1023	2047	4095
2	0	1	3	5	9	13	17	25	33	41	49	65	81
3	0	1	3	5	7	11	15	19	23	27	31	39	47
4	0	1	3	5	7	9	13	17	21	25	29	33	37

Table 5.8: Frame-Stewart numbers FS^n_{h+2}

Corollary 5.29. *For $\Delta_{h,\nu} \le n \le \Delta_{h,\nu+1}$ there are $\Delta_{h-1,k+1}$ discs making precisely 2^k moves for $k \in [\nu]_0$ and $n - \Delta_{h,\nu}$ discs that make precisely 2^ν moves in Stewart's algorithm.*

Proof. We proceed by double induction on h and ν. Let $h = 1$; then $\nu \le n \le \nu + 1$. For $k \in [\nu]_0$, one disc makes exactly 2^k moves and $n - \nu$ discs make 2^ν moves. This is known from the classical case of the TH with three pegs; cf. Proposition 2.4.

Now let $h \in \mathbb{N}_2$. If $\nu = 0$, then $n \in B$ and there are n discs that make exactly one move. If $\nu \in \mathbb{N}$, then $n = \Delta_{h,\nu} + x$ with $x \le \Delta_{h-1,\nu+1}$.

Consider the case $x \le \Delta_{h-1,\nu}$, where $\Delta_{h,\nu-1} \le m = n - \Delta_{h-1,\nu} = \Delta_{h,\nu-1} + x \le \Delta_{h,\nu}$, so that by induction assumption there are $\Delta_{h-1,k+1}$ discs that make 2^k moves for $k \in [\nu - 1]_0$ and x discs making $2^{\nu-1}$ moves in the first step of the algorithm and again in the third step. In the second step $\ell = \Delta_{h-1,\nu}$ discs are moved, so that by induction assumption $\Delta_{h-2,k+1}$ discs make 2^k moves for $k \in [\nu]_0$. Together we have

- 1 disc that makes 1 move ($k = 0$),

- $\Delta_{h-1,k} + \Delta_{h-2,k+1} = \Delta_{h-1,k+1}$ discs that make 2^k moves for $k \in [\nu - 1]$,

- x discs that make 2^ν moves ($k = \nu$).

Now let $\Delta_{h-1,\nu} \le x \le \Delta_{h-1,\nu+1}$, where $m = \Delta_{h,\nu}$ and again $\Delta_{h-1,k+1}$ discs make 2^k moves for $k \in [\nu - 1]_0$ and $\Delta_{h,\nu} - \Delta_{h,\nu-1} = \Delta_{h-1,\nu}$ discs make $2^{\nu-1}$ moves in the first and in the third step, respectively. In the second step $\ell = n - m = \Delta_{h,\nu} + x - \Delta_{h,\nu} = x$ discs are moved, of which $\Delta_{h-2,k+1}$ make precisely 2^k moves for $k \in [\nu]_0$ and $x - \Delta_{h-1,\nu}$ make 2^ν moves by induction assumption. This adds up in the same way as in the case before. □

Remark. Note that for $n \in \mathbb{N}$ there is precisely one disc making exactly one move in Stewart's algorithm. This must be the largest disc n so that the lower subtower is never empty, reflecting the fact that $m - n \in \mathbb{N}$; cf. the remark after Lemma 5.25.

In his attempt to prove the *Frame-Stewart conjecture*, i.e. $d_p(0^n, 1^n) = FS^n_p$, for the case $p = 4$, i.e. $h = 2$, or in other words to solve The Reve's Puzzle, Bousch defines the function, or rather integer sequence, Φ as our Φ_2, i.e.

$$\forall n \in \mathbb{N}_0 : \Phi(n) = \sum_{k=0}^{n-1} 2^{\nabla_{2,k}}.$$

But instead of making use of Theorem 5.27 to show that $[62, (1.2)]$

$$\forall\, n \in \mathbb{N}: \; \Phi(n) = \min\left\{2\Phi(m) + M_{n-m} \mid m \in [n]_0\right\},$$

he uses only the weaker statement $[62, (2.1)]$

$$\forall\, m, \ell \in \mathbb{N}_0: \; \Phi(m + \ell) \le 2\Phi(m) + M_\ell, \tag{5.25}$$

i.e. the inequality from Lemma 5.25 (case $h = 2$).

The task to get from perfect state 0^{1+n} to perfect state $(h+1)^{1+n}$ can be fulfilled in $2\mathrm{d}_{h+2}(0^n, t) + 1$ moves, where $0t$ with $t \in [h]^n$ is any half-way state. To prove the Frame-Stewart conjecture, we are therefore left with showing that

$$\forall\, t \in [h]^n: \; \mathrm{d}_{h+2}(0^n, t) \ge \overline{\Phi}_h(n),$$

where $\overline{\Phi}_h(n) = \dfrac{1}{2}\left(\Phi_h(n+1) - 1\right) = \dfrac{1}{2}\displaystyle\sum_{k=1}^{n} 2^{\nabla_{h,k}}$.

Note that for $h = 1$, i.e. $p = 3$, we have $t = 1^n$ and $\overline{\Phi}_1(n) = M_n$.

5.5.4   The Reve's Puzzle Solved

We will now concentrate on the case of $p = 4$ pegs, i.e. $h = 2$. For brevity we put $\mathrm{d} := \mathrm{d}_4$. We already saw that, considered as sequences, $\Delta_2 = \Delta$ and $\Phi = \Phi_2$; we will also write ∇ for ∇_2 and $\overline{\Phi}$ for $\overline{\Phi}_2$. Our goal is to prove that

$$\forall\, t \in (Q \setminus B)^n: \; \mathrm{d}(0^n, t) \ge \overline{\Phi}(n). \tag{5.26}$$

Lemma 5.24 translates to

Lemma 5.30. (Cf. $[62, \text{Lemme } 2.1]$) *For all $x \le \nu + 1$:*

$$\Phi(\Delta_\nu + x) = c_\nu + x \cdot 2^\nu = \sum_{k=0}^{\nu-1}(k+1)2^k + x \cdot 2^\nu = (\nu - 1 + x)2^\nu + 1.$$

Let $2_0^{\mathbb{N}}$ be the set of all finite subsets of $\mathbb{N}$ and define for $E \in 2_0^{\mathbb{N}}$:

$$\forall\, \mu \in \mathbb{N}_0: \; \Psi_\mu(E) = \sum_{d \in E} 2^{\min\{\nabla_{d-1},\mu\}} - c_\mu,$$

where c is again Dudeney's sequence, such that $\Psi_\mu(E) \le \displaystyle\sum_{d \in E} 2^{\nabla_{d-1}}$; let

$$\Psi(E) = \max_{\mu \in \mathbb{N}_0} \Psi_\mu(E).$$

Note that $\Psi(E) \ge \Psi_0(E) = |E|$.

The most fundamental applications of Ψ are on segments of $\mathbb{N}$.

Lemma 5.31. (Cf. $[62, \text{Lemme } 2.2]$) *For all $n \in \mathbb{N}_0$:* $\Psi[n] := \Psi([n]) = \overline{\Phi}(n)$.

Proof. Without loss of generality $n \in \mathbb{N}$, i.e. $n = \Delta_\nu + x$, $\nu \in \mathbb{N}$, $x \in [\nu + 1]_0$. Then

$$
\begin{aligned}
\Psi_{\mu+1}[n] - \Psi_\mu[n] &= \sum_{\delta=0}^{n-1} \left(2^{\min\{\nabla_\delta, \mu+1\}} - 2^{\min\{\nabla_\delta, \mu\}} \right) - (\mu + 1)2^\mu \\
&= 2^\mu \left(|[n]_0 \setminus [\Delta_{\mu+1}]_0| - (\mu + 1) \right) \\
&= 2^\mu \left((n - \Delta_{\mu+1})(n > \Delta_{\mu+1}) - (\mu + 1) \right) > 0 \\
&\Leftrightarrow (n - \Delta_{\mu+1})(n > \Delta_{\mu+1}) > \mu + 1 \\
&\Leftrightarrow n \geq \Delta_{\mu+1} + \mu + 2 = \Delta_{\mu+2} \\
&\Leftrightarrow \nu = \nabla_n \geq \mu + 2 \Leftrightarrow \mu < \nu - 1.
\end{aligned}
$$

Therefore, $\Psi_\mu[n]$ is maximal at $\mu = \nu - 1$, so that

$$
\begin{aligned}
\Psi[n] = \Psi_{\nu-1}([n]) &= \sum_{\delta=0}^{n-1} 2^{\min\{\nabla_\delta, \nu-1\}} - c_{\nu-1} \\
&= \sum_{\delta=0}^{\Delta_\nu - 1} 2^{\nabla_\delta} + \sum_{\delta=\Delta_\nu}^{n-1} 2^{\nu-1} - c_{\nu-1} \\
&= \Phi(\Delta_\nu) + (n - \Delta_\nu)2^{\nu-1} - c_{\nu-1} \\
&= (\nu + x)2^{\nu-1} = \overline{\Phi}(n),
\end{aligned}
$$

the latter two equalities coming from Lemma 5.30. □

It follows with (5.25) that ([62, (2.2)])

$$
\forall\, k, \ell \in \mathbb{N}_0 : \quad \Psi[k + \ell] \leq 2\Psi[k] + 2^{\ell-1} . \tag{5.27}
$$

A lower bound is given in

Lemma 5.32. (Cf. [62, Lemme 2.3]) *For all $m \in \mathbb{N}$:* $\Psi[m + 1] \geq 2^{\nabla_{m-1}+1}$.

Proof. Let $\mu := \nabla_{m-1}$. Then it suffices to show that

$$
\forall\, \mu \in \mathbb{N}_0 : \quad \Psi[\Delta_\mu + 2] \geq 2^{\mu+1} \tag{5.28}
$$

because $\Delta_\mu \leq m - 1$ and $\Psi[\cdot]$ is monotone increasing by Lemma 5.31. From the same lemma we get $\Psi[\Delta_\mu + 2] = \overline{\Phi}(\Delta_\mu + 2) = \frac{1}{2}(\Phi(\Delta_\mu + 3) - 1)$. Then (5.28) follows (with equality) if $\mu \in B$. Otherwise the right hand side is greater than or equal to $(\mu + 2)2^{\mu-1}$ by Lemma 5.30, so that (5.28) follows as well. □

The subsequent Lemmas 5.33 and 5.35 to 5.37 on properties of Ψ for more general finite subsets of $\mathbb{N}$ are also needed for the proof of Theorem 5.38.

Lemma 5.33. (Cf. [62, Lemme 2.4]) *Let $E \in 2_0^\mathbb{N}$ and define $k := |E|$. Then for every $\mu \in \mathbb{N}_0$:*

$$
\Psi_\mu[k] \leq \Psi_\mu(E) \leq M_k
$$

in particular, $|E| \leq \Psi[|E|] \leq \Psi(E) \leq M_{|E|} .$

Proof. With a strictly monotone increasing mapping $\iota \in E^k$ we have $\nabla_{\ell-1} \leq \nabla_{\iota\ell-1}$; this proves the first inequality. For the second, note that $\forall\, m \in \mathbb{Z}: 1+m \leq 2^m$; therefore, putting $m = k - \mu$, we get $\Psi_\mu(E) \leq k2^\mu - c_\mu = (k-\mu+1)2^\mu - 1 \leq M_k$. □

Corollary 5.34. *In Lemma 5.33 we have $\Psi(E) = M_{|E|}$ only if $|E| \leq 1 + \nabla_{\min E-1}$.*

Proof. We may assume that $k := |E| \in \mathbb{N}$. Let $\mu \in \mathbb{N}_0$ with $\Psi_\mu(E) = \Psi(E) = M_k$. As $1 + m = 2^m \Leftrightarrow m \in B$, the last line of the proof of Lemma 5.33 tells us that either $\mu = k$ and $\Psi_k(E) = k2^k - c_k$, or $\mu = k-1$ and $\Psi_{k-1}(E) = k2^{k-1} - c_{k-1}$. The former case is equivalent to

$$\sum_{d \in E} 2^{\min\{\nabla_{d-1},k\}} = k2^k \,,$$

i.e. $\forall\, d \in E: \nabla_{d-1} \geq k$; the latter case means

$$\sum_{d \in E} 2^{\min\{\nabla_{d-1},k-1\}} = k2^{k-1} \,,$$

i.e. $\forall\, d \in E: \nabla_{d-1} \geq k - 1$, so that this condition is necessary and sufficient in any case. □

Lemma 5.35. (Cf. [62, Lemme 2.6]) *Let $A \in 2_0^{\mathbb{N}}$ and $\kappa \in \mathbb{N}$ with $|A \smallsetminus [\Delta_\kappa]| \leq \kappa$. Then*

$$\forall\, a \in A: \Psi(A) - \Psi(A \smallsetminus \{a\}) \leq 2^{\kappa-1} \,.$$

Proof. For all $\mu \in \mathbb{N}_{\kappa-1}$ we have

$$\Psi_{\mu+1}(A) - \Psi_\mu(A) = 2^\mu \left(|A \smallsetminus [\Delta_{\mu+1}]| - (\mu+1)\right) \leq 2^\mu \left(|A \smallsetminus [\Delta_\kappa]| - \kappa\right) \leq 0\,.$$

This shows that $\Psi_\mu(A)$ is non-increasing from $\mu = \kappa - 1$, such that there is a $\mu \leq \kappa - 1$ with $\Psi(A) = \Psi_\mu(A)$, whence

$$\Psi(A) - \Psi(A\smallsetminus\{a\}) \leq \Psi_\mu(A) - \Psi_\mu(A\smallsetminus\{a\}) = 2^{\min\{\nabla_{a-1},\mu\}} \leq 2^\mu \leq 2^{\kappa-1}\,.   □$$

Lemma 5.36. (Cf. [62, Lemme 2.7]) *Let $\ell \in \mathbb{N}$, $k \in \mathbb{N}_{\Delta_{\ell-1}}$, and $A \subseteq [k]$, $A' \subseteq \mathbb{N}$, $|A'| \leq \ell$. Then*
$$\Psi(A \cup A') - \Psi(A) \leq \Psi[k+\ell] - \Psi[k]\,.$$

Proof. Since $\Psi[\cdot]$ is monotone increasing, we may assume that $|A'| = \ell$. Let $\iota \in A'^\ell$ be injective and put $A_\rho := A \cup \{\iota_\sigma \mid \sigma \in [\rho]\}$ for $\rho \in [\ell+1]_0$, such that $A_0 = A$ and $A_\ell = A \cup A'$. By a telescoping argument it is then sufficient to show that $\Psi(A_\rho) - \Psi(A_{\rho-1}) \leq \Psi[k+\rho] - \Psi[k+\rho-1]$ for all $\rho \in [\ell]$. By Lemma 5.31 the right hand side is $2^{\kappa-1}$, where $\kappa := \nabla_{k+\rho}$, such that the inequality follows from Lemma 5.35 if we can show that $|A_\rho \smallsetminus [\Delta_\kappa]| \leq \kappa$.

We first observe that

$$\Delta\rho - \rho = \Delta_{\rho-1} \leq \Delta_{\ell-1} \leq k$$

and consequently $\rho \le \kappa$. Moreover, from $\Delta_{\kappa+1} > k + \rho$ we get $\Delta_\kappa + \kappa \ge k + \rho$. Putting these facts together, we obtain[72]

$$
\begin{aligned}
|A_\rho \smallsetminus [\Delta_\kappa]| \le \rho + |[k] \smallsetminus [\Delta_\kappa]| &= \rho + [k - \Delta_\kappa]_{\Delta_\kappa+1} \\
&= \rho + (k - \Delta_\kappa)(k > \Delta_\kappa) \\
&\le \rho + (\kappa - \rho)(k > \Delta_\kappa) \\
&\le \max\{\rho, \kappa\} = \kappa.
\end{aligned}
$$

□

Lemma 5.37. (Cf. [62, Lemme 2.8])

$$
\forall\, A, A' \in 2_0^{\mathbb{N}}:\ \Psi(A) + \Psi(A') \ge \frac{1}{2}\Psi\big[|A \cup A'| + 2\big] - 1.
$$

Proof. Let $E := A \cup A'$ and $k := |E|$. Then for $\mu \in \mathbb{N}_0$ we have

$$
\begin{aligned}
\Psi(A) + \Psi(A') &\ge \Psi_\mu(A) + \Psi_\mu(A') \\
&= \Psi_\mu(A \cup A') + \Psi_\mu(A \cap A') \\
&\ge \Psi_\mu(E) + \Psi_\mu(\varnothing) \\
&\ge \Psi_\mu[k] - c_\mu,
\end{aligned}
\tag{5.29}
$$

where the last inequality follows from Lemma 5.33.

We write $k+3$ as $\Delta_\nu + x$ with $x \in [\nu+1]_0$, i.e. $\nu = \nabla_{k+3} \ge 2$. Put $\mu := \nu - 2 \in \mathbb{N}_0$; then $\Delta_\mu = \Delta_{\mu+1} - (\mu+1) = \Delta_\nu - (2\mu+3) \le k - 2\mu \le k$ and with Lemma 5.30

$$
\begin{aligned}
\Phi(\Delta_\mu) &= 1 + (\mu-1)2^\mu, \\
\overline{\Phi}(k+2) &= (\mu+1+x)2^{\mu+1}.
\end{aligned}
$$

It follows that

$$
\begin{aligned}
\Psi_\mu[k] - c_\mu &= \sum_{\delta=0}^{k-1} 2^{\min\{\nabla_\delta, \mu\}} - 2c_\mu \\
&= \sum_{\delta=0}^{\Delta_\mu - 1} 2^{\nabla_\delta} + (n - \Delta_\mu)2^\mu - 2c_\mu \\
&= \Phi(\Delta_\mu) + (k - \Delta_\mu)2^\mu - 2c_\mu \\
&= 1 + (\mu - 1 + k - \Delta_\mu)2^\mu - 2c_\mu \\
&= (k - \Delta_\mu - \mu + 1)2^\mu - 1 \\
&= (\Delta_{\mu+2} - \Delta_\mu - \mu - 2 + x)2^\mu - 1 \\
&= (\mu + 1 + x)2^\mu - 1 = \frac{1}{2}\overline{\Phi}(k+2) - 1,
\end{aligned}
$$

so that with (5.29) and Lemma 5.31, for $n = k + 2$, the proof is complete. □

Finally, Bousch announces his main result (cf. [62, Théorème 2.9])

---

[72] For $z \in \mathbb{Z} \smallsetminus \mathbb{N}_0$ we use the convention that $[z]_\lambda = \varnothing$ for all $\lambda \in \mathbb{N}_0$.

Theorem 5.38. *Let* $s, t \in Q^n$, $|t([n])| \leq 2$. *Then, for* $i \notin t([n])$,

$$d(s, t) \geq \Psi \left(s^{-1}(\{i\}) \right) ;$$

this inequality is [62, (2.3)].

For example, let 2 and 3 be two pegs that contain all discs in the goal state t, i.e. $t([n]) \subseteq \{2, 3\}$; then the lower bound in Theorem 5.38 depends only on the set of those discs which are, e.g., on peg $i = 0$ in the initial state s.

In view of (5.26) and Lemma 5.31 we can conclude (cf. [62, Corollaire 4.1]):

Corollary 5.39. *For all* $n \in \mathbb{N}_0$: $d_4(0^n, 1^n) = FS_4^n$.

We propose to call the sequence given by FS_4^n the *Bousch sequence*; cf. A007664.

5.5.5   The Proof of Theorem 5.38

We first give a brief outline. The proof is by induction on n and demonstrates the importance of choosing the statement in a smart way. The Reve's puzzle does *not* allow for a simple induction argument because, contrary to the case of three pegs, the reduction to one disc less does not lead to a task of the same type, namely to move a whole tower. On the other hand, the fact that the largest disc will only move once in a shortest path (see Proposition 5.7) reduces the P0 task to the P1 task to distribute the discs of a tower to two pegs different from the starting peg of the tower, i.e. to get from a perfect state to a *critical* state. As already suggested earlier (p. 225) it might even be necessary to consider P2-type tasks $s \rightarrow t$, and this is what Bousch does by providing a lower bound on the length of solutions for a class of such tasks.

In the induction step, the largest disc $n \in \mathbb{N}$ and possibly another one $m \in [n]$ play the decisive role together with (up to) four moves on a shortest path γ from $\gamma_0 = s$ to $\gamma_{d(s,t)} = t$, namely

- τ_0, the first move of disc m;

- τ_1, the first move of disc m to the peg in $t([n])$ which is not $t(n)$;

- τ_2, the first move of disc n;

- τ_3, the last move of disc n.

Considering a series of nested cases, unavoidable moves of discs, in particular from $s^{-1}(\{i\})$, in the invervals between s, the γ_{τ_ℓ}s and t are then counted and their addition leads to the desired lower bound for $d(s, t)$.

In [62, Section 5] Bousch makes some remarks on how he arrived at the proof and in particular the function Ψ.

We will now present the details of the proof.

0. The case $n = 0$ is trivial. (The theorem can easily be demonstrated for some small values of n.)

Now let $n \in \mathbb{N}$ and the statement be true for all smaller numbers. For simplicity, we may assume that $i = 0$, $t_n = 2$, and $t([n-1]) \subseteq \{2,3\}$. The case $E := s^{-1}(\{0\}) = \emptyset$ can trivially be dealt with, because the right hand side of the inequality would equal 0.

1. $(E \neq \emptyset)$ For $r \in Q^n$ we write $\bar{r} := r \upharpoonright [n-1] \in Q^{n-1}$, and we may apply the induction hypothesis to obtain

$$L := \mathrm{d}(s,t) \geq \mathrm{d}(\bar{s},\bar{t}) \geq \Psi(\overline{E}),$$

where $\overline{E} := E \cap [n-1]$. We now distinguish the cases that $n \notin E$ or $n \in E$.

1.0. $(n \notin E \neq \emptyset)$ Then $\overline{E} = E$ and we are done.

1.1. $(n \in E)$ Let $\gamma : [L+1]_0 \to Q^n$ be a shortest s,t-path in H_4^n. We will write γ_τ for $\gamma(\tau)$; in particular, $\gamma_0 = s$ and $\gamma_L = t$. Now let

$$E' := \{d \in E \mid \exists \tau \in [L+1]_0 : \gamma_{\tau,d} = 3\},$$

i.e. the set of those discs originally on peg 0 which eventually pass through peg 3 on γ. We now distinguish the cases where this set is empty or not.

2.0. $(n \in E, E' = \emptyset)$ Now $t(E) = \{2\}$, and the discs from E, suitably relabelled, make a transition from peg 0 to peg 2 in $H_3^{|E|}$ which takes at least $M_{|E|}$ moves. Together with Lemma 5.33 we get $L \geq M_{|E|} \geq \Psi(E)$, and we are done. Note that by Corollary 5.34 we have $L > \Psi(E)$ if $|E| > 1 + \nabla_{\min E-1}$.

We now continue with $E' \neq \emptyset$ and define $m := \max E' \in [n]$.

2.1. $(n \in E, m \in E')$ We now give the formal definitions of the special move numbers τ_ℓ, $\ell \in [4]_0$, mentioned above and the corresponding states $x_\ell := \gamma_{\tau_\ell}$ on the path γ; we will also consider the states before $x_{\ell-} := \gamma_{\tau_\ell - 1}$.

Let $\tau_0 := \min\{\tau \in [L] \mid \gamma_{\tau,m} \neq 0\}$. This means that disc m moves for the first time in move number τ_0 and this move is from peg 0 to peg $x_{0,m} \in [3]$. Therefore, all discs smaller than m are on $Q \setminus \{0, x_{0,m}\}$ in state x_0 and in x_{0-}. By induction hypothesis we know that, together with the move of disc m,

$$\mathrm{d}(s, x_0) \geq 1 + \Psi(\overline{\overline{E}}), \tag{5.30}$$

where $\overline{\overline{E}} := E \cap [m-1]$.

Let $\tau_1 := \min\{\tau \in [L] \mid \gamma_{\tau,m} = 3\}$. Then in x_1 all discs smaller than m are on $Q \setminus \{3, x_{1-,m}\}$.

Let $\tau_2 := \min\{\tau \in [L] \mid \gamma_{\tau,n} \neq 0\}$. In x_2 only disc n lies on peg $x_{2,n} \in [3]$ and peg $x_{2-,n} = 0$ is empty. Therefore,

$$\mathrm{d}(s, x_2) \geq 1 + \Psi(\overline{E}). \tag{5.31}$$

Note that (5.30) and (5.31) are the same if $m = n$.

Finally, let $\tau_3 := \max\{\tau \in [L]_0 \mid \gamma_{\tau,n} \neq 2\} + 1 \in [L]$. Then in x_3 the only disc on peg 2 is disc n and peg $x_{3-,n} \neq 2$ is empty.

It is obvious that $\tau_0 \leq \tau_1$ and $\tau_0 \leq \tau_2 \leq \tau_3$, but we do not know the position of τ_1 with respect to τ_2 or τ_3. In any case, $\tau_1 \neq \tau_3$, because the goals of these moves are different. If $\tau_\ell = \tau_2$ for some $\ell \in B$, then $m = n$, because otherwise the discs moved are different. We will now turn to the distinction between $m = n$ and $m \neq n$.

3.0. ($n = m \in E' \subseteq E$) This means that on its way from peg 0 to peg 2 the largest disc passes through peg 3. We have $0 < \tau_0 = \tau_2 \leq \tau_1 < \tau_3 \leq L$, so that γ runs in the order $s \to x_0 = x_2 \to x_1 \to x_3 \to t$.

If we put $A = E$ and $\kappa = \nabla_n$ in Lemma 5.35, then $|A \setminus [\Delta_\kappa]| \leq n - \Delta_\kappa \leq \kappa$, and we get $\Psi(E) - \Psi(\overline{E}) \leq 2^{\nabla_n - 1}$ whence from (5.31) we obtain

$$d(s, x_0) = d(s, x_2) \geq \Psi(E) - M_{\nabla_n - 1} . \tag{5.32}$$

Recall that $\overline{x}_3 \in (Q \setminus \{2, x_{3-,n}\})^{n-1}$. Let $a := 3(x_{3-,n} \neq 3)$, $a' := (x_{3-,n} \neq 1)$ and $A := \overline{x}_3^{-1}(\{a\})$, $A' := \overline{x}_3^{-1}(\{a'\})$. Then $A \cup A' = [n-1]$, so that by Lemmas 5.37 and 5.31:

$$
\begin{aligned}
\Psi(A) + \Psi(A') &\geq \frac{1}{2} \Psi[n+1] - 1 \\
&= \frac{1}{2} \overline{\Phi}(n+1) - 1 \\
&= \frac{1}{4} \left(2^{\nabla_{n+1}} + 2^{\nabla_n} \right) + \frac{1}{2} \Psi[n-1] - 1 \\
&\geq M_{\nabla_n - 1} + \frac{1}{2} \Psi[n-1] .
\end{aligned}
\tag{5.33}
$$

If $a = 0$, then $\overline{x}_0 \in (Q \setminus \{0, x_{0,n}\})^{n-1}$; if $a = 3$, then $\overline{x}_1 \in (Q \setminus \{x_{1-,n}, 3\})^{n-1}$, so that induction hypothesis yields, together with at least one move of disc n,

$$d(x_0, x_3) \geq 1 + d(\overline{x}_{(a=3)}, \overline{x}_3) \geq 1 + \Psi(A) . \tag{5.34}$$

Similarly, since $\overline{t} \in (Q \setminus B)^{n-1}$,

$$d(x_3, t) = d(\overline{x}_3, \overline{t}) \geq \Psi(A') . \tag{5.35}$$

Putting together (5.32), (5.34), and (5.35), we arrive at

$$L = d(s, x_0) + d(x_0, x_3) + d(x_3, t) \geq \Psi(E) + \Psi(A) + \Psi(A') + 1 - M_{\nabla_n - 1}$$

and finally with (5.33) at

$$L \geq \Psi(E) + \frac{1}{2} \Psi[n-1] + 1 > \Psi(E) .$$

3.1. ($E' \ni m < n \in E$) Now let $E'' := E \cap ([n-1] \setminus [m])$ and $\mu := |E''| + 1 \in [n-m]$. We will have to distinguish the cases $\Delta_\mu < m$ and $\Delta_\mu \geq m$.

4.0. ($E' \ni m < n \in E$, $\Delta_\mu < m$) Define $\kappa := \nabla_{m+\mu}$. Since $E \subseteq [m] \cup E'' \cup \{n\}$, the difference $E \setminus [\Delta_\kappa]$ contains at most $\kappa' := (m - \Delta_\kappa)(m > \Delta_\kappa) + \mu$ elements. If $m \le \Delta_\kappa$, then $\kappa' = \mu \le \nabla_m \le \kappa$. Otherwise, by the definition of κ we have

$$m + \mu < \Delta_{\kappa+1} = \Delta_\kappa + \kappa + 1,$$

whence $\kappa' = m - \Delta_\kappa + \mu \le \kappa$. In any case, we can apply Lemma 5.35 with $A = E$ and $a = n$ to obtain

$$\Psi(E) - \Psi(\overline{E}) \le 2^{\nabla_{m+\mu} - 1}$$

and (5.31) yields

$$d(s, x_2) \ge \Psi(E) - M_{\nabla_{m+\mu} - 1}. \tag{5.36}$$

We now have to distinguish the cases $\tau_1 < \tau_3$ and $\tau_1 > \tau_3$.

4.0.0. ($E' \ni m < n \in E$, $\Delta_\mu < m$, $\tau_1 < \tau_3$) Now γ passes through $s \to x_0 \to x_1 \to x_2 \to x_3 \to t$ or $s \to x_0 \to x_2 \to x_1 \to x_3 \to t$. We have $\overline{x}_1 \in (Q \setminus \{3, x_{1-,m}\})^{m-1}$, where we write $\overline{r} := r \upharpoonright [m-1]$. Induction assumption gives

$$d\left(\overline{x}_1, \overline{x}_3\right) \ge \Psi(A) \tag{5.37}$$

for $A := \overline{x}_3^{-1}(\{3\})$.

Moreover, $\overline{x}_3 \in (Q \setminus \{2, x_{3-,n}\})^{n-1}$ with $x_{3-,n} \in B$; in particular $\overline{x}_3(E'') = \{1 - x_{3-,n}\}$. Therefore, $\overline{t} \in (Q \setminus \{x_{3-,n}, 1 - x_{3-,n}\})^{n-1}$, so that

$$d(x_3, t) = d\left(\overline{x}_3, \overline{t}\right) \ge \Psi(A') \tag{5.38}$$

for $A' := \overline{x}_3^{-1}(\{1 - x_{3-,n}\}) \supset \overline{x}_3^{-1}(\{1 - x_{3-,n}\}) \dot\cup E''$. Thus $A \cup A' \supset [m-1] \dot\cup E''$, such that $|A \cup A'| \ge m + \mu - 2$. By Lemma 5.37 we have

$$\Psi(A) + \Psi(A') \ge \frac{1}{2}\Psi[m+\mu] - 1. \tag{5.39}$$

Moreover, $s(E'') = \{0\}$ and since these discs avoid peg 3, we have $x_2(E'') = \{3 - x_{2,n}\} \subseteq [2]$, so that they have to perform at least $M_{|E''|} = 2^{\mu-1} - 1$ moves between s and x_2 (cf. 2.0). By (5.30) the discs from $[m]$ perform at least $1 + \Psi\left(\overline{\overline{E}}\right)$ moves between s and x_0, and finally disc n moves precisely once from s to x_2. Putting this together with (5.37) and (5.38), we get

$$
\begin{aligned}
L \;&=\; d(s, x_3) + d(x_3, t) \\
&\ge\; 2^{\mu-1} - 1 + 1 + \Psi\left(\overline{\overline{E}}\right) + 1 + \Psi(A) + \Psi(A') \\
&\ge\; 2^{\mu-1} + \Psi\left(\overline{\overline{E}}\right) + \frac{1}{2}\Psi[m+\mu],
\end{aligned}
$$

the latter inequality following from (5.39). Because $m > \Delta_\mu$, i.e. $m - 1 \ge \Delta_\mu$, Lemma 5.36 (with $\ell = \mu + 1$, $k = m - 1$, $A = \overline{\overline{E}}$, $A' = E \setminus [m-1]$) tells us that

$$\Psi(E) - \Psi\left(\overline{\overline{E}}\right) \le \Psi[m+\mu] - \Psi[m-1],$$

whence, using (5.27) with $k = m - 1$ and $\ell = \mu + 1$,

$$L - \Psi(E) \geq 2^{\mu - 1} + \Psi[m - 1] - \frac{1}{2}\Psi[m + \mu] \geq 0.$$

Note that we counted only one move of disc n. In the case of the second ordering of the x_ℓs we may add one move, because then that disc has to move at least twice, i.e. $L > \Psi(E)$. Similarly, if $x_0 < x_1$ in the first ordering, we may add one move of disc m between x_0 and x_1, leading again to $L > \Psi(E)$.

4.0.1. $(E' \ni m < n \in E, \Delta_\mu < m, \tau_1 > \tau_3)$ Now the path γ goes $s \to x_0 \to x_2 \to x_3 \to x_1 \to t$. We have $\overline{\overline{x}}_1 \in (Q \smallsetminus \{3, x_{1-,m}\})^{m-1}$. If $x_{1-,m} \in B$, let $a := 2$ and $a' := 1 - x_{1-,m} \in B$; otherwise, i.e. if $x_{1-,m} = 2$, let $a := x_{3-,n} \in B$ and $a' := 1 - a$. With $A := \overline{\overline{x}}_1^{-1}(\{a\})$ and $A' := \overline{\overline{x}}_1^{-1}(\{a'\})$ we have $A \,\dot\cup\, A' = [m - 1]$ in both cases. Moreover, $\overline{\overline{x}}_3 \in (Q \smallsetminus \{2, x_{3-,n}\})^{m-1}$, so that peg a is empty in any case, whence induction assumption tells us that, together with at least one move of disc m between x_3 and x_1, we have

$$d(x_3, x_1) \geq 1 + d\left(\overline{\overline{x}}_3, \overline{\overline{x}}_1\right) \geq 1 + \Psi(A). \tag{5.40}$$

Since $\overline{\overline{t}} \in (Q \smallsetminus \{a', 1 - a'\})^{m-1}$, we get

$$d(x_1, t) \geq d\left(\overline{\overline{x}}_1, \overline{\overline{t}}\right) \geq \Psi(A'). \tag{5.41}$$

We note that x_2 may be equal to x_3, summarize (5.36), (5.40), (5.41) and make use of Lemma 5.37 to obtain

$$
\begin{aligned}
L &= d(s, x_2) + d(x_2, x_3) + d(x_3, x_1) + d(x_1, t) \\
&\geq \Psi(E) - M_{\nabla_{m+\mu}-1} + 0 + 1 + \Psi(A) + \Psi(A') \\
&\geq \Psi(E) - M_{\nabla_{m+\mu}-1} + \frac{1}{2}\Psi[m + 1]. \tag{5.42}
\end{aligned}
$$

Since the assumption $\Delta_\mu < m$ is equivalent to $m + \mu \geq \Delta_{\mu+1}$, we have $\kappa := \nabla_{m+\mu} \geq \mu + 1$. With $\Delta_{\nabla_{m+\mu}} \leq m + \mu$ it follows that $m - 1 = (m + \mu) - (\mu + 1) \geq \Delta_\kappa - \kappa = \Delta_{\kappa-1}$, so that $\nabla_{m-1} \geq \kappa - 1$. Lemma 5.32 yields $\Psi[m + 1] \geq 2^{\nabla_{m-1}+1} \geq 2^\kappa$ and (5.42) reads $L > \Psi(E)$.

4.1. $(E' \ni m < n \in E, \Delta_\mu \geq m)$ The path γ obviously runs like $s \to x_2 \to x_3 \to t$, where $x_2 = x_3$ or $x_3 = t$ are a priori not excluded. So $L = d(s, x_2) + d(x_2, x_3) + d(x_3, t) \geq d(s, x_2) + d(x_3, t)$.

We can employ Lemma 5.35 with $A = E$, $a = n$, and $\kappa = \mu$, because $\{d \in E \mid d > \Delta_\mu\} \subseteq \{d \in E \mid d > m\} = E'' \cup \{n\}$. We obtain

$$\Psi(E) - \Psi(\overline{E}) \leq 2^{\mu-1}.$$

Together with (5.31) we get

$$d(s, x_2) \geq \Psi(E) - M_{\mu-1}. \tag{5.43}$$

Since $x_3(E'') = \{1 - x_{3-,n}\} \subseteq B$ and because these discs do not pass through peg 3, we get $t(E'') = \{2\}$ and $d(x_3, t) \geq M_{\mu-1}$. With (5.43) we arrive at $L \geq \Psi(E)$.

Note that if $\tau_1 > \tau_2$, then we may add a move of disc m between x_2 and t to the right hand side, so that $L > \Psi(E)$ in that case.

This completes the proof which, in fact, tells us a little more (cf. Theorem 5.18) if an upper bound is known for $d(s, t)$, like, e.g., from the Dudeney-Frame-Stewart algorithms.

Proposition 5.40. *If* $n \in \mathbb{N}$ *and* $s(n) = i$ *in the situation of Theorem 5.38 and* $d(s, t) \leq \Psi(E)$ *for* $E := s^{-1}(\{i\})$, *then the largest disc* n *moves exactly once in a shortest* s, t-*path* γ, *i.e.* $x_2 = x_3$. *Moreover, for* γ *only the cases 2.0, 4.0.0 (with* $x_0 = x_1 < x_2$) *and 4.1 (with* $x_0 \leq x_1 < x_2$) *can occur. Case 2.0 is excluded if* $|E| > 1 + \nabla_{\min E-1}$, *i.e. in particular if* $E = [n]$ *with* $n \neq 1$.

Proof. The named cases contain exactly one move of disc n and the proof of Theorem 5.38 shows that in all other cases we would have $d(s, t) > \Psi(E)$. □

Corollary 5.41. *In a shortest solution for the P0 task in* H_4^{n+1}, $n \in \mathbb{N}$, *the largest disc moves exactly once, the second largest exactly twice.*

This is, of course, also a consequence of Theorem 5.18.

5.6 Hanoi Graphs H_p^n

Obviously, the TH with four or more pegs can be modeled just like the classical one by state graphs. Let $p \in \mathbb{N}_3$; then the *Hanoi graph (on p pegs)* H_p^n has all regular states as vertices, two vertices being adjacent if they are obtained from each other by a legal move of one disc. See Figure 5.12 for a graphical representation of H_4^4. For more illustrations of Hanoi graphs see [338].

The vertex set of H_p^n is thus

$$V(H_p^n) = [p]_0^n \, ,$$

and we will again denote its elements by $s = s_n \ldots s_1$ with $s_d \in [p]_0$ signifying the peg where disc $d \in [n]$ is lying in state $s \in V(H_p^n)$. In contrast to the definition of Sierpiński graphs S_p^n, $p \in \mathbb{N}_4$, the extension of the edge sets from H_3^n to H_p^n, $p > 3$, is not based on the mathematical structure of the graphs, but on the rules of the TH puzzle. Formally, for any $p \in \mathbb{N}_3$ and any $n \in \mathbb{N}_0$,

$$E(H_p^n) = \{\{\underline{s}i\overline{s}, \underline{s}j\overline{s}\} \mid i, j \in [p]_0, i \neq j, d \in [n], \underline{s} \in [p]_0^{n-d}, \overline{s} \in ([p]_0 \setminus \{i, j\})^{d-1}\} \, .$$
$$(5.44)$$

Here, for each edge, d is again the moving disc, such that all smaller discs have to be in a state $\overline{s}$ on pegs different from both i and j. Since $V(H_p^n) = [p]_0^n$ and two vertices of H_p^n can be adjacent only if they differ in exactly one coordinate, Hanoi graphs H_p^n can be interpreted as spanning subgraphs of Hamming graphs, that is,

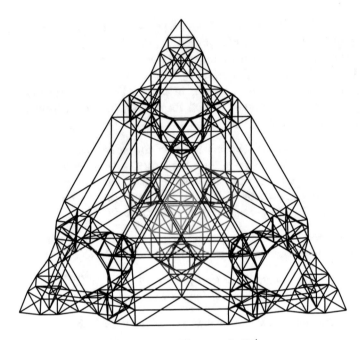

Figure 5.12: The graph H_4^4

of Cartesian products of complete graphs; see [221, Section 2.2] for this point of view to Hanoi graphs.

The edge sets of Hanoi graphs can be expressed, as done before for $p = 3$ (cf. (2.12)) in a recursive definition:

$$E(H_p^0) = \varnothing,$$
$$\forall\, n \in \mathbb{N}_0: E(H_p^{1+n}) = \big\{\{ir, is\} \mid i \in [p]_0,\ \{r, s\} \in E(H_p^n)\big\}$$
$$\cup \big\{\{ir, jr\} \mid i, j \in [p]_0,\ i \neq j,\ r \in ([p]_0 \smallsetminus \{i, j\})^n\big\}. \quad (5.45)$$

The first term of (5.45) corresponds to the edges between the states in which the largest disc is on a fixed peg i. Any edge from the second set is between two states in which disc $n+1$ is on different pegs, while the remaining discs lie on the other $p-2$ pegs. If follows that

$$\|H_p^0\| = 0\,,\ \|H_p^{1+n}\| = p\|H_p^n\| + \binom{p}{2}|H_{p-2}^n| = p\|H_p^n\| + \binom{p}{2}\cdot(p-2)^n. \quad (5.46)$$

An easy consequence is

Proposition 5.42. *For $p \in \mathbb{N}_4$ and $n \in \mathbb{N}_2$, $\|S_p^n\| < \|H_p^n\|$.*
In particular,
$$S_p^n \cong H_p^n \Leftrightarrow p = 3 \text{ or } n \leq 1\,.$$

Proof. Obviously, $S_p^0 = H_p^0$ and $S_p^1 = H_p^1$, and we have already seen that $S_3^n \cong H_3^n$ for all n. Pairs of subgraphs iS_p^n, $i \in [p]_0$, being linked by a single edge, we have the recurrence relation

$$\|S_p^{1+n}\| = p\|S_p^n\| + \binom{p}{2},$$

and $\|S_p^n\| < \|H_p^n\|$ for $p > 3$ and $n > 1$ follows from (5.46). □

Solving the recurrence (5.46) with Lemma 2.18, we get (cf. [237, Corollary 3.3]):

Proposition 5.43. *For any $n \in \mathbb{N}_0$ and $p \in \mathbb{N}_3$,*

$$\|H_p^n\| = \frac{p(p-1)}{4}\left(p^n - (p-2)^n\right).$$

Proposition 5.43 can, of course, also be proved directly by recourse to (5.44) or even more directly as follows (cf. D. Arett and S. Dorée [20, p. 202f]). Fix two pegs i and j. Then a move between them is legal if and only if they are not both empty. The number of states in which both i and j are empty is $(p-2)^n$, hence there are exactly $p^n - (p-2)^n$ states that allow a move between pegs i and j. Since there are $\binom{p}{2}$ pairs of pegs and each edge has been counted twice, namely as moves from i to j and from j to i, the number of edges is thus $\frac{1}{2}\binom{p}{2}\left(p^n - (p-2)^n\right)$.

Another expression for $\|H_p^n\|$ involving **Stirling numbers of the second kind** is given in Exercise 5.9. A formula for $\|H_p^n\|$ cannot easily be obtained from the Handshaking Lemma because of the diversity of degrees in H_p^n. For instance, while disc 1 can always move to $p-1$ pegs, the second smallest top disc may still go to $p-2$ targets, such that for $n > 1$ any non-perfect vertex has degree at least $2p-3$, which, for $p > 3$, is larger than p, the maximal degree of S_p^n; this is an even simpler argument for the second statement in Proposition 5.42. An extension of this argument leads to the following result (see [20, p. 203] and [211, Proposition 1]), where $\delta(G)$ is the minimum degree of G.

Proposition 5.44. *Let $p \in \mathbb{N}_3$, $n \in \mathbb{N}$, and let s be a vertex of H_p^n corresponding to a regular state with k pegs occupied, where $k \in [p]$. Then the degree of s is*

$$\binom{p}{2} - \binom{p-k}{2}.$$

Consequently,

$$\delta\left(H_p^n\right) = p - 1 \quad \text{and} \quad \Delta\left(H_p^n\right) = \binom{p}{2} - \binom{p-n}{2},$$

with the understanding that $\binom{p-n}{2} = 0$ for $n > p$.

Proof. Let the k top discs in state s be arranged according to size from d_1 (smallest) to d_k (largest). Then d_ℓ can move to $p - \ell$ pegs, $\ell \in [k]$. This adds up to

$$\sum_{\ell=1}^{k} (p - \ell) = \sum_{\ell=1}^{p} (p - \ell) - \sum_{\ell=k+1}^{p} (p - \ell) = \sum_{\ell=0}^{p-1} \ell - \sum_{\ell=0}^{p-k-1} \ell = \binom{p}{2} - \binom{p-k}{2}.$$

The minimum degree is obtained for $k = 1$, that is, for perfect states. For the maximum degree, if $n \geq p - 1$ then there are regular states with $k = p - 1$ and hence $\Delta(H_p^n) = \binom{p}{2}$; otherwise k can only be as large as n. □

Remark 5.45. *If $n \in \mathbb{N}_2$, then there are more than 2 vertices of the form $i^{n-1}j$, $i \neq j$, which all have odd degree $2p - 3$. Therefore the corresponding Hanoi graphs are not semi-eulerian.*

Even if S_p^n is not isomorphic to H_p^n anymore for $p > 3$, we can still imbed it, at least for any odd p. More precisely:

Proposition 5.46. *For every odd $p \in \mathbb{N}_3$ and all $n \in \mathbb{N}_0$, S_p^n can be imbedded isomorphically into H_p^n in such a way that k^n is mapped onto itself for all $k \in [p]_0$.*

Proof. The case $n = 0$ is trivial. For $n \in \mathbb{N}_0$ let ι_n be the isomorphic imbedding from S_p^n into H_p^n fulfilling $\iota_n(k^n) = k^n$ by induction assumption. We construct the mapping $\iota_{1+n} : S_p^{1+n} \to H_p^{1+n}$ in the following way. For $k \in [p]_0$ define the permutation π_k on $[p]_0$ as follows:[73]

$$\forall i \in [p]_0 : \pi_k(i) = \tfrac{1}{2}\left(k(p+1) - i(p-1)\right) \bmod p;$$

it has precisely one fixed point, namely k. Then let $\pi_k^{(n)}$ denote the bijection on $[p]_0^n$ with $\pi_k^{(n)}(s_n \ldots s_1) = \pi_k(s_n) \ldots \pi_k(s_1)$. Define

$$\forall k \in [p]_0 \ \forall s \in [p]_0^n : \iota_{1+n}(ks) = k\pi_k^{(n)}\left(\iota_n(s)\right).$$

This obviously constitutes a bijection with

$$\iota_{1+n}(k^{1+n}) = k\pi_k^{(n)}\left(\iota_n(k^n)\right) = k\pi_k^{(n)}(k^n) = k^{1+n}.$$

It remains to show that $\{\iota_{1+n}(ij^n), \iota_{1+n}(ji^n)\} \in E(H_p^{1+n})$ for $i, j \in [p]_0$, $i \neq j$. We have $\iota_{1+n}(ij^n) = i\pi_i^{(n)}\left(\iota_n(j^n)\right) = i\pi_i(j)^n$ and similarly $\iota_{1+n}(ji^n) = j\pi_j(i)^n$. Moreover,

$$i \neq \pi_i(j) = \frac{1}{2}(ip + i - jp + j) \bmod p = \frac{1}{2}(jp + j - ip + i) \bmod p = \pi_j(i) \neq j,$$

and so the two vertices are adjacent in H_p^n. □

---

[73]The reader might wonder where this definition comes from: it derives from the canonical total coloring of the complete graph on p vertices.

In the case $p = 3$ the imbedding of Proposition 5.46 is an isomorphism, because the graphs have the same order. This isomorphism is the one represented as an automaton in Figure 4.4.

The proof of Proposition 5.46 does not work for even p, because the permutations π_k are not well defined then. This reflects the fact that K_p cannot be totally colored with p colors for even p. In fact, Proposition 5.46 is not true for even p. For instance, S_4^2 can not be imbedded isomorphically into H_4^2: for degree reasons, extreme vertices have to be mapped onto perfect states; but then any choice of single connections between the subgraphs $\iota_2(kS_4^1)$, $k \in [4]_0$, will lead to a contradiction (cf. the pictures of S_4^2 and H_4^2 in Figures 9.9 and 5.6, respectively). This kind of argument can be extended to every even $p \geq 4$ and all $n \geq 2$ by considering cliques in H_p^n. In fact, the following holds (see [206, Lemma 2.1]):

Lemma 5.47. *Every complete subgraph of H_p^n, $p \in \mathbb{N}_3$, $n \in \mathbb{N}$, is induced by edges corresponding to moves of one and the same disc. In particular, $\omega(H_p^n) = p$ and the only p-cliques of H_p^n are of the form $\underline{s}H_p^1$ with $\underline{s} \in [p]_0^{n-1}$.*

Proof. Take any vertex s joined to two vertices s' and s'' by edges corresponding to the moves of two different discs. Then the positions of these discs differ in s' and s''. Since vertices in H_p^n can only be adjacent if they differ in precisely one coordinate, s' and s'' can not be adjacent. This proves the first assertion. In any state s, the smallest disc can move to $p - 1$ pegs, so that s is contained in the p-clique $s_n \ldots s_2 H_p^1$. On the other hand, a disc $d \neq 1$ can be transferred to at most $p - 2$ pegs, namely those not occupied by disc 1.   □

Proposition 5.46 and Lemma 5.47 in combination with Theorem 4.3 lead to (see [206, Theorem 3.1]):

Theorem 5.48. *Let $p \in \mathbb{N}_3$, $n \in \mathbb{N}$. Then S_p^n can be embedded isomorphically into H_p^n if and only if p is odd or $n = 1$.*

For details of the proof, see [206]. Recalling that Hanoi graphs can be interpreted as spanning subgraphs of Hamming graphs, we obtain the same result for Sierpiński graphs S_p^n with odd p.

We have seen in Proposition 2.25 that the connectivity κ of the graphs H_3^n is 2. This fact extends to all Hanoi graphs as follows.

Proposition 5.49. *For any $p \in \mathbb{N}_3$ and any $n \in \mathbb{N}$, $\kappa\left(H_p^n\right) = p - 1$.*

Proof. By definition, $\kappa(K_p) = p - 1$, hence the assertion holds for $n = 1$ because $H_p^1 \cong K_p$. Since $\delta\left(H_p^n\right) = p - 1$ by Proposition 5.44, we have $\kappa(H_p^n) \leq p - 1$ for all n. To show that no $p - 2$ vertices disconnect H_p^n, we proceed by induction on n, the base case being already treated. If we delete $p - 2$ vertices from H_p^{1+n}, at most $(p-2)^2$ edges between subgraphs kH_p^n, induced by vertices of the form $k\bar{s}$ with $k \in [p]_0$ and $\bar{s} \in [p]_0^n$, are destroyed. But every pair of these subgraphs has

$(p-2)^n \geq p-2$ connecting edges, such that at most $p-2$ connections between the subgraphs are completely lost, which leaves the whole graph still connected. □

Hamiltonicity also extends to all Hanoi graphs; see Exercise 5.10. This is not the case, however, for the existence of perfect codes. While there are, of course, perfect codes for all Hanoi graphs isomorphic to Sierpiński graphs and also for H_p^2 (see Exercise 5.11), no other Hanoi graph has this property, as found out by Q. Stierstorfer [399, Hauptsatz 4.4]. (An attempt at H_4^3 had already been made by D. Arett in [19].) So the quest for the domination number of Hanoi graphs H_p^n for $p \in \mathbb{N}_4$ and $n \in \mathbb{N}_3$ will be a true challenge. From the structure of the graphs one is tempted to conjecture that $\gamma(H_p^n) \leq \gamma(S_p^n)$; cf. (4.14). This is true for odd p by virtue of Theorem 5.48 because the domination number of a spanning subgraph can not be smaller than the domination number of the graph itself. The only case known so far for even p, namely $\gamma(H_4^3) = 13 = \gamma(S_4^3)$ (see [209]), supports the conjecture further (and even $\gamma(H_p^n) = \gamma(S_p^n)$). Note that for the special case of Hanoi graphs H_p^2 even power domination and propagation radius could be approached; see [425, Theorems 3.1 and 3.2].

We next list several additional graph-theoretical properties of the graphs H_p^n that have been dealt with by now.

Planarity Recall that the graphs H_3^n are planar for any $n \in \mathbb{N}_0$. From Figure 5.6 we see that H_4^0, H_4^1, and H_4^2 are planar. On the other hand, H_4^3 is *not* planar; see Exercise 5.12. Since this is a subgraph of every H_4^n with $n > 4$, they too are non-planar. Finally, $H_5^1 \cong K_5$ is contained in all H_p^ns for $p > 4$. In conclusion, the only planar Hanoi graphs are H_p^0, H_3^n, H_4^1, and H_4^2, a result first obtained by Hinz and Parisse [210, Theorem 2].

Not a single crossing number of non-planar Hanoi graphs for more than one disc is known explicitly! Whereas $\mathrm{cr}(S_4^3)$ had been found to be 12 (see Section 4.2.2), the best upper bound for $\mathrm{cr}(H_4^3)$ is 72 [371, Theorem 3.12]. The corresponding drawing of R. S. Schmid is shown in Figure 5.13. Note that for this drawing it turned out to be advantageous *not* to employ planar drawings of the 4 subgraphs iH_4^2, but ones with 9 crossings each. Among the 24 edges connecting these subgraphs, 12 have 4 crossings each and 12 contain 2; since they were counted twice, this makes up for another 36 crossings.

Schmid's drawing and the question of its optimality with respect to the number of crossings has even made it to a French popular science magazine; see [97, p. 112]. Nobody has ever dared to approach the question of genera of non-complete non-planar Hanoi graphs.

Chromatic number Since H_p^n contains complete subgraphs on p vertices, $\chi(H_p^n) \geq p$. On the other hand, Arett and Dorée [20, p. 206] observed that the function

$$c(s) = \left(\sum_{d=1}^{n} s_d \right) \bmod p$$

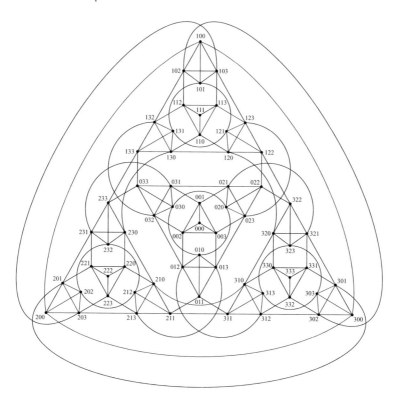

Figure 5.13: Drawing in the plane of H_4^3 with 72 crossings

defines a vertex coloring of H_p^n. Indeed, if s and s' are adjacent states, then they differ in exactly one position and consequently $c(s) \neq c(s')$. So $\chi\left(H_p^n\right) \leq p$ and we conclude that $\chi\left(H_p^n\right) = p$.

Chromatic index The arguments used to determine $\chi\left(H_p^n\right)$ form a direct extension of Proposition 2.24 and its proof for the case of vertex colorings. The same proposition also gives the chromatic index $\chi'\left(H_3^n\right) = 3$, where the label of the idle peg of the move associated with an edge defines an edge coloring of H_3^n with three colors (cf. [378, p. 97]). However, the quest for the chromatic index $\chi'\left(H_p^n\right)$ for bases higher than 3 turned out to be quite intriguing.[74] The intrinsic difficulty seems to be the fact that an answer appeared to be more demanding for small n and large p than for higher values of n. Hinz and Parisse [211] overcame all the difficulties and proved:

Theorem 5.50. *For any $p \in \mathbb{N}_3$ and any $n \in \mathbb{N}_2$, $\chi'\left(H_p^n\right) = \Delta\left(H_p^n\right)$.*

---

[74]The quest was initiated by Stockmeyer during the *Workshop on the Tower of Hanoi and Related Problems*, Maribor, Slovenia, 2005.

Recall (from p. 48) that by Vizing's theorem for any graph G, $\chi'(G) = \Delta(G)$ or $\chi'(G) = \Delta(G) + 1$. In the respective cases G is called *Class 1* and *Class 2*. Since $H_p^1 = K_p$, Theorem 5.50 can thus be rephrased by saying that all Hanoi graphs are Class 1 except the instances of H_p^1 with odd p.

Total chromatic number In the same paper [211] where the chromatic index was settled, the total chromatic numbers are also determined:

Theorem 5.51. *For any $p \in \mathbb{N}_3$ and any $n \in \mathbb{N}_2$, $\chi''\left(H_p^n\right) = \Delta\left(H_p^n\right) + 1$.*

This result in particular implies that the Total coloring conjecture (see p. 48) is true on Hanoi graphs.

We conclude the section by describing the symmetries of Hanoi graphs, a result due to S. E. Park [333]. Recall from Theorem 2.29 that H_3^n, $n \in \mathbb{N}$, has exactly six symmetries, more precisely, $\mathrm{Aut}(H_3^n) \cong \mathrm{Sym}(T)$. We gave a short proof by applying the fact (Lemma 2.8) that a state is uniquely determined by the vector of its distances to the extreme vertices. In Theorem 4.13 we further noticed that the argument can be extended to all Sierpiński graphs and concluded that $\mathrm{Aut}(S_p^n) \cong \mathrm{Sym}([p]_0)$. On the other hand, the distances in H_p^n, $p \in \mathbb{N}_4$, being much less well understood, we follow different arguments here to obtain the analogous result for Hanoi graphs (cf. [333, Main Theorem]). We begin with a little lemma [333, Lemma 5].

Lemma 5.52. *Let $n \in \mathbb{N}$ and $s \in [p]_0^n$ with $s_n = i$. Then for any $j \in [p]_0 \smallsetminus \{i\}$, $\mathrm{d}(s, i^n) < \mathrm{d}(s, j^n)$.*

Proof. We already know from Theorem 5.18 that on a shortest path P from s to j^n, disc n moves exactly once. The corresponding edge in H_p^n is $e = \{ir, jr\}$ with $r \in ([p]_0 \smallsetminus \{i, j\})^{n-1}$. Let P be the union of the path Q from s to ir, the edge e and the path R from jr to j^n. Now merge Q with the path obtained from R by switching the values of i and j in its vertices to obtain a path from s to i^n which has a length of 1 less than P. $\qquad\square$

Theorem 5.53. *For any $p \in \mathbb{N}_3$ and any $n \in \mathbb{N}$, $\mathrm{Aut}(H_p^n) \cong \mathrm{Sym}([p]_0)$.*

Proof. We proceed in several steps.
(i) $\forall\, g \in \mathrm{Aut}(H_p^n)\ \exists_1\, \sigma \in \mathrm{Sym}([p]_0)\ \forall\, k \in [p]_0 : g(k^n) = \sigma(k)^n$.
The existence is clear, because g preserves degrees and perfect states are the only edges with degree $p - 1$. Uniqueness is also obvious.

This defines a mapping $\iota :\ \mathrm{Aut}(H_p^n) \to \mathrm{Sym}([p]_0)$, $g \mapsto \sigma_g$. We will show now that it is a group isomorphism.
(ii) The mapping ι is surjective, since for every $\sigma \in \mathrm{Sym}([p]_0)$:

$$g :\ [p]_0^n \to [p]_0^n,\ s_n \ldots s_1 \mapsto \sigma(s_n) \ldots \sigma(s_1)$$

is in $\mathrm{Aut}(H_p^n)$ and $\iota(g) = \sigma$.

(iii) $\forall\, g \in \mathrm{Aut}(H_p^{1+n})\ \forall\, k \in [p]_0 : g(ks) = \sigma_g(k)g_k(s)$ with $g_k \in \mathrm{Aut}(H_p^n)$.
Let $s \in [p]_0^n$ and $j \in [p]_0 \smallsetminus \{k\}$; then, by Lemma 5.52,

$$\mathrm{d}\big(g(ks), \sigma_g(k)^{1+n}\big) = \mathrm{d}(ks, k^{1+n}) < \mathrm{d}(ks, j^{1+n}) = \mathrm{d}\big(g(ks), \sigma_g(j)^{1+n}\big),$$

which, again by Lemma 5.52, is only true if $g(ks)_{n+1} = \sigma_g(k)$. Hence $g \upharpoonright k[p]_0^n \in \mathrm{Aut}\big(\sigma_g(k)H_p^n\big)$.

(iv) $\forall\, g \in \mathrm{Aut}(H_p^n)\ \forall\, s \in [p]_0^n : g(s) = \sigma_g(s_n)\ldots\sigma_g(s_1)$.
This is proved by induction on n. The statement is true for $n = 1$, because H_p^1 contains only perfect states. Now let $g \in \mathrm{Aut}(H_p^{1+n})$, $n \in \mathbb{N}$. Then, by (iii) and induction assumption, $g(ks) = \sigma_g(k)\sigma_{g_k}(s_n)\ldots\sigma_{g_k}(s_1)$, and we have to show that $\sigma_{g_k}(j) = \sigma_g(j)$ for every $j \in [p]_0$.
 From $g(ki^n) = \sigma_g(k)g_k(i^n) = \sigma_g(k)\sigma_{g_k}(i)^n$ it follows immediately, putting $i = k$, that $\sigma_{g_k}(k) = \sigma_g(k)$. Moreover, for $i \neq k$ and any $i \neq j \neq k$, we know that the vertices $\sigma_g(k)\sigma_{g_k}(i)^n$ and $\sigma_g(j)\sigma_{g_j}(i)^n$ are adjacent because they are images under g of adjacent vertices. This is possible only if $\sigma_{g_k}(i) = \sigma_{g_j}(i)$, since neighboring vertices differ in just one pit. But then, the σs being permutations, $\{\sigma_{g_k}(j), \sigma_{g_k}(k)\} = \{\sigma_{g_j}(j), \sigma_{g_j}(k)\}$, whence $\sigma_{g_k}(j) = \sigma_g(j)$.
 It follows that ι is injective.

(v) The consistency of ι with the group structures is straightforward. $\qquad \square$

At this point we should mention that there is also the concept of Hanoi Towers groups, introduced by R. Grigorchuk and Šunik in [167]. Roughly speaking, the *Hanoi Towers group* on p pegs, denoted by $H^{(p)}$, is the group of p-ary tree automorphisms generated by the transposition automorphisms. These transpositions correspond to the moves between associated pegs which in turn implies that the so-called *Schreier graph* of the action of $H^{(p)}$ on the nth level of the p-ary tree is obtained from H_p^n by adding the appropriate number of loops to vertices of H_p^n such that each of them has $\binom{p}{2}$ incident edges, thus making the resulting graph more regular. Going into details would be beyond the scope of this book; let us just mention that in [166] Grigorchuk and Šunić determined the spectrum of the Schreier graphs corresponding to H_3^n.

Many properties of all Hanoi graphs have thus been clarified by now. On the other hand, while metric properties of H_3^n are well understood, it should be clear from the previous sections that metric properties of H_p^n are notoriously difficult for $p > 3$. In preparation for more numerical experiments, we will now investigate equivalence classes of states and tasks. Referring to Section 0.8.2 again, we now consider the group $(\Gamma, \cdot, 1) = \big(\mathrm{Aut}(H_p^n), \circ, \mathrm{id}\big)$ and let it act on $X = V(H_p^n) = [p]_0^n$ by the action $g.s = \sigma_g(s_n)\ldots\sigma_g(s_1)$. Then, by Theorem 5.53, $|\Gamma| = p!$ and in order to make use of Theorem 0.23 and Corollary 0.24, we have to look at the sets X^g and Γ_s.
 Recall that $\mathrm{FP}(\sigma)$ is the set of fixed points of $\sigma \in \mathrm{Sym}([p]_0)$. Then $q := |\mathrm{FP}(\sigma)| \in [p+1]_0$ and σ as restricted to $[p]_0 \smallsetminus \mathrm{FP}(\sigma)$ is a derangement on $p - q$ elements. A state s is a fixed point under the corresponding automorphism g if

and only if (we may, of course, assume that $n \in \mathbb{N}$) it "lives" on $\mathrm{FP}(\sigma)$, i.e. $s([n]) \subseteq \mathrm{FP}(\sigma)$; there are $|\Gamma_s| = (p - |s([n])|)!$ such automorphisms. Theorem 0.23 yields $|\Gamma.s| = \dfrac{p!}{(p - |s([n])|)!}$, such that the size of an equiset $[s]$ can have one of the values $\dfrac{p!}{(p-q)!}$, $q \in [p]$. There are $\left\{ {n \atop q} \right\}$ states s whose n discs are distributed *onto* precisely q pegs, whence the number of equisets in H_p^n is

$$\left|[X]\right| = \sum_{q=1}^{p} \left\{ {n \atop q} \right\}.$$

On the other hand, there are $\binom{p}{q}(p-q)\mathrm{i}$ permutations σ with $|\mathrm{FP}(\sigma)| = q$, each leading to q^n fixed points of the corresponding automorphism g, such that

$$\sum_{g \in \Gamma} X^g = \sum_{q=1}^{p} \binom{p}{q}(p-q)\mathrm{i}\, q^n$$

whence from Corollary 0.24:

$$\left|[X]\right| = \sum_{q=1}^{p} \frac{(p-q)\mathrm{i}}{(p-q)!}\, \frac{q^n}{q!}.$$

We summarize these facts, including the case $n = 0$ for completeness.

Theorem 5.54. *The vertex set $[p]_0^n$ of H_p^n, $n \in \mathbb{N}_0$, decomposes into*

$$\sum_{q=0}^{p} \left\{ {n \atop q} \right\} = \sum_{q=0}^{p} \frac{(p-q)\mathrm{i}}{(p-q)!}\, \frac{q^n}{q!}$$

equisets with respect to the symmetries of H_p^n.

For $q \in [p-1]_0$, $\left\{ {n \atop q} \right\}$ equisets have size $\frac{p!}{(p-q)!}$; $\left\{ {n \atop p-1} \right\} + \left\{ {n \atop p} \right\}$ equisets have size $p!$.

If we let the automorphism group of H_p^n act on the set $X = [p]_0^n \dot\times [p]_0^n$ of non-trivial tasks for the TH with p pegs and $n \in \mathbb{N}$ discs, we obtain (cf. [161, Theorems 1 and 2]), by the same arguments, but observing that we now have to consider the union $s([n]) \cup t([n])$ of the images of the initial and goal states to be distributed onto q pegs,

Theorem 5.55. *The set of non-trivial tasks $[p]_0^n \dot\times [p]_0^n$ on H_p^n, $n \in \mathbb{N}$, decomposes into*

$$\sum_{q=2}^{p} \left(\left\{ {2n \atop q} \right\} - \left\{ {n \atop q} \right\} \right) = \sum_{q=2}^{p} \frac{(p-q)\mathrm{i}}{(p-q)!}\, \frac{q^n(q^n - 1)}{q!}$$

equisets with respect to the symmetries of H_p^n.

For $q \in \{2, \ldots, p-2\}$, $\left\{ {2n \atop q} \right\} - \left\{ {n \atop q} \right\}$ equisets have size $\frac{p!}{(p-q)!}$; $\left\{ {2n \atop p-1} \right\} + \left\{ {2n \atop p} \right\} - \left\{ {n \atop p-1} \right\} - \left\{ {n \atop p} \right\}$ equisets have size $p!$.

Some numerical values can be found in [208, Table 3].

5.7 Numerical Results and Largest Disc Moves

Only very few mathematically provable statements are known for P2-type tasks of the TH with more than three pegs. In particular, this applies to the diameter of H_p^n. The first general result is

Proposition 5.56. *For all $p \in \mathbb{N}_3$ and $n \in \mathbb{N}_0$:*

$$\mathrm{diam}(H_p^n) \le 2^n - 1.$$

Proof. Induction on n, where in the induction step we can choose a special path between states is and jt, namely passing ik^n and jk^n for some $k \in [p]_0$ with $i \ne k \ne j$. It can be made, by Lemma 5.16, of length $\mathrm{d}(s, k^n) + 1 + \mathrm{d}(t, k^n)$ which, by induction assumption, is less than or equal to $2^{n+1} - 1$. □

Although we know from Theorem 2.31 that the upper bound in Proposition 5.56 can not be improved for $p = 3$, the numerical results from Section 5.3 show that in contrast to the Sierpiński situation (cf. Corollary 4.10), this upper bound is very bad for general p. In particular, this is true for small n, where we have the following

Theorem 5.57. *For all $p \in \mathbb{N}_3$ and $n \in [p - 1]$:*

$$\mathrm{diam}(H_p^n) = 2n - 1.$$

The lower bound for $\mathrm{diam}(H_p^n)$ follows from the remark preceding Proposition 5.20, whereas the upper bound is an immediate consequence of

Proposition 5.58. *Let $s, t \in [p]_0^n$, $p \in \mathbb{N}_3$, $n \in [p - 1]$. Then there is an s,t-path in H_p^n in which at least one disc moves at most once and all other discs move at most twice; discs in end position[75] do not move at all. In particular, $\mathrm{d}(s,t) \le 2n - 1$.*

For the proof of Proposition 5.58 we need the following lemma (cf. [23, Lemma 5.13]).

Lemma 5.59. *Let $s, t \in [p]_0^n$, $p \in \mathbb{N}_3$, $n \in [p - 1]$. Then there exists a peg which is empty in both states, s and t, or from state s some disc can move directly to its position on the bottom of some peg in state t.*

Proof. Assume that no peg is empty simultaneously in both, s and t. Let there be $e \in [p - 1]$ empty and $o = p - e$ occupied pegs in state s. Then state t contains e bottom discs on the pegs empty in s. At least one of these must be a top disc in state s, because otherwise there would be $o + e = p$ or more discs. □

Now the proof of Proposition 5.58 follows by induction on n.

---

[75] A disc is in its end position if it and all discs underneath it in t are on their goal peg.

Proof of Proposition 5.58. For $n = 1$, we have $H_p^n \cong K_p$, such that at most one move of the only disc is needed. Now let $s, t \in [p]_0^{n+1}$, $n \in [p-2]$. Move as many discs as possible directly to their end positions. Then, by virtue of Lemma 5.59, some peg k is empty simultaneously in the resulting state s' and in the goal state t. Let d be the smallest disc which is not already in its end position. (If there is no such disc, then $s' = t$ and each disc was moved at most once.) Then d must be a top disc in both s' and t. (Color discs in end position green, all others red, then no red disc lies underneath a green one in s' and t.) Now move disc d to peg k and solve the task $s'_{n+1} \ldots s'_{d+1} s'_{d-1} \ldots s'_1 \to t_{n+1} \ldots t_{d+1} t_{d-1} \ldots t_1$ on the pegs from $[p]_0 \smallsetminus \{k\}$. By induction assumption this can be done by moving at least one disc from $[n+1] \smallsetminus \{d\}$ at most once and all others at most twice; discs in end position in $s^{(\prime)}$ do not move. (If $p = 3$ at most one disc is moved, which is possible even on 2 pegs.) Disc d is finally moved from k to t_d in its second move. $\square$

Note that an algorithm could be based on the procedure just described which solves a task in H_p^n in at most $2n - 1$ moves, but that this solution is not necessarily optimal.

With a similar approach one can improve the upper bound on distances for flat-ending tasks.

Proposition 5.60. *Let $s, t \in [p]_0^n$, $p \in \mathbb{N}_3$, $n \in [p]_0$, with $\forall d \in [n] : t_d = d - 1$. Then there is an s,t-path in H_p^n in which at most $\left\lfloor \frac{n}{2} \right\rfloor$ discs move precisely twice and all other discs move at most once, those in end position not at all. In particular, $\mathrm{d}(s,t) \le n + \left\lfloor \frac{n}{2} \right\rfloor$.*

The proof will be given in Exercise 5.13, together with the radius of small Hanoi graphs.

Apart from the trivial upper bound $\mathrm{diam}(H_p^n) \le 2^n - 1$ (see Proposition 5.56) there is only one further, equally trivial, statement, namely $\mathrm{diam}(H_p^n) = 2n - 1$, if $n \in [p-1]$ (see Theorem 5.57). For $p = 3$ equality also holds for the first formula (see Theorem 2.31), but for larger p there has been no argument so far whether the inequality is sharp or not. Numerical evidence shows that indeed the diameter is *much* smaller than M_n, but this is still an interesting theoretical question in comparison to Sierpiński graphs S_p^n with $\mathrm{diam}(S_p^n) = M_n$ (see Corollary 4.10). We will give such an argument here.

Let the integer sequence $(x_n)_{n \in \mathbb{N}_0}$ be defined by

$$x_0 = 0, \ x_1 = 1, \ \forall n \in \mathbb{N}_2 : x_n = x_{n-1} + x_{n-2} + 2. \tag{5.47}$$

Then $x_n = F_{n+3} - 2$ (A001911). Moreover, $x_2 = M_2$ and

$$\forall n \in \mathbb{N}_3 : x_n < M_n. \tag{5.48}$$

Theorem 5.61. $\forall p \in \mathbb{N}_4 \ \forall n \in \mathbb{N}_0 : \mathrm{diam}(H_p^n) \le F_{n+3} - 2.$

Proof. Clearly, $\operatorname{diam}(H_p^0) = 0$ and $\operatorname{diam}(H_p^1) = 1$, so we may assume that $n \in \mathbb{N}_2$. Let $is, jt \in [p]_0^n$. If $i = j$, then

$$\operatorname{d}(is, jt) = \operatorname{d}(s, t) \le \operatorname{diam}(H_p^{n-1}).$$

Let $i \ne j$. Choose $\ell \in [p]_0 \setminus \{i, j, s_{n-1}\}$ and $k \in [p]_0 \setminus \{i, j, \ell\}$. (Here we need $p \ge 4$.) Then

$$\begin{aligned}
\operatorname{d}(is, jt) &\le \operatorname{d}(is, is_{n-1}\ell^{n-2}) + \operatorname{d}(is_{n-1}\ell^{n-2}, ik\ell^{n-2}) \\
&\quad + \operatorname{d}(ik\ell^{n-2}, jk\ell^{n-2}) + \operatorname{d}(jk\ell^{n-2}, jt) \\
&\le \operatorname{diam}(H_p^{n-2}) + 2 + \operatorname{diam}(H_p^{n-1}).
\end{aligned}$$

The statement of the theorem now follows from (5.47). □

Together with (5.48) we get:

Corollary 5.62. $\forall\, p \in \mathbb{N}_4 \ \forall\, n \in \mathbb{N}_3: \ \operatorname{diam}(H_p^n) < M_n$.

Instead of only considering s_{n-1} separately, one could also, for $n \in \mathbb{N}_3$, treat s_{n-1} and s_{n-2} as special, in which case $\operatorname{diam}(H_p^n)$ can be estimated by the sequence $(y_n)_{n \in \mathbb{N}_0}$ defined by

$$y_0 = 0, \ y_1 = 1, \ y_2 = 3, \ \forall\, n \in \mathbb{N}_3: \ y_n = y_{n-1} + y_{n-3} + 4, \tag{5.49}$$

which unfortunately does not seem to have such a nice representation as the sequence x, but is strictly smaller than the latter for $n \in \mathbb{N}_6$. An even better bound, which in addition does depend on p, can be found by a kind of Stewart strategy. Here we go on all p pegs from is to $i\underline{s}\ell^m$ for some $\ell \in [p]_0 \setminus \{i, j\}$ and with $\underline{s} = s_{n-1} \ldots s_{m+1}, m \in [n]_0$, then on pegs different from ℓ to $i\underline{s}'\ell^m$ with $\underline{s}'_d \in [p]_0 \setminus \{i, j\}$ for all $d \in [n-1] \setminus [m]$ and minimize with respect to m. Finally one can move disc n to peg j and transfer the other discs to state t on all pegs. With $z_{3,n} = M_n$, $z_{p,0} = 0$ and $\forall\, n \in [p-1]: z_{p,n} = 2n - 1$ this leads to the recurrence

$$\forall\, n \in \mathbb{N}_p: \ z_{p,n} = z_{p,n-1} + \min\{z_{p,m} + z_{p-1,n-1-m} \mid m \in [n]_0\} + 1. \tag{5.50}$$

In Table 5.9 the upper bounds for $\operatorname{diam}(H_4^n)$ are compared with calculated values.

Making use of the known diameters and (5.50) we get the estimate $\operatorname{diam}(H_4^{16}) \le 211$. This is even better than the estimate obtained from going through semi-perfect states which is 213.

Although maybe not as fast as 2^n, radii, eccentricities, and diameters *are* increasing for larger n. For $i, j \in [p]_0$, $i \ne j$, and $s, t \in [p]_0^n$, $n \in \mathbb{N}_0$, consider an is, jt-path of length $\operatorname{d}(is, jt)$. It engenders an is, it-walk by just ignoring all moves of disc $n + 1$. Since $i \ne j$, this walk must be shorter by at least one edge than the original path. This shows that

$$\forall\, i, j \in [p]_0, \ i \ne j \ \forall\, s, t \in [p]_0^n, \ n \in \mathbb{N}_0: \ \operatorname{d}(s, t) = \operatorname{d}(is, it) < \operatorname{d}(is, jt),$$

n	0	1	2	3	4	5	6	7	8	9	10
M_n	0	1	3	7	15	31	63	127	255	511	1023
x_n	0	1	3	6	11	19	32	53	87	142	231
y_n	0	!	3	7	12	19	30	46	69	103	153
$z_{4,n}$	0	1	3	5	10	17	26	39	57	82	115
$\mathrm{diam}(H_4^n)$	0	1	3	5	9	13	17	25	33	41	49

n	11	12	13	14	15	16	17
M_n	2047	4095	8191	16383	32767	65535	131071
x_n	375	608	985	1595	2582	4179	6763
y_n	226	333	490	720	1057	1551	2275
$z_{4,n}$	157	212	283	372	486	632	811
$\mathrm{diam}(H_4^n)$	65	81	97	113	130	161	193

Table 5.9: Comparison of the known diameter of H_4^n with the theoretical upper bounds

such that in particular

$$\forall j \in [p]_0 \ \forall t \in [p]_0^n, \ n \in \mathbb{N}_0 : \ \varepsilon(t) < \varepsilon(jt) .$$

Hence, the radius, eccentricity of perfect states $\varepsilon(0^n)$, and the diameter of H_p^n are strictly increasing with respect to n for fixed p (cf. [369, Lemma 6]). The relatively sophisticated arguments even for small Hanoi graphs indicate how difficult it will be to obtain more quantitative results in general. We therefore have to resort to computational experiments.

5.7.1   Path Algorithms

In order to construct solutions for P2-type tasks for the TH with p pegs by computer, e.g., to calculate the diameter of H_p^n, it would, of course, be most economic just to consider one representative of each equiset of tasks. There is an elegant algorithm by K. A. M. Götz to obtain a representative set of states (cf. [161]). We construct a tree with 0 as a starting vertex, signifying that the largest disc lies on peg 0. In every next level we append new disc configurations by placing the smaller disc onto all already occupied pegs and only on the first non-occupied one. In the nth row we get representatives for all equisets; see Figure 5.14.

In case one wants to perform calculations for a couple of problems with different numbers of pegs, it should be more effective, at first not to constrict the entries in the tree to $p$its, but to construct it admitting $q$its, where q is the largest p one is interested in and afterwards to sort out those configurations where more than p pegs are involved.

A couple of computations for eccentricities have been carried out by Meier [308] for $p > 4$ to the effect that Korf's phenomenon could not be detected

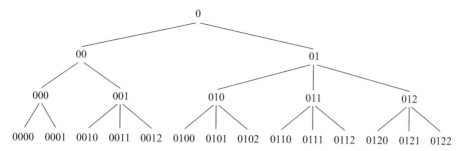

Figure 5.14: Generating representatives of equivalence classes in $[V(H_3^4)]$

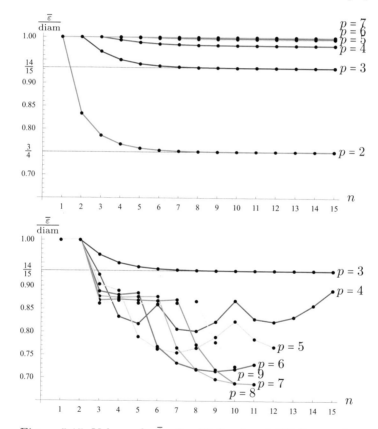

Figure 5.15: Values of $\frac{\bar{\varepsilon}}{\text{diam}}$ for S_p^n (top) and H_p^n (bottom)

for disc numbers up to 17 ($p = 5$), 15 ($p = 6$), and 14 ($p = 7, 8$), respectively. With recourse to the restriction to representatives it was also possible to tackle the calculation of what we called EX for $p = 4$, namely $EX(n) = \text{diam}(H_p^n) - \text{d}(0^n, (p-1)^n)$. The values turned out to be 0 for $n \leq 11$ ($p = 5, 6$), 10 ($p = 7$), and 8 ($p = 8$), the limits being caused by availability of computer power.

An impressive indication of how much more intricate metric properties are for Hanoi graphs H_p^n in comparison with Sierpiński graphs S_p^n is given in Figure 5.15, where the average eccentricity, normalized with respect to diameter, is plotted against n.

5.7.2 Largest Disc Moves

With the ordinary BFS algorithm we get a rooted tree and can therefore obtain, for each vertex of the graph, only one shortest path to the root. In order to keep track of *all* shortest paths, we have to modify the search. Instead of immediately marking a vertex as visited, we register all its immediate predecessors. Only after the new level is completed are all its vertices marked as visited. With the resulting Algorithm 19 we can compute *distance layered graphs* like the one for H_4^7 in Figure 5.16; cf. also Exercise 5.14.

Algorithm 19 All shortest paths algorithm

Procedure $\mathrm{ASP}(G, r)$
Parameter G: graph
Parameter r: root vertex $\{r \in V(G)\}$
    for $u \in V(G)$ **do**
        $P(u) \leftarrow \{\}$ {set of predecessors of u}
        mark u as unvisited
    end for
    $level \leftarrow -1$
    $nextLevel \leftarrow \{r\}$
    mark r as visited
    while $nextLevel \neq \{\}$
      $level \leftarrow level + 1$
      $currentLevel \leftarrow nextLevel$
      $nextLevel \leftarrow \{\}$
      for $u \in currentLevel$ **do**
        for $v \in N(u)$ **do**   {set of neighbors of u}
          if v is unvisited **then**
            put vertex v into $nextLevel$
            $P(v) \leftarrow P(v) \cup \{u\}$
          end if
        end for
        for $v \in nextLevel$ **do**
          mark v as visited
        end for
      end for
    end while

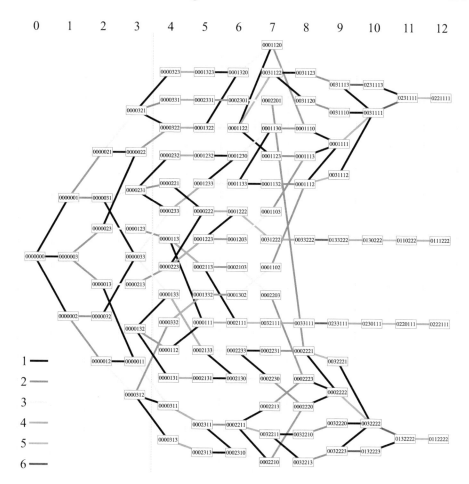

Figure 5.16: Distance layered graph H_4^7 from perfect state 0^7 till half-path states

Figure 5.16 shows that there can be no simple formula for Hanoi graphs like (4.12) in Corollary 4.7 in the case of Sierpiński graphs; for instance, disc 5 can move already in the 7th step, but the 4 smallest discs are reunited on one peg only after 9 moves. Moreover, we see that on half-path, when the largest disc 7 can move for the first time, there are the following states: 0221111, 0112222, 0222111, and 0111222. To each of them a lot of paths lead which makes up for many shortest paths between perfect states. Note that all these half-path states are of subtower type. It seems that $\binom{p}{2}$ is the smallest number of discs where this happens.

With these computations we are also able to address the subtower assumption of Frame further. We look for optimal non-subtower solutions in H_p^{n+1}, i.e. to distribute $n \in \mathbb{N}_0$ discs from peg 0 to pegs 1 to $p-2$ (to allow disc $n+1$ to move from peg 0 peg $p-1$). To be more specific, we ask for those *half-way states*

$s \in [p-2]^n$ possessing a *misplaced disc* $d \in [n-2]$ which lies directly upon disc $D \in [n] \setminus [d+1]$ on peg $s_d = s_D \in [p-2]$. In other words, the appearances of s_d in s are *not* consecutive. A specific example for $p = 5$ and $n = 6$, namely $s = 331212$, had been given in [87, p. 237]. (For $p = 3$ or $n \leq 2$ no such misplaced discs can exist in s.) Our numerical computations, which were summarized in Table 5.5 of the first edition of this book, led us to propose, in Conjecture 5.41 of said edition, that optimal non-subtower solutions exist if and only if $p + 1 \leq n + 2 \leq \binom{p}{2}$. This conjecture has been supported by further numerical experiments performed by Petr, the results of which are presented in our new Table 5.10; distributions differing only by a permutation of pegs 1 to $p - 2$ are counted as one.

$n \setminus p$	4	5	6	7
0	0	0	0	0
1	0	0	0	0
2	0	0	0	0
3	1	0	0	0
4	2	3	0	0
5	0	14	6	0
6	0	28	46	10
7	0	20	184	111
8	0	5	450	696
9	0	0	614	3012
10	0	0	489	9120
11	0	0	234	18585
12	0	0	67	25864
13	0	0	9	25230
14	0	0	0	17677
15	0	0	0	9024
16	0	0	0	3354
17	0	0	0	885
18	0	0	0	154
19	0	0	0	14
20	0	0	0	0

Table 5.10: The number of essentially different non-subtower solutions in H_p^{n+1}

For the trivial cases where $n + 2 \leq p$, each of the n smaller discs must and will move precisely once, such that the optimal solution is unique (up to permutations of pegs in $[p-2]$) and obviously a *subtower solution* containing no misplaced disc, because starting in 0^n the first move of any disc is to the bottom of some peg. For the remaining cases we make the following observation:

Lemma 5.63. *If $n + 2 \geq p + 1$, then in an optimal half-way state of n discs, none of the pegs from $[p-2]$ is empty.*

The proof is left to the reader as Exercise 5.15.

We now show that there *are* optimal non-subtower solutions for $p+1 \leq n+2 \leq \binom{p}{2}$. The following theorem is essentially due to Parisse.[76]

Theorem 5.64. *Optimal non-subtower solutions between two perfect states in H_p^{n+1} exist if $p + 1 \leq n + 2 \leq \binom{p}{2}$.*

Proof. The case $p = 3$ is obviously true, so we may assume that $p \geq 4$. Based on Exercise 5.7 it suffices to find an $s \in [p-2]^n$ with a misplaced disc and such that $\mathrm{d}(0^n, s) \leq 2n - p + 2$.

Let us first introduce some notation. We define $\ell := \min\{n+2-(p-1), p-1\}$ and $m := \min\{n+2-(p-2), 2(p-2)\}$. Then

$$\ell = \begin{cases} n-p+3, & n+2 \leq 2(p-1); \\ p-1, & n+2 \geq 2(p-1); \end{cases}$$

and

$$m = \begin{cases} n-p+4, & n+2 \leq 3(p-2); \\ 2p-4, & n+2 \geq 3(p-2). \end{cases}$$

We will need the following:

a) $2 \leq \ell \leq p-1$,

b) $1 \leq m - \ell \leq p-3$,

c) $n - m \begin{cases} = p-4, & n+2 \leq 3(p-2); \\ \in [\frac{1}{2}(p-1)(p-4)] \setminus [p-4], & n+2 > 3(p-2). \end{cases}$

To prove (a), let $n \leq 2(p-2)$. Then $\ell = n + 2 - (p-1) \in [p-1] \setminus [1]$. If $n \geq 2(p-1)$, then $\ell = p - 1 \geq 3$.

For (b) we have $m - \ell = 1 \leq p-3$, if $n+2 \leq 2(p-1)$. If $2(p-1) \leq n+2 \leq 3(p-2)$, then $m - \ell = n+2-(2p-3) \in [p-3]$. Finally, if $n+2 \geq 3(p-2)$, then $m - \ell = p-3 \geq 1$.

If $n + 2 \leq 3(p-2)$, then $n - m = p - 4$ follows directly from the definition of m. For $n + 2 > 3(p-2)$ we have

$$n - m = n - 2p + 4 = n + 2 - 2(p-1) \in [\tfrac{1}{2}(p-1)(p-4)] \setminus [p-4].$$

Together this proves (c).

Note that for $p = 4$ we have $3 \leq \ell + 1 = m = n \leq 4$.

We now construct a half-way state s containing a misplaced disc. Starting from perfect state 0^n, we proceed in several steps.

1. Move disc 1 to peg $p - 1$.

2. Transfer the subtower consisting of all discs in $[\ell] \setminus [1]$ from peg 0 to peg 1, avoiding peg $p - 1$.

---

[76] D. Parisse, private communication, 2014.

3. Transfer the subtower consisting of all discs in $[m] \setminus [\ell]$ from peg 0 to peg 2, avoiding pegs 1 and $p - 1$.

4. Move disc 1 from peg $p - 1$ onto disc $\ell + 1$ on peg 2.

5. Distribute all discs in $[n] \setminus [m]$ (still on peg 0), avoiding pegs 1 and 2, among the empty pegs in $[p - 2] \setminus [2]$.

In steps 2 and 3 we need the conditions $2 \le \ell \le n$ and $\ell < m \le n$, respectively, in view of Lemma 5.63 and to misplace disc 1 on peg 2. These conditions are fulfilled by the above properties of ℓ and m. Note that in the case $p = 4$ step 3 is possible because only $m - \ell = 1$ disc is transferred. (The TH with 2 pegs can be solved only for at most 1 disc!) Moreover, the last step is void for $p = 4$.

We now have to determine the number of moves performed in steps 1 to 5 to reach the resulting state s from 0^n. They are: 1 move in step 1, $FS_{p-1}^{\ell-1}$ moves in step 2, $FS_{p-2}^{m-\ell}$ moves in step 3,[77] 1 move in step 4, and finally in step 5 we made μ_{p-2}^{n-m} moves; for the definition of the latter notation see Exercise 5.7. This gives us, together with the properties of ℓ and m which also guarantee that the conditions in the exercise are fulfilled, the following values:

$$
\begin{aligned}
FS_{p-1}^{\ell-1} &= 1 + 2\mu_{p-1}^{\ell-2} = 1 + 2\left(\ell - 2 + (\ell - p + 1)(\ell \ge p)\right) = 2\ell - 3, \\
FS_{p-2}^{m-\ell} &= 1 + 2\mu_{p-2}^{m-\ell-1} = 1 + 2\left(m - \ell - 1 + (m - \ell - p + 3)(m - \ell \ge p - 2)\right) \\
&= 2(m - \ell) - 1, \\
\mu_{p-2}^{n-m} &= n - m + (n - m - p + 4)(n - m \ge p - 3) \\
&= n - m + (n - m - p + 4)(n + 2 > 3(p - 2)) \\
&= \begin{cases} n - m, & n + 2 \le 3(p - 2); \\ 2n - 2m - p + 4, & n + 2 > 3(p - 2). \end{cases}
\end{aligned}
$$

Altogether we have

$$
\begin{aligned}
1 + FS_{p-1}^{\ell-1} + FS_{p-2}^{m-\ell} + 1 + \mu_{p-2}^{n-m} &= 2m - 2 + \begin{cases} n - m, & n + 2 \le 3(p - 2); \\ 2n - 2m - p + 4, & n + 2 > 3(p - 2) \end{cases} \\
&= \begin{cases} m + n - 2 = 2n - p + 2, & n + 2 \le 3(p - 2); \\ 2n - p + 4 - 2 = 2n - p + 2, & n + 2 > 3(p - 2) \end{cases} \\
&= 2n - p + 2.
\end{aligned}
$$

So we have obtained the desired state s. $\square$

A typical example for such a state would be $s = 3^2 4 2^3 1^4 2$ in H_6^{11}, i.e. to solve the canonical task for $n + 1 = 12$ discs on $p = 6$ pegs, such that $\ell = 5$ and $m = 8$. The steps performed in the proof of Theorem 5.64 are shown in Figure 5.17, where disc 1, the misplaced one, is colored in red, the discs from $[\ell] \setminus [1]$ are in cyan,

---

[77]This demonstrates a big advantage of the $[\cdot]$ notation: it avoids the *fencepost error*, a typical *off-by-one error (OBOE)*, because $\|[m] \setminus [\ell]\| = m - \ell$ for $m \in \mathbb{N}_\ell$.

those from $[m] \smallsetminus [\ell]$ are green, those from $[n] \smallsetminus [m]$ are purple, and finally the inactive largest disc 12 is black. Steps 1 and 4 are the single moves of disc 1, in step 2 the cyan subtower is transferred in $FS_5^4 = 7$ moves, step 3 is the transfer of the green subtower in $FS_4^3 = 5$ moves, and in the final step 5 the purple discs make $F_4^3 = \frac{1}{2}(FS_4^4 - 1) = 4$ moves. Altogether we make $18 = \frac{1}{2}(FS_6^{12} - 1) = F_6^{11}$ moves (cf. Table 5.8), whence $0s$ is a non-subtower half-way state. Of course, we would have reached a subtower half-way state in the same number of moves had we placed disc 1 onto peg 1 in step 4, but the point of Theorem 5.64 is *not* to show that there are no such states, but to guarantee the existence of non-subtower ones as long as $n + 1 < \binom{p}{2}$. If you try to apply the method of their construction to $n + 1 = \binom{p}{2}$, say, you will see that it fails. This does, of course, not mean that they do not exist!

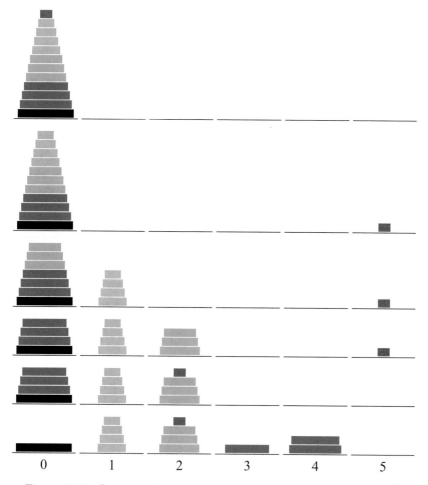

Figure 5.17: Construction of a non-subtower half-way state in H_6^{12}

For the special case $n + 2 = p + 1$ one can determine the number of non-isometric non-subtower solutions; see Exercise 5.16. By numerical evidence from Table 5.10 we may also conjecture that this number is $\frac{1}{2}p(p-3)$ for $n+2 = \binom{p}{2}$. (It even seems that for $p \geq 4$ this is the number of non-isomorphic subtower solutions for $n + 2 = \binom{p}{2} - 1$.)

What remains to be decided from Conjecture 5.41 of the first edition is "only" the absence of optimal non-subtower solutions for $n+2 > \binom{p}{2}$. It would be proved if one can show that a half-way non-subtower state can not be reached in μ_p^n moves. By Exercise 5.7, we may consider a special case:

Conjecture 5.65. *Let $p \in \mathbb{N}_4$ and $n + 1 = \binom{p}{2}$. If $s \in [p-2]^n$ contains a misplaced disc, then there is no $0^n, s$-path of length $p(p-2)$ in H_p^n, i.e. $\mathrm{d}(0^n, s) > p(p-2)$.*[78]

The remaining cases are then contained in

Conjecture 5.66. *Let $p \in \mathbb{N}_4$ and $n \in \mathbb{N}_{\binom{p}{2}}$. If $s \in [p-2]^n$ contains a misplaced disc, then there is no $0^n, s$-path of length μ_p^n in H_p^n, i.e. $\mathrm{d}(0^n, s) > \mu_p^n$.*

The smallest case of Conjecture 5.66 corresponds to Figure 5.16. If confirmed, Conjectures 5.65 and 5.66, together with Proposition 5.20, would justify Frame's subtower assumption, because solutions engendered by the construction of the FS numbers are always subtower solutions.

Recall from Chapter 2 (cf. p. 141) that a long standing myth for three pegs was that the largest disc will not move more than once in a shortest solution. It was shown in Theorem 5.18 that this is in fact true for P1-type tasks even for any number of pegs. However, already H_3^2 demonstrated that the one-move-assumption is false in general (cf. Figure 2.25). On the other hand, Corollary 2.37 asserts that on a geodesic of $H_3^n \cong S_3^n$ the largest disc moves at most twice, and we were able to extend this statement to *all* Sierpiński graphs S_p^n in Theorem 4.9. As for Hanoi graphs H_p^n, we already proved in Proposition 5.58 that two largest-disc-moves, which we will again refer to as LDMs, are *sufficient* to solve any task if $n < p$, and Aumann showed in [23, Theorem 5.16] (cf. [24, Theorem 2.2]) that in this case the largest disc (in fact, any disc; see [24, Theorem 2.2]) never moves thrice in an optimal solution. Intuition says that this should be the case for all p and n, but, alas, intuition is (again) completely wrong!

The graph H_4^4 in Figure 5.12 looks so tangly that it comes as no surprise that it contains examples of tasks whose shortest solutions include three LDMs. One such task is $0233 \rightarrow 3001$. It can be solved optimally with 1, 2 or 3 LDMs; see Exercise 5.17. More LDMs are not possible for H_4^n though. The general upper bound is an immediate consequence of the boxer rule Lemma 5.16:

---

[78]In [284, p. 141–143] Édouard Lucas shows that $2a_q$, where $a_q = q(q+2)$, is the number of moves of $2q$ pieces in *Un jeu de pions*, later to morph in the second edition into frogs in *Le jeu des grenouilles*, who in turn were joined by toads in *Le bal des crapauds et des grenouilles* (cf. [291, p. 117–124]). (The sequence a is A005563.)

Theorem 5.67. *The number of LDMs in a shortest path in H_p^n is at most $p - 1$.*

Proof. Any k LDMs will involve $k + 1$ different pegs, such that $k \le p - 1$. □

But are there examples that this upper bound of LDMs may be *necessary* for an optimal solution? Of course, for $p = 3$, we already saw one in Figure 2.26 for H_3^3. For $p = 4$ we have to provide 6 discs to find a task requiring 3 LDMs to be solved optimally, e.g., $022333 \to 300101$, which has a unique optimal solution of 10 moves, 3 of which are LDMs. (With 2 LDMs 11 moves are needed and even 12 if disc 6 is moved only once; cf. Exercise 5.18.)

The upper bound of Theorem 5.67 seems very rough and inconsistent with the intuition that the largest disc will not move more than twice. However, the following surprising theorem due to Aumann [23, Korollar 6.2] asserts that the bound of Theorem 5.67 is best possible for any p!

Theorem 5.68. *If $n \in \mathbb{N}_{p(p-2)}$, then $p - 1$ LDMs are necessary for some P2-type tasks.*

The proof is by constructing such tasks explicitly; see [24, p. 7–12].

It is now perfectly natural to ask, which patterns of LDMs do occur. For instance, are there tasks in H_p^n whose shortest path may have two or five, but not one, three, four or more than five LDMs? This pattern will be encoded as the bit vector 10010. What are the minimum numbers of pegs and discs for such a pattern of LDMs to occur? For the classical case $p = 3$ we got a complete picture of the number of tasks with (the only) non-trivial LDM patterns 01, 11, and 10 in Section 2.4. Since there are only few analytical methods to construct such tasks or to prove their (non-)existence for $p > 3$, the occurrences of LDM patterns were computed numerically.

In Algorithm 19 each vertex will be accompanied by the bit vector describing its LDMs which is updated each time the vertex is visited. If the largest disc is moved in the transition from vertex u to vertex v, then the new bit vector of v is calculated as follows

if $LDM(u) = 0$ **then**
    $LDM(v) \leftarrow LDM(v) \vee 1$ {"$\vee$" means bitwise (inclusive) or}
else $LDM(v) \leftarrow LDM(v) \vee 2 \cdot LDM(u)$ {"$2\cdot$" means left-shift of binary vector};

otherwise, if a smaller disc is moved, we update the bit vector of v as follows

    $LDM(v) \leftarrow LDM(v) \vee LDM(u)$.

A comprehensive account of these computations is [24]. Analytical and numerical results for 3 and up to 7 pegs are collected in Figure 5.18. For each particular p a table is drawn using small graphical symbols, the meanings of which are briefly explained on top of the figure. Possible patterns of LDMs for a given number of pegs are coded on the left side of the corresponding table with black and white dots. The number at the head of these first few columns stands for the number of LDMs. The values above the other columns are the respective numbers of discs.

●○●● Pattern of LDMs, e.g. 4, 2, and 1, but not 3 LDMs.
  ✓   There exist some tasks with the given pattern.
  ✗   There does not exist any task with the given pattern.
      The result is mathematically proved.
      The result is numerically computed.
      The result is conjectured.
  ?   No conjecture for the result.

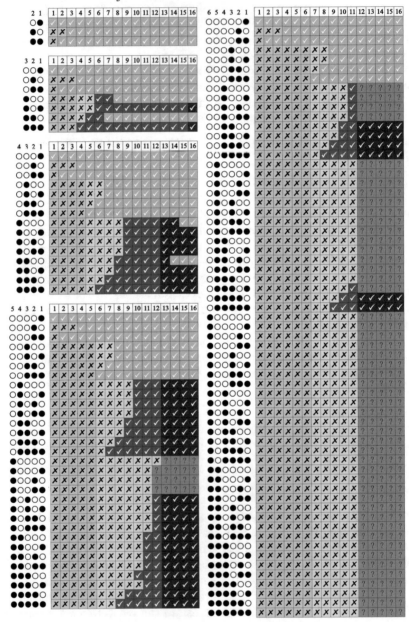

Figure 5.18: Patterns of LDMs

This right-hand side of every table displays colored squares each containing a cross mark, a check mark, or a question mark. The first two of these marks indicate the (non-)existence of the corresponding kind of task. Squares with a check mark are filled with a dark color and squares with a cross mark with a light color in order to discern possible regularities. Green fields testify that the (non-)existence of an LDM pattern was proved by analytical arguments, blue stands for numerical results and brown for a conjecture. With the question mark we disclose that we do not even have a clue for what the correct answer in these cases might be.

The non-regular structure of the entries in the tables in Figure 5.18 enforces the impression of high complexity of the TH with more than three pegs, as it was imposed already by other considerations in this chapter. The interpretation of these tables is not easy, because seemingly recurring structures are annihilated by counter-examples. One conjecture, however, is strongly suggested by the facts:

Conjecture 5.69. *If an LDM pattern $b \in [2^{p-1}]_0$ occurs in H_p^n, then it occurs in every H_p^m for $m \in \mathbb{N}_n$.*

5.8 Exercises

5.1. Fill in the details of the bookkeeping in the proof of Theorem 5.4.

5.2. Show that the FSC* holds for $\nu \in [3]_0$.

5.3. Let G be a connected graph with $|G| \geq 2$.

　　a) If $d(u,v) = 1$ for some $u, v \in V(G)$, then $|\varepsilon(u) - \varepsilon(v)| \leq 1$.

　　b) Prove that $\forall \varepsilon \in [\mathrm{diam}(G)]_0 \setminus [\mathrm{rad}(G)]_0 \exists v \in V : \varepsilon(v) = \varepsilon$.

5.4. (Klavžar, Milutinović, and Petr [240]) Show that the TH with $n \in \mathbb{N}$ discs and $p \in \mathbb{N}$ pegs has exactly

$$\sum_{k=1}^{p} \binom{n-1}{k-1} p^{\underline{k}}$$

regular states in which discs on every peg form a superdisc. Here $p^{\underline{k}} = \prod_{i=0}^{k-1}(p-i)$ is the k-th *falling power* of p, and the empty set of discs is treated as a superdisc.

5.5. Show that $\widetilde{P}_q(\nu) = P_q(\nu)$ for all $q, \nu \in \mathbb{N}_0$.

5.6. Show that any $0^8, 1^4 3^4$-path in H_4^8 has length at least 24 and that any optimal path must pass through the vertex $12^3 3^4$. (Hence, an optimal solution can be obtained from transferring 4 smallest discs using 4 pegs and then 4 largest discs using 3 pegs.)

5.7. Let $p \in \mathbb{N}_3$ and $n \in \left[\binom{p}{2}\right]_0$. Then

$$\mu_p^n := \min\{d(0^n, s) \mid s \in [p-2]^n\} = n + (n-p+2)(n \geq p-1)$$

and

$$FS_p^{n+1} = \mathrm{d}(0^{1+n}, (p-1)^{1+n}) = 1 + 2\mu_p^n.$$

5.8. Recall from (5.8) that $g(3,n)$ is the minimum number of moves, taken over all $s \in T^n$, needed to displace each disc at least once when starting at s. Show that $g(3,n) = 2^{n-2} + 1$ for $n \in \mathbb{N}_2$.

5.9. (Klavžar, Milutinović, and Petr [237, p. 62]) Show that for any $n \in \mathbb{N}_0$ and $p \in \mathbb{N}_3$,

$$\|H_p^n\| = \frac{1}{4} \sum_{k=1}^{p} k\,(2p-k-1) \left\{ {n \atop k} \right\} p^{\underline{k}}.$$

5.10. (Hinz and Parisse [210, Theorem 1]) Show that H_p^n is hamiltonian for any $p \in \mathbb{N}_3$ and any $n \in \mathbb{N}$.

5.11. Prove that H_p^2 has a unique perfect code and that its codewords are the perfect states.

5.12. Show that the graph H_4^3 is *not* planar.

5.13. Give a proof for Proposition 5.60. Then show the following for $p \in \mathbb{N}_3$:

    a) $\forall\, n \in \mathbb{N}_0 : \mathrm{rad}(H_p^n) \geq n + \left\lfloor \frac{n}{2} \right\rfloor$,

    b) $\forall\, n \in [p]_0 : \mathrm{rad}(H_p^n) = n + \left\lfloor \frac{n}{2} \right\rfloor$.

5.14. (Klavžar, Milutinović, and Petr [240]) Consider the TH with $n \in \mathbb{N}_3$ pegs and n discs, and the task to move the discs from a perfect state i^n to a different perfect state j^n. Show that there are precisely $\binom{n-1}{2}(n-2)!$ (cf. A180119) states that are at distance n from i^n on optimal paths from i^n to j^n.

5.15. Prove Lemma 5.63.

5.16. Show that there are exactly Δ_{p-3} non-isomorphic non-subtower half-way states if $n = p - 1$.

5.17. Find optimal solutions for the task $0233 \to 3001$ in H_4^4 which employ 1, 2, and 3 LDMs, respectively.

5.18. Show that to solve the task $022333 \to 300101$ in H_4^6 employing precisely $k \in [3]$ LDMs, $13 - k$ moves are necessary and sufficient.

Chapter 6

Variations of the Puzzle

TH is an example of a one person game; such games are known as *solitaire games*. There are plenty of other mathematical solitaire games, the Icosian game, the Fifteen puzzle, and Rubik's Cube are just a few prominent examples. Numerous variations of the TH can also be defined, some natural and some not that natural. In fact, Lucas himself in [286, p. 303] pointed out the following: "Le nombre des problèmes que l'on peut se poser sur la *nouvelle* Tour d'Hanoï est incalculable."[79] Many variations were indeed studied and some of them we already encountered in previous chapters: the Linear TH in Chapter 2, problems in Chapter 3 allowing for irregular states, the Switching TH in Chapter 4, and the tasks in Chapter 5 where more than three pegs are available.

In the next section we make clear what is understood as a variant of the TH. We first illustrate this by a brief look at a variety called Exchanging Discs TH, and then by introducing and solving the Black and White TH. Numerous "colored" variants are listed in the subsequent section, including the Tower of Antwerpen. In the concluding section we present the Bottleneck TH which allows for larger discs above smaller ones up to a certain discrepancy. We describe an optimal algorithm and note that the solution for the Bottleneck TH may not be unique. We close this chapter by briefly mentioning a related version, the Sinner's TH.

6.1 What is a Tower of Hanoi Variant?

In order to make precise which variants of the TH are of interest, we define the framework as follows.

Any variant of the TH consists of pegs and discs such that discs can be stacked onto pegs. In addition, it obeys the following common rules:

1. Pegs are distinguishable.

---

[79]The number of problems which one can pose oneself on the *new* Tower of Hanoi is incalculable.

© Springer International Publishing AG, part of Springer Nature 2018
A.M. Hinz et al., *The Tower of Hanoi – Myths and Maths*,
https://doi.org/10.1007/978-3-319-73779-9_7

2. Discs are distinguishable.

3. Discs are on pegs all the time except during moves.

4. One or more discs can only be moved from the top of a stack.

5. Task: given an initial distribution of discs among pegs (*initial state*) and a goal distribution of discs among pegs (*final state*), find a shortest sequence of moves that transfers discs from the initial state to the final state obeying the rules.

Although the above conventions appear rather restrictive, they offer a tremendous number of different variations. For instance,

- there can be an arbitrary number of pegs (as we have seen already);

- pegs can be distinguished also by their heights, that is, by the number of discs they can hold;

- discs can be distinguished in size and/or color;

- (certain) irregular (with respect to TH rules) states may be admitted;

- more than one top disc may be moved in a single move;

- there can be additional restrictions or relaxations on moves, the latter even violating the divine rule;

- and, of course, any combination of the above.

The interest for such variants goes all the way back to Lucas, who in [286] proposed a variant with five pegs and four groups of discs of different colors. Only the central peg can hold all discs; see Figure 0.11. Every group contains four discs and the 16 discs have pairwise different sizes. The group of color $c \in [4]$ consists of the four discs $d \in [16]$ with $1 + (16 - d) \bmod 4 = c$. All this can be deduced from the figure of the goal state as produced by Lucas on the first page of his article. Figure 6.1 shows the initial configuration of this type of puzzle. A possible task is to transfer all discs onto the middle peg obeying the divine rule; see [285, Quatrième problème].

Note that since the 16 discs are of mutually different sizes, this puzzle is equivalent to an instance of a type P1 problem (that is, to reach a perfect state from a given regular state) with $p = 5$ and $n = 16$, namely, e.g., $(1234)^4 \rightarrow 0^{16}$. In [285], Lucas also proposed type P2 problems (to reach a regular from another regular state). Although his formulation of the tasks leaves room for interpretation, we think that he is requiring two [285, Première problème] or three [285, Troisième problème] colors to be united on the center peg, respectively, with the other colors remaining in their initial position. Finally, Lucas proposes similar tasks with the initial state changed to $1^4 2^4 3^4 4^4$, i.e. the discs grouped in four towers according to size, and the inverse problems.

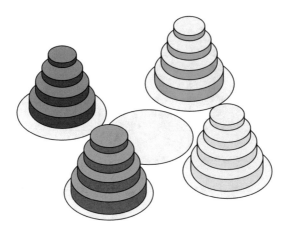

Figure 6.1: An initial state of the Lucas variant with four colored stacks

It is interesting to compare, for the task to reunite two colors, what would be the optimal number of moves if they were restricted to 3 or 4 pegs, respectively. The former task is equivalent to $(12)^4 \to 0^8$ in H_3^8 and has been solved already in Proposition 2.20 with the result that it takes 182 moves to accomplish. Using 4 pegs, i.e. in H_4^{12}, a new phenomenon occurs, namely that it makes a difference which pair of colors is chosen to be united. If the relatively smallest discs end where they were, i.e. for the task $(123)^4 \to (003)^4$, it takes 39 moves (see Figure 6.2 and Figure 6.3, where all shortest paths are shown with the special solution highlighted by the black dots). If the relatively largest discs remain fixed, i.e. $(123)^4 \to (100)^4$, the optimal number of moves is 32, and for the intermediate case $(123)^4 \to (020)^4$ it is 37. The minimum number of moves needed to conjoin two colors using all 5 pegs ranges from 31 with the relatively smallest discs fixed to 23 for the largest ones. In the variant with four size-groups, i.e. starting from $1^4 2^4 3^4 4^4$, the range is wider, from 16 to 38. These results can be obtained either by a combinatorial analysis or by computation.

A dependence on the relative size of the color groups of discs also occurs in [285, Deuxième problème], the P2-type task mentioned in Section 2.4, namely to switch the position of two colors. If only one auxiliary peg is employed, as Lucas imposed, then, as we will see below in Section 6.3, 183 moves are necessary. If, however, we solve the task on all 5 pegs, we will only need between 29 if the relative largest groups are swiched and 21 for the smallest two. Again the range is larger for size-grouped tasks, namely from 17 to 33.

The tasks in [285, Troisième problème] and [285, Quatrième problème] could only be approached by numerical experiments at the time. To unite three colors starting from $(1234)^4$ (or $1^4 2^4 3^4 4^4$) needs from 39 (31) to 43 (47) moves depending on which color group from 1 to 4 ends in their starting position. For the goal 0^{16} our calculations found 63 (54) as the optimal number of moves. Moreover,

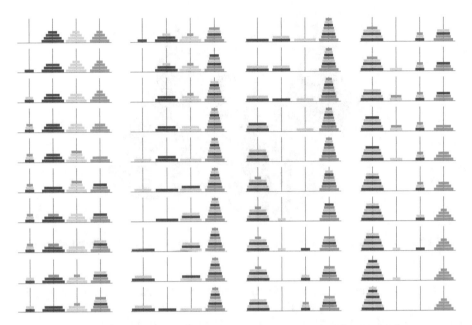

Figure 6.2: Optimal solution for the task $(123)^4 \to (003)^4$

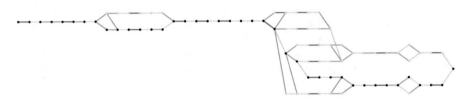

Figure 6.3: All solutions for the task $(123)^4 \to (003)^4$; the color of an edge corresponds to the color of the disc moved

$\varepsilon((1234)^4) = 63$ and $\varepsilon(1^4 2^4 3^4 4^4) = 60$. In the latter case, the goal state 3^{16} seems to be the most eccentric challenge.

Although all problems from [285] can be interpreted formally in the framework of Chapters 2 and 5, the coloring of the discs changes the character of the game because it transmits additional information during execution; cf. Exercise 2.5. In other variants, however, colors enter into the rules of the game.

We next describe another old variant in order to illustrate what is meant by moving more than one top disc in a single move and what is understood as additional restrictions on moves. This variant, named the *Exchanging Discs Tower of Hanoi* was proposed back in 1944 [378] and solved 50 years later by Stockmeyer et al. [408]. The variant is of type P0—transfer a perfect tower of n discs from one peg to another—where the only moves allowed are:

- disc 1 can move from any peg to any peg, and

- if for some $d \in [n-1]$, discs d and $d+1$ are top discs on different pegs, then they can be exchanged in a single move.

In [408] an algorithm is designed that solves the Exchanging discs TH. The number of moves performed can be expressed as a non-homogeneous linear recurrence of order 4 which in turn implies that asymptotically the number of moves made is approximately

$$1.19188 \cdot 1.85356^n.$$

Moreover, it is proved that the solution given by the algorithm is the unique optimal solution.

In the rest of this section we take a closer look at another color variant of the TH that was proposed in 2010 by Stockmeyer and F. Lunnon in [409]. Just as Lucas's variant, this one involves a combination of more pegs and colored discs, but it also gives an additional restriction on moves. We will call this variant the *Black and white TH*.

The Black and white TH consists of four pegs, a set of n white discs, and a set of n black discs. Each of the disc sets contains—just as in the classical TH— discs of mutually different diameters and the two sets of discs are identical with the exception of their color. The rules are as in the classical TH, except that the divine rule is complemented by "you must not place a disc on the other disc of the same size". Initially the n white discs form a tower on peg i and the n black discs form a tower on peg j, where $\{i,j\} \in \binom{Q}{2}$, $i \neq j$. The goal is to reach the state in which the roles of pegs i and j are interchanged: white discs form a tower on peg j and black discs on peg i. Figure 6.4 shows, for $n = 5$, the initial state for the case $i = 0$, $j = 2$, and another legal state of the puzzle. It follows from the extended divine rule that a peg can hold only one disc of a given size. Therefore, it is not necessary that a peg can hold more than n discs at a time.

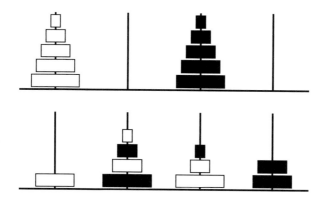

Figure 6.4: An initial and a legal state of the Black and white TH

Algorithm 20 gives a solution for the Black and white TH. It reduces to several instances of the standard TH with three pegs, so any procedure from Chapter 2 that finds the unique solution of the TH can be used. Since we are dealing with a puzzle with four pegs, let us again make use of Algorithm 16 whose call $FS3(m, 1, i, j, k)$ transfers a tower of m smallest discs from peg i to peg j using auxiliary peg k (and the fourth peg is not involved at all). This is independent of the color of discs.

Algorithm 20 Black and white

Procedure $bw(n, i, j)$
Parameter n: number of white/black discs $\{n \in \mathbb{N}_0\}$
Parameter i: source peg for white discs $\{i \in [4]_0\}$
Parameter j: source peg for black discs $\{j \in [4]_0\}$
  if $n \neq 0$ and $i \neq j$ **then**
      $\{i, j, k, \ell\} \leftarrow \{0, 1, 2, 3\}$ $\{k$ and ℓ are the pegs different from i and $j\}$
      $FS3(n - 1, 1, i, k, \ell)$ $\{$transfers $n - 1$ smallest white discs to peg $k\}$
      move (white) disc n from i to ℓ $\{$moves largest white disc to peg $\ell\}$
      $FS3(n, 1, j, i, \ell)$ $\{$transfers the black discs to their goal peg$\}$
      move (white) disc n from ℓ to j $\{$moves largest white disc to its goal$\}$
      $FS3(n - 1, 1, k, j, \ell)$ $\{$transfers $n - 1$ smallest white discs to their goal$\}$
  end if

Note that in the call $FS3(n, 1, j, i, \ell)$ the black disc n is moved only once (from peg j to peg i), so the white disc n on peg ℓ is not an obstruction for the call. Since procedure FS3 needs $2^m - 1$ moves to transfer a tower of m discs from one peg to another, it is straightforward to see that Algorithm 20 completes its task in $2^{n+1} - 1$ moves. Moreover, it can be shown that no other algorithm can do better (a proof of which is left for Exercise 6.1), hence we have arrived at the following result from [409]:

Theorem 6.1. *Algorithm 20 solves the Black and white TH with n black discs and n white discs in the optimal number of $2^{n+1} - 1$ moves.*

We close the section by listing variations of Black and white TH that were studied in [409].

Variant A Consider the Black and white TH with an additional restriction on the moves: white discs cannot be placed on auxiliary peg k and black discs cannot be placed on auxiliary peg ℓ.

The minimum number of moves needed to solve variant A is

$$3(2^n - 1),\tag{6.1}$$

as shown in [409], where two optimal algorithms for solving this variant are described.

Variant B Suppose we have $2s$, $s \in \mathbb{N}_3$, pegs numbered by integers from $[2s]_0$ and s stacks each consisting of n discs. The stacks are identical except for their color, that is, there are s pairwise differently colored stacks. Initially the stacks are on pegs $0, 2, \ldots, 2s - 2$, respectively. The goal is that for $i \in [s]_0$, the stack from peg $2i$ is transferred to peg $(2i + 2) \bmod (2s)$, where discs from peg $2i$ can, besides pegs $2i$ and $2i + 2$, only use peg $2i + 1$. Peg $2i + 1$ can thus be considered as a private peg for discs that are initially on peg $2i$. In Figure 6.5 this variant is presented for the case with four colored stacks ($s = 4$) each containing four discs ($n = 4$).

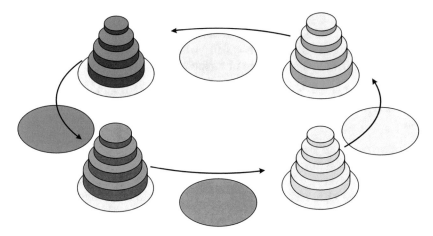

Figure 6.5: Variant B of the Black and white TH with $s = 4 = n$

The minimum number of moves needed to solve variant B is

$$s (2^n - 1) + 2^{n-1}. \tag{6.2}$$

For an example see Exercise 6.2. An optimal algorithm for the general case can be found in [409].

Extending the definition of variant B to $s = 2$, a puzzle equivalent to variant A is obtained. However, different approaches for $s \geq 3$ and $s = 2$ are needed and this is the reason why they are treated as different puzzles. In particular, for $n \geq 2$, the minimum number of moves for variant A given in (6.1) is larger than the number obtained from (6.2) for $s = 2$. This means that the above mentioned optimal algorithm for variant B cannot be applied for puzzle A (because it would lead to forbidden moves).

Variant C This variant is obtained from variant A by adding an additional peg to it and allowing white and black discs to use this peg.

Variant C is of similar nature as the TH with four pegs, Stockmeyer and Lunnon give an algorithm that needs $FS_4^n + 2FS_4^{n-1} + 2$ moves to reach the final state. Having in mind our considerations from Chapter 5 it seems unlikely that a provably optimal algorithm for variant C can be found easily.

6.2 Ambiguous Goal

Let in a P1-type task $s \to j^n$ in H_3^n, $n \in \mathbb{N}$, the goal peg j be *not* specified. For instance, starting a P0-type task from 0^n, the target might be either 1^n or 2^n. If a player gets lost in $s \in T^n$, the question arises for which $j \in [2]$ we have $d(s, j^n) < d(s, (3-j)^n)$. (The case of equality can only occur for $s = 0^n$; cf. Remark 2.12.)

Observation. Let $s \in T^n \setminus \{0^n\}$. If the number of leading 0s is even, then j is the first (from the left) non-zero tit i in s; otherwise $j = 3 - i$. (In particular, $j = s_n$, if $s_n \neq 0$; this is already a consequence of formulas (2.6) and (2.7).) More formally,

Proposition 6.2. *Let $n \in \mathbb{N}$ and $s = 0^{n-d} s_d \bar{s}$ with $d \in [n]$, $s_d \in [2]$, $\bar{s} \in T^{d-1}$. For $j \in [2]$ with $d(s, j^n) < d(s, (3-j)^n)$ we have: $j = s_d$, if $n - d$ is even, and $j = 3 - s_d$ otherwise.*

Proof. Entering s into two P1-automata with initial states $j = 1$ and $j = 2$, respectively, will find these automata after $n - d$ inputs in states 1 and 2, or 2 and 1, depending on whether $n - d$ is even or odd; cf. p. 117. In the first case, entering $s_d \in [2]$ will leave either the first or the second state unchanged depending on whether $s_d = 1$ or $s_d = 2$, such that the corresponding j^n is the closer perfect state from s; cf. formula (2.8). Similarly for the odd case. $\qquad\square$

The procedure is realized in Algorithm 21.

Algorithm 21 Comparison of goals

Procedure $\mathrm{cg}(n, s)$
Parameter n: number of discs $\{n \in \mathbb{N}\}$
Parameter s: regular state $\{s \in T^n\}$

     $d \leftarrow n$ {index of tit}
     $b \leftarrow 0$ {parity of $n - d$}
     while $s_d = 0$
         $b \leftarrow 1 - b$, $d \leftarrow d - 1$
     end while
     $j = 1 + (b = 2 - s_d)(= 1 + b \veebar (s_d - 1))$

There are exactly $2 \cdot 3^{n-k}$ states $s \in T^n \setminus \{0^n\}$ for which Algorithm 21 stops after $k \in [n]$ inputs. To find the average performance of the algorithm we therefore have to calculate

$$\sum_{k=1}^{n} k \cdot 3^{n-k} = \sum_{k=0}^{n-1} (n-k) \cdot 3^k = n \sum_{k=0}^{n-1} 3^k - \sum_{k=1}^{n-1} k \cdot 3^k.$$

The first sum on the right-hand side is equal to $\dfrac{3^n - 1}{2}$ by Lemma 2.18b, and for the second sum we get $\dfrac{3 + (2n - 3) \cdot 3^n}{4}$ from [162, (2.26)]. So the number of tits entered for all s adds up to $\frac{3}{2}(3^n - 1) - n$. This means that on the average $\frac{3}{2} - n(3^n - 1)^{-1}$ tits are checked, i.e. asymptotically, for large n, just $\frac{3}{2}$.

The situation is much easier for Sierpiński graphs S_p^n. From formula (4.10) it follows that the order in which $j \in [p]_0$ occurs first in $s \in [p]_0^n$ (from left to right) is the order of distance from s to extreme vertex j^n (strictly from small to large); (only) for j that do *not* occur in s we have $d(s, j^n) = 2^n - 1$.

No such result is known for Hanoi graphs H_p^n if $p \geq 4$ though. For instance, from Figure 5.6 it is clear that in H_4^2

$$d(03, 00) < d(03, 11) = d(03, 22) < d(03, 33).$$

Only the first inequality is supported by a general statement, namely Lemma 5.52. Otherwise, the situation seems to be more complex.

6.3 The Tower of Antwerpen

We already discussed a variation proposed by Lucas himself in 1889 containing discs of different colors. Of course, the variety of variations one can pose with colored discs is beyond any control. In Section 6.1 we have considered the Black and white TH and three offsprings; in this section we present additional variants of TH with colored discs. We first stage the Tower of Antwerpen, then list four variations of this puzzle, and conclude with the Hanoi Rainbow problem which in turn leads to numerous additional variants.

The *Tower of Antwerpen* was proposed in 1981/82 by Wood [443, Variant 2] and presumably invented in that city. It consists of three pegs, a set of n black discs with pairwise different diameters, a set of n yellow discs, and a set of n red discs.[80] These three sets are identical except for their colors. Each peg can hold all discs. Initially each of the three colored sets of discs forming a tower on one peg, the goal is to reach the state in which the three colored sets again form towers but such that each tower rests on a different peg than originally.[81] It is understood that the classical TH rules have to be observed, which means, in particular, that discs of the same size may be put on top of each other during execution. See Figure 6.6 for the Tower of Antwerpen with $n = 5$.

The problem was solved eight years later by Minsker in [312]. Clearly, for $n = 1$ five moves are optimal, otherwise the following holds:

---

[80] Wood used different colors.

[81] One could say explicitly from which peg to which peg a given colored stack has to be moved, as done in [443], but the puzzle clearly remains the same as presented here.

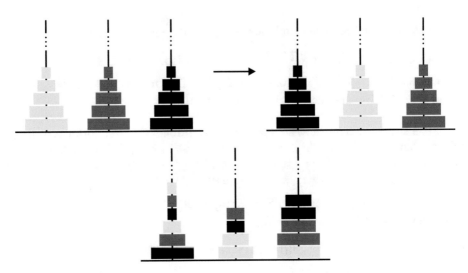

Figure 6.6: Initial and final and a general regular state of the Tower of Antwerpen

Theorem 6.3. *Let $n \in \mathbb{N}_2$. Then the Tower of Antwerpen puzzle with $3n$ discs can be solved in the optimal number of $3 \cdot 2^{n+2} - 8n - 10$ moves.*

An optimal solution of the Tower of Antwerpen need not be unique. Already for $n = 1$ the reader can easily verify that there are three optimal solutions; for $n = 2$ see Exercise 6.3. (In fact, for $n = 2$ there are 30 optimal sequences of moves, see [312].)

In the rest of the section we present variations of the Tower of Antwerpen that also received attention in the literature.

Little Tower of Antwerpen

The *Little Tower of Antwerpen* is defined just as the Tower of Antwerpen, except that there are two colored towers (say golden and silver) each consisting of n discs and the goal is to interchange these two towers. We remind the reader that a disc of one color is allowed to be put onto a disc of the same size of the other color, which makes this puzzle inherently different from the Black and white TH discussed in Section 6.1. The Little Tower of Antwerpen was introduced and discussed by Minsker in [313]. He obtains the minumum number of moves $\frac{1}{3}\left(7 \cdot 2^{n+1} - 9n - 11 + (n \text{ odd})\right)$. Moreover, he finds that there are exactly $2^{\lfloor (n+1)/2 \rfloor}$ different optimal sequences of moves.

In [325], Obara and Hirayama considered the following variant of the Little Tower of Antwerpen. As before, the initial state consists of two colored towers each composed of n discs, say n golden discs on peg 0 and n silver discs on peg 2. The goal is now to reach the state where on peg 0 one finds the silver disc n, the golden disc $n-1$, the silver disc $n-2$, the golden disc $n-3$, ..., and the disc 1 of the corresponding color (depending on the parity of n), all the other discs lie on peg 2. A recursive solution for this puzzle is proposed with no attempt to prove its optimality.

In Figure 6.7, the above two tasks are summarized (bottom arrow and left-most arrow upwards, respectively), together with still some other ones one might pose.

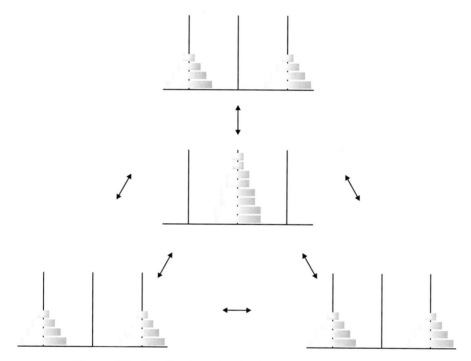

Figure 6.7: A collection of tasks for rules from Antwerpen

Twin- and Triple-Tower problems

All the tasks of Figure 6.7 can also be solved in the setting of a variation very similar to the Little Tower of Antwerpen which was proposed in 1991 by van Zanten [448]. It was named the *Twin-Tower problem* and differs from the Little Tower of Antwerpen by allowing a silver disc to be put on the golden disc of the same size, but not the other way round. All the rest is the same, in particular the object of the puzzle is to exchange a perfect tower of $n \in \mathbb{N}$ silver discs with a perfect tower of n golden discs. Denoting silver discs by $1, 3, \ldots, 2n-1$ and the golden ones by $2, 4, \ldots, 2n$, i.e. shrinking the silver discs, it is easy to observe that the Twin-Tower problem is the following instance of a type P2 problem of the classical TH:

$(01)^n \to (10)^n$. Based on our experience from Sections 2.4 and 4.2, the solution is rather straightforward. We first show that moving the largest disc only once will lead to the unique optimal solution. Applying the H_3-to-S_3-automaton to the first three pairs of input $(0,1)$, $(1,0)$, and $(0,1)$ yields the sequence of pairs $(0,1), (2,2), (1,0)$ and starting the P2 decision automaton with $i = 0$, $j = 1$, and $k = 2$ (from the first pair) in state A, we enter the second pair (k,k) leading to state B; the third pair (j,i) already gives the final result that this is a type I task. We can now decompose the optimal solution into the transfer of $2n-1$ discs to peg 2, the single move of disc $2n$ to peg 1, and the sorting of the $(2n-1)$-tower according to parity. It can be argued whether this is a humane solution: although the human problem solver might skip the decision problem and assume that the largest disc moves only once because the initial and final states make the alternative rather unlikely, and even if the P1-type part of the solution can be handled, the sorting back seems to be a rather difficult task for a human. By Proposition 2.20 the optimal solution will need altogether $2 \left\lfloor \frac{5}{7} 2^{2n-1} \right\rfloor + 1 = \left\lfloor \frac{5}{7} 4^n \right\rfloor + (n \bmod 3 = 1)$ moves (cf. [448, Theorem 3.3]). Note that this is the same number of moves as needed for Noland's problem (cf. Section 2.2.1), except for the case "$n \bmod 3 = 1$", where one more step is necessary. Apart from the trivial case $n = 1$, the latter occurs, e.g., for the task in Figure 6.8, where the optimal solution comprises 183 moves; cf. Section 6.1.

Figure 6.8: Interchanging gold and silver

In Exercise 6.4 the reader is asked to solve the problem when we add a single person to the twins. Some solutions to the tasks in Figure 6.7 may be surprising; see Exercise 6.5.

Van Zanten also posed a *Triple-Tower problem* [447, p. 7f], which starts from dealing $n \in \mathbb{N}$ discs cyclically among all three pegs and asks for a permutation of the situation on the pegs. Again, this can be solved efficiently with the methods of Section 4.2. For an example, see Exercise 6.6.

Linear Twin Hanoi

Minsker [314] proposed and investigated an additional variant of the Little Tower of Antwerpen: it differs by disallowing moves between pegs 0 and 2. The task is to interchange a perfect tower consisting of n white discs stacked initially on peg 0 with a perfect tower consisting of n black discs stacked initially on peg 2. (We could say that the *Linear Twin Hanoi* is a combination of the Little Tower of Antwerpen with the Linear TH of Section 2.3.1.) The minimum number of moves needed to solve the Linear Twin Hanoi as found by Minsker is $10 \cdot 3^{n-1} - 4n - 2$, and for $n \geq 2$ he identifies $2^{5 \cdot 3^{n-2}-2}$ different optimal sequences of moves.

Classical-Linear Hybrid problem

Yet another variation of the Linear Twin Hanoi (and consequently of the (Little) Tower of Antwerpen) was suggested by Minsker [315]. Now there are n discs, each colored black or white in an arbitrary way, which initially form a perfect tower on peg 0. The goal is to transfer the tower to peg 2, where the black discs must follow the linear rule (no moves between pegs 0 and 2) while there is no such restriction for the white discs. Minsker gives an optimal algorithm for this puzzle in [315], while in [316] he extends the study of this puzzle to P1- and P2-type problems. Type P1 problems have unique optimal solutions, just like in the classical TH, but there are situations when the largest disc moves twice in the optimal sequence of moves, contrary to the case of the classical TH; cf. Theorem 2.7.

To conclude the section colorfully we describe the *Hanoi rainbow problem*. This variant was proposed by R. Neale in [321] and subsequently studied by Minsker in [311]. The puzzle is a type P0 problem that differs from the classical TH as follows. The discs are colored such that in the initial state adjacent discs are assigned different colors, while the divine rule is extended by the requirement that a disc can never be placed on a disc of the same color. Minsker first gives a natural extension, called Procedure Rainbow, of the classical recursive Algorithm 4. This procedure always solves the problem, however, it need not be optimal. A more involved procedure is also given that always finds an optimal solution. The special case when the discs of the initial configuration are alternately colored with m colors is of particular interest. Here, for $m = 3$, Procedure Rainbow is proved to be optimal, and it was conjectured that this holds for any odd $m \geq 5$. As far as we know, this conjecture remains open.

6.4 The Bottleneck Tower of Hanoi

A plausible way to vary the TH is to relax the divine rule as follows. Consider the standard setup with three pegs and discs $1, 2, \ldots, n$, and let t be a fixed positive integer, the *discrepancy* of the puzzle. Then a top disc d of some peg is allowed to move to some other peg j whenever

$$d - d' < t \tag{6.3}$$

holds for any disc d' that lies on peg j. In other words, if in a regular state of the puzzle disc d is placed higher than disc d', then d has to be less than $d' + t$. The task is to transfer discs from a perfect state to a perfect state on another peg with the minimum number of moves under this relaxed divine rule. Figure 6.9 displays a regular state of the new puzzle with $t = 4$. It is a regular state for all puzzles with discrepancy at least 4, but it is not regular for smaller discrepancies.

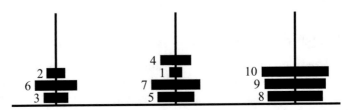

Figure 6.9: A regular state of the Bottleneck Tower of Hanoi with discrepancy 4

Note that when $t = 1$, condition (6.3) reads $d \le d'$ and consequently $d < d'$. Therefore discrepancy $t = 1$ is the usual divine rule. However, for each fixed $t \ge 2$ a new puzzle is obtained.

These puzzles were introduced in 1981 by Wood [443, Variant 1] who also posed the problem to determine the minimum number of moves for any discrepancy t and any number of discs n. In 1992 Poole [341] named these puzzles *Bottleneck Tower of Hanoi*. More specifically, let BTH_t stand for the Bottleneck TH with discrepancy t. Poole also provided a solution to Wood's problem and implicitly described an algorithm that uses the minimum number of moves. We now describe this algorithm, where we may assume that $n \ge 2$.

The main idea is to start by moving the smallest $n-1$ discs to the auxiliary peg, then move disc n to the goal peg, and finally move the smallest $n-1$ discs from the auxiliary peg to the goal peg. Hence the idea goes along the same lines as the classical recursive solution of the TH (Algorithm 4). However, while moving the smallest $n-1$ discs one should use the discrepancy as much as possible. To do so, divide the $(n-1)$-tower of the initial perfect state from bottom to top into subtowers of t discs each, except the topmost one, called B_1, which may contain between 1 and t discs. The partition is schematically shown in Figure 6.10, where N comes from the decomposition $n-1 = (N-1)t + r$, with $N \in \mathbb{N}$ and $r \in [t]$.

Let B be a subtower as constructed above and suppose that at some stage of the algorithm the discs of B are the top discs on peg i. Assume that all discs on peg j, $j \ne i$, are bigger than any of the discs from B. Then the discrepancy t allows us to move all discs of B one by one from peg i to peg j. Clearly, we need only t moves (or r, if $B = B_1$) for this transfer and the resulting state is a legal state with respect to the BTH_t. Since after this operation the discs of B lie on peg j in the upside-down order with respect to their previous position on peg i, we denote the sequence of these moves by

$$\text{upside-down}(B, i, j).$$

Let the move of a disc d from peg i to peg j be written as $i \overset{d}{\to} j$ and recall

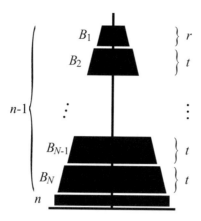

Figure 6.10: Setup for the Bottleneck Tower of Hanoi algorithm

that $p0(n, i, j)$ (Olive's algorithm—Algorithm 2) transfers n discs from peg i to peg j. Then Poole's approach can be encoded as presented in Algorithm 22.

Algorithm 22 Solution of the Bottleneck Tower of Hanoi

Procedure BTH(t, n, i, j)
Parameter t: discrepancy of the Bottleneck TH $\{t \in \mathbb{N}\}$
Parameter n: number of discs $\{n \in \mathbb{N}_0\}$
Parameter i: source peg $\{i \in T\}$
Parameter j: goal peg $\{j \in T\}$
   if $n = 0$ or $i = j$ **then** STOP
   if $n = 1$ **then** move disc n from i to j and STOP
   let $n - 1 = (N - 1)t + r$, where $r \in [t]$
   let B_1 contain discs $1, \ldots, r$
   for $\ell = 1, \ldots, N - 1$, let $B_{\ell+1}$ contain discs $r + (\ell - 1)t + 1, \ldots, r + \ell t$
   $k \leftarrow 3 - i - j$ {the auxiliary peg different from i and j}
   run $p0(N, i, k)$ and replace its moves $k' \overset{d}{\to} k''$ with upside-down(B_d, k', k'')
         {using discrepancy t, transfers $n - 1$ smallest discs to auxiliary peg}
   move disc n from i to j {moves largest disc to goal peg}
   run $p0(N, k, j)$ and replace its moves $k' \overset{d}{\to} k''$ with upside-down(B_d, k', k'')
         {using discrepancy t, transfers $n - 1$ smallest discs to goal peg}

Correctness of Algorithm 22 Since the procedure $p0(n, i, j)$ obeys the standard divine rule and each of the blocks B_ℓ, $\ell \in [N]$, consists of at most t discs, each of the states reached by Algorithm 22 is legal with respect to BTH$_t$. Moreover, recall from Proposition 2.4 that during the run of $p0(n, i, j)$ disc $d \in [n]$ is moved 2^{n-d} times. It follows that Algorithm 22 moves each block

B_ℓ, $\ell \in [N]$, exactly $2 \cdot 2^{N-\ell}$ times. Since this is an even number, all blocks have their smallest disc on top at the end of the execution. In other words, Algorithm 22 terminates with the perfect state on peg j.

Number of moves made by Algorithm 22 For $n \leq 1$, the algorithm makes n disc moves. If $n > 1$, then in each of the calls p0(N, i, k) and p0(N, k, j), block B_1 is moved 2^{N-1} times. Since each move of B_1 consists of r individual disc moves, the total number of moves of discs from B_1 is

$$2 \cdot r \cdot 2^{N-1} = r2^N .$$

Similarly, in each of the calls p0(N, i, k) and p0(N, k, j) block $B_{\ell+1}$, $\ell \in [N-1]$, is moved $2^{N-\ell-1}$ times. Since each move of $B_{\ell+1}$ consists of t moves of discs, the total number of moves of discs from $B_2, \ldots, B_N$ is

$$2 \cdot t \cdot (2^{N-2} + 2^{N-3} + \cdots + 2^0) = 2t(2^{N-1} - 1) .$$

Together with the single move of disc n, the total number of moves is

$$r2^N + 2t(2^{N-1} - 1) + 1 = (r + t)2^N - 2t + 1 .$$

Optimality of Algorithm 22 In [341] Poole correctly claimed that Algorithm 22 uses the minimum number of moves needed to solve the Bottleneck TH but his argument was not complete because it assumed that before the last move of disc n (to the goal peg), all the other discs lie on the spare peg.[82] However, during the last move of disc n, discrepancy t permits that discs $n-t+1, \ldots, n-1$ are distributed arbitrarily among source peg and auxiliary peg. About 15 years later, Xiaomin Chen, B. Tian and L. Wang [81] and Y. Dinitz and S. Solomon [106] pointed out that Poole's arguments were insufficient and presented complete proofs of optimality of Algorithm 22. The two groups of authors worked on the problem independently, at almost the same time,[83] and with different approaches. In both cases the arguments are long and non-trivial.

In conclusion, we have the following result:

Theorem 6.4. *Let $t \in \mathbb{N}$ be the discrepancy of the Bottleneck TH and let $n \in \mathbb{N}_2$. Set $n - 1 = (N - 1)t + r$, where $r \in [t]$. Then Algorithm 22 solves BTH_t in the optimal number of moves $(r + t)2^N - 2t + 1$.*

---

[82]This is the corresponding quotation from [341, p. 206]: "Since the tower must finish in standard position, disk n must be placed on the (empty) destination peg. In order to do this, the other disks must first be moved into a legal position on the spare peg."

[83]The paper of Dinitz and Solomon was submitted in February 2006, the one of Chen, Tian and Wang in May 2006. On the other hand, the latter paper was published in 2007 and the former in 2008.

Note that when $t = 1$ the decomposition $n - 1 = (N - 1)t + r$ gives $r = 1$ and we have $N = n - 1$. Hence the number of moves made by Algorithm 22 is $2^n - 1$ which is the number of moves of the optimal solution of TH (Theorem 2.1).

Poole's arguments from [341] did not consider solutions in which the largest disc moves more than once. Dinitz and Solomon [106] then proved that in every optimal sequence of moves the largest disc indeed moves only once. On the other hand, in [105] they showed that an optimal solution of the Bottleneck TH is not unique in general (cf. Exercise 6.7); earlier uniqueness was wrongly claimed in [341, Corollary 3]. Moreover, they described the set of all optimal solutions for the BTH_t and proved that their number is

$$\left(\binom{t + \tau}{t} - \binom{t + \tau}{t + 1} \right)^{\left\lceil 2^{\lceil \frac{n-1}{t} \rceil - 2} \right\rceil - 1},$$

where $\tau = (n - 1) \bmod t$.

The Bottleneck TH on four pegs was considered by A. A. K. Majumdar and A. Halder in [298].

To close the section we add that Chen, Tian, and Wang [81] considered another related variant of TH and named it *Sinner's Tower of Hanoi*. It differs from the standard TH in that for a given s, one can disobey the divine rule at most s times. More precisely, at most s times we may put a disc directly onto a smaller disc. Chen et al. solved the Sinner's TH and proved that the minimum number of moves for n discs is

$$\begin{cases} 2n - 1, & n \leq s + 2; \\ 4n - 2s - 5, & s + 2 \leq n \leq 2s + 2; \\ 2^{n-2s} + 6s - 1, & 2s + 2 \leq n. \end{cases}$$

6.5 Exercises

6.1. Show that any solution of the Black and white TH with $2n$ discs requires at least $2^{n+1} - 1$ moves.

6.2. Give an optimal solution for the variant B of the Black and white TH for $s = 3$ and $n = 2$.

6.3. Show that the solution for the Tower of Antwerpen with 3×2 discs is not unique.

6.4. (van Zanten [447, p. 6f]) Solve the Twin-Tower problem for an odd total number of discs, i.e. with an extra disc $2n + 1$.

6.5. Which of the two largest discs (gold or silver) will move more often in the (unique) optimal solution for the task given by the leftmost arrow in Figure 6.7, but with the rules of the Twin-Tower problem?

6.6. What is the first move and the length of the optimal path for the task $01201201 \to 10210210$ in H_3^8?

6.7. (Dinitz and Solomon [105]) Verify that the sequence of transfers

$$
\begin{array}{llllllll}
1234567|| & \to & 34567|21| & \to & 567|21|43 & \to & 567||1243 & \to \\
7|65|1243 & \to & 217|65|43 & \to & 217|465|3 & \to & 17|2465|3 & \to \\
17|32465| & \to & 7|132465| & \to & |132465|7 & \to & |32465|17 & \to \\
3|2465|17 & \to & 3|465|217 & \to & 43|65|217 & \to & 1243|65|7 & \to \\
1243||567 & \to & 43|21|567 & \to & |21|34567 & \to & ||1234567 &
\end{array}
$$

is an optimal solution for the BTH$_2$ and that it is different from the sequence of moves produced by Algorithm 22.

Chapter 7

The Tower of London

The *Tower of London (TL)* was invented in 1982 by Shallice [379] and has received an astonishing attention in the psychology of problem solving and in neuropsychology. Just for an illustration, we point out that in the paper [208], which sets up the mathematical framework for the TL, no less than 79 references are listed! The success of the TL is due to the fact that on one hand it is an easy-to-observe psychological test tool, while on the other hand it can be applied in different situations and for numerous clinical goals. It is hence not surprising that several additional variations of the TL were proposed to which we will turn in Section 7.2. Here we only mention the *Tower of Oxford* introduced by G. Ward and A. Allport in [430] and named in [208, p. 2936], but which is mathematically the same puzzle as the TH without the divine rule or either the Bottleneck TH with maximal discrepancy.[84]

7.1 Shallice's Tower of London

The classical TL consists of three differently colored balls (instead of discs) of equal size and of three pegs that can only hold up to 1, 2, and 3 balls, respectively. See Figure 7.1 for two states of the puzzle.

The goal of the puzzle is to reach a specified state from another designated state in the minimum number of moves. The best way to visualize all tasks is to draw the state graph of the puzzle. Let the balls be numbered 1, 2, 3, where, say, 1=blue, 2=red, 3=yellow. In addition, use the symbol "|" to delimit balls among pegs and list balls on a given peg from top to bottom, just as it was done in Chapter 3. For instance, the states from Figure 7.1 are

$$||123 \quad \text{and} \quad 1|3|2,$$

---

[84]However, the Tower of Oxford and the TH without the divine rule are arguably different in the psychological sense.

© Springer International Publishing AG, part of Springer Nature 2018
A.M. Hinz et al., *The Tower of Hanoi – Myths and Maths*,
https://doi.org/10.1007/978-3-319-73779-9_8

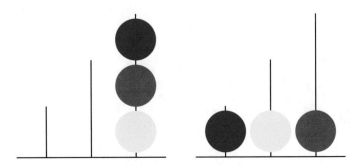

Figure 7.1: Two states of the Tower of London

respectively.

By inspection it is straightforward to deduce that the state graph of the TL has order 36 and that it can be drawn as in Figure 7.2. We denote it by L. Clearly, L is planar and *degree balanced* in the sense that it has 12 vertices of each of the degrees 2, 3, and 4.

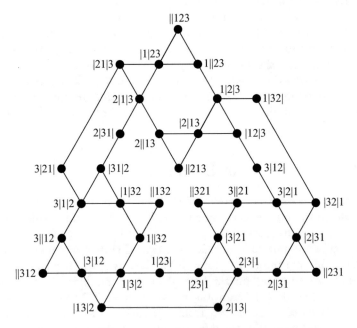

Figure 7.2: The state graph of the Tower of London

Note the similarity between L and the Hanoi graph H_3^3, but also their differences. It is therefore clear that the TL and TH as used in psychological tests are *not* equivalent.

With L at hand it is easy to check that the minimum number of moves from the left-hand side state to the right-hand side state of Figure 7.1 is 7. Moreover, there are three different solutions with that many moves. It is also easy to check that the diameter of the TL graph is 8. The following property of L is also of interest and is not completely obvious from its drawing.

Proposition 7.1. *L contains a hamiltonian path but no hamiltonian cycle.*

Proof. L contains a hamiltonian path as can be seen from Figure 7.3.

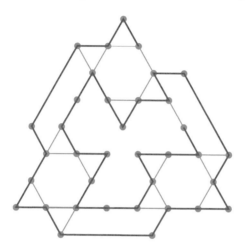

Figure 7.3: A hamiltonian path in L

To show that L is *not* hamiltonian though, we make the following two observations on the graph (cf. Figure 7.4).

- The 12 vertices of degree 2 form a perfect code on L; their neighborhoods, consisting of one vertex of degree 3 and one vertex of degree 4, respectively, are therefore not overlapping.

- Any two vertices of degree 3 are not adjacent in L.

From this it follows that on a cycle C in L containing all its vertices also no two vertices of degree 4 (in L) are adjacent. (See Exercise 7.1 for an alternative proof of this fact.)

Now have a look at Figure 7.4. Since the vertex $||123$ is of degree two, C necessarily contains the subpath $1||23 \rightarrow ||123 \rightarrow |1|23$. Since $|1|23$ is of degree four, the above argument implies that the next vertex on C must be $|21|3$. Since $3|21|$ is of degree two, C continues with $3|21|$ and $3|1|2$. The latter vertex is of degree four, and there are two possibilities how to continue C. If we would proceed with $|31|2$, we would not be able to leave the inner circle in Figure 7.4 anymore. So we must continue with $3||12$. Proceeding with the argument we infer that C

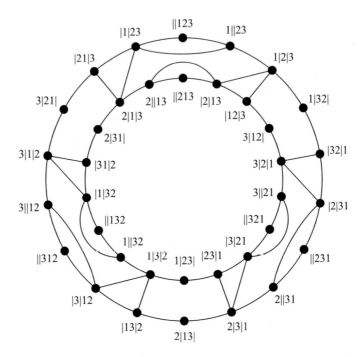

Figure 7.4: Equivalent drawing of the state graph of the Tower of London

stays on the outer circle without ever visiting any of the vertices on the inner circle, the final contradiction.   □

Let us mention in passing that the fact, used in the preceding proof, that the 12 vertices of degree 2 form a perfect code in L results in $\gamma(L) = 12$ by Corollary 2.28. The sets of vertices of degree 3 and 4, respectively, are also minimal L-dominating sets; they are not 1-error correcting though (cf. Figure 7.4).

We now give a brief example of how the TL is used in (neuro)psychology; as already mentioned, this will be just the tip of an iceberg of related extensive investigations.

Again, a *task* is an ordered pair of different states, namely the start and the goal state. Hence there are 36×35 different tasks one can pose on the TL. They come in blocks of six so-called "iso-problems" (cf. W. K. Berg and D. L. Byrd in [51]) differing only in a permutation of the colors. So there are only 210 essentially different tasks representing the equivalence classes of the set of tasks with respect to the six color permutations.

Of course, from the mathematical point of view, iso-problems are of the same difficulty. On the other hand they look different to a test person. Starting out from this, A. H. Faber et. al. [144] performed an experiment on 81 volunteers. It turned out that only one third of the participants were aware of the repetition of iso-problems which, together with the fact that the test persons showed a linear trend of improvements in solution times, led to the conclusion that they learned without realizing it. The authors summarize by saying that "this study might open a new functional domain, namely implicit memory, that can be assessed with the Tower of London."

7.2 More London Towers

For normal test persons the classical TL is by far too simple. With only 36 vertices and a diameter of 8, the London graph L can almost be explored in one flash. Only the number of alternative moves, caused by the maximal degree of 4, poses a little difficulty in finding *shortest* paths. J. R. Tunstall [420] was the first to propose a TL with four balls and all three pegs extended to be able to hold one more ball each, i.e. 2, 3 and 4, respectively (cf. Figure 7.5). This version has been investigated mathematically in [208] embedded in a more general notion of London graphs. It turns out that this class of graphs offers a number of challenging mathematical problems which we will explore in the present section.

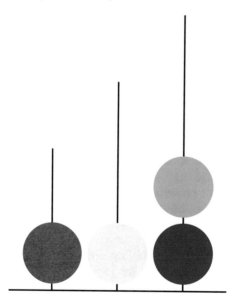

Figure 7.5: State $2|3|41$ of Tunstall's TL version L_{234}^4

The generalized TL has $p \in \mathbb{N}_3$ pegs and $n \in \mathbb{N}_2$ balls (to avoid trivialities). Each peg labeled[85] $k \in [p]$ can hold at most a given number $h_k \in \mathbb{N}$, its *height*, of balls. Then $n \le \sum_{k=1}^{p} h_k$ balls, colored, i.e. numbered, from 1 to n, can be distributed

---

[85]For the TL this labeling has an advantage over the one for the TH.

among the pegs. A legal move consists of transferring the topmost ball on one peg to the top of another one, space permitting. Again there is no other limitation like the divine rule to be observed here.

Each state can be represented uniquely by a permutation $s \in \mathrm{Sym}_{n+p}$, where s_i is the position of ball $i \in [n]$ or either of the bottom of peg $i-n$, if $i \in [n+p] \setminus [n]$. These positions are read from top to bottom on a peg and from left to right for the pegs just like in the classical TL. As there, we will write state s in the form $\Sigma_1 | \dots | \Sigma_p$, where Σ_k is the string of ball labels (colors) in positions $s_{n+k-1} + 1$ to $s_{n+k} - 1$, representing the balls (from top to bottom) on peg $k \in [p]$; here s_{n+0} is not s_n, but 0. Again, the upright bars stand for the bottoms of the pegs, peg p being omitted. For example, Figure 7.5 shows the state $2|3|41$ in Tunstall's TL, where $p = 3$, $n = 4$, and $h = 234$ (i.e. $h_1 = 2$, $h_2 = 3$, $h_3 = 4$). In this state, ball 2 is in position 1, such that $s_2 = 1$, the bottom of peg 1 is in position 2, meaning $s_5 = 2$, and so on until we reach $s_1 = 6$ and $s_7 = 7$ (the omitted bar); therefore, in the notation for a permutation as on p. 42, the state is written as $s = 6135247$. The corresponding state graph will be denoted by L_{234}^4. More generally, we have the following.

Definition 7.2. *For $p \in \mathbb{N}_3$, $n \in \mathbb{N}_2$, and $h \in [n]^p$ with $\sum\limits_{k=1}^{p} h_k \geq n$, the London graph L_h^n has the vertex set consisting of all those $s \in \mathrm{Sym}_{n+p}$, which fulfill the condition*

$$\forall\, k \in [p]: \ 1 \leq s_{n+k} - s_{n+k-1} \leq h_k + 1, \ s_{n+p} = n + p.$$

The edge set of L_h^n is composed of those pairs of vertices whose corresponding states are linked by an individual legal move of one ball. The special case $O_p^n := L_{n^p}^n$ is called an Oxford graph; *cf. [208, p. 2937].*

Note that the restriction to values in $[n]$ for the h_ks is no loss of generality, because there are not more than n balls to be distributed.

Quite obviously, the classical London graph L is equal to L_{123}^3. However, London graphs pose a lot of interesting mathematical questions which also touch their usefulness in cognitive tests. It is even not simple to determine their orders in general (see Exercise 7.2), except for the Oxford graphs (see Exercise 7.3), for which even the size is accessible (see Exercise 7.4). More importantly, two properties are crucial in applications for psychology tests, namely connectedness and planarity.

An obvious condition for connectedness of a London graph is that it must be possible to distribute all balls such that the tallest peg remains empty, because no ball on that peg can be moved anymore if all the other pegs are filled completely. In what follows, we will always assume that the pegs are arranged in increasing order of their heights, i.e. $h_1 \leq h_2 \leq \cdots \leq h_p$. From the mathematical point of view this is no loss of generality, but we emphasize again that for the human problem solver the *Olympic Tower of London* L_{231}^3 (see Figure 7.6), for instance, might have a different aspect from Shallice's classical TL L_{123}^3!

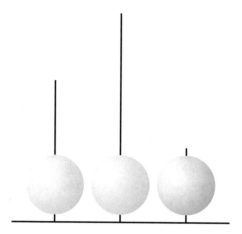

Figure 7.6: The Olympic Tower of London during victory ceremony

The necessary condition for connectedness then reads

$$n \leq \sum_{k=1}^{p-1} h_k. \tag{7.1}$$

This is a nice example of an obviously necessary condition being (non-obviously) sufficient as well! Here is the result of Götz [161, Lemma 18] (cf. also Hinz [201, Theorem]):

Theorem 7.3. *The London graph L_h^n is connected if and only if condition (7.1) holds.*

Proof. We only have to show sufficiency of condition (7.1) for any two vertices to be joined by a path. This will be done in a sequence of reductions.

0. We may assume that $n = \sum_{k=1}^{p-1} h_k$, because otherwise we can introduce virtual balls, whose moves can afterwards be ignored.

1. We distribute all balls which are on peg p in the initial state onto the other pegs. In a similar way, we prepare the goal state by moving away all balls from peg p, an action which can be undone at the very end of the procedure. So we may assume that peg p is empty in the initial and final states.

2. We now choose successively a ball which is not yet on its goal position (peg *and* height on the peg) in order to interchange it with the ball currently present there. If this can be done without altering any other ball's position, then there is at least one more ball in the right place. With peg p empty and all the others filled, it cannot happen that only one ball remains in false position during this procedure.

3. We may assume that one of the two balls to be interchanged lies on top of peg 1, because three switches of this type cover the general case; for instance,

$$cx|xbx|xax| \ \to \ bx|xcx|xax| \ \to \ ax|xcx|xbx| \ \to \ cx|xax|xbx| \ ,$$

where every x replaces a (possibly empty) collection of balls.

4. Let a be the top ball on peg 1. Then there are two cases:

   4.1 Ball b is on peg 1 as well. Move the top ball c on peg 2 to peg p, then a to peg 2 and all other balls above and including b from peg 1 to peg p. Then a can move back to peg 1, b to peg 2, all balls from peg p except c return to 1, then b as well, and finally c to its original position.

   4.2 Ball b is on a different peg. Move all balls above it on the same peg one after the other to peg p, then b on top of p, a to the position b just left, b to peg 1, and all balls from peg p to where they came from.

This completes the proof, which the reader is invited to recapitulate in Exercise 7.5 a.　　　　　　　　　　　　　　　　　　　　　　　　　　　　□

Note that, in general, the procedure in the proof of Theorem 7.3 does not yield an optimal path; see Exercise 7.5 b. From now on, we will only consider connected London graphs L_h^n, i.e. those fulfilling (7.1).

The other interesting question relating to them is planarity. Apart from its mathematical interest this is because psychologists like to use the representation of a test person's performance on the underlying graph of the tower puzzle in use and therefore crossings of edges might lead to confusion. We will limit our discussion of planarity of London graphs to the case of $p = 3$ pegs employed in psychological test tools.

From Figure 7.7 it is clear that all L_h^2 are planar.

In fact, with $\subseteq$ denoting "subgraph",

$$L_{111}^2 \subseteq L_{112}^2 \subseteq L_{122}^2 \subseteq L_{222}^2 = O_3^2$$

are shown here adding successively red, green, blue, and yellow. Note from the figure that $H_3^2 \subseteq O_3^2$. Actually, the *underlying graph*, i.e. obtained by replacing arcs with edges, of the mixed graph $\overrightarrow{H}_p^n$ of Lucas's second problem from Chapter 3 has the same vertex set as and is a subgraph of O_p^n; cf. Figure 3.2. Let us also mention that the graphs L_h^2 provide a nice exercise for domination. We can easily find 2-element perfect codes for $h = 111$, $h = 112$, and $h = 122$, such that the corresponding domination number is 2 by Corollary 2.28. The problem for $L_{222}^2 = O_3^2$ is slightly more delicate and left as Exercise 7.6. For general London graphs it will be very difficult to address the question of domination number given that even the order of these graphs can not be determined explicitly in many cases (cf. Exercise 7.2).

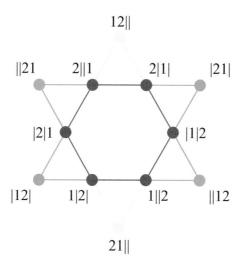

Figure 7.7: London graphs for 2 balls

Turning to London graphs L_h^3, we have the following scheme.

$$L$$
$$\|$$
$$
\begin{array}{ccccccc}
L_{122}^3 & \subseteq & L_{123}^3 & \subseteq & L_{133}^3 & & \\
\cap & & \cap & & \cap & & \\
L_{222}^3 & \subseteq & L_{223}^3 & \subseteq & L_{233}^3 & \subseteq & L_{333}^3
\end{array}
$$
$$\|$$
$$O_3^3$$

We have already seen, for instance in Figure 7.4, that L is planar. Deleting vertices (and adjacent edges) for states with three balls on peg 3 gives a planar drawing of L_{122}^3. On the other hand, adding the six vertices $|ijk|$, $\{i,j,k\} = [3]$, to the drawing in Figure 7.4, together with edges linking them to $i|jk|$ and $|jk|i$, respectively, shows that L_{133}^3 is planar too.

To analyze L_{222}^3, we remind the reader of two facts about the complete bipartite graph $K_{3,3}$. First, it is not planar and second, it can be drawn without crossings on the torus, i.e. it is *toroidal*. The latter property can be seen in the left picture of Figure 7.8. The right-hand picture in that figure shows the same situation on a plane representation of the torus. Here, the edge leaving the top boundary has to be identified with the edge entering at the bottom, and similarly for the two lines extending to the left and right side of the rectangle which has been obtained by cutting the torus at two appropriate orthogonal circles.

With this preparation we are now able to interpret the drawing of London graph L_{222}^3 in Figure 7.9.

The figure shows that L_{222}^3 can be drawn on a torus without crossing. On the

Figure 7.8: The graph $K_{3,3}$ on the torus

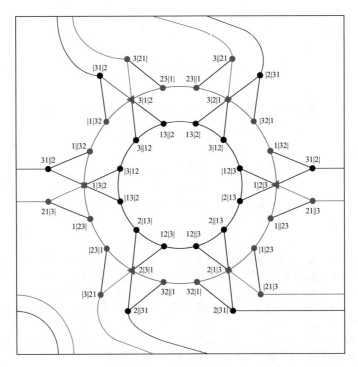

Figure 7.9: The graph L_{222}^3 on the torus

other hand, the subgraph drawn in red is a subdivision of a $K_{3,3}$ whose vertices are the flat states, i.e. those with one ball per peg. (The elements of the two independent parts of the vertex set of this $K_{3,3}$ are represented in Figure 7.9 as squares and triangles, respectively.) Hence L_{222}^3 is not planar, but toroidal; in other words, its genus is 1. However, it is not clear what its crossing number is; cf. Exercise 7.7. All the other L_h^3 are toroidal because O_3^3 is as is shown in

Exercise 7.8. It is worth noting that these are the only non-complete non-planar tower graphs (Hanoi, Sierpiński, or London) whose genus is known!

Adding a ball to L_{222}^3 shows that L_{222}^4 is a subdivision of that graph (the extra ball can always be moved out of the way) and therefore not planar either. This argument can not be used for L_{133}^4, because L_{133}^3 is planar. So one has to find, e.g., a subdivision of $K_{3,3}$ in L_{133}^4 to prove that it is not planar. This is a rather tedious task given that $|L_{133}^4| = 168$, so we will not give the details here. We summarize these results as follows.

Proposition 7.4 (Hinz, A. Spitzer). *Among the connected London graphs for* $p = 3$, *the only planar ones are* L_h^2, L_{122}^3, L_{123}^3, *and* L_{133}^3.

Proof. It only remains to show that all London graphs for 3 pegs and $n \geq 4$ balls are non-planar. This follows for $n = 4$ from the non-planarity of L_{133}^4 and L_{222}^4, because every L_h^4 must contain one of these as a subgraph. For $n > 4$ one can argue as above in the case of L_{222}^4 by adding balls to L_h^4. □

The already somewhat demanding graph L_{222}^3 will also provide a good example for the rich symmetrical structure of London graphs. The reason for this variability is that we now have two types of permutations, namely of ball colors and of pegs. For those puzzles, like Shallice's and Tunstall's TL, where the peg heights are pairwise different, only color permutations are symmetries of the graph. For instance, although Shallice's TL and the Olympic TL have isomorphic state graphs, their vertex sets are different such that the isomorphisms are no automorphisms. We also already pointed out the difference of mathematical and psychological equivalence of these two variants. So we do *not* claim to consider all automorphisms of London graphs, but the above mentioned "obvious" symmetries. For Tunstall's TL L_{234}^4 this means that the 69432 (non-trivial) tasks simply fall into 2893 equisets of 24 iso-problems each. This already provides a sufficient variety for the design of psychological tests, but things become more exciting when some peg heights are equal, i.e. when symmetries can overlap. For instance, the Oxford graph O_3^2 shown in Figure 7.7 has only two non-equivalent types of states, namely flat or tower-like, and the, admittedly not difficult, tasks $2||1 \rightarrow |1|2$ and $|2|1 \rightarrow 2|1|$, for instance, are equivalent since they have been obtained from each other by a rotation of the 3 pegs combined with a switch of the 2 colors. In fact, applying all combinations of symmetries, their equiset contains 12 iso-problems. This is the case for all but the two equivalence classes corresponding to tasks with just a switch of the ball colors, i.e. represented by $12|| \rightarrow 21||$ and $1|2| \rightarrow 2|1|$, and which contain only 6 iso-problems each, engendered by rotations of pegs and switches of colors. The 132 tasks of O_3^2 can therefore be partitioned into 12 equisets, 10 of which containing 12 elements.

We now can make use of these facts to deduce metric properties of the graph O_3^2 in a most efficient way. For instance, in order to obtain the average eccentricity, we just have to read-off the eccentricities of one representative of each of the two equally sized equisets of states, e.g., $\varepsilon(12||) = 4$ and $\varepsilon(1|2|) = 3$ to find

$\bar{\varepsilon}(O_3^2) = 3.5$. Similarly, we can find the average distance in O_3^2 by first summing up all distances from each of the two representatives, namely $\displaystyle\sum_{s \in V(O_3^2)} d(12||, s) = 26$

and $\displaystyle\sum_{s \in V(O_3^2)} d(1|2|, s) = 21$, such that $\bar{d}(O_3^2) = \dfrac{6 \cdot 26 + 6 \cdot 21}{132} \approx 2.136$. Note that, contrary to our custom in earlier chapters, we divided by the number of non-trivial tasks, just to please psychologists and to be able to compare with lists of values for other tower graphs like in [208, Tables 1 and 2]; that we included trivial tasks in the sum of distances is, of course, irrelevant. Other graph parameters like minimal, average, and maximal degree, which are 2, 3, and 4, respectively, for O_3^2, are also relevant in psychological test design, because they are measures for the amount of alternatives a subject has for his decisions during a solution of a task. For a discussion of graph theoretical measures for tower tasks in (neuro-)psychology, see [201].

To employ symmetry techniques for more complicated London graphs, we have to get a bit more formal and come back to the ideas of Section 0.8.2. For $n, p \in \mathbb{N}$, we define the group $(\Gamma_{np}, \cdot, 1_{np})$ by

$$\Gamma_{np} = \operatorname{Sym}_n \times \operatorname{Sym}_p, \ (\chi_1, \pi_1) \cdot (\chi_2, \pi_2) = (\chi_1 \circ \chi_2, \pi_1 \circ \pi_2), \ 1_{np} = (\operatorname{id}_n, \operatorname{id}_p).$$

It obviously has order $|\Gamma_{np}| = n! \cdot p!$. Recalling that we write a permutation σ from Sym_q in the form $\sigma(1) \ldots \sigma(q)$, we now describe the action of Γ_{np} on the set $X = V(L_h^n)$, where $h \in [n]^p$. A state $s = \Sigma_1 | \ldots | \Sigma_p$ is transformed into $(\chi, \pi).s = \mathrm{X}\big(\Sigma_{\pi^{-1}(1)} | \ldots | \Sigma_{\pi^{-1}(p)}\big)$, where X permutes all ball colors according to χ. For instance, in the above example in O_3^2, the state $2||1$ is transformed by the rotation of pegs $\pi = 312$ (with $\pi^{-1} = 231$) into $|1|2$ and then by the switch of colors $\chi = 21$ into $|2|1$, such that $(21, 312).2||1 = |2|1$.

To analyze equisets of states it now suffices to find the fixed points under this action. Similarly, information about equisets of tasks are obtained by the action of Γ_{np} on the set $V(L_h^n) \dot{\times} V(L_h^n)$, defined in the analogous way. We will ask the reader to do the details for L_{222}^3 in Exercise 7.9 to arrive at

Proposition 7.5. *With respect to the symmetries engendered by permutations of pegs and colors, the 42 states of L_{222}^3 fall into 2 equisets of sizes 6 and 36, respectively. Minimal, average, and maximal degrees of vertices are 3, $\frac{24}{7} \approx 3.43$, and 6, respectively. The eccentricity of any vertex (and consequently the diameter of the graph) is 5. The average distance on L_{222}^3 is $\frac{139}{41} \approx 3.39$. The 1722 non-trivial tasks come in 49 equisets, 47 of which have size 36, the other two having size 18 and 12, respectively.*

Those who think that L_{222}^3 is too easy, should try O_4^4 (with $|O_4^4| = 840$ and $\|O_4^4\| = 2880$), where the 704760 tasks fall into 1400 equisets, 1068 of size 576, 298 of size 288, 1 of size 192, 9 of size 144, 23 of size 96, and 1 of size 72; the diameter is 9, the average distance about 4.82, and the minimal, average, and maximal degrees are 3, $\frac{48}{7} \approx 6.86$, and 12, respectively (cf. the tables in [208]). Note that

H_4^4 has the same diameter, but puzzlers trying to solve tasks from O_4^4 (optimally!) will appreciate the divine rule of the TH.

7.3 Exercises

7.1. Show that if G is a hamiltonian graph and X a subset of its vertices such that $G - X$ consists of $|X|$ connected components, then no two vertices of X are consecutive vertices of a hamiltonian cycle of G. Deduce from this that no two vertices of degree 4 of L can be consecutive vertices of a hamiltonian cycle.

7.2. a) Show that the condition on a permutation s in Definition 7.2 is equivalent to the existence of an h-*partition* σ of n, i.e. a $\sigma \in \mathbb{N}_0^p$ with $\sum_{k=1}^{p} \sigma_k = n$, fulfilling $\forall\, k \in [p] : \sigma_k \le h_k$. (The value of σ_k is the number of balls on peg k in state s, i.e. the length of string Σ_k.)

b) Find the number l_h^n of such partitions for $h \in \mathbb{N}_0^p$, $p \in \mathbb{N}_0$, and $n \in \mathbb{N}_0$.

c) Show that $|L_h^n| = n! \cdot l_h^n$.

7.3. a) Show that $l_{n^p}^n = \binom{p+n-1}{n}$ for $n \in \mathbb{N}_0$ and $p \in \mathbb{N}$.

b) Give a more direct argument for $|O_p^n| = \dfrac{(p-1+n)!}{(p-1)!}$.

7.4. Show that

$$\|O_p^n\| = \frac{np}{2} \frac{(p-2+n)!}{(p-2)!}$$

for $p \in \mathbb{N}_2$, $n \in \mathbb{N}_0$.

7.5. Consider the task to get from state $1|2|43$ to state $|4|231$ in Tunstall's TL L_{234}^4.
a) Construct a solution according to the algorithm in the proof of Theorem 7.3.
b) Find a shortest solution for the task.

7.6. Determine $\gamma(O_3^2)$ and show that O_3^2 has no perfect code.

7.7. Show that $1 \le \mathrm{cr}\left(L_{222}^3\right) \le 8$ or give tighter bounds.

7.8. Construct a drawing of O_3^3 on the torus.

7.9. Give a detailed proof for Proposition 7.5.

Chapter 8

Tower of Hanoi Variants with Restricted Disc Moves

In Chapter 6 we have introduced the concept of a TH variant and presented several such puzzles: the Black and white TH, additional variants with colored discs, and the Bottleneck TH. We continued in Chapter 7 where the TL (and its variations) were treated in detail. In this chapter we turn our attention to the TH with restricted disc moves. These restrictions may consist of disallowing moves between certain pegs or either of imposing a direction of moves. (Note that Variants A, B and C of the Black and white TH introduced on p. 288ff are also defined prohibiting certain moves depending on disc colors.) The variants are pretty well understood if only three pegs are present as we will see in Section 8.2. It will turn out that for these, exactly five, puzzles a unified algorithm to generate their, respectively unique, optimal solutions can be constructed. To make things precise, we will discuss *solvability* of a variant in Section 8.1.

The variety of TH with restricted disc moves grows dramatically to 83 puzzles on four pegs, hence in Section 8.3 we carefully select which of them deserve special attention, mainly those where moves are *not* oriented. The most prominent among these will be the so-called Star TH and the Linear puzzle, but the intrinsic difficulty of the (oriented) Cyclic TH with more than three pegs is also emphasized. In the final part of this chapter, Section 8.5, we give a classification of the solvable problems with restricted disc moves into exponential and sub-exponential variants.

8.1 Solvability

Any TH variant dealt with in this chapter is uniquely specified by a digraph D whose vertices correspond to the pegs and in which an arc $(i,j) \in A(D)$ indicates that a disc may be moved from peg i to peg j. D is also called the *move graph* of the respective variant which is denoted by $\mathrm{TH}(D)$. The move graph D is actually the state graph for $n = 1$ disc of the corresponding TH variant $\text{TH}(D)$.[86] (The state graph for $n = 0$ discs is always given by $V = \{e\}$ and $A = \varnothing$, where e is the empty word; it is therefore not specific for any variant.) More formally, let D be a (di)graph of order $|D| =: p \in \mathbb{N}_2$; we may assume that $V(D) = [p]_0$. Then, for $n \in \mathbb{N}_0$, we define the (di)graph H_D^n by

Definition 8.1. $V(H_D^n) = V(D)^n$ *(an $s \in V(D)^n$ is written as $s_n \ldots s_1$),*

$$A(H_D^0) = \varnothing,$$
$$A\left(H_D^{1+n}\right) = \{(is, it) \mid (s, t) \in A\left(H_D^n\right)\}$$
$$\cup \left\{(is, js) \mid (i, j) \in A(D),\ s \in (V(D) \setminus \{i, j\})^n\right\}. \qquad (8.1)$$

Remark. Definition 8.1 makes sense for $p = 0$ and $p = 1$ as well, but leads to trivial graphs only. One may also restrict $V(D)$ to a subset, but we will not pursue this here.

Proposition 8.2. $H_D^1 = D$, $|H_D^n| = |D|^n$, *and* $\|H_D^n\| = \frac{1}{2}\|D\|\,(p^n - (p-2)^n)$.

Proof. The first two statements follow directly from Definition 8.1. By virtue of (8.1) in Definition 8.1 we obtain the recurrence

$$\|H_D^0\| = 0,\ \forall\, n \in \mathbb{N}_0 : \|H_D^{1+n}\| = p\|H_D^n\| + \|D\|(p-2)^n,$$

which can be resolved by Lemma 2.18. $\square$

There is also a direct definition of the graphs H_D^n:

Definition 8.3.

$$V(H_D^n) = V(D)^n,$$
$$A(H_D^n) = \left\{(\underline{s}i\overline{s}, \underline{s}j\overline{s}) \mid (i,j) \in A(D),\ \overline{s} \in (V(D) \setminus \{i,j\})^{d-1},\ d \in [n]\right\}. \qquad (8.2)$$

For undirected D we replace A by E and ordered pairs by unordered ones throughout; of course, $|E| = \frac{1}{2}|A|$ then. We will write $(V, A) \simeq (V, E)$. For instance, let $\overset{\leftrightarrow}{K}_p$ denote the *complete digraph* on p vertices, that is, a digraph with two arcs of opposite orientation between each pair of its p vertices: $A(\overset{\leftrightarrow}{K}_p) = \{(i,j) \mid i, j \in [p]_0,\ i \ne j\}$. (See the leftmost picture in Figure 8.1 for $\overset{\leftrightarrow}{K}_3$.) Then $\text{TH}(\overset{\leftrightarrow}{K}_p)$ is the standard TH with p pegs: $H_{\overset{\leftrightarrow}{K}_p}^n \simeq H_p^n$.

We will restrict our considerations to type P0 problems—to transfer a tower of discs from one peg to another in the minimum number of moves, i.e. to find a (shortest) path from a perfect state i^n to a perfect state j^n in H_D^n. Here a u, v-*path* $P_{1+\ell}$ (of *length* $\ell \in \mathbb{N}$), is a digraph on $V(P_{1+\ell}) = \{v_0, \ldots, v_\ell\}$, $|V(P_{1+\ell})| = 1 + \ell$, with

---

[86]This may have led some authors to call the digraphs D "Hanoi graphs"; this should not be mixed up with the standard definition of Hanoi graphs as employed in the present book.

$v_0 = u$, $v_\ell = w$, and $A(P_{1+\ell}) = \{(v_{k-1}, v_k) \mid k \in [\ell]\}$. The situation can be different for different (ordered) pairs of source and goal pegs. We therefore say that a TH with oriented disc moves is *solvable* if for any choice of source and target pegs and for every number of discs there exists a sequence of legal moves joining initial and goal states. Formally, a TH variant TH(D) is *solvable*, if

$$\forall\, n \in \mathbb{N}_0 \; \forall\, \{i, j\} \in \binom{[p]_0}{2} \; \exists\, (i^n, j^n)\text{-path in } H_D^n. \tag{8.3}$$

Apart from the classical complete graphs $D \simeq K_p$, there are two prototype cases for $p = 3$. One is the Linear TH, where $D = \overset{\leftrightarrow}{L}_3$ is the digraph with the vertex set $V(\overset{\leftrightarrow}{L}_3) = T$ and the arc set $A(\overset{\leftrightarrow}{L}_3) = \{(0,1), (0,2), (1,0), (2,0)\}$; it is drawn as the central picture of the top row in Figure 8.1. Of course, this graph is undirected, i.e. $D \simeq H_{3,\mathrm{lin}}^n$, for which we have proved solvability in Section 2.3.1. In fact, an immediate consequence is the solvability for all TH variants whose underlying graph D is undirected and connected; more generally:

Proposition 8.4. *If the undirected graph D, $|D| \geq 3$, contains an i, j-path, $i, j \in [p]_0$, then there is an i^n, j^n-path in H_D^n.*

Proof. [87] Induction on the length $\ell \in \mathbb{N}$ of the path in D. If $\ell = 1$, then i and j lie on a path in D of length 2, because D is connected and has at least 3 vertices. Then there is an i^n, j^n-path in H_D^n by the solvability of the Linear TH. Now assume the statement to be true for lengths up to and including $\ell \in \mathbb{N}$ and let P be an i, j-path in D of length $\ell + 1 \geq 2$. Let $k \neq i$ be the neighbor of j in P. Then there are an i^n, k^n-path and a k^n, j^n-path in H_D^n, which glued together lead to an i^n, j^n-path. $\qquad\square$

State graphs for undirected move graphs have other interesting properties. As a curiosity we have:

Proposition 8.5. *Let D be an undirected move graph of a TH puzzle. Then the state graph H_D^n of the puzzle with $n \in \mathbb{N}$ discs is bipartite if and only if D is bipartite.*

Proof. Recall that a graph is bipartite if and only if it does not contain an odd cycle.

Suppose first that D is not bipartite and let $i_1 i_2 \dots i_{2k+1}$, $k \in \mathbb{N}$, be an odd cycle in D. Then the vertices $i_1^{n-1} i_1, i_1^{n-1} i_2, \dots, i_1^{n-1} i_{2k+1}$ induce an odd cycle in H_D^n, which thus is also not bipartite.

Conversely, assume that D is bipartite. Let C be an arbitrary cycle in H_D^n, $n \in \mathbb{N}$, and let s be an arbitrary vertex of C. Traversing C starting in s (representing a regular state of the puzzle) and returning to s, each disc is moved an even number of times (possibly zero times) because D is bipartite. It follows that the total number of moves to get from state s back to state s along C is even. Consequently C is an even cycle. Hence H_D^n is bipartite. $\qquad\square$

---

[87] Jasmina Ferme, private communication, 2017.

The prototype case for move restrictions which are indeed oriented, is the *Cyclic Tower of Hanoi* with p pegs, $p \in \mathbb{N}_3$. The digraph $\overrightarrow{C}_p$ of this puzzle has the vertex set $V(\overrightarrow{C}_p) = [p]_0$, and the arc set is given by

$$A(\overrightarrow{C}_p) = \{(0,1), (1,2), \ldots, (p-2, p-1), (p-1, 0)\}\,.$$

$\overrightarrow{C}_p$ is called the *directed cycle on p vertices*, see the rightmost picture in Figure 8.1 for $\overrightarrow{C}_3$. The proof of the next result is left for Exercise 8.1.

Proposition 8.6. *$TH(\overrightarrow{C}_p)$ is solvable for any $p \in \mathbb{N}_3$.*

This example shows that the proof of Proposition 8.4 can not be employed for directed D because of the reference to the Linear TH on three pegs. The following definition is needed. A digraph $D = (V, A)$ is called *strongly connected* or *strong* for short, if for any distinct vertices $u, v \in V$ there is a directed path from u to v and a directed path from v to u. Now we can state the result which is essentially due to E. L. Leiss [260, Theorem]. As the author admits, his proof is "quite long and tedious". Therefore we present a more general statement with a less involved proof.

Theorem 8.7. *Let D be a digraph with $V(D) = [p]_0$, $p \in \mathbb{N}_3$. Then the Tower of Hanoi variant $TH(D)$ is solvable if and only if D is strong.*

Proof. If $TH(D)$ is solvable, then (8.3) applies, in particular, for $n = 1$, which means that D is strong.

Now let D be strong. We prove (8.3) by induction on n. The case $n = 0$ is trivial. Fix $\{i, j\} \in \binom{[p]_0}{2}$ and let P be an (i,j)-path in D (exists because D is strong) whose first move is from i to k, say. Let $\ell \in [p]_0 \setminus \{i, k\}$ (exists because $|D| \geq 3$). Starting in i^{1+n}, we proceed as follows:

1. Transfer the n-tower consisting of discs 1 to n from peg i to peg ℓ.

2. Move disc $n + 1$ from peg i to peg k.

3. Transfer the n-tower from peg ℓ to peg i.

4. Move disc $n + 1$ from peg k along the rest of P to peg j.

5. Transfer the n-tower to peg j.

This takes us to j^{1+n}. Note that step 3 is only needed if ℓ lies on P and can therefore be avoided if P does not contain all vertices of D. Also step 4 is void if and only if $k = j$. So the resulting path contains either two or three transfers of the n-tower made possible by induction assumption.   □

8.2   An Algorithm for Three Pegs

In this section we consider the variants defined on three pegs. By Theorem 8.7 we only need to consider those TH(D) for which D, with $V(D) = T$, is strong. Following the approach of A. Sapir [364] we show that there is a common framework for all of them. More precisely, a single algorithm can be designed that finds the unique optimal solution in each of the cases. Stockmeyer conjectures in [407, p. 69] that all these solutions "can be solved with appropriate automata." At the end of the section we also discuss the number of moves in these solutions.

It is straightforward to verify that there are exactly five non-isomorphic strongly connected digraphs on three vertices; see Figure 8.1. Among the variants corresponding to these digraphs we have already met the classical TH, the Linear TH, and the Cyclic TH (top row in Figure 8.1). Algorithm 23 will solve all variants. In particular it generalizes the classical recursive Algorithm 4 as well as the solutions for the Linar TH given in Section 2.3.1.

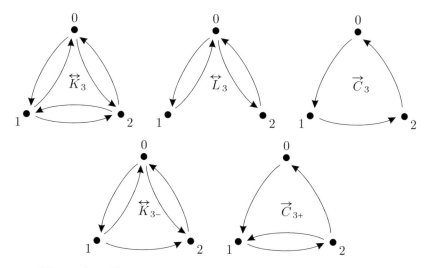

Figure 8.1: The strongly connected digraphs on three vertices

Theorem 8.8. *For every strongly connected digraph D on three vertices, Algorithm 23 returns the unique solution of TH(D) with the minimum number of moves.*

Proof. For the proof it will be useful to rename Algorithm 23 to Algorithm A. The move of a disc d from peg i to peg j, its first move from peg i to peg j, and its last move from peg i to peg j will be denoted by

$$i \xrightarrow{d} j, \quad f : i \xrightarrow{d} j, \quad \text{and} \quad l : i \xrightarrow{d} j,$$

Algorithm 23 Recursive general algorithm for three pegs

Procedure $gp0(D, n, i, j)$
Parameter D: strongly connected digraph with $V(D) = T$
Parameter n: number of discs $\{n \in \mathbb{N}_0\}$
Parameter i: source peg $\{i \in T\}$
Parameter j: goal peg $\{j \in T\}$

    if $n \neq 0$ and $i \neq j$ **then**
        $k \leftarrow 3 - i - j$ $\{$the auxiliary peg different from i and $j\}$
        if $(i, j) \in A(D)$ **then**
            $gp0(D, n - 1, i, k)$ $\{$transfers $n - 1$ smallest discs to auxiliary peg$\}$
            move disc n from i to j $\{$moves largest disc to goal peg$\}$
            $gp0(D, n - 1, k, j)$ $\{$transfers $n - 1$ smallest discs to goal peg$\}$
        else
            $gp0(D, n - 1, i, j)$ $\{$transfers $n - 1$ smallest discs to goal peg$\}$
            move disc n from i to k $\{$moves largest disc to auxiliary peg$\}$
            $gp0(D, n - 1, j, i)$ $\{$transfers $n - 1$ smallest discs to source peg$\}$
            move disc n from k to j $\{$moves largest disc to goal peg$\}$
            $gp0(D, n - 1, i, j)$ $\{$transfers $n - 1$ smallest discs to goal peg$\}$
        end if
    end if

---

respectively.

To see that Algorithm A returns a solution we only need to observe that whenever $(i, j) \notin A(D)$, we have $(i, k) \in A(D)$ and $(k, j) \in A(D)$ because D is strongly connected. By induction, the algorithm indeed returns a solution.

It remains to verify that Algorithm A returns the only solution of $\mathrm{TH}(D)$ with the minimum number of moves. This is again done by induction and is clear for $n = 0$ and $n = 1$. Let Algorithm X be an arbitrary algorithm that solves $\mathrm{TH}(D)$ in the minimum number of moves for $n + 1$ discs, $n \in \mathbb{N}$. Note that the analogue of the boxer rule (Lemma 2.32) holds and therefore disc $n + 1$ is moved either once or twice.

A sequence of moves performed by Algorithm A and by Algorithm X to transfer discs $1, 2, \ldots, d$ (that is, all discs from the set $[d]$) from peg i to peg j will be denoted by

$$A : i \xrightarrow{[d]} j \quad \text{and} \quad X : i \xrightarrow{[d]} j,$$

respectively. We now separately consider the possibilities that (i, j) is an arc of D or not. We begin with the easier case when it is not an arc. (Interestingly, Algorithm A is more involved in this case.)

So suppose that $(i, j) \notin A(D)$. Then disc $n+1$ necessarily moves twice, namely from peg i via peg k to peg j. Hence the moves of Algorithm X can be decomposed into

$$X : i \xrightarrow{[n]} j, \quad f : i \xrightarrow{n+1} k, \quad X : j \xrightarrow{[n]} i, \quad l : k \xrightarrow{n+1} j, \quad X : i \xrightarrow{[n]} j.$$

By induction assumption,

$$X : i \xrightarrow{[n]} j \; = \; A : i \xrightarrow{[n]} j \quad \text{and} \quad X : j \xrightarrow{[n]} i \; = \; A : j \xrightarrow{[n]} i \,,$$

which implies that Algorithm X makes the same moves as Algorithm A, also for $n + 1$ discs.

Now suppose that $(i, j) \in A(D)$. We will show that in this case the largest disc moves only once. Assume it moves twice, which means in particular that $(i, k) \in A(D)$ and $(k, j) \in A(D)$. Then, by induction assumption, the sequence of moves of Algorithm X can be decomposed into:

$$\mathcal{X} = A : i \xrightarrow{[n]} j, \; f : i \xrightarrow{n+1} k, \; A : j \xrightarrow{[n]} i, \; l : k \xrightarrow{n+1} j, \; A : i \xrightarrow{[n]} j \,.$$

Let $\mathcal{A}$ be the sequence of moves of Algorithm A, that is,

$$\mathcal{A} = A : i \xrightarrow{[n]} k, \quad f : i \xrightarrow{n+1} j, \quad A : k \xrightarrow{[n]} j \,.$$

We are going to show that $|\mathcal{A}| < |\mathcal{X}|$. Since in both $\mathcal{A}$ and $\mathcal{X}$ the sequences of moves of discs from $[n]$ are with respect to Algorithm A, we will in the rest omit the prefix A. Employing the induction assumption again, we have

$$i \xrightarrow{[n]} k \; = \; i \xrightarrow{[n-1]} j, \; i \xrightarrow{n} k, \; j \xrightarrow{[n-1]} k,$$
$$k \xrightarrow{[n]} j \; = \; k \xrightarrow{[n-1]} i, \; k \xrightarrow{n} j, \; i \xrightarrow{[n-1]} j,$$
$$i \xrightarrow{[n]} j \; = \; i \xrightarrow{[n-1]} k, \; i \xrightarrow{n} j, \; k \xrightarrow{[n-1]} j \,.$$

To be able to compare $|\mathcal{A}|$ with $|\mathcal{X}|$ we also need to deal with $j \xrightarrow{[n]} i$. For this, we distinguish two cases.

Case 1. $(j, i) \in A(D)$.
In this case $j \xrightarrow{[n]} i = j \xrightarrow{[n-1]} k, \; j \xrightarrow{n} i, \; k \xrightarrow{[n-1]} i$. Now,

$$|\mathcal{X}| = |i \xrightarrow{[n]} j| + 1 + |j \xrightarrow{[n]} i| + 1 + |i \xrightarrow{[n]} j|$$
$$= 2 \cdot \left(|i \xrightarrow{[n-1]} k| + |k \xrightarrow{[n-1]} j| \right) + |j \xrightarrow{[n-1]} k| + |k \xrightarrow{[n-1]} i| + 5$$
$$\geq 2 \cdot |i \xrightarrow{[n-1]} j| + |j \xrightarrow{[n-1]} k| + |k \xrightarrow{[n-1]} i| + 5$$
$$> 2 \cdot |i \xrightarrow{[n-1]} j| + |j \xrightarrow{[n-1]} k| + |k \xrightarrow{[n-1]} i| + 3$$
$$= |\mathcal{A}| \,.$$

In the above computation we have used identities provided above and the obvious triangle inequality $|i \xrightarrow{[n-1]} j| \leq |i \xrightarrow{[n-1]} k| + |k \xrightarrow{[n-1]} j|$.

Case 2. $(j,i) \notin A(D)$.

In this case $j \xrightarrow{[n]} i = j \xrightarrow{[n-1]} i, j \xrightarrow{n} k, i \xrightarrow{[n-1]} j, k \xrightarrow{n} i, j \xrightarrow{[n-1]} i$. Therefore,

$$
\begin{aligned}
|\mathcal{X}| &= \left|i \xrightarrow{[n]} j\right| + 1 + \left|j \xrightarrow{[n]} i\right| + 1 + \left|i \xrightarrow{[n]} j\right| \\
&= 2 \cdot \left(\left|i \xrightarrow{[n-1]} k\right| + \left|k \xrightarrow{[n-1]} j\right| + \left|j \xrightarrow{[n-1]} i\right|\right) + \left|i \xrightarrow{[n-1]} j\right| + 6 \\
&= \left(\left|i \xrightarrow{[n-1]} k\right| + \left|k \xrightarrow{[n-1]} j\right|\right) + \left(\left|j \xrightarrow{[n-1]} i\right| + \left|i \xrightarrow{[n-1]} k\right|\right) + \left(\left|k \xrightarrow{[n-1]} j\right| + \left|j \xrightarrow{[n-1]} i\right|\right) \\
&\quad + \left|i \xrightarrow{[n-1]} j\right| + 6 \\
&\geq 2 \cdot \left|i \xrightarrow{[n-1]} j\right| + \left|j \xrightarrow{[n-1]} k\right| + \left|k \xrightarrow{[n-1]} i\right| + 6 \\
&> |\mathcal{A}| \, .
\end{aligned}
$$

Hence Algorithm X cannot be optimal. So we have shown that disc $n + 1$ moves directly from peg i to peg j. Again by induction Algorithm X makes the same moves as Algorithm A.   □

Allouche and Sapir [10] used Algorithm 23 to construct optimal sequences of moves for the five solvable TH variants with the so-called *infinite morphic sequences*. Earlier [8, 9] this had been done for the classical and the Cyclic TH. Instead of giving formal definitions (the interested reader can find them in [10]) and details for all of them, let us present here a particularly appealing sequence for the linear puzzle $TH(\overleftrightarrow{L}_3)$. Let a, $\bar{a}$, b, and $\bar{b}$, denote the move of a topmost disc from peg 1 to peg 0, from peg 0 to 1, from 0 to 2, and from 2 to 0, respectively. To obtain the required infinite sequence, start with the term a and subsequently apply the following substitution rules:

$$
a \to a\, b\, a, \quad \bar{a} \to a\, b\, \bar{a}, \quad b \to \bar{b}\, \bar{a}\, b, \quad \bar{b} \to \bar{b}\, \bar{a}\, \bar{b}.
$$

The first three steps of this procedure are:

$$
\begin{aligned}
a \;\longrightarrow\; & a\, b\, a \\
\longrightarrow\; & a\, b\, a\, \bar{b}\, \bar{a}\, b\, a\, b\, a \\
\longrightarrow\; & a\, b\, a\, \bar{b}\, \bar{a}\, b\, a\, b\, a\, \bar{b}\, \bar{a}\, \bar{b}\, a\, b\, a\, \bar{b}\, \bar{a}\, b\, a\, b\, a\, \bar{b}\, \bar{a}\, b\, a\, b\, a.
\end{aligned}
$$

The reader is invited to verify that the first $3^3 - 1 = 26$ terms from the last line (that is, all terms but the last one) give the solution for the Linear TH with three discs, source peg 1 and goal peg 2.

The infinite sequence just constructed is 3-automatic; the classical TH admits a 2-automatic sequence; the Cyclic TH sequence is not k-automatic for any k.[88]

---

[88] A precise definition of this term is out of the scope of the present book. It would be something like "A sequence is k-automatic if it is the image, under a coding, of an iterative fixed point of a k-uniform morphism."; see [217, Definition 2.10].

It was conjectured in [10] and taken up as a challenge in the first edition of the present book (Chapter 9, Allouche-Sapir Conjecture) that the latter holds for the remaining two sequences as well. Meanwhile, this has been confirmed by C. Holz auf der Heide for $\overrightarrow{C}_{3+}$ in [217, Theorem 2.27] by considering the *frequency of letters* in the sequence. Although this method seems not to apply for the remaining open question for $\overleftrightarrow{K}_{3-}$, Holz auf der Heide was able, based on a different promising procedure, to reduce it to a computational problem in [217, Section 2.3].

We next look at the number of moves performed by Algorithm 23. For the classical TH we know it from Theorem 2.1, while the Linear TH has been covered in Section 2.3.1. Consider next the Cyclic TH. By symmetry we infer that

$$|0 \xrightarrow{[n]} 1| = |1 \xrightarrow{[n]} 2| = |2 \xrightarrow{[n]} 0|$$

and

$$|1 \xrightarrow{[n]} 0| = |2 \xrightarrow{[n]} 1| = |0 \xrightarrow{[n]} 2|.$$

Setting $a_n = |0 \xrightarrow{[n]} 1|$ and $b_n = |1 \xrightarrow{[n]} 0|$, Algorithm 23 gives us $a_0 = 0 = b_0$ and

$$a_{n+1} = 2b_n + 1,$$
$$b_{n+1} = a_n + 2b_n + 2.$$

Inserting the first recurrence relation into the second we get $b_{n+2} = 2b_{n+1} + 2b_n + 3$, which, together with $b_0 = 0$ and $b_1 = 2$ results in

$$b_n = \frac{3 + 2\sqrt{3}}{6}\left(1 + \sqrt{3}\right)^n + \frac{3 - 2\sqrt{3}}{6}\left(1 - \sqrt{3}\right)^n - 1$$

and therefore

$$a_{n+1} = 2b_n + 1 = \frac{3 + 2\sqrt{3}}{3}\left(1 + \sqrt{3}\right)^n + \frac{3 - 2\sqrt{3}}{3}\left(1 - \sqrt{3}\right)^n - 1.$$

The values a_n and b_n (cf. A005665/6) can be rewritten as follows:

$$a_n = \frac{\sqrt{3}}{6}\left((1 + \sqrt{3})^{n+1} - (1 - \sqrt{3})^{n+1}\right) - 1,$$
$$b_n = \frac{\sqrt{3}}{12}\left((1 + \sqrt{3})^{n+2} - (1 - \sqrt{3})^{n+2}\right) - 1.$$

The remaining digraphs to be considered are the digraph $\overleftrightarrow{K}_{3-}$ obtained from the complete digraph $\overleftrightarrow{K}_3$ by removing one arc (the lower-left picture in Figure 8.1) and the digraph $\overrightarrow{C}_{3+}$ obtained from the cyclic digraph $\overrightarrow{C}_3$ by adding one arc (the lower-right picture in Figure 8.1). To determine the number of moves for TH($\overleftrightarrow{K}_{3-}$) and TH($\overrightarrow{C}_{3+}$) one proceeds similarly as we did above for TH($\overrightarrow{C}_3$). In Exercises 8.2

and 8.3 the reader is asked to develop a system of recurrence equations for these two variants. In all the cases, the growth of the minimum number of moves with respect to the number of discs n is asymptotically as $c\lambda^n$. Here λ depends only on the given variant as represented by D, whereas c depends on D and possibly on the given initial and goal pegs. Table 8.1 collects the values of λ.

D	$\overleftrightarrow{K}_3$	$\overleftrightarrow{K}_{3-}$	$\overrightarrow{C}_{3+}$	$\overrightarrow{C}_3$	$\overleftrightarrow{L}_3$
λ	2	≈ 2.343	$\frac{1+\sqrt{17}}{2} \approx 2.562$	$1+\sqrt{3} \approx 2.732$	3

Table 8.1: Growth parameters for the number of moves in $\mathrm{TH}(D)$

D. Berend and Sapir [48] proved another result that holds for all TH variants on three pegs:

Theorem 8.9. *For $n \in \mathbb{N}$ and any strongly connected digraph D on three vertices, the diameter of the state graph H_D^n is realized by the distance between some two perfect states.*

The above theorem thus asserts that the maximum number of moves of an optimal path among all type P2 problems is realized by a P0 problem. Of course, in some cases, like the linear one, the two perfect states need to be selected appropriately. In the linear case $D = \overleftrightarrow{L}_3$ (see the central drawing in the top row of Figure 8.1) an extremal task is to move a perfect tower from peg 1 to peg 2.

In [48] Berend and Sapir proved that the conclusion of Theorem 8.9 holds for more than three pegs as well in the case of the Cyclic TH. Recall from Section 5.3 that this is not so for the TH based on complete digraphs. In Section 8.3 we will say more about TH variants on more than three pegs. We conclude the current section with several remarks about $\mathrm{TH}(\overrightarrow{C}_3)$.

$\mathrm{TH}(\overrightarrow{C}_3)$ was independently proposed in 1979 by Hering [191] and in 1981 by M. D. Atkinson [22]. They both presented algorithms to solve the tasks $0^n \to 1^n$ and $1^n \to 0^n$ and also computed the number of moves a_n and b_n, respectively. Hering's note includes a minimality proof, assuming, as usual, the largest disc to move only once. Iterative solutions of $\mathrm{TH}(\overrightarrow{C}_3)$ were given by T. R. Walsh [429], Er [130], and T. D. Gedeon [158]. In [134], Er defines a *pseudo ternary code* for every move number in the optimal path from which the disc used in that move can be deduced similarly to the Gros sequence for the classical unrestricted TH. On the other hand, Allouche [5] proved that the infinite sequence obtained from the moves of the solution as n goes to infinity is not k-automatic for any $k \in \mathbb{N}_2$; in other words, no finite automaton can solve the problem based on move number. Er [131] also considered a "semi-cyclic" variant in which some discs are allowed to move only clockwise and the other discs only counterclockwise.

Stockmeyer [402] considered the state digraph of $\text{TH}(\overrightarrow{C}_3)$ and proved several interesting properties of it, Exercise 8.4 gives a basic one. A more advanced property is the fact that for any pair of different vertices there is a unique shortest directed path between them. In other words, an arbitrary type P2 problem (regular to regular) has a unique solution. Based on these results Stockmeyer computed the average distance between the states of the $\text{TH}(\overrightarrow{C}_3)$. Earlier Er [132] obtained the average distance for the P1-type problem (regular to perfect). Their results are collected in the next theorem.

Theorem 8.10. *Let* $n \in \mathbb{N}_0$. *Then the average distance between a regular state and a fixed perfect state of the* $\text{TH}(\overrightarrow{C}_3)$ *is*

$$\frac{5 + 3\sqrt{3}}{18}(1 + \sqrt{3})^n - \frac{5}{9} + \frac{5 - 3\sqrt{3}}{18}(1 - \sqrt{3})^n,$$

while the average distance between regular states is

$$\frac{77 + 57\sqrt{3}}{414}(1 + \sqrt{3})^n - \frac{1}{9} + \frac{77 - 57\sqrt{3}}{414}(1 - \sqrt{3})^n - \frac{6}{23}\left(\frac{1}{3}\right)^n.$$

The Cyclic Tower of Antwerpen

Recall from Section 6.3 that the Tower of Antwerpen consists of three pegs, a set of n black discs with pairwise different diameters, a set of n yellow discs, and a set of n red discs, and that these three sets are identical except for their colors. Minsker [317] investigated the *Cyclic Tower of Antwerpen*, that is, the Tower of Antwerpen in which disc moves are specified by the digraph $\overrightarrow{C}_3$. Up to the symmetries of the puzzle, there are three type P0 problems:

- *Transpose Cyclic Tower of Antwerpen*: interchange two towers, say the yellow tower and the red tower, and end up with the third (black) tower on its original peg.
- *Shift Cyclic Tower of Antwerpen*: transfer each of the three towers along one arc of $\overrightarrow{C}_3$.
- *Double Shift Cyclic Tower of Antwerpen*: transfer each of the three towers along two arcs of $\overrightarrow{C}_3$.

For the first of these three problems Minsker designed an algorithm of length

$$\sqrt{3}\left((1 + \sqrt{3})^{n+1} - (1 - \sqrt{3})^{n+1}\right) - 6n - 3$$

and proved it to be optimal [317, Theorem 7]. Interestingly, the number of optimal solutions is growing exponentially w.r.t. n; see [317, Theorem 17].

It is not difficult to see that the Shift Cyclic Tower of Antwerpen and the Double Shift Cyclic Tower of Antwerpen are both solvable (Exercise 8.5). On the other hand, not even some "natural" algorithms for these two problems are known for which one could conjecture optimality; cf. [317, p. 52].

8.3   Undirected Move Graphs on More Than Three Pegs

From the previous section we know that there are exactly five solvable three-peg TH with oriented disc moves. Equivalently (cf. Theorem 8.7), there exist precisely five non-isomorphic strong digraphs on three vertices. There are 83 such digraphs on four vertices ([178, p. 218]; see also A035512 which can be extracted from the figures of all the 218 digraphs on four vertices given in [177, p. 227–230]. Hence we are faced with 83 different TH variants on four pegs. Therefore very good reasons should be given why one would be interested in a specific case on four or more pegs. By now we have encountered the following interesting examples:

- The TH with $p \in \mathbb{N}_4$ pegs. It corresponds to the complete digraph $\overset{\leftrightarrow}{K}_p$.

- The Cyclic TH with $p \in \mathbb{N}_4$ pegs. It corresponds to the digraph $\vec{C}_p$.

- Let $\overset{\leftrightarrow}{L}_p$, $p \in \mathbb{N}_3$, be the digraph that is obtained from the path on p vertices by replacing each of its edges by a pair of opposite arcs. (Recall that $\overset{\leftrightarrow}{L}_3$ is the digraph of the Three-in-a-row puzzle.) Let us call the puzzle corresponding to $\overset{\leftrightarrow}{L}_p$ the *Linear TH on p pegs* or *p-in-a-row* puzzle.

We have already treated in detail the classical TH with p pegs. Before considering the Cyclic TH and the Linear TH with p pegs, we introduce another version which is due to Stockmeyer [401]. As we will see, this puzzle interlaces with The Reve's puzzle and the Linear TH, leads to some interesting mathematics, and offers some challenging conjectures, so we are well justified to treat it in detail. It is one of only six undirected connected graphs of order 4, so we will now concentrate on these.

Assume that the undirected graph D is connected. Then TH(D) is solvable by Proposition 8.4, i.e. the classical task to transfer a perfect tower of discs from one peg to another has always a (shortest) solution, no matter which pegs are considered and how many discs are involved.

For $p = 3$ there are only two such rules, namely those where the state graph H_D^1 is either a complete graph $K_3 = C_3$ or a path graph $P_{1+2} = K_{1,2}$. The six undirected move graphs on 4 pegs are: the complete graph K_4, the *diamond* $K_4 - e$, the cycle C_4, the *paw* $K_{1,3} + e$, the *claw* (or *star*) $K_{1,3}$, and the path P_{1+3}; see Figure 8.2. Among these, apart from the classical unrestricted one associated with *The Reve's Puzzle*, the Star Tower of Hanoi plays a special role. We are therefore motivated to look deeper into the details.

Let $p \in \mathbb{N}_3$ and $n \in \mathbb{N}_0$. Then we define the *Star Tower of Hanoi* with n discs $d \in [n]$ on p pegs labelled as $i \in [p]_0$ with the same rules as the corresponding Tower of Hanoi, but with the restriction that disc moves are only allowed between the *central* peg 0 and a *peripheral* peg $i \in [p-1]$. In the diction of Section 8.1 the disc moves are restricted on the (undirected) star graph with p vertices, i.e. $D \simeq K_{1,p-1}$. Special cases are the Linear TH for $p = 3$ and the variant for $p = 4$ introduced by

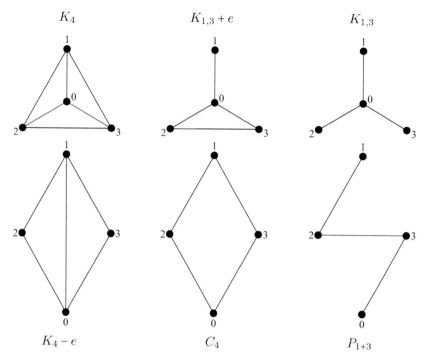

Figure 8.2: Undirected connected graphs of order 4 ($e = \{2,3\}$)

Stockmeyer in 1994 (see [401, Section 4]), which we will call, following [63], *Stockmeyer's Tower*. While the former case has been solved completely in Section 2.3.1, Stockmeyer in the latter case derived an upper bound for the number of moves needed to transfer a perfect tower from one peripheral peg to another one and conjectured that this bound is in fact sharp. *Stockmeyer's conjecture* was one of the challenges in Chapter 9 of the first edition and has only recently been confirmed in the article [63] by Bousch. We will outline his approach below, in Section 8.3.3.

On a more formal base we will consider the state graphs $St_p^n := H_{K_{1,p-1}}^n$ corresponding to the Star Tower of Hanoi and call them *Star Hanoi graphs*. They are given by (cf. Definition 8.3)

Definition 8.11. *For $p \in \mathbb{N}_3$ and $n \in \mathbb{N}_0$,*

$$V(St_p^n) = [p]_0^n ,$$
$$E(St_p^n) = \left\{ \{\underline{s}0\overline{s}, \underline{s}i\overline{s}\} \mid i \in [p-1], \overline{s} \in ([p-1] \smallsetminus \{i\})^{d-1}, d \in [n] \right\} . \tag{8.4}$$

Note that in (8.4) we have $\underline{s} \in [p]_0^{n-d}$; therefore, it immediately follows that $|St_p^n| = p^n (= |H_p^n|)$ and $\|St_p^n\| = \frac{p-1}{2} (p^n - (p-2)^n) (= \frac{2}{p} \|H_p^n\|)$. In particular, $\|St_3^n\| = 3^n - 1$ and $\|St_4^n\| = 3 \cdot 2^{n-1}(2^n - 1)$. Degree properties are a little more involved. Of course, $\Delta(St_p^0) = 0$.

Proposition 8.12. *Let $p \in \mathbb{N}_3$, $n \in \mathbb{N}$, and $s \in [p]_0^n$. Then*

1. $\deg(s) = 1 \iff s = i^n$, $i \in [p-1]$.

2. *If $s \notin \{i^n \mid i \in [p-1]\}$, then $2 \leq \deg(s) \leq p-1$.*

3. $\deg(0^n) = p-1$.

In particular, $\Delta(St_p^n) = p-1$.

Proof. Vertices i^n, $i \in [p-1]$, corresponding to perfect states on peripherial pegs, have degree 1 because only disc 1 can move (from peg i to peg 0). If disc 1 lies on peg 0, i.e. $s_1 = 0$, it can move to any peg i, $i \in [p-1]$; here no other disc can move, whence $\deg(s) = p-1$ in that case. In particular this proves *3*. If disc 1 lies on a peripheral peg i, i.e. $s_1 = i \in [p-1]$, then it can move to peg 0 (only) and, provided that $s \neq i^n$, the smallest disc d not on peg i can move as well—to any peg different from i, if it lies on peg 0, and to peg 0 otherwise. This proves *1* and the lower bound in *2*. For the upper bound we remark that if the top disc on peg 0, if any, can move to some peg in $[p-1] \setminus \{i\}$, then the top disc on that peg, if any, cannot move.   □

Remark. The lower bound in *2* of Proposition 8.12 is sharp, because $\deg(s) = 2$, e.g., for $s = 21^{n-1}$ with the only neighbors 01^{n-1} and $21^{n-2}0$, if $n \in \mathbb{N}_2$.

There is an equivalent recursive definition of the edge sets of St_p^n (cf. Definition 8.1).

Definition 8.13. *For $p \in \mathbb{N}_3$ and $n \in \mathbb{N}_0$,*

$$E(St_p^0) = \varnothing,$$
$$E(St_p^{1+n}) = \big\{\{is, it\} \mid i \in [p]_0, \; \{s,t\} \in E(St_p^n)\big\}$$
$$\cup \big\{\{0s, is\} \mid i \in [p-1], \; s \in ([p-1] \setminus \{i\})^n\big\}. \qquad (8.5)$$

The elements from the first set in (8.5) form the edge sets of the subgraphs iSt_p^n of St_p^{1+n} induced by vertices is, $s \in [p]_0^n$. The edges from the last set in (8.5) are called *critical*. There are $(p-1)(p-2)^n$ of them, i.e. 2 for $p = 3$ and $3 \cdot 2^n$ for $p = 4$, and they correspond to moves of the largest disc allowing to change in St_p^{1+n} from subgraph $0St_p^n$ to subgraph iSt_p^n or vice versa. This shows that the Star Hanoi graphs are connected and therefore equipped with the canonical distance function d_p. (If no doubt about p is possible, we will simply write d; the dependence on n will be clear from the length of the arguments.)

The case $p = 3$ is now completely covered: since St_3^n, $n \in \mathbb{N}$, is a connected graph with the two pendant vertices 1^n and 2^n and all other vertices having degree 2, it must be a path graph of length $3^n - 1$. The only two critical edges are $\{12^{n-1}, 02^{n-1}\}$ and $\{01^{n-1}, 21^{n-1}\}$. See Figure 8.3 for drawings of St_3^1, St_3^2, and St_3^3, where the horizontal edges are the critical ones, respectively.

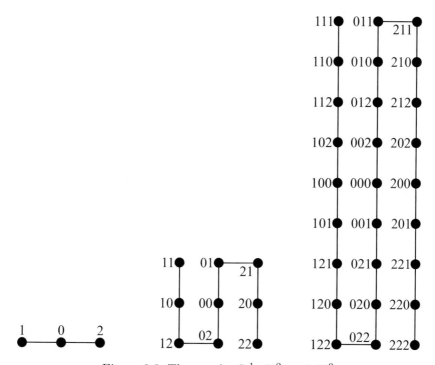

Figure 8.3: The graphs St_3^1, St_3^2, and St_3^3

It is clear that St_3^n is the same as the state graph of the Linear TH with end vertices 1^n and 2^n. Examples of drawings of graphs St_4^n can be found in Figure 8.4. In particular, the 12 critical edges in St_4^3 are represented in yellow; they are $\{022, 122\}$, $\{033, 133\}$, $\{023, 123\}$, $\{032, 132\}$, $\{011, 211\}$, $\{033, 233\}$, $\{013, 213\}$, $\{031, 231\}$, $\{011, 311\}$, $\{022, 322\}$, $\{012, 312\}$, and $\{021, 321\}$.

Let us mention in passing that the alternative drawing of St_4^3 in Figure 8.5 shows that $\mathrm{cr}(St_4^3) \le 3$; as the sets of vertices $\{012, 031, 023\}$ and $\{013, 032, 021\}$ form a $K_{3,3}$ subdivision (see [149, p. 51]), we know that $1 \le g(St_4^3) \le \mathrm{cr}(St_4^3) \le 3$. More on planarity of Star Hanoi graphs can be found in Exercise 8.6.

The obvious fact in St_3^n that on any geodesic in the graph there are at most two critical edges is shared by all St_p^n; this is a consequence of the *boxer rule*:

Lemma 8.14. *The largest disc that is moved on a (directed) shortest path between two states in St_p^n will never come back to a peg it had left before.*
In particular, this disc moves at most twice, namely from peg i to peg 0 only, or from peg 0 to peg j only, or from peg i through peg 0 to peg j with $\{i, j\} \in \binom{[p-1]}{2}$.

Proof. If the largest moved disc returns to a peg it had left before, one may leave out all its moves in between because it does not restrict the moves of the other discs. This results in a strictly shorter path.

If the largest moved disc starts on a peripheral peg, then its first move is to the central peg 0. Once moved away from there it cannot move anymore because the next move would be back to peg 0.   □

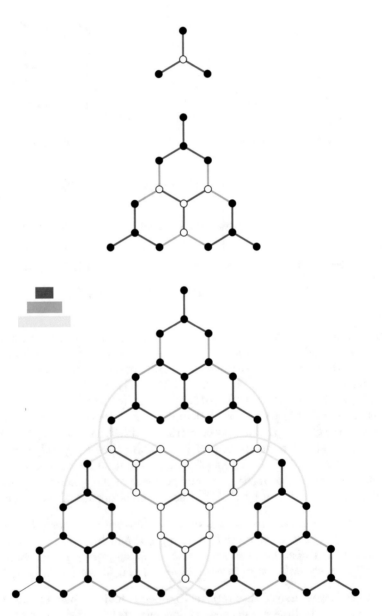

Figure 8.4: The graphs St_4^n for $n \in [3]$

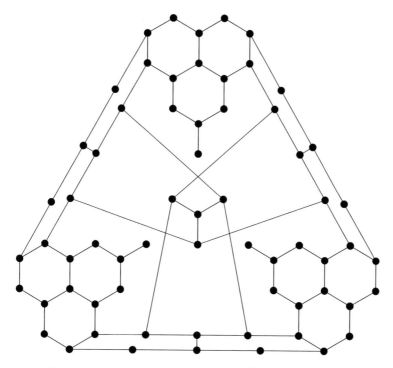

Figure 8.5: Drawing of the graph St_4^3 with 3 crossings

The moves of the largest disc being fixed, we have the opportunity to reduce a typical task for $n + 1$ discs to tasks with only n discs as a motivation for an induction proof; cf. [63, Lemme 2.1].

Proposition 8.15. *Let* $\{i, j\} \in \binom{[p-1]}{2}$, $p \in \mathbb{N}_3$, *and* $s, t \in [p]_0^n$, $n \in \mathbb{N}_0$. *Then there are* $s' \in ([p-1] \setminus \{i\})^n$ *and* $t' \in ([p-1] \setminus \{j\})^n$ *such that*

$$d(is, jt) = 2 + d(s, s') + d(s', t') + d(t', t) =: 2 + d(s, s', t', t).$$

Proof. Since $i \neq j$, the largest disc $n+1$ has to move, on a shortest is, jt-path, from i through 0 to j (only). Before the first move a state is' must have been reached where no disc in s' lies on either peg 0 or peg i. Similarly, the second move of disc $n + 1$ is only possible from a state $0t'$ with all discs in t' outside pegs 0 and j. Clearly, all transits of the smaller n discs from s to s', from s' to t' and from t' to t have to be on shortest paths in St_p^n in order to obtain a shortest is, jt-path in St_p^{1+n}. □

The classical task for all TH variants is to transfer a perfect tower from one peg to another, i.e. to get from i^n to j^n on a shortest path in the state graph and to determine its length and possibly to provide an algorithm for the optimal solution(s). For the Star TH there are two types of these tasks, depending on whether one of the towers is on the center peg at the beginning (or at the end) or both are peripheral. We will call this *task* 0 and *task* 1 and denote the lengths of shortest paths by $d_p^0(n) := d_p(0^n, 1^n)$ and $d_p^1(n) := d_p(1^n, 2^n)$, respectively.

For the prototype case $p = 3$, it is clear from the above that

$$d_3^1(n) = 3^n - 1 \text{ and } d_3^0(n) = \frac{1}{2}d_3^1(n) = \frac{1}{2}(3^n - 1) = \Sigma T_n$$

with $\Sigma T = 0, 1, 4, 13, 40, 121, 364, 1093, \ldots$ (A003462), where the sequence $T = 0, 1, 3, 9, 27, 81, 243, 729, \ldots$ is defined by $T_n = (n \in \mathbb{N}) \cdot 3^{n-1}$ (A140429). It will turn out that the question is more intricate for $p \geq 4$, which is not surprising because already the case $p = 4$ produced a lot of trouble for the classical rules ("The Reve's Puzzle"). In the latter case, the strategy of Stewart (1941) turned out to be successful and has, at last, been proved to be optimal. So it was quite natural for Stockmeyer, anticipating this optimality, to apply the same strategy for task 1 of the Star TH with four pegs in [401].

We present this *Stewart-Stockmeyer strategy* here for the general case $p \geq 4$: divide the n-tower on peg 1 into $m \in [n]_0$ smaller discs and $n - m$ larger ones; move the m smaller discs to peg $p - 1$, then the $n - m$ larger ones to the goal peg 2 avoiding peg $p - 1$, and finally the smaller discs to the goal as well. The parameter m is then chosen in such a way that the resulting number of moves is minimal (with respect to this strategy). This number will be denoted by $st_p^1(n)$ (for either Star, Stewart or Stockmeyer). The algorithm applied to n discs amounts to using the same algorithm for m discs twice and the (presumed) optimal solution for task 1 in St_{p-1}^{n-m} once. It should be mentioned that there might be other solutions using the same number of moves. E.g., for the task to get from 111 to 222 in St_4^3 one may, after the forced first move to 110, either pass the vertex 113, as in the algorithm just described, or go through vertex 112 instead to reach 222 in $st_4^1(3) = 12$ moves on the (only) alternative optimal path (see the paths between two pendant vertices in Figure 8.5; they are 111, 110, 113, 103, 123, 023, 003, 013, 213, 203, 223, 220, 222 and 111, 110, 112, 102, 132, 032, 030, 031, 231, 201, 221, 220, 222, respectively). The situation may be compared with the non-standard solutions for the classical problem in H_p^n; cf. above, Theorem 5.64 and Conjectures 5.65 and 5.66.

Similarly one may proceed for task 0 (which has not been considered in [401, 63] and neither in the first edition of this book). Here we transfer m discs from peg 0 to peg $p - 1$, then $n - m$ discs to the goal peg 1 avoiding peg $p - 1$ and finally the m smaller discs to the goal, but the algorithm is less symmetric, because it uses itself only in the first step, then the algorithm for task 0 in St_{p-1}^{n-m} in the second step, and finally the task 1 algorithm for St_p^m. Denoting the number of moves made by this algorithm by $st_p^0(n)$, we get for $p \in \mathbb{N}_4$ and $\iota \in \{0, 1\}$:

$$\forall n \in \mathbb{N}: \quad st_p^\iota(n) = \min \left\{ st_p^\iota(m) + st_p^1(m) + st_{p-1}^\iota(n - m) \mid m \in [n]_0 \right\}, \quad (8.6)$$

with $st_p^\iota(0) = 0$ and $\forall n \in \mathbb{N}_0: \ st_3^\iota(n) = 2^\iota \Sigma T_n$. Note that an immediate consequence is $st^0_p \le st^1_p$; see Exercise 8.7. According to [76, Theorem 1][89] (put $p_3 = 3$, $q_3 = 2$ and $p_i = 2$, $q_i = 1$ otherwise), the solution for $\iota = 1$ is $\frac{1}{2}st^1_p(n) = \Sigma a_p(n)$, where a_p is the sequence starting with $a_p(0) = 0$ and listing in non-decreasing order all numbers of the form $3^{\alpha_0} \cdot 2^{\Sigma^{p-3}_{i=1}\alpha_i}$ with $\alpha \in \mathbb{N}^{1+p-3}_0$.

Recall from Definition 5.26 that the Frame-Stewart numbers for the presumed optimal solution of the classical TH with $p \in \mathbb{N}_4$ pegs and $n \in \mathbb{N}_0$ discs (which we write as $FS_p(n)$ for comparison now) are given by the recurrence

$$\forall n \in \mathbb{N}: \ FS_p(n) = \min\{2FS_p(m) + FS_{p-1}(n-m) \mid m \in [n]_0\}\ , \qquad (8.7)$$

with $FS_p(0) = 0$ and $\forall n \in \mathbb{N}_0: \ FS_3(n) = M_n = \Sigma B_n$, the Mersenne numbers, where $B_n = \overline{M}_n = (n \in \mathbb{N}) \cdot 2^{n-1} = 0, 1, 2, 4, 8, 16, 32, 64, \ldots$ (A131577). Making use of (5.22) and Theorem 5.27, the recurrence relation in (8.7) can also be written as

$$\forall n \in \mathbb{N}: \ FS_p(n) = \min\{2FS_p(m) + \Sigma F_p(n-m) \mid m \in [n]_0\}\ , \qquad (8.8)$$

where $F_p(n) = (n \in \mathbb{N}) \cdot 2^{\nabla_{p-3}(n-1)}$. Again by [76, Theorem 1] (this time put $p_i = 2$, $q_i = 1$ throughout) this sequence can also be obtained from the non-decreasing ordering of the numbers of the form $2^{\Sigma^{p-3}_{i=0}\alpha_i}$ with $\alpha \in \mathbb{N}^{1+p-3}_0$. Moreover, we can analyse the number of moves made asymptotically by individual discs. For the classical TH cf. Corollary 5.29. Here, by Exercise 7.3 there are $\binom{p-3+k}{p-3} = \binom{p-3+k}{k}$ ways to express $k \in \mathbb{N}_0$ as a sum $\Sigma^{p-3}_{i=0}\alpha_i$. They will all occur, if n is large enough, namely exceeds the sum $\Sigma^k_{\kappa=0}\binom{p-3+\kappa}{p-3} = \Sigma^k_{\kappa=0}\Delta_{p-3,\kappa+1} = \Sigma(\Delta_{p-3})_{k+1} = \Delta_{p-2,k+1}$ (see the line following (5.9)).

Proposition 8.16. *Let $p \in \mathbb{N}_3$ and $k \in \mathbb{N}_0$. Then exactly $\binom{p-3+k}{p-3} = \binom{p-3+k}{k}$ disc(s) move(s) exactly 2^k times in the Frame-Stewart algorithm for H^n_p, if $n \ge \Delta_{p-2,k+1}$.*

A similar result should be true for the Star Tower of Hanoi. We conjecture the following:[90]

Conjecture 8.17. *Let $p \in \mathbb{N}_3$ and $k, \ell \in \mathbb{N}_0$. Then exactly $\binom{p-4+k}{p-4} = \binom{p-4+k}{k}$ disc(s) move(s) exactly $2^{1+k} \cdot 3^\ell$ times in the Stewart-Stockmeyer algorithm for task 1, if n is large enough. (Here we use the convention $\binom{k-1}{-1} = (k = 0) = \binom{k-1}{k}$.)*

We will come back to this conjecture after we will have understood Stockmeyer's Tower.

The *(generalized) Stockmeyer conjecture* is: $d^\iota_p = st^\iota_p$.

8.3.1   Stockmeyer's Tower

Stockmeyer's Tower is the case $p = 4$ of the Star TH for the task $\iota = 1$ (see Figure 8.6 for the underlying (di)graph $\overleftrightarrow{K}_{1+3} \simeq K_{1+3}$).

---

[89]The numbers occuring in that theorem were called *generalized Frame-Stewart numbers*. In that article; they form a wide generalization of Frame-Stewart numbers.
[90]P. K. Stockmeyer, private communication, 2014.

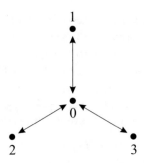

Figure 8.6: Move graph of Stockmeyer's Tower

The Stewart-Stockmeyer algorithm described above for the general Star puzzle is shown for Stockmeyer's Tower in Figure 8.7.

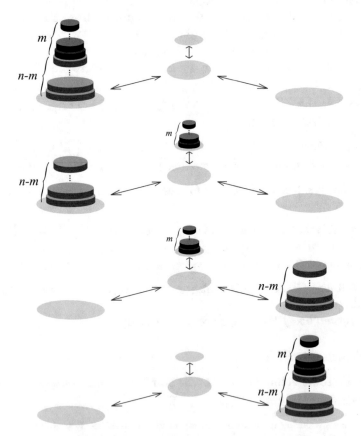

Figure 8.7: Stewart-Stockmeyer strategy for $TH(K_{1,3})$

The recurrence (8.6) reduces to

$$st_4^1 = 0, \ \forall n \in \mathbb{N}: \ st_4^1(n) = \min\left\{2st_4^1(m) + st_3^1(n - m) \mid m \in [n]_0\right\}.$$

If we define the *Stockmeyer sequence* Σ by $\Sigma_n = \frac{1}{2}st_4^1(n)$ we get

$$\Sigma_0 = 0, \ \forall n \in \mathbb{N}: \ \Sigma_n = \min\left\{2\Sigma_m + \Sigma T_{n-m} \mid m \in [n]_0\right\}. \tag{8.9}$$

The sequence starts $0, 1, 3, 6, 10, 16, 24, 33, \ldots$ and Stockmeyer's original conjecture was that for all $n \in \mathbb{N}_0$: $d_4^1(n) = 2\Sigma_n$. In [401] it was shown that the sequence of differences $A := \overline{\Sigma}$ is the sequence $A = 0, 1, 2, 3, 4, 6, 8, 9, \ldots$, starting with $A_0 = 0$, followed by the list of 3-smooth numbers[91] arranged in strictly increasing order (A003586), and that the minimum in (8.9) is achieved for[92] $m_n = n - 1 - \lfloor \mathrm{lt}(A_n) \rfloor$ only. This may be compared with the classical Tower of Hanoi, where $\overline{FS_3}(n) = B_n$ which form the sequence starting with 0, followed by the 2-smooth numbers.

Theorem 8.18. *The value $m_n = n - 1 - \lfloor \mathrm{lt}(A_n) \rfloor$ is the unique m that defines Σ_n in (8.9). Moreover, for every $i \in [n]$, there is a disc that makes exactly $2A_i$ moves in the corresponding algorithm and hence*

$$st_4^1(n) = 2\Sigma_n = 2\sum_{i=1}^{n} A_i.$$

Proof. We perform (strong) induction on $n \in \mathbb{N}$. For $n = 1$ only $m = 0 = n - 1 - \lfloor \mathrm{lt}(A_n) \rfloor$ is admissible and the only disc makes $2 = 2A_n$ moves.

For $n \geq 2$ let $\ell := n - m (\in [n])$. Discs $n - \mu$, $\mu \in [\ell]_0$, are transferred according to the Linear TH, hence by Exercise 2.21 they move $2 \cdot 3^\mu = 2T_{\mu+1}$ times, respectively. On the other hand, by induction assumption the smallest m discs make $4A_{\nu+1}$ moves, where ν runs through $[m]_0$.

Note that the sequences given by $a_\mu = T_{\mu+1}$ and $b_\nu = 2A_{\nu+1}$ are disjoint because a_μ is not even. Moreover, they partition the sequence $A \upharpoonright \mathbb{N}$. Indeed, let $A_i = 2^j 3^k$, $j, k \in \mathbb{N}_0$. If $j = 0$ then $A_i = 3^k$ is a term of a and if $j \geq 1$ then $A_i = 2 \cdot 2^{j-1} 3^k$ is a term of b. The sum of the first ℓ terms of a and the first m terms of b will be minimized if their union consists of the terms of $A \upharpoonright [n]$. This will happen if and only if for the terms of a we have $a_{\ell-1} \leq A_n < a_\ell$. For the minimum to be attained we thus get $3^{\ell-1} \leq A_n < 3^\ell$, i.e. $(\ell - 1)\ln(3) \leq \ln(A_n) < \ell \ln(3)$, which in turn implies that $\ell = 1 + \lfloor \mathrm{lt}(A_n) \rfloor$.

We have thus proved that $m_n = n - 1 - \lfloor \mathrm{lt}(A_n) \rfloor$ is the unique value that minimizes the Stewart-Stockmeyer strategy for $\mathrm{TH}(K_{1,3})$. Moreover, the argument also shows that discs move $2A_i$, $i \in [n]$, times, respectively, hence $st_4^1(n) = 2\sum_{i=1}^{n} A_i$. $\square$

Combining Theorem 8.18 with Theorem 5.5 we infer:

Corollary 8.19. *For any $n \in \mathbb{N}_0$, $st_4^1(n) = T(n, 3, 2)$.*

---

[91] A *b-smooth number* is a positive integer with no prime factor strictly larger than $b \in \mathbb{N}_2$.
[92] The *ternary logarithm* is given by $\mathrm{lt}(x) = \ln(x)/\ln(3)$.

With Theorem 8.18 in hand we can state Algorithm 24, where $\mathrm{Linear}(\ell;i,j)$ denotes the call of the procedure in which ℓ discs are moved from peg i via peg 0 to peg j, $i,j \in [3]$, $i \neq j$; recall that 0 is the central peg of the Star TH.

Algorithm 24 A solution for Stockmeyer's Tower

Procedure $\mathrm{ST}(n;i,j)$
Parameter n: number of discs $\{n \in \mathbb{N}_0\}$
Parameter i: source peg $\{i \in [3]\}$
Parameter j: goal peg $\{j \in [3] \smallsetminus \{i\}\}$
   if $n \neq 0$ **then**
       $\ell \leftarrow 1 + \lfloor \mathrm{lt}(\mathrm{A}_n) \rfloor$, $m \leftarrow n - \ell$
       $\mathrm{ST}(m;i, 6-i-j)$ $\{6-i-j$ is the auxiliary peg different from 0, i and $j\}$
       $\mathrm{Linear}(\ell;i,j)$
       $\mathrm{ST}(m; 6-i-j, j)$
   end if

We point out that Theorem 8.18 asserts only that Algorithm 24 makes the minimum number of moves among the algorithms that follow the Stewart-Stockmeyer strategy. This does not mean that Algorithm 24 makes the smallest number of moves among all procedures that solve Stockmeyer's Tower. Stockmeyer said in [401]: "We strongly suspect, though, that this algorithm is indeed optimal."

Before we can come to Bousch's confirmation of this conjecture and in order to be able to realize Algorithm 24, we have to learn more about 3-smooth numbers, i.e. the sequence A.

Let us also mention that the center-to-periphery case of Stockmeyer's Tower has been treated by him computationally for small n. It turned out[93] that

$$\mathrm{d}_4^0 \upharpoonright [16]_0 = 0, 1, 4, 7, 14, 23, 32, 47, 68, 93, 120, 153, 198, 255, 318, 399 = st_4^0 \upharpoonright [16]_0 \,.$$

Moreover, in all these cases, the eccentricity of 0^n in St_4^n is realized by the distance to i^n, $i \in [3]$, only. There is no approach known to attack recurrences like (8.6) for $\iota = 0$ analytically though. A list of calculated values of $st_4^0(n)$ for $n \in [56]_0$ can be found in [149, p. 46]. Recent computations of Borut Lužar[94] confirmed $\mathrm{d}_4^0(n) = st_4^0(n)$ for $n = 16, 17, 18$ as well. In the range of $n \in [19]_0$ the sequence of differences of $\mathrm{d}_4^0 = st_4^0$ is

$$0, 1, 3, 3, 7, 9, 9, 15, 21, 25, 27, 33, 45, 57, 63, 81, 81, 99, 121 \,.$$

8.3.2   3-Smooth Numbers

A first useful property of the sequence A is given in the following lemma (cf. [63, (2.2)]).

---

[93]P. K. Stockmeyer, private communication, 2014.
[94]B. Lužar, private communication, 2017.

Lemma 8.20. *For all $m \in \mathbb{N}_0$ and all $n \in \mathbb{N}$ we have $A_{m+n} \leq A_{m+1} \cdot A_n$; in particular*

$$A_n < A_{n+1} \leq 2A_n . \tag{8.10}$$

Proof. We proceed by induction on n. The case $n = 1$ is trivial, so let us assume that $A_{m+n} \leq A_{m+1} \cdot A_n$ for some $n \in \mathbb{N}$ (and all $m \in \mathbb{N}_0$). Then by monotonicity, $A_{m+n} < A_{m+1} \cdot A_{n+1}$. As the product of two 3-smooth numbers is again a 3-smooth number, the value on the right-hand side must be greater than or equal to the minimum of 3-smooth numbers strictly larger than A_{m+n}, i.e. A_{m+n+1}.

For the second inequality in (8.10) put $m = 1$. □

According to [305, Corollary 2.1] the sequence Σ also fulfills the recurrence

$$\widetilde{\Sigma}_0 = 0, \ \forall n \in \mathbb{N}: \ \widetilde{\Sigma}_n = \min\left\{3\widetilde{\Sigma}_m + \Sigma B_{n-m} \mid m \in [n]_0\right\} . \tag{8.11}$$

This is the recurrence given for A259823, where we expect the minimum to be assumed for $m_n = n - 1 - \lfloor \mathrm{lb}(A_n) \rfloor$ only. A. Matsuura used it in [305, Section 4] to obtain the number of moves made in a Stewart-type algorithm to solve a task for the *Diamond Tower of Hanoi*, where the pegs are living on the diamond graph $K_4 - e$, namely to get from a degree-3 vertex to a degree-2 vertex, i.e. from peg 1 to peg 2 in Figure 8.2. It is not obvious, however, why an optimal solution for this task should have precisely $\frac{1}{2}$ of the length of an optimal solution for Stockmeyer's Tower. Matsuura's algorithm is shown in Algorithm 25. Here Classical($\ell; i, j, k$) is the classical transfer of an ℓ-tower from peg i to peg j using the (only) auxiliary peg k and $\overline{\mathrm{DT}}$ is DT in reverse.

Algorithm 25 A solution for the Diamond TH

Procedure $\mathrm{DT}(n; i, j)$
Parameter n: number of discs $\{n \in \mathbb{N}_0\}$
Parameter i: source peg $\{i \in \{0, 1\}\}$
Parameter j: goal peg $\{j \in \{2, 3\}\}$
  if $n \neq 0$ then
    $\ell \leftarrow 1 + \lfloor \mathrm{lb}(A_n) \rfloor$, $m \leftarrow n - \ell$
    $\mathrm{DT}(m; i, h)$ $\{h \in \{2, 3\} \setminus \{j\}\}$
    Classical($\ell; i, j, k$) $\{k \in \{0, 1\} \setminus \{i\}\}$
    $\overline{\mathrm{DT}}(m; k, h)$
    $\mathrm{DT}(m; k, j)$
  end if

The reader is invited in Exercise 8.8 to prove that the following analog of Theorem 8.18 holds, namely

Theorem 8.21. *The value $m_n = n - 1 - \lfloor \mathrm{lb}(A_n) \rfloor$ is the unique m that defines $\widetilde{\Sigma}_n$ in (8.11). Moreover, for every $i \in [n]$, there is a disc that makes exactly A_i moves*

in Algorithm 25 and hence the total number of moves made by the algorithm is

$$\widetilde{\Sigma}_n = \sum_{i=1}^{n} A_i = \Sigma_n.$$

Note that although Theorem 8.21 shows that $d_{K_4-e}(1^n, 2^n) \leq \Sigma_n = \frac{1}{2}d_{K_{1,3}}(1^n, 2^n)$, Algorithm 25 does not in general lead to an optimal solution; see Exercise 8.9. (Optimality was neither claimed nor conjectured in [305].) The equivalence of (8.9) and (8.11) can also be deduced from [76, Theorem 1]. (Clearly, the sequence A is symmetric in 2 and 3.) In [76, Theorem 3] this is also applied to the case $\iota = 1$ of (8.6). So it seems that Matsuura has chosen his Diamond TH strategy in order to obtain the solvable recurrence (8.11).

Let us have a closer look at the sequence A. While it is easy to arrange the 2-smooth numbers according to increasing size to obtain the sequence B, this ordering is not so straightforward for 3-smooth numbers. The corresponding sequence A can be defined by the recurrence

$$A_0 = 0, \ \forall n \in \mathbb{N}_0 : A_{n+1} = \min\{2^k \cdot 3^\ell > A_n \mid k, \ell \in \mathbb{N}_0\}. \tag{8.12}$$

(This is properly defined if we put $\varphi(m) = \min\{2^k \cdot 3^\ell > m \mid k, \ell \in \mathbb{N}_0\}$ in the Fundamental Theorem of recursion. A prototype of such an arrangement is the sequence π of prime numbers with $\pi_0 = 2$ followed by the list of odd primes $\pi_1 = 3$, $\pi_2 = 5$, $\pi_3 = 7$ &c.) From this definition it immediately follows that $A_n < A_{n+1}$. Therefore

$$\forall (k, \ell) \in \mathbb{N}_0^2 \ \exists_1 n \in \mathbb{N}_0 : A_n < 2^k \cdot 3^\ell \leq A_{n+1}. \tag{8.13}$$

But then, by (8.12), we have that $A_{n+1} = 2^k \cdot 3^\ell$ in (8.13). This defines a mapping $n_0 \in \mathbb{N}_0^{\mathbb{N}_0^2}$ with $(k, \ell) \mapsto n_0(k, \ell)$, $A_{n_0(k,\ell)+1} = 2^k \cdot 3^\ell$. The mapping is injective by the uniqueness of prime decomposition and it is surjective, because all elements A_{n+1} of A are of the form $2^k \cdot 3^\ell$, such that $n = n_0(k, \ell)$.

So n_0 is bijective (and hence $\mathbb{N}_0^2$ equipotent with $\mathbb{N}_0$; compare with Cantor's First Diagonalization Method), but not easily accessible. The sequence

$$n_0(\cdot, 0) = 0, 1, 3, 5, 8, 12, 16, 21, 27, 33, 40, 47, \ldots$$

contains all indices n for which A_{n+1} is a pure power of 2 (A022331 - 1) and the sequence

$$n_0(0, \cdot) = 0, 2, 6, 11, 18, 26, 36, 48, \ldots$$

contains all indices n for which A_{n+1} is a pure power of 3 (A022330 - 1). Looking at Table 8.2, which contains the first 55 values of n_0, one is led to the following:

Proposition 8.22. $\forall k, \ell \in \mathbb{N}_0 : n_0(k, \ell) = k \cdot \ell + n_0(k, 0) + n_0(0, \ell).$

This is [318, Theorem 1] (cf. also [63, Lemme 3.3]). Since the arguments in [318] are somewhat heuristic, we present here a more formal proof.

ℓ / k	0	1	2	3	4	5	6	7
0	0	2	6	11	18	26	36	48
1	1	4	9	15	23	32	43	
2	3	7	13	20	29	39	51	
3	5	10	17	25	35	46		
4	8	14	22	31	42	54		
5	12	19	28	38				
6	16	24	34	45				
7	21	30	41	53				
8	27	37	49					
9	33	44						
10	40	52						
11	47							

Table 8.2: Values for $n_0(k,\ell) \in [55]_0$ (from [149])

Proof of Proposition 8.22. Since

$$
\begin{aligned}
n_0(k,\ell) &= |\{n \in \mathbb{N}_0 \mid n < n_0(k,\ell)\}| \\
&= |\{(\kappa,\lambda) \in \mathbb{N}_0^2 \mid n_0(\kappa,\lambda) < n_0(k,\ell)\}| \\
&= |\{(\kappa,\lambda) \in \mathbb{N}_0^2 \mid A_{n_0(\kappa,\lambda)+1} < A_{n_0(k,\ell)+1}\}| \\
&= |\{(\kappa,\lambda) \in \mathbb{N}_0^2 \mid 2^\kappa \cdot 3^\lambda < 2^k \cdot 3^\ell\}|,
\end{aligned}
\tag{8.14}
$$

we can divide $\mathbb{N}^2$ into four mutually disjoint parts:

0. $[k]_0 \times [\ell]_0$,
1. $[k]_0 \times \mathbb{N}_\ell$,
2. $\mathbb{N}_k \times [\ell]_0$,
3. $\mathbb{N}_k \times \mathbb{N}_\ell$.

All (κ,λ) from the rectangle in 0 have to be counted, since obviously $2^\kappa \cdot 3^\lambda < 2^k \cdot 3^\ell$ for them. This makes up for the term $k \cdot \ell$.

Similarly, no term from 3 must be counted since there we have $2^\kappa \cdot 3^\lambda \geq 2^k \cdot 3^\ell$.

So we are left with cases 1 and 2. We only treat case 1; case 2 is done analogously. We show by induction on ℓ that

$$|\{(\kappa,\lambda) \in [k]_0 \times \mathbb{N}_\ell \mid 2^\kappa \cdot 3^\lambda < 2^k \cdot 3^\ell\}| = |\{(\kappa,\lambda) \in \mathbb{N}_0^2 \mid 2^\kappa \cdot 3^\lambda < 2^k\}|\ (= n_0(k,0)) \, .$$

This is true for $\ell = 0$, because $\kappa \in \mathbb{N}_k$ does not count. Moreover, replace λ by $\lambda - 1$,

$$\{(\kappa,\lambda) \in [k]_0 \times \mathbb{N}_{\ell+1} \mid 2^\kappa \cdot 3^\lambda < 2^k \cdot 3^{\ell+1}\} = \{(\kappa,\lambda) \in [k]_0 \times \mathbb{N}_\ell \mid 2^\kappa \cdot 3^\lambda < 2^k \cdot 3^\ell\} \, . \quad \square$$

The arguments in the preceeding proof can be checked, as in [318], with the example $k = 5$ and $\ell = 4$, where we have $n_0(5,4) = 50$ (yellow in Table 8.2), i.e. $A_{51} = 2^5 \cdot 3^4 = 2592$. We have to find the entries in Table 8.2 that are less than 50. They appear in three blocks: the rectangle of $k \cdot \ell = 20$ entries, made up from $\kappa \in [k]_0$ and $\lambda \in [\ell]_0$ (red), and the polygons with the blue and cyan entries in Table 8.2. If we move the block of blue entries to the left until the vertical base line (this amounts in dividing by powers of 3), we see that the block corresponds to the value $12 = n_0(5,0)$ and hence contains 12 elements. Similarly, shifting the cyan block up to the horizontal base line, this block contains $18 = n_0(0,4)$ elements.

If we define mappings $k_0, \ell_0 \in \mathbb{N}_0^{\mathbb{N}}$ by

$$k_0(n) = \min\{k \in \mathbb{N}_0 \mid n \le 2^k\}, \ \ell_0(n) = \min\{\ell \in \mathbb{N}_0 \mid n \le 3^\ell\} \, ,$$

[318, Theorems 2 and 3] claim that

$$n_0(k,0) = \sum_{\kappa=1}^{k} \ell_0(2^\kappa) \ \text{and} \ n_0(0,\ell) = \sum_{\lambda=1}^{\ell} k_0(3^\lambda) \, . \tag{8.15}$$

Again, we want to present a much more straightforward proof here, where for symmetry reasons we restrict ourselves to the first identity in (8.15), which we prove by induction on k. The case $k = 0$ is clear. Now

$$
\begin{aligned}
n_0(k+1,0) &= |\{(\kappa,\lambda) \in \mathbb{N}_0^2 \mid 2^\kappa \cdot 3^\lambda < 2^{k+1}\}| \\
&= |\{(0,\lambda) \in \{0\} \times \mathbb{N}_0 \mid 3^\lambda < 2^{k+1}\}| + |\{(\kappa,\lambda) \in \mathbb{N} \times \mathbb{N}_0 \mid 2^\kappa \cdot 3^\lambda < 2^{k+1}\}| \\
&= |\{\lambda \in \mathbb{N}_0 \mid 3^\lambda < 2^{k+1}\}| + |\{(\kappa,\lambda) \in \mathbb{N}_0^2 \mid 2^\kappa \cdot 3^\lambda < 2^k\}| \\
&= \min\{\ell \in \mathbb{N}_0 \mid 2^{k+1} \le 3^\ell\} + |\{(\kappa,\lambda) \in \mathbb{N}_0^2 \mid 2^\kappa \cdot 3^\lambda < 2^k\}| \\
&= \ell_0(k+1) + \sum_{\kappa=1}^{k} \ell_0(2^\kappa) = \sum_{\kappa=1}^{k+1} \ell_0(2^\kappa) \, .
\end{aligned}
$$

This completes the proof of (8.15).

As $2^{k_0(3^\lambda)-1} < 3^\lambda \le 2^{k_0(3^\lambda)}$ for $\lambda \in \mathbb{N}$, we get $(k_0(3^\lambda) - 1)\ln(2) < \lambda \ln(3) \le k_0(3^\lambda)\ln(2)$, whence $k_0(3^\lambda) - 1 < \lambda S \le k_0(3^\lambda)$ and consequently $k_0(3^\lambda) = \lceil \lambda S \rceil$, where, as before, $S := \mathrm{lb}(3)$ (the Hausdorff dimension of the Sierpiński triangle). Similarly, $\ell_0(2^\kappa) = \lceil \kappa C \rceil$, where again $C := \mathrm{lt}(2)$ (the Hausdorff dimension of the

Cantor set). (This can also be obtained from the fact that for $\kappa \in \mathbb{N}$ we have $\ell_0(2^\kappa) = a_0(\kappa, 0) = 1 + \lfloor \kappa C \rfloor = \lceil \kappa C \rceil$, the latter because $\kappa C \notin \mathbb{N}_0$.)

So we arrive at

$$n_0(k, \ell) = k \cdot \ell + \sum_{\kappa=1}^{k} \lceil \kappa C \rceil + \sum_{\lambda=1}^{\ell} \lceil \lambda S \rceil . \tag{8.16}$$

Equation (8.16) is sufficiently handy to provide the position $n_0(k, \ell) + 1$ of a 3-smooth number, which can easily be decomposed into k powers of 2 and ℓ powers of 3, in the sequence A. If we want to know, for a given $n \in \mathbb{N}_0$, which is the 3-smooth number A_{n+1}, we learn from equation (8.16) at least that we only have to check for a finite number of pairs (k, ℓ).

As an application of (8.16) we will give a proof of

Lemma 8.23. $\forall n \in \mathbb{N}_2 : 2 \cdot A_{n+1} \leq 3^n$.

Proof. We have $2 \cdot A_{n+1} = 2^{1+k} \cdot 3^\ell \leq 3^{1+k+\ell}$ for $n = n_0(k, \ell)$. From (8.16) we get $1 + k + \ell \leq n_0(k, \ell)$ if $(k, \ell) \notin \{(0,0), (1,0)\}$, but these exceptions correspond to $n \in \{0, 1\}$. □

We can now approach the question whether m_n in Stockmeyer's solution for his star puzzle is growing with n. As a preparation we show:

Lemma 8.24. $\forall 2 \leq m \leq \ell : 2\Sigma_\ell - 2\Sigma_m \leq \Sigma T_\ell - \Sigma T_m$, i.e. $2\Sigma_\ell + \Sigma T_m \leq 2\Sigma_m + \Sigma T_\ell$.

Proof. From Lemma 8.23 we get

$$\forall k \in \mathbb{N}_2 : 2(\Sigma_{k+1} - \Sigma_k) = 2 \cdot A_{k+1} \leq 3^k .$$

Therefore,

$$2(\Sigma_\ell - \Sigma_m) = 2 \sum_{k=m}^{\ell-1} A_{k+1} \leq \sum_{k=m}^{\ell-1} 3^k = \Sigma T_\ell - \Sigma T_m . \qquad \square$$

We need one more preparation.

Lemma 8.25. $m_1 = 0$, $m_2 = 1 = m_3$, $\forall n \in \mathbb{N}_4 : m_n \geq 2$.

Proof. Since $\lfloor \mathrm{lt}(A_1) \rfloor = 0 = \lfloor \mathrm{lt}(A_2) \rfloor$ and $\lfloor \mathrm{lt}(A_3) \rfloor = 1$, we get the equalities from the formula $m_n = n - 1 - \lfloor \mathrm{lt}(A_n) \rfloor$.

For $n \geq 4$ we show that in (8.9) the minimum is *not* assumed for $m = 0$ or $m = 1$, i.e. that

$$2\Sigma_0 + \Sigma T_n > 2\Sigma_1 + \Sigma T_{n-1} > 2\Sigma_2 + \Sigma T_{n-2} ,$$

or $\Sigma T_n > 2 + \Sigma T_{n-1} > 6 + \Sigma T_{n-2}$. This in turn follows from $3^{n-1} > 2$ for $n \geq 3$ and $3^{n-2} > 4$ for $n \geq 4$. □

We now have

Proposition 8.26. $\forall n \in \mathbb{N} : 2m_n \geq n - 1$.

Proof. The cases for $n \in [5]$ are covered by Lemma 8.25. So let us assume that $2m_n < n - 1$ for some $n \geq 6$. Put $m = m_n$ and $\ell = n - m_n \geq m_n \geq 2$ in Lemma 8.24. Then $\ell \in [n]_0$ and by uniqueness of m_n we get $\ell = m_n$, i.e. $2m_n = n$, a contradiction. $\square$

We are now prepared to come back to Conjecture 8.17. For the case $p = 4$ we have the following when we look at the proof of Theorem 8.18.

1. If $2\sigma_n(d)$ denotes the number of moves made by disc $d \in [n]$ during execution of Stockmeyer's algorithm for n discs, then $\sigma_n(d) = 2\sigma_{m_n}(d)$, if $d \in [m_n]$ and otherwise $\sigma_n(d) = 3^{n-d}$; cf. [149, Lemma 2.5].

2. Let $n_0 \in \mathbb{N}$. Then for every $n \in \mathbb{N}_{n_0}$ there is exactly one disc $d \in [n]$ with $\sigma_n(d) = A_{n_0}$.

3. Let $k, \ell \in \mathbb{N}_0$. Then exactly one disc $d \in [n]$ moves exactly $2\sigma_n(d) = 2^{1+k} \cdot 3^\ell$ times in the Stewart-Stockmeyer algorithm for task 1, if $n > n_0(k,\ell)$, where $A_{n_0(k,\ell)+1} = 2^k \cdot 3^\ell$.

Proof of Conjecture 8.17. [95] We employ double induction on $p \in \mathbb{N}_3$ and $n \in \mathbb{N}_0$. For $p = 3$ exactly one disc moves exactly $2 \cdot 3^\ell$ times if $n > \ell$; this is case $k = 0$. The statement is also true for $k \in \mathbb{N}$, but no disc is moving.

Let $p \geq 4$ and the statement be true for up to $p - 1$ pegs. If $n = 0$, then no disc is moving. In the formula
$$st_p^1(n) = 2st_p^1(m_n) + st_{p-1}^1(n - m_n)$$
the first summand on the right hand side refers to the top m_n discs and by induction assumption (on n) exactly $\binom{p-4+k-1}{p-4} = \binom{p-5+k}{p-4}$ of them move exactly $2 \cdot 2^{1+(k-1)} \cdot 3^\ell = 2^{1+k} \cdot 3^\ell$ times. Among the $n - m_n$ larger discs from the second summand exactly $\binom{p-1-4+k}{p-1-4} = \binom{p-5+k}{p-5}$ move exactly $2^{1+k} \cdot 3^\ell$ times by induction assumption (on p). Altogether exactly $\binom{p-5+k}{p-4} + \binom{p-5+k}{p-5} = \binom{p-4+k}{p-4}$ discs move exactly $2^{1+k} \cdot 3^\ell$ times. We have to make use of the fact that both m_n and $n-m_n$ are growing with n; for m_n this follows from Proposition 8.26 and for $n - m_n = 1 + \lfloor \mathrm{lt}(A_n)\rfloor$ it is obvious from the monotonicity of A (and lt). $\square$

So we may now talk about Theorem 8.17. For the lower bound on n let us define
$$\forall \kappa \in \mathbb{N}_0: a_\kappa(k,\ell) = \left|\{\lambda \in \mathbb{N}_0 \mid 2^\kappa \cdot 3^\lambda \leq 2^k \cdot 3^\ell\}\right|$$
$$= (\kappa = k) + \left|\{\lambda \in \mathbb{N}_0 \mid 2^\kappa \cdot 3^\lambda < 2^k \cdot 3^\ell\}\right|. \quad (8.17)$$

Note that $a_\kappa(k,\ell) = 0 \Leftrightarrow \kappa > k + \lfloor \ell S\rfloor$, where $S := \mathrm{lb}(3)$ (the Hausdorff dimension of the Sierpiński triangle). *Ferme's conjecture* is
$$n \geq \sum_{\kappa=0}^{k+\lfloor \ell S\rfloor} a_\kappa(k,\ell)\binom{p-4+\kappa}{p-4}.$$

---

[95] J. Ferme, private communication, 2017.

For $p = 3$ and $p = 4$ (where according to (8.17)

$$\sum_{\kappa=0}^{k+\lfloor \ell S \rfloor} a_\kappa(k, \ell) = 1 + n_0(k, \ell))$$

this is true and also for a couple of computed cases for $p = 5$. The numbers $a_\kappa(k, \ell)$ can be calculated from $a_0(k, \ell) = \ell + 1 + \lfloor kC \rfloor$, where $C := \mathrm{lt}(2)$ (the Hausdorff dimension of the Cantor set), $a_{\kappa+1}(0, \ell) = 0$ for $\ell \leq \kappa$ and $= \ell - \kappa$ otherwise, and with the recurrence $a_{\kappa+1}(k, \ell) = a_\kappa(k - 1, \ell)$ for $k \in \mathbb{N}$.

8.3.3   Bousch's Proof of Stockmeyer's Conjecture

We will now give an outline of Bousch's confirmation of the Stockmeyer conjecture for $p = 4$. As in [62] the main theorem of [63] is a lower bound for a more general situation than the classical tower-to-tower task. Lacking an alternative approach its proof will be based on induction on the number of discs as with so many statements around the Tower of Hanoi. And as with any induction proof there will be a trade off between a weak statement that is more likely to be true, but leads to a weak induction assumption, and a stronger statement, which may be more difficult to prove, but offers a stronger tool in the induction step. Of course, one has to avoid the statement envisaged to be false! Finally, in view of the ultimate goal, in our case the inequality $d_4^1 \geq 2\Sigma$, the result has to be sharp in this special case. In [63, Section 5] Bousch explains carefully how he got to the lower bound he eventually sets off to prove.

To set the stage let $\{i, j, k\} = [3]$ be the set of labels for the peripheral pegs and define $\forall \lambda \in \mathbb{N}_0 : j_\lambda = i + (\lambda \bmod 2)(j - i)$. Moreover, for $n, \ell \in \mathbb{N}_0$ and $s = (s_0, \ldots, s_\ell) \in ([4]_0^n)^{[1+\ell]_0}$ we define the *multi-distance*

$$d(s) = \sum_{\lambda=1}^{\ell} d(s_{\lambda-1}, s_\lambda).$$

It can be interpreted as the length of an s_0, s_ℓ-walk moving forward on shortest paths through the other vertices in natural order. Recall from Proposition 8.15 that in intermediate states s_λ, $\lambda \in [\ell - 1]$, the allowed peg other than k will alternate between j and i, i.e. it will be j_λ.

Next we define a sequence of integer sequences S_ℓ by

$$\forall \ell \in \mathbb{N}_0 \ \forall n \in \mathbb{N}_0 : \ S_\ell(n) = \min \{2\Sigma_m + \ell \cdot \Sigma T_{n-m} \mid m \in [n + 1]_0\} . \qquad (8.18)$$

Note that $S_1 = \Sigma$, because $S_1(0) = 0 = \Sigma_0$ and for $n \in \mathbb{N}$ the minimum in (8.18) for $\ell = 1$ is not assumed for $m = n$ since by (8.9) there is an $m < n$ with $2\Sigma_m + \Sigma T_{n-m} = \Sigma_n < 2\Sigma_n$.

The goal is now a lower bound for the distance between any two arbitrary states of Stockmeyer's Tower of the form

$$\forall s, t \in [4]_0^n : \ \tfrac{1}{2}d(s, t) + \Sigma_n \geq \sum_{d=1}^{n} \{(s_d = i) + (t_d = j)\}A_d . \qquad (8.19)$$

Note that the sum on the right is $2\Sigma_n$, if and only if $s = i^n$ and $t = j^n$, whence in particular $\mathrm{d}(i^n, j^n) \geq 2\Sigma_n$, and consequently Stockmeyer's conjecture would be proved.

Unfortunately, (8.19) does not lend itself for a straightforward induction, because the introduction of a new (largest) disc will not lead to the same type of statement. Therefore, we have to make use of alternating intermediate states as described above. Let us first assume that $s_0 = i^n$ and $s_\ell = j^n$ (with an odd ℓ) and that for $\lambda \in [\ell - 1]$ we have $s_\lambda = j_\lambda^{n-m} k^m \in \{j_\lambda, k\}^n$. Inspired by Stewart's strategy we transfer $m \in [n+1]_0$ discs from peg i to peg k in $2\Sigma_m$ moves. Then the other $n - m$ discs alternate ℓ times between pegs i and j, which takes $2\ell \cdot \Sigma \mathrm{T}_{n-m}$ moves, and finally the m smaller discs are brought from k to j. Altogether this takes $4\Sigma_m + 2\ell \cdot \Sigma \mathrm{T}_{n-m}$ moves, i.e. $2S_\ell(n)$ moves if m is properly chosen according to (8.18). So we should expect

$$\mathrm{d}(s) \geq 2S_\ell(n) \tag{8.20}$$

in this situation.

For general s_0 and s_ℓ, which will occur in the induction proof envisaged, we will have to introduce "penalties" for those discs which are initially not on peg i and at the end not on peg j_ℓ. In [63, Section 5] Bousch explains carefully how he got to the suitable penalties, namely, for $n \in \mathbb{N}$,

$$2\sum_{d=1}^{n} \{(s_0(d) \neq i) + (s_\ell(d) \neq j_\ell)\}\mathrm{A}_d - (\ell \geq 2)\{(s_0(n) \neq i) + (s_\ell(n) \neq j_\ell)\}\left(2\mathrm{A}_n - \overline{S_\ell}(n)\right).$$

This can also be written as

$$4\Sigma_n \quad - \quad 2\sum_{d=1}^{n}\{(s_0(d) = i) + (s_\ell(d) = j_\ell)\}\mathrm{A}_d$$
$$- \quad (\ell \geq 2)\{2 - (s_0(n) = i) - (s_\ell(n) = j_\ell)\}\left(2\mathrm{A}_n - \overline{S_\ell}(n)\right).$$

Note that for $\ell = 1$ this reduces to

$$4\Sigma_n - 2\sum_{d=1}^{n}\{(s_0(d) = i) + (s_1(d) = j)\}\mathrm{A}_d,$$

which, together with (8.20), corresponds exactly to (8.19). Therefore, the envisaged lower bound on multi-distances is (see [63, Théorème 2.2]):

Theorem 8.27. *Let $n, \ell \in \mathbb{N}_0$ and $s \in ([4]_0^n)^{[1+\ell]_0}$ with $s_\lambda \in \{j_\lambda, k\}^n$ for $\lambda \in [\ell - 1]$. Then*

$$\tfrac{1}{2}\mathrm{d}(s) \quad \geq \quad S_\ell(n) - 2\Sigma_n + \sum_{d=1}^{n}\{(s_0(d) = i) + (s_\ell(d) = j_\ell)\}\mathrm{A}_d$$
$$+ \quad (n \geq 2)(\ell \geq 2)\{2 - (s_0(n) = i) - (s_\ell(n) = j_\ell)\}\left(\mathrm{A}_n - \tfrac{1}{2}\overline{S_\ell}(n)\right).$$

Remark 8.28. *One might wonder where the ugly last term comes from if $n, \ell \geq 2$. Let us assume it is not there and apply Theorem 8.27 in the situation where $s_0(n) = k = s_\ell(n)$. Then the right-hand side becomes*

$$S_\ell(n) - 2\Sigma_n + \sum_{d=1}^{n-1} \{(s_0(d) = i) + (s_\ell(d) = j_\ell)\} A_d \, ;$$

this is the same as

$$S_\ell(n-1) - 2\Sigma_{n-1} + \sum_{d=1}^{n-1} \{(s_0(d) = i) + (s_\ell(d) = j_\ell)\} A_d - 2A_n + \overline{S_\ell}(n),$$

which may be strictly smaller ($\overline{S_\ell}(n)$ is non-decreasing in ℓ and goes to $2A_n$ as $\ell \to \infty$, whence $\overline{S_\ell}(n) \leq 2A_n$; see [63, Section 3]) than the right-hand side in Theorem 8.27 when applied to $\overline{s} := s \upharpoonright [n-1]$, if $s_0(n-1) = i$ and $s_\ell(n-1) = j_\ell$; but clearly $\mathrm{d}(s) = \mathrm{d}(\overline{s})$.

As said before, an immediate consequence of Theorem 8.27 is the truth of Stockmeyer's conjecture:

Corollary 8.29. *For all $n \in \mathbb{N}_0$ and $\{i, j\} \in \binom{[3]}{2}$: $\mathrm{d}(i^n, j^n) = 2\Sigma_n$ in St_4^n.*

Proof. Put $\ell = 1$, $s_0 = i^n$ and $s_1 = j^n$ in Theorem 8.27 to get $\mathrm{d}(i^n, j^n) \geq 2\Sigma_n$. The opposite inequality follows from Stockmeyer's algorithm. □

Remark. If we try to apply the same argument to the question whether $\mathrm{d}_4^0 = st_4^0$, Theorem 8.27 is of no help, because the right-hand side in that case is just 0.

But how is Theorem 8.27 proved? Bousch gives, in [63, Section 3] seven lemmas about the properties of the sequences S_ℓ which are too technical to be reproduced here. Based on these, in [63, Section 4] the proof of Theorem 8.27 is presented. As the case $\ell = 0$ is trivial, the right-hand side being ≤ 0, we assume a fixed $\ell \in \mathbb{N}$ and proceed by induction on n, where again the case $n = 0$ is obviously true, both sides of the inequality being 0. For $n = 1$ we observe that $A_1 = \Sigma_1 = 1$ and $\overline{S_\ell}(1) = S_\ell(1) = \min\{\ell, 2\} = 1 + (\ell \geq 2)$. The inequality to be proved reads $\frac{1}{2}\mathrm{d}(s) \geq -(\ell = 1) + (s_0(1) = i) + (s_\ell(1) = j_\ell)$. For $\ell = 1$, the right-hand side is positive, namely equal to 1, only if $s_0(1) = i$ and $s_1(1) = j$, but then $\mathrm{d}(s) = 2$. If $\ell \geq 2$, then $\mathrm{d}(s_0, s_1) \geq 2(s_0(1) = i)$ and $\mathrm{d}(s_{\ell-1}, s_\ell) \geq 2(s_\ell(1) = j_\ell)$.

For the induction step we may now assume that the statement is true up to and including some $n \in \mathbb{N}$. Let $s_\lambda = s_\lambda(n+1)\overline{s}_\lambda \in [4]_0^{1+n}$ for $\lambda \in [\ell + 1]_0$. Then $\mathrm{d}(s) \geq \mathrm{d}(\overline{s})$, because we may leave out all moves of the largest disc $n + 1$ when going from the left expression to the right one. Moreover, the induction assumption applies for $\bar{s}$, i.e.

$$\tfrac{1}{2}\mathrm{d}(\bar{s}) \;\geq\; S_\ell(n) - 2\Sigma_n + \sum_{d=1}^{n}\{(s_0(d) = i) + (s_\ell(d) = j_\ell)\}\mathrm{A}_d$$

$$+ \;(n \geq 2)(\ell \geq 2)\{2 - (s_0(n) = i) - (s_\ell(n) = j_\ell)\}\left(\mathrm{A}_n - \tfrac{1}{2}\overline{S_\ell}(n)\right)$$

$$\geq\; S_\ell(n) - 2\Sigma_n + \sum_{d=1}^{n}\{(s_0(d) = i) + (s_\ell(d) = j_\ell)\}\mathrm{A}_d\,,$$

where in the last estimate we used $2\mathrm{A}_n \geq \overline{S_\ell}(n)$ again. Altogether

$$\tfrac{1}{2}\mathrm{d}(s) \;-\; S_\ell(n+1) + 2\Sigma_{n+1} - \sum_{d=1}^{n+1}\{(s_0(d) = i) + (s_\ell(d) = j_\ell)\}\mathrm{A}_d$$

$$-\;(\ell \geq 2)\{2 - (s_0(n+1) = i) - (s_\ell(n+1) = j_\ell)\}\left(\mathrm{A}_{n+1} - \tfrac{1}{2}\overline{S_\ell}(n+1)\right)$$

$$\geq\; S_\ell(n) - S_\ell(n+1) - 2\Sigma_n + 2\Sigma_{n+1}$$

$$-\;\{(s_0(n+1) = i) + (s_\ell(n+1) = j_\ell)\}\mathrm{A}_{n+1}$$

$$-\;(\ell \geq 2)\{2 - (s_0(n+1) = i) - (s_\ell(n+1) = j_\ell)\}\left(\mathrm{A}_{n+1} - \tfrac{1}{2}\overline{S_\ell}(n+1)\right)$$

$$=\; K\left(\mathrm{A}_{n+1} - (\ell \geq 2)\left(\mathrm{A}_{n+1} - \tfrac{1}{2}\overline{S_\ell}(n+1)\right)\right) - \overline{S_\ell}(n+1)\,, \qquad (8.21)$$

where K is short for $2 - (s_0(n+1) = i) - (s_\ell(n+1) = j_\ell)$ (or $(s_0(n+1) \neq i) + (s_\ell(n+1) \neq j_\ell)$); obviously, $K \in [3]_0$. We have to show that the expression in (8.21) is non-negative. If $K = 2$, this follows from $2\mathrm{A}_{n+1} \geq \mathrm{A}_{n+1} = \overline{S_1}(n+1)$ for $\ell = 1$ and otherwise from $S_\ell(n+1) \geq S_\ell(n+1) - S_\ell(n) = \overline{S_\ell}(n+1)$. The case $K = 1 = \ell$ is also immediate, the expression in (8.21) being 0. Left over are the cases $K = 1 < \ell$ and $K = 0$. They are treated in a typical, but extremely tedious combinatorial manner. We cannot do the details here, but want to show how Bousch cuts the problem into digestible pieces.

The first case to be treated is $K = 1 < \ell$ [63, Section 4.1], where for symmetry reasons it is assumed that $s_0(n+1) = i$ and consequently $s_\ell(n+1) \neq \lambda_\ell$. Then $\alpha := \min\{\lambda \in [\ell] \mid s_\lambda(n+1) \neq j_\lambda\}$ and $\beta := \ell - \alpha \in [\ell]_0$ are considered. This leads to subcases "$\alpha, \beta \geq 2$", "$\alpha \geq 2, \beta = 1$", "$\alpha = 1 = \beta$", "$\alpha = 1, \beta = 2$", "$\alpha = 1, \beta \geq 3$", and "$\beta = 0$", respectively, all treated in a similar manner, making use of the lemmas from [63, Section 3].

The final case, covered in [63, Section 4.2], is $K = 0$, where we define $E := \{\lambda \in [\ell-1] \mid s_\lambda(n+1) \neq j_\lambda\}$ with the subcases $E = \varnothing$ and $E \neq \varnothing$. In the latter case we define $\alpha := \min E$, $\alpha^* := \max E$, $\beta := \alpha^* - \alpha$, and $\gamma := \ell - \alpha^*$, such that $\alpha + \beta + \gamma = \ell$. This is the most exhausting part of the combinatorics in the proof. The main subcases are

1. $\alpha, \gamma \geq 2$ ([63, Sous-section 4.2.1]),

2. $\alpha = 1 = \gamma$ ([63, Sous-section 4.2.2]),

3. $\alpha = 1 < \gamma$ ([63, Sous-section 4.2.3]),

4. $\alpha > 1 = \gamma$ (by symmetry from 3).

Subcase 1 is subdivided into $\beta = 0$, $\beta = 1$, and $\beta \geq 2$. In subcase 2 we have to distinguish $\beta = 0$, $\beta = 1$, $\beta = 2$, and $\beta \geq 3$, as we have for subcase 3, where for $\beta = 0$ we even have to separate $\gamma = 2$ and $\gamma \geq 3$. All in all this is a masterpiece of a combination of a clever idea (the inequality in Theorem 8.27) and a careful combinatorial analysis to make the idea work.

8.3.4 Stewart-Type Algorithms

Except for the classical case $D = K_4$ there are always essentially different tasks to be analyzed depending on the choice of initial and final pegs i and j, $\{i, j\} \in \binom{Q}{2}$, when going from i^n to j^n in H_D^n for the other 5 undirected graphs D on 4 vertices. In particular, we do not know $d_D(i^n, j^n)$ for any of these non-classical cases but for $D = K_{1,3}$, where we know from Exercise 8.7 and Theorem 8.18 that

$$d_D(0^n, 1^n) \leq st_4^0(n) \leq st_4^1(n) = d_D(1^n, 2^n) = 2\Sigma_n.$$

Therefore, we have to rely on the approach Stockmeyer had, as shown by Bousch successfully, adopted in the latter case to all the others, namely to employ a Stewart-type strategy: cut the n-tower into two subtowers of m smaller and $n - m$ larger discs and transfer these subtowers appropriately. The latter means that even for fixed i and j there might be alternative Stewart-type algorithms and we have to choose the best among these, which still does *not* guarantee overall minimality of the resulting path(s). We will call the solutions from the best Stewart-type algorithm *relatively optimal* and denote their lengths by x_n for $\{i, j\} = \{0, 1\}$, y_n for $\{i, j\} = \{1, 2\}$, and z_n for $\{i, j\} = \{2, 3\}$, respectively, e.g., in the case of the Diamond TH with $D = K_4 - e$, $e = \{2, 3\}$. Here we have the following options (with the seeds for $n = 0$ being 0 in all cases):

$$
\begin{aligned}
x_n &= \min\{2y_m + \Sigma B_{n-m} \mid m \in [n]_0\}, \\
y_n' &= \min\{x_m + y_m + \Sigma T_{n-m} \mid m \in [n]_0\}, \\
y_n'' &= \min\{y_m + z_m + \Sigma B_{n-m} \mid m \in [n]_0\}, \\
y_n &= \min\{y_n', y_n''\}, \\
z_n &= \min\{2y_m + 2\Sigma T_{n-m} \mid m \in [n]_0\}.
\end{aligned}
\tag{8.22}
$$

We have seen that Matsuura adopted the not even relatively optimal strategy equivalent to

$$y_n''' = \min\{3x_m + \Sigma B_{n-m} \mid m \in [n]_0\}$$

and leading to $y_n''' = \Sigma_n$; see Theorem 8.21.

Note that from (8.22) it is immediate that $x_n \leq z_n$ and we are tempted to conjecture that $x_n \leq y_n = y_n'' \leq z_n \leq 2x_n$. To test such guesses, Lužar has

task $\setminus n$	0	1	2	3	4	5	6	7	8
$d_{K_4}(0^n,1^n)$	0	1	3	5	9	13	17	25	33
$d_{K_{1,3}+e}(0^n,1^n)$	0	1	4	7	12	20	29	39	54
$d_{K_{1,3}+e}(1^n,2^n)$	0	2	5	10	16	25	35	50	68
$d_{K_{1,3}+e}(2^n,3^n)$	0	1	3	7	11	17	25	35	47
$d_{K_{1,3}+e}(3^n,0^n)$	0	1	3	6	10	16	24	32	43
$d_{K_{1,3}}(0^n,1^n)$	0	1	4	7	14	23	32	47	68
$d_{K_{1,3}}(1^n,2^n)$	0	2	6	12	20	32	48	66	90
$d_{K_4-e}(0^n,1^n)$	0	1	3	5	9	13	19	27	35
$d_{K_4-e}(1^n,2^n)$	0	1	3	6	10	14	21	29	39
$d_{K_4-e}(2^n,3^n)$	0	2	4	8	14	20	28	36	50
$d_{C_4}(0^n,1^n)$	0	2	4	10	16	22	32	50	68
$d_{C_4}(1^n,2^n)$	0	1	4	7	12	21	30	41	56
$d_{P_{1+3}}(0^n,1^n)$	0	3	10	19	34	57	88	123	176
$d_{P_{1+3}}(1^n,2^n)$	0	1	4	9	18	29	44	69	96
$d_{P_{1+3}}(2^n,3^n)$	0	1	4	7	14	23	34	53	78
$d_{P_{1+3}}(3^n,1^n)$	0	2	6	12	22	36	54	78	112

task $\setminus n$	9	10	11	12	13	14	15
$d_{K_4}(0^n,1^n)$	41	49	65	81	97	113	129
$d_{K_{1,3}+e}(0^n,1^n)$	72	95	122	151	185	227	276
$d_{K_{1,3}+e}(1^n,2^n)$	86	111	141	175	218	268	322
$d_{K_{1,3}+e}(2^n,3^n)$	63	81	101	131	163	199	235
$d_{K_{1,3}+e}(3^n,0^n)$	59	76	95	120	152	185	221
$d_{K_{1,3}}(0^n,1^n)$	93	120	153	198	255	318	399
$d_{K_{1,3}}(1^n,2^n)$	122	158	206	260	324	396	492
$d_{K_4-e}(0^n,1^n)$	43	57	73	89	109	129	159
$d_{K_4-e}(1^n,2^n)$	49	64	80	96	120	146	178
$d_{K_4-e}(2^n,3^n)$	66	84	104	124	154	186	218
$d_{C_4}(0^n,1^n)$	86	108	138	176	230	284	342
$d_{C_4}(1^n,2^n)$	75	102	131	164	201	250	309
$d_{P_{1+3}}(0^n,1^n)$	253	342	449	572	749	980	1261
$d_{P_{1+3}}(1^n,2^n)$	133	182	241	322	409	532	675
$d_{P_{1+3}}(2^n,3^n)$	105	138	187	248	329	412	515
$d_{P_{1+3}}(3^n,1^n)$	158	212	272	352	466	604	766

Table 8.3: Distances $d_D(i^n,j^n)$

performed a series of computations based on BFS algorithms; the results are shown in Table 8.3.

All numbers from Table 8.3 for $D \neq P_{1+3}$ coincide with those obtained from solving numerically recurrences for the corresponding lengths of (relatively optimal) Stewart-type solutions like (8.22) for the Diamond TH. The same applies for y_n, z_n, and v_n, the latter corresponding to the task $3^n \to 1^n$, in $D = P_{1+3}$ if we employ for x the (beginning of the) sequence of *optimal* move numbers from Table 8.3. However, the sequence x obtained from the recurrence

$$
\begin{aligned}
x_n &= \min\{3x_m + 3\Sigma T_{n-m} \mid m \in [n]_0\}, \\
y_n &= \min\{x_m + v_m + \Sigma T_{n-m} \mid m \in [n]_0\}, \\
z_n &= \min\{y_m + v_m + \Sigma T_{n-m} \mid m \in [n]_0\}, \\
v_n &= \min\{x_m + y_m + 2\Sigma T_{n-m} \mid m \in [n]_0\}
\end{aligned} \tag{8.23}
$$

differs from the optimal values already from $n = 3$. So we are lead to state the following

Conjecture 8.30. *Stewart-type algorithms lead to the optimal move number for the tasks $i^n \to j^n$ in undirected and connected move graphs D, if $d_D(i, j) \le 2$, i.e. in particular if* diam$(D) \le 2$.

The conjecture is supported by the obvious next case $D = K_{1,4}$ of the star with four leaves, where calculations of the optimal path lengths by BFS and the values obtained by solving the corresponding recurrence for the lengths of Stewart-type solutions lead to the same numbers.

8.3.5   The Linear Tower of Hanoi

The simplest case of an undirected connected D with diam$(D) > 2$ is the path $D = P_{1+3}$ associated with the four-peg Linear TH. This problem has been addressed in [401, Section 3] and studied intensely in literature. But even from calculating minimal paths for $n \in [8]_0$, realized by Lužar (2017), no optimal strategy could be discerned yet for the task to transfer an n-tower from one extreme peg to the other. Recall that for reasons of comparability, we arrange the pegs in the order 0–3–2–1 (cf. Figure 8.2). As with the classical TH, there is a striking difference between the puzzle on three pegs and the ones on four or more pegs; the case $p = 3$ is easy, the cases $p \ge 4$ seem notoriously difficult. (See Exercise 8.10 for the more general p-in-a-row, or *Linear TH* on p pegs with 2 discs.)

Stockmeyer [401] initiated the study of the Four-in-a-row TH and made some starting observations. In particular he proposed an algorithm that solves the puzzle with n discs in $3^n + n - 1$ moves. More specifically, his algorithm transfers n discs from the perfect state on peg 0 to the perfect state on peg 1. Berend, Sapir and Solomon [50] followed by proving that the task to transfer discs from peg 0 to peg 1 is the most time consuming among the tasks to transfer discs from peg i to peg j, where $i, j \in Q$ and proved that the task can be done in less than

$$
1.6\sqrt{n}\,3^{\sqrt{2n}}
$$

moves. The main idea to derive this result is to split the tower of n discs into a subtower of small upper discs, an intermediate subtower of big discs, and the third subtower consisting of disc n only, where the intermediate subtower consists of about $\sqrt{2n}$ discs. Then moving subtowers "naturally" solves the puzzle, while a careful analysis of the number of moves yields the estimate. Table 8.4 lists minimum numbers of moves for up to $n = 10$ discs for all different tasks: moving the tower from peg 3 to peg 2, from peg 0 to peg 3, from peg 0 to peg 2, and from peg 0 to peg 1, respectively. For additional numerical results on the Four-in-a-row puzzle see J. W. Emert, R. B. Nelson and F. W. Owens [126].

	minimum number of moves for:			
n	$3 \overset{[n]}{\longrightarrow} 2$	$0 \overset{[n]}{\longrightarrow} 3$	$0 \overset{[n]}{\longrightarrow} 2$	$0 \overset{[n]}{\longrightarrow} 1$
1	1	1	2	3
2	4	4	6	10
3	7	9	12	19
4	14	18	22	34
5	23	29	36	57
6	34	44	54	88
7	53	69	78	123
8	78	96	112	176
9	105	133	158	253
10	138	182	212	342

Table 8.4: Minimum number of moves for the Four-in-a-row TH [50, Table 1]

Berend, Sapir and Solomon [50] proved in addition that for every Linear TH and for any combination of source and goal pegs, the number of moves needed to transfer a perfect tower grows sub-exponentially (see the next section for a formal definition) as a function of n. To obtain this result, they partition the discs into $p - 1$ subtowers, where the last subtower consists of disc n only. This procedure is hence like Frame's approach to the classical multipeg TH (cf. Section 5.4). However, contrary to the latter, where it turned out that Frame's and Stewart's approaches are equivalent for any $p \geq 4$ (cf. Theorem 5.14), the situation is different for the p-in-a-row TH. Here splitting into $p - 1$ subtowers leads to an algorithm with a smaller number of moves; in fact, there is rarely a situation where smaller subtowers unify to form larger ones.

8.4 The Cyclic Tower of Hanoi

If in the Linear TH on p pegs we impose a one-sided orientation on the moves, then clearly the task to transfer a tower from one peg to another can not be solved for more than one disc. In [262], Leiss proposed a partially oriented move graph that would allow for at least *some* of the tasks. For $p \in \mathbb{N}_3$ let D_p be given by $V(D_p) = [p]_0$ and $A(D_p) = \{(i, i+1) \mid i \in [p-1]_0\} \cup \{(1,0),(2,1)\}$. (The pegs are arranged in a line in natural order $0-1-\cdots-(p-1)$.) $D_3 \simeq P_{1+2}$ is the only solvable in this class of digraphs. In $H^n_{D_p}$ the (oriented) task $i^n \to j^n$ can be solved if and only if $i \in [3]_0$ and $j \in [p]_0$. Let us call such (ordered) pairs (i, j) *admissible*. In Exercise 8.11 the reader will show that $(0^n, (p-1)^n)$ leads to the largest value of $d_{D_p}(i^n, j^n)$ among all admissible pairs. In fact, Leiss' goal was to characterize D_p as the "worst" in a generous class of move graphs (containing all solvable ones) of order p with respect to such a distance, with the possible exception of a finite number of graphs, all containing directed cycles of order at most p.

Among the oriented cycles, $\mathrm{TH}(\overrightarrow{C}_4)$ was first considered in 1944 by Scorer, Grundy and Smith in [378]. They discussed the task $0^n \to 2^n$ and proposed the following natural recursive algorithm to solve it:

$$0 \xrightarrow{[n-1]} 2, \ 0 \xrightarrow{n} 1, \ 2 \xrightarrow{[n-1]} 0, \ 1 \xrightarrow{n} 2, \ 0 \xrightarrow{[n-1]} 2 . \tag{8.24}$$

Stockmeyer [401] observed that this algorithm is not optimal, see Exercise 8.12. However, the minimum number of moves has not yet been determined. Even worse, there is no natural algorithm which one might conjecture to be optimal. One of the reasons why $\mathrm{TH}(\overrightarrow{C}_4)$ is so difficult lies in the fact that there is a large number of minimal move sequences. For instance, the task $0^n \to 2^n$ for $n = 4$, 5 and 6 has 640, 2 688 and 54 839 936 different optimal solutions, respectively. (See [401].) Additional numerical values for $\mathrm{TH}(\overrightarrow{C}_4)$ were produced by Emert, Nelson and Owens in [127].

Berend and Sapir [47] made a profound study of $\mathrm{TH}(\overrightarrow{C}_p)$, $p \in \mathbb{N}_4$, and proved several explicit lower and upper bounds for the number of moves in optimal solutions. From their many results we extract the following:

Theorem 8.31. *For every $p \in \mathbb{N}_3$ and $n \in \mathbb{N}_{p-1}$, the number of moves needed to solve the $0^n \to 1^n$ task of $\mathrm{TH}(\overrightarrow{C}_p)$ is at least*

$$\frac{(p-1)^3}{3}\left(1 + \frac{6}{7(p-1)^2}\right)^n .$$

The proof of this theorem is quite involved and therefore not given here. Note that the bound of Theorem 8.31 is rather bad for $p = 3$.

8.5 Exponential and Sub-Exponential Variants

In the previous sections we considered several variants of the TH with restricted disc moves. Recall that each task of the Four-in-a-row TH can be completed in less than

$$1.6\sqrt{n}\, 3^{\sqrt{2n}}$$

moves, while on the other hand the (computationally least complex) task $0^n \to 1^n$ of $\mathrm{TH}(\overrightarrow{C}_4)$ requires, for $n \geq 3$, at least

$$9\left(1 + \frac{2}{21}\right)^n$$

moves. We have also said that the first estimate is sub-exponential. On the other hand, the lower bound for the number of moves for the $\mathrm{TH}(\overrightarrow{C}_4)$ is exponential. In this section we are going to make these statements precise and, moreover, present a classification of the solvable variants with restricted disc moves into exponential and sub-exponential variants. The main result of this section (Theorem 8.32) is a statement that covers all solvable $\mathrm{TH}(D)$.

Let D be a strong digraph and H_D^n the state digraph of $\mathrm{TH}(D)$ with n discs. Then $\mathrm{TH}(D)$ is called *exponential* if there exist constants $C > 0$ and $\lambda > 1$ such that

$$\mathrm{diam}(H_D^n) \geq C\lambda^n$$

holds for all $n \in \mathbb{N}$. On the other hand, $\mathrm{TH}(D)$ is called *sub-exponential* if for every $\varepsilon > 0$ there exists a constant C (C may depend on ε) such that

$$\mathrm{diam}(H_D^n) \leq C(1 + \varepsilon)^n$$

holds for all $n \in \mathbb{N}_0$.

It could in principle be possible that for some strong digraph D the $\mathrm{TH}(D)$ is neither exponential nor sub-exponential. However, as proved in [49] this is not the case, that is, $\mathrm{TH}(D)$ is either exponential or sub-exponential. In order to formulate the corresponding classification we say that a vertex of a strong digraph D is a *shed* if the digraph $D - v$ contains a strong subdigraph on at least three vertices. Then we have the following appealing result.

Theorem 8.32. (Berend and Sapir [49, Theorem 2]) *If $\mathrm{TH}(D)$ is solvable, then $\mathrm{TH}(D)$ is sub-exponential if and only if D contains a shed.*

If D has three vertices, then no vertex-deleted subdigraph can contain a strong subdigraph on at least three vertices. Hence Theorem 8.32 in particular implies that $\mathrm{TH}(D)$ is exponential for any D on three vertices, a fact which we already implicitly observed in the last part of Section 8.2. To clarify the structure of digraphs with sheds on at least four vertices, the following concept is useful. A digraph D is called a *moderately enhanced cycle* if it is obtained from a directed cycle by adding some arcs in such a way that no two added arcs are adjacent. Figure 8.8 displays an example of such a graph.

Then we have:

Proposition 8.33. ([49, Proposition 2]) *A strong digraph D on at least four vertices contains a shed if and only if D is not a moderately enhanced cycle.*

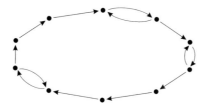

Figure 8.8: A moderately enhanced cycle

The reader is asked to prove Proposition 8.33 in Exercise 8.13.

We conclude the chapter with a further question to be posed on a TH with restricted disc moves. It relates to situations where the task of moving a perfect tower from one peg to another might be feasible for a small number of discs and not solvable otherwise.

Let D be a digraph, and let i (source peg) and j (target peg) be two specific vertices of D. Suppose in addition that for every other vertex k, there exists a directed walk in D from i through k to j. We will call D an i,j-digraph, abbreviated D_{ij}. Then $i^1 \rightarrow j^1$ is solvable. However, if (for instance) D is the directed i,j-path on $p \geq 2$ vertices, denoted by $\overrightarrow{P}_{ij}^{(p)}$, then $i^n \rightarrow j^n$ can not be solved for $n \geq 2$. Hence $\mathrm{TH}(D_{ij})$ will be called *weakly solvable* if for every $n \in \mathbb{N}$ a tower of n discs can be transferred from i to j. This issue was formulated by Leiss in [260].

If $\mathrm{TH}(D_{ij})$ is not weakly solvable, then there is a largest number of discs for which the task $i^n \rightarrow j^n$ can be carried out; let $max\,(D_{ij})$ denote this number. For instance, as mentioned above, $max\left(\overrightarrow{P}_{ij}^{(p)}\right) = 1$. In addition, let

$$M(p) = \max\{max\,(D_{ij}) \mid \mathrm{TH}(D_{ij}) \text{ is not weakly solvable}, |V(D_{ij})| = p\}\,.$$

As $max\,(D_{ij})$ is finite and since the number of digraphs on p vertices is also finite, $M(p)$ is well-defined. In fact, Leiss [261, Section 2.3] proved that $M(p)$ grows super-polynomially when p tends to infinity and asked whether the growth is exponential. The first part of the following theorem due to D. Azriel and Berend [26, Theorem 1.3] answers his question in the negative.

Theorem 8.34. *There exists a constant $C > 0$ such that $M(p) \leq Cp^{\frac{1}{2}\mathrm{lb}(p)}$ holds for all $p \in \mathbb{N}_2$. On the other hand, for every $\varepsilon > 0$ there exists a constant $C > 0$ (depending on ε) such that $M(p) \geq Cp^{\left(\frac{1}{2}-\varepsilon\right)\mathrm{lb}(p)}$ holds for all $p \in \mathbb{N}_2$.*

Azriel, N. Solomon and S. Solomon [27] gave an algorithm that solves $\mathrm{TH}(D_{st})$ for two infinite families of graphs D and also prove its optimality. That is, they solve infinitely many variants of TH! In the introduction to Chapter 6 we have quoted Lucas that the number of problems which one can pose oneself on the TH is incalculable. His ingenious remark is thus literally justified by the investigations in [27]!

8.6   Exercises

8.1. Prove Proposition 8.6.

8.2. Determine a system of recurrence relations that define the number of moves in $\mathrm{TH}(\overrightarrow{C}_{3+})$.

8.3. Determine a system of recurrence relations that define the number of moves in $\mathrm{TH}(\overleftrightarrow{K}_{3-})$.

8.4. Show that the state digraph of $\mathrm{TH}(\overrightarrow{C}_3)$ with n discs is an *orientation* of the Hanoi graph H_3^n, that is, a digraph obtained from H_3^n by giving an orientation to each of its edges. Show also that the state digraph of $\mathrm{TH}(\overrightarrow{C}_3)$ is strong.

8.5. Show that the Shift Cyclic Tower of Antwerpen and the Double Shift Cyclic Tower of Antwerpen are solvable.

8.6. Although not directly related to the Star puzzle, planarity of Star Hanoi graphs St_p^n is of some interest. Show that the only planar Star Hanoi graphs are St_p^0, St_p^1, St_3^n, and St_4^2.

8.7. Show that $\forall\, p \in \mathbb{N}_3 \;\forall\, n \in \mathbb{N}_0 : \; st_p^0(n) \le st_p^1(n)$.

8.8. Prove Theorem 8.21.

8.9. Show that $d_{K_4-e}(1^5, 2^5) < \widetilde{\Sigma}_5 \,(= \Sigma_5)$.

8.10. (Klavžar, Milutinović, and Petr [240]) Consider the p-in-a-row puzzle $\mathrm{TH}(P_{1+(p-1)})$, $p \in \mathbb{N}_3$. Show that the number of optimal solutions to transfer two discs from one extreme peg to the other is $C_p{-}2C_{p-1}$, where $C_n = \frac{1}{n+1}\binom{2n}{n}$, $n \in \mathbb{N}_0$, are the *Catalan numbers* (cf. [387]).

8.11. Let $p \in \mathbb{N}_3$ and $n \in \mathbb{N}_0$. Show that for all admissible pairs (i,j) of the digraph D_p (cf. p. 351) we have

$$d_{D_p}(i^n, j^n) \le d_{D_p}(0^n, (p-1)^n) = 3^n - 1 + n(p-3).$$

8.12. Determine the number of moves performed by the algorithm described in (8.24) that solves the task $0^n \to 2^n$ of the $\mathrm{TH}(\overrightarrow{C}_4)$. In addition find a solution for $n = 3$ discs with 18 moves and conclude that the algorithm of (8.24) is not optimal.

8.13. Let $T(D)$ be solvable. Show that D is a moderately enhanced cycle if and only if D contains no shed (that is, there is no vertex v of D such that the digraph $D - v$ contains a strong subdigraph on at least three vertices).

Chapter 9

Hints, Solutions and Supplements to Exercises

Chapter 0

Exercise 0.1   The ansatz leads to $\xi^{n+2} = \xi^{n+1} + \xi^n$, such that $\xi^2 = \xi + 1$, which is solved by $\tau := (1 + \sqrt{5})/2 > 0$ and $\sigma := (1 - \sqrt{5})/2 = -1/\tau < 0$.

Let $F_n = a\tau^n + b\sigma^n$ with $0 = F_0 = a + b$ and $1 = F_1 = a\tau + b\sigma = (a - b)\sqrt{5}/2$. Hence $a = 1/\sqrt{5} = -b$ and consequently $F_n = \dfrac{\tau^n - \sigma^n}{\sqrt{5}}$. (This formula is named for J. P. M. Binet.) Similarly, if $L_n = a\tau^n + b\sigma^n$ with $2 = L_0 = a + b$ and $1 = L_1 = 1 + (a - b)\sqrt{5}/2$, we get $L_n = \tau^n + \sigma^n$.

For $n \in \mathbb{N}$, we have

$$\frac{F_{n+1}}{F_n} = \frac{\tau^{n+1} - \sigma^{n+1}}{\tau^n - \sigma^n} = \frac{\tau - \sigma\left(\frac{\sigma}{\tau}\right)^n}{1 - \left(\frac{\sigma}{\tau}\right)^n}.$$

But $\sigma/\tau = -1/\tau^2$, whence $|\sigma/\tau| < 1$, such that $F_{n+1}/F_n \to \tau$, as $n \to \infty$. Similarly for L_n.

Remark 9.1. *While it is by an easy induction proof that one can show that there is only one mapping $F \in \mathbb{N}_0^{\mathbb{N}_0}$ which fulfills the recurrence (0.4) and we proved the existence by deducing Binet's formula, the question whether any such recurrence leads to a well-defined mathematical object is rather subtle and usually ignored in the literature. Because of the latter fact and since the* Fundamental Theorem of recursion *is of outstanding importance even to define such elementary things like addition of the natural numbers, we will state and prove it here for the connoisseur (and all those who want to become one). Its first application will be the definition of factorials in the next exercise.*

Fundamental Theorem of recursion. *Let M be a set, $\eta_0 \in M$, and $\varphi \in M^{\mathbb{N}_0 \times M}$. Then there exists exactly one mapping $\eta \in M^{\mathbb{N}_0}$ which fulfills the recurrence*

© Springer International Publishing AG, part of Springer Nature 2018
A.M. Hinz et al., *The Tower of Hanoi – Myths and Maths*,
https://doi.org/10.1007/978-3-319-73779-9_10

$$\eta(0) = \eta_0, \ \forall \, k \in \mathbb{N}_0 : \ \eta(k+1) = \varphi\,(k, \eta(k))\,.$$

Proof. Uniqueness. Let $\eta, \mu \in M^{\mathbb{N}_0}$ both fulfill the recurrence. Then $\eta(0) = \eta_0 = \mu(0)$ and if $\eta(k) = \mu(k)$ for $k \in \mathbb{N}_0$, then $\eta(k+1) = \varphi\,(k, \eta(k)) = \varphi\,(k, \mu(k)) = \mu(k+1)$. By induction we get $\eta = \mu$.

Existence. The wanted function $\eta \subseteq \mathbb{N}_0 \times M$ must have the following properties:

$$\forall \, k \in \mathbb{N}_0 \ \exists \, \eta_k \in M : \ (k, \eta_k) \in \eta, \tag{A}$$

$$\forall \, k \in \mathbb{N}_0 \ \forall \, m_1, m_2 \in M : \ (k, m_1), (k, m_2) \in \eta \Rightarrow m_1 = m_2, \tag{B}$$

$$(0, \eta_0) \in \eta \ \wedge \ \forall \, k \in \mathbb{N}_0 \ \forall \, m \in M : \ (k, m) \in \eta \Rightarrow (k+1, \varphi(k, m)) \in \eta. \tag{C}$$

Condition (C) is satisfied, e.g., by $\mathbb{N}_0 \times M$. Therefore we choose

$$\eta = \bigcap \{\mu \subseteq \mathbb{N}_0 \times M \mid \mu \text{ fulfills (C)}\}\,.$$

We now prove (A) and (B) simultaneously by induction on k. By definition, $(0, \eta_0) \in \eta$. Assume that $(0, \mu_0) \in \eta$ for some $M \ni \mu_0 \neq \eta_0$. Let $\mu := \eta \setminus \{(0, \mu_0)\}$. Then $(0, \eta_0) \in \mu$ and if $(\kappa, m) \in \mu$, then $(\kappa+1, \varphi(\kappa, m)) \in \mu$ as well, since no element of the form $(\kappa+1, \ell)$ has been taken away from η. Therefore, μ has property (C); but then $\eta \subseteq \mu$, in contradiction to the definition of the latter set.

For the induction step, put $\eta_{k+1} := \varphi(k, \eta_k)$. If $(k+1, m) \in \eta$ for some $M \ni m \neq \eta_{k+1}$, define $\mu := \eta \setminus \{(k+1, m)\}$. Again (C) is fulfilled for μ ($(0, \eta_0)$ has not been excluded and $(k+1, m) \neq (k+1, \varphi(k, \eta_k)))$, and we get the contradiction $\eta \subseteq \mu$. $\quad\square$

Exercise 0.2   a) Induction on ℓ. There is exactly one (injective) mapping from $[0] = \varnothing$ to $[k]$. An injection ι from $[\ell+1]$ to $[k]$, $\ell \in [k]_0$, can have k values for $\iota(\ell+1)$, and $\iota \upharpoonright [\ell]$ ($\upharpoonright$ stands for "restricted to") runs through all injections from $[\ell]$ to $[k] \setminus \{\iota(\ell+1)\}$, such that there are all in all $k\dfrac{(k-1)!}{((k-1)-\ell)!} = \dfrac{k!}{(k-(\ell+1))!}$ injections from $[\ell+1]$ to $[k]$.

The statement about bijections follows from the special case $\ell = k$, together with the Pigeonhole principle which excludes the possibility of an injection from $[k]$ to a smaller set.

b) Again by the Pigeonhole principle, there is no subset of $[k]$ with $\ell > k$ elements. For $k \geq \ell$, we have from (a) $\dfrac{k!}{(k-\ell)!}$ injective mappings from $[\ell]$ to $[k]$. Each element of $\dbinom{[k]}{\ell}$ occurs $\ell!$ times as a permuted image set of one of these injections.

c) For $k < \ell$, all three terms are 0. For $k = \ell$, we have $\dbinom{k+1}{\ell+1} = 1 = \dbinom{k}{\ell}$ and $\dbinom{k}{\ell+1} = 0$. If $k > \ell$, then $k+1 \geq \ell+1$, $k \geq \ell$, and $k \geq \ell+1$, such that from (b) we get

$$\binom{k}{\ell} + \binom{k}{\ell+1} = \frac{k!}{\ell!(k-\ell)!} + \frac{k!}{(\ell+1)!(k-\ell-1)!}$$

$$= \frac{(k+1)!}{(\ell+1)!(k-\ell)!} = \binom{k+1}{\ell+1}.$$

Exercise 0.3 a) The case $k = 0$ is just for definiteness: of course, the empty set $[0]$ can neither be permuted nor deranged in the ordinary sense of these words, but there is precisely one mapping from $[0]$ to $[0]$, and this mapping does not have a fixed point, such that $\overline{f}_0 = 1$. On the other hand, the only mapping from $[1]$ to $[1]$ clearly has a fixed point, whence $\overline{f}_1 = 0$.

Let $k \in \mathbb{N}$ and $\ell \in [k]$. Any derangement σ on $[k]$ can be modified to yield a derangement $\widetilde{\sigma}$ on $[k+1]$, namely by defining $\widetilde{\sigma} = \sigma$ on $[k] \smallsetminus \{\ell\}$, $\widetilde{\sigma}(\ell) = k+1$, and $\widetilde{\sigma}(k+1) = \sigma(\ell)$, i.e. by mapping ℓ on $k+1$ and the latter to the former image of ℓ. This makes up for $k \cdot \overline{f}_k$ derangements on $[k+1]$.

A derangement σ on $[k-1]$ can be manipulated as follows. Define $\tau : [k-1] \to [k] \smallsetminus \{\ell\}$ by $\tau(i) = i$ for $i < \ell$ and $\tau(i) = i+1$ otherwise. This yields a derangement $\widetilde{\sigma}$ on $[k+1]$ by putting $\widetilde{\sigma} = \tau \circ \sigma \circ \tau^{-1}$ on $[k] \smallsetminus \{\ell\}$, $\widetilde{\sigma}(\ell) = k+1$, and $\widetilde{\sigma}(k+1) = \ell$. This leads to another $k \cdot \overline{f}_{k-1}$ derangements on $[k+1]$ which are obviously different from the ones constructed earlier by the fact that they just switch $k+1$ with some $\ell \in [k]$. Conversely, a derangement on $[k+1]$ either has this property or not, such that it is easy to convince oneself that all of them are among the $k \left(\overline{f}_k + \overline{f}_{k-1} \right)$ constructed ones. We thus observed the two basic rules of counting, namely not to forget any object and not to count any twice.

b) To show that (0.25) follows from (0.26) is easy: clearly, $x_1 = 0$ and for $k \in \mathbb{N}$, we can add the two equations

$$x_{k+1} = (k+1)x_k - (-1)^k \,,$$
$$x_k = kx_{k-1} + (-1)^k \,,$$

to obtain $x_{k+1} + x_k = (k+1)x_k + kx_{k-1}$, whence $x_{k+1} = k\left(x_k + x_{k-1}\right)$.

As Euler admits himself, "however, it is not as easy" to derive (0.26) from (0.25). Therefore, we first prove formula (0.27) assuming (0.25) by induction, where the induction step for $k \in \mathbb{N}$ reads

$$x_{k+1} = k(x_k + x_{k-1})$$
$$= (k+1)! \left(\sum_{\ell=0}^{k-1} \frac{(-1)^\ell}{\ell!} + \frac{(-1)^k k}{(k+1)!} \right)$$
$$= (k+1)! \sum_{\ell=0}^{k+1} \frac{(-1)^\ell}{\ell!} \,.$$

Finally, the implication "(0.27) $\Rightarrow$ (0.26)" is trivial.

Let us mention that (0.26) is derived from the definition of a derangement in [45] by considering the set of permutations with exactly one fixed point.

Exercise 0.4 $1039_{10} = 10000001111_2$ and $11_{10} = 1011_2$, such that every bit of 11 is smaller than or equal to the corresponding one of 1039 and therefore $\binom{1039}{11}$ is odd according to (0.17).

Exercise 0.5 a) By virtue of Corollary 0.8 it suffices to prove the statement including the text in square brackets. Arrange the $2k$ odd vertices into pairs and declare these to be k new edges. Then all vertices are even, such that a constructive proof, for instance based on Fleury's algorithm, will show that the resulting graph is eulerian.

b) Delete one edge from the eulerian graph obtained in (a).

Remark. Since it is undeniably easier to build just one bridge instead of two, Wilson's claim about (closed) eulerian trails in Königsberg (cf. p. 38) is verified.

Exercise 0.6 Employ Fleury's algorithm: start in a vertex of odd degree, A say, avoid using a bridge (edge) that separates the remaining graph into two components, and delete used edges. For example: ABCDABDAC.

Exercise 0.7 We can employ Fleury's algorithm for the graph in Figure 9.1, whose vertices are all even. Starting with edge a, we follow Fleury to get the trail abcdef, here denoted by the labels of the edges.

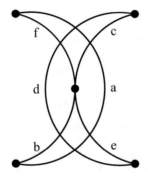

Figure 9.1: Muhammad's sign manual as a graph

Exercise 0.8 We will describe the algorithm informally. To check an edge ab for being a bridge in G, we search for a path from b to a in $G - ab$. Starting in b, we screen all neighbors of b, then neighbors of neighbors and so on, but skipping those already encountered. (Since we stay as close to the **root** b as possible, this is called a *breadth-first search (BFS)*). The result is a tree rooted in b and **spanning** its connected component in $G - ab$. If a is not on that tree, then ab is a bridge.

For a practical realization, we may construct a queue of vertices to be checked for their neighbors. Initially we define $q_v = n := |G|$ for all $v \in V(G)$. The length ℓ of the queue is put to 1 and $q_b = 1$, such that vertex b is now first in the line. Then we search successively for its neighbors u in $G - ab$ which have not yet been visited, i.e. with $q_u = n$. If we encounter a, we stop: ab is not a bridge. Otherwise we increase ℓ by 1 and put $q_u = \ell$. After all neighbors of b have been scanned, we decrease ℓ by 1 and also all $q_u \in [n-1]$ and restart the search for neighbors of the first vertex v in the queue, that is with $q_v = 1$, as long as there is any, i.e. $\ell = |\{u \in V(G) \mid q_u \in [n-1]\}| \neq 0$.

Exercise 0.9 Recall the group axioms, where $\rho, \sigma, \tau \in S_n$:

0. $\sigma \circ \tau \in S_n$,

1. $\rho \circ (\sigma \circ \tau) = (\rho \circ \sigma) \circ \tau$,

2. $\sigma \circ \iota = \sigma = \iota \circ \sigma$,

3. $\exists \sigma^{-1} : \sigma \circ \sigma^{-1} = \iota = \sigma^{-1} \circ \sigma$.

All these axioms apply by virtue of the corresponding properties of the composition of functions. Commutativity is clear for $n \leq 2$. Otherwise, let, e.g., $\sigma = 2314\ldots n$ and $\tau = 2134\ldots n$; then $\sigma \circ \tau = 3214\ldots n \neq 1324\ldots n = \tau \circ \sigma$.

Exercise 0.10   a) Coming from L T, the next three moves are $\mu \lambda \mu$, which can be found twice in both equations (0.18) and (0.19), such that there are the four solutions

$$\begin{array}{l} \text{L T S R Q Z B G H X W V J K F D C P N M}, \\ \text{L T S R Q Z X W V J H G B C P N M D F K}, \\ \text{L T S R Q Z B C P N M D F G H X W V J K}, \\ \text{L T S R Q P N M D C B Z X W V J H G F K}. \end{array}$$

b) Here the triple is $\mu \lambda^2$, occurring only in one kind in each equation, such that we have the two solutions

$$\begin{array}{l} \text{J V T S R W X Z Q P N M L K F D C B G H}, \\ \text{J V T S R W X H G F D C B Z Q P N M L K}. \end{array}$$

**Exercise 0.11   ** Let us assume that $K_{3,3}$ is planar. Since it is 2-colorable, it does not contain any triangle. Therefore $4\|\|K_{3,3}\|\| \leq 2\|K_{3,3}\|$, such that with (0.20), we arrive at $9 = \|K_{3,3}\| \leq 2(|K_{3,3}| - 2) = 8$, a contradiction.

**Exercise 0.12   ** Hint: Reduce the careful brothers problem to the *missionaries and cannibals problem*: three missionaries and three cannibals want to cross the river with the same boat as before; the cannibals must never outnumber the missionaries on a river bank.

Identifying women with cannibals and men with missionaries, it is clear that every solution of the careful brothers problem leads to a solution of the missionaries and cannibals problem, because individuals do not play a role in the latter and women can never outnumber men on a bank since otherwise a woman would be without her brother. So if we solve missionaries and cannibals problem, we only have to verify that the solution will also satisfy the careful brothers problem. We will stay with the women/men notation and denote by mw a constellation of $m \in \{0, 1, 2, 3\}$ men and $w \in \{0, 1, 2, 3\}$ women on the bank where the boat is present. By the jealousy condition, only 10 of the 16 combinations are admissible: 00, 01, 02, 03, 11, 22, 30, 31, 32, 33. Of these 00 (no transfer possible), 01 (can only be reached from 33 and leads back there) and 30 (more women than men on the other bank after transit) can be excluded immediately. The remaining 7 constellations lead to the graph in Figure 9.2, whose edges are boat transfers labelled with the passenger(s).

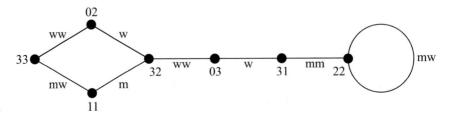

Figure 9.2: The graph of the missionaries and cannibals problem

Note that this graph contains a **loop** at vertex 22. Quite obviously, there are precisely four optimal solutions of length 11. One of the solutions has been given in Latin verses (cf. [345, p. 74]): *Binae, sola, duae, mulier, duo, vir mulierque, Bini, sola, duae, solus, vir cum mulier.*

In fact, all solutions lead to solutions of the careful brothers problem, where now individual identities increase the number of different solutions. For instance the transfer $33 \to 02$ can be performed with any of the three pairs of women, whereas the reverse transfer has to be done with the only pair present. A combinatorial analysis of all cases based on Figure 9.3 leads to 486 solutions altogether. It seems that this number has never been given in the abundant literature of the problem.

Variations include higher numbers of couples and (remember Königsberg!) an island in the middle of the river; cf. [345]. The problem has also been employed as an example for problem representations, i.e. models, in the Artificial Intelligence literature; cf. [15].

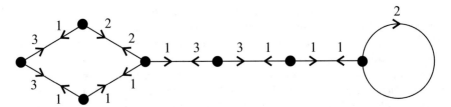

Figure 9.3: Numbers of alternatives in the careful brothers problem

Exercise 0.13

A solution, which as usual assumed the largest disc to move only once and which contains a flaw in the definition of the "related problem", was provided by L. Badger and M. Williams in [29].

Let x_{n+1}, $n \in \mathbb{N}_0$, be the necessary (and sufficient) number of moves to transfer a perfect tower of $n + 1$ discs, i.e. all stacked on one peg in order of decreasing size from bottom to top, following the relaxed rule. Then, as in the proof of Theorem 0.26, before the first move of the largest disc, the discs of a perfect n-tower have to be transferred to another peg, but this time with the even more relaxed rule that on the starting peg there is no restiction because it is always occupied by the overall largest disc. The minimal number of moves to achieve this subgoal defines the sequence y_n. Then, after the last move of the largest disc all other discs have to be ordered to form a perfect n-tower above the largest disc. By symmetry, this takes another unavoidable y_n moves, whence first and last move of the largest disc are the same and $x_{n+1} = 2y_n + 1$.

Obviously, $y_0 = 0$ and to transfer $n \in \mathbb{N}$ discs of a perfect tower to some arrangement on another peg, we first have to bring $n - 1$ discs to an intermediate peg, which takes y_{n-1} moves because we are starting from the same peg. Then move the largest disc and finally the $n - 1$ smaller ones directly, one by one, on top of it. Since all these moves are necessary (and sufficient), we get $y_n = y_{n-1} + 1 + (n - 1) = y_{n-1} + n$. It follows from (0.22) that $y = \Delta$ and consequently $x_{n+1} = 2\Delta_n + 1 = n^2 + n + 1$. (The sequence $x = 0, 1, 3, 7, 13, 21, 31, 43, 57, \ldots$ is essentially A002061.)

Chapter 1

Exercise 1.1 For $n \in \mathbb{N}$ the path of length $M_n = 2^n - 1$ from $\alpha^{(n)}$ to $\omega^{(n)}$ in R^n can be dissected into the two paths from 0^n to 1^n and from $1^n = 11^{n-1}$ to $\omega^{(n)} = 10^{n-1}$ of lengths ℓ_n and ℓ_{n-1}, respectively.

Exercise 1.2 In every move precisely one bit is switched. This yields the vertex coloring. It can be proved by induction that moves of type 0 and 1 are alternating along a shortest path from $\alpha^{(n)}$ to $\omega^{(n)}$, obviously starting and ending with move type 0.

Exercise 1.3 **a)** The state graph is constructed by adding edges $\{\underline{s}00, \underline{s}11\}$ with $\underline{s} \in B^{n-2}$ to R^n, just like the red edges in Figure 9.4.

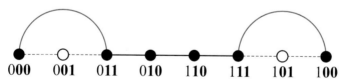

000   001   011   010   110   111   101   100

Figure 9.4: The graph R^3 with double moves (red edges)

b) We may define the graph $\widetilde{R}^n$ from the previous graph by deleting all vertices of the form $\underline{s}01$, because they subdivide the new edges. It is a path graph again, whose length is $2^n - 1 + 2^{n-2} - 2 \cdot 2^{n-2} = 3 \cdot 2^{n-2} - 1$.

c) As in (1.2), we have $p_1 = 1$, $\forall n \in \mathbb{N}: p_{n+1} + p_n = 3 \cdot 2^{n-1} - 1$. For $\gamma_n := 2^{n-1} - p_n$ this means that $\gamma_1 = 0$ and $\forall n \in \mathbb{N}: \gamma_{n+1} = 1 - \gamma_n$, whence

$$p_n = 2^{n-1} - (n \text{ even}).$$

Exercise 1.4   Hint: $(2^n - 1) \mod 3 = n \mod 2$. For $n = 0$, $C = V$ and $|C| = 1$. For $n \in \mathbb{N}$, either 0^n or $0^{n-1}1$ is a codeword. Consequently, the codewords are those vertices $s \in B^n$ with $\mathrm{d}(s) \mod 3 = 0$ or 1, respectively. From the hint it follows that for even n necessarily $\alpha^{(n)}$ and $\omega^{(n)}$ are codewords and that $|C| = \frac{1}{3}(2^n + 2)$; for odd n there are two perfect codes, containing either $\alpha^{(n)}$ or $\omega^{(n)}$ and with $|C| = \frac{1}{3}(2^n + 1)$.

Exercise 1.5 We are looking for the state $s \in B^{14}$ which lies on the path R^{14} between 1^{14} and 0^{14} at a distance 9999 from the former. Since $\beta_{14} = \mathrm{d}(1^{14}, s) + \mathrm{d}(s, 0^{14})$, we get

$$\mathrm{d}(s, 0^{14}) = \left[\frac{2}{3}(2^{14} - 1)\right] - \mathrm{d}(1^{14}, s)$$
$$= 10922 - 9999 = 923$$
$$= 00001110011011_2.$$

We can now apply the Gray code automaton in Figure 1.5 which yields $s = 00001001010110$. Therefore, rings 2, 3, 5, 7, and 10 are on the bar, all others are off.

Exercise 1.6   Obviously, $\mathrm{W}(P_1) = 0$, and for a pendant vertex e of P_k, $k > 1$, we have

$$\mathrm{W}(P_k) = \sum_{v \in V(P_k)} \mathrm{d}(e, v) + \mathrm{W}(P_{k-1})$$
$$= \Delta_{k-1} + \mathrm{W}(P_{k-1}) = \binom{k}{2} + \mathrm{W}(P_{k-1}).$$

Comparing this with Exercise 0.2 c, we realize that $\mathrm{W}(P_{k-1})$ fulfills the same recurrence as $\binom{k}{3}$, such that $\mathrm{W}(P_k) = \binom{k+1}{3}$.

Exercise 1.7 The formula $g_{2^n} = n + 1$ follows directly from Corollary 1.10. Let $k = 2^r(2s + 1)$ with $r \in [n]_0$ and $s \in [2^{n-r-1}]_0$. Then

$$2^n \pm k = 2^r(2^{n-r} \pm 2s \pm 1) = 2^r(2s' + 1)$$

with $s' = 2^{n-r-1} + s$ or $s' = 2^{n-r-1} - s - 1$, depending on whether $\pm$ is plus or minus; in any case $s' \geq 0$, because $s \leq M_{n-r-1}$. Again by Corollary 1.10 we get $g_{2^n \pm k} = r + 1 = g_k$.

Exercise 1.8   Write $k = 2^r(2s + 1)$, $r, s \in \mathbb{N}_0$. Then observe that in the binary representation $(b_n b_{n-1} \ldots b_1 b_0)_2$ of k we have $b_0 = b_1 = \cdots = b_{r-1} = 0$ and $b_r = 1$. Hence $\widetilde{g}_k = r$. Since $g_k = r + 1$ by Corollary 1.10, the conclusion follows.

Exercise 1.9   The recursion for q follows immediately from the binary representation of k. But then $g_k = q(k-1) - q(k) + 2$, because the right-hand side fulfills the recursion of the Gros sequence in Proposition 1.9. Summing $\widetilde{g}_k = q(k-1) - q(k) + 1$ for k from 1 to n yields Legendre's formula.

Exercise 1.10 Hint: Starting at 0^n, walk along the edges of the cube by switching b_{g_k} in the k-th step for $k \in [2^n - 1]$. This path will end in $0^{n-1}1$, corresponding to a corner of the n-cube which is linked to our starting corner by an extra edge.

Exercise 1.11 We will show that Corollary 1.10 remains valid with g replaced by a. We first claim that for any $r \in \mathbb{N}_0$, (i) $a_{2^r} = r + 1$, and that (ii) $a_k < r + 1$ for any $k < 2^r$.

We proceed by induction on r, the cases $r = 0$ and $r = 1$ being trivial. Let $r \geq 2$ and suppose that $a_{2^{r-1}} = r$ and that the terms before $a_{2^{r-1}}$ are smaller than r. The $2^{r-1} - 1$ terms after the term $a_{2^{r-1}}$ constructed by the greedy algorithm coincide with the first $2^{r-1} - 1$ terms of the sequence. Indeed, if we would construct a square, it would not contain $a_{2^{r-1}}$ because by the induction hypothesis the terms before it are smaller. So a square would have to appear before (or, equivalently, after) $a_{2^{r-1}}$, which is not possible. Hence we have a sequence containing $(2^{r-1}-1) + 1 + (2^{r-1}-1) = 2^r - 1$ terms of the form $1, 2, \ldots, 1, r-1, 1, 2, \ldots, 1, r, 1, 2, \ldots, 1, r-1, 1, 2, \ldots, 1$. Now we see that $a_{2^r} = r + 1$, for otherwise a square would be constructed. Moreover, the above argument also implies part (ii) of the claim.

Suppose now that $k = 2^r(2s+1)$, where $s \in \mathbb{N}_1$. Let $t \in \mathbb{N}$ be selected such that $2^t < k < 2^{t+1}$. To show that $a_k = r + 1$ in this case too, we proceed by induction on t, where the case $t = 1$ is straightforward. So let $t \geq 2$. Then by the above, $a_k = a_{k-2^t}$. Since $k - 2^t = 2^r(2s+1) - 2^t = 2^r((2s+1) - 2^{t-r})$ and $(2s+1) - 2^{t-r}$ is odd, the induction hypothesis implies $a_{k-2^t} = r + 1$. But then $a_k = r + 1$ as well.

As an alternative, we may prove Proposition 1.11 by using induction on $l \in \mathbb{N}$ to show that $a_{2l-1} = 1$, $a_{2l} = a_l + 1$ (cf. Proposition 1.9).

Exercise 1.12 Hint: Use the fact that g is the greedy square-free sequence and argue that g is also the greedy strongly square-free sequence.

Chapter 2

Exercise 2.1 Observe that for any $r, s \in T^m$, $r \neq s$, $rT^{n-m} \cap sT^{n-m} = \varnothing$. The rest follows from the fact that sT^{n-m} is the set of regular states with the largest m discs fixed in distribution s.

Exercise 2.2 We may assume that the transfer of the tower is from peg 0 to peg 2. Replace the induction step in the proof of Theorem 2.1 by the following argument. Before the first move of disc d, to peg i, say, state $0^{n-d}0(3-i)^{d-1}$ must be reached moving only discs from $[d-1]$ in the minimum number of moves, which by (strong) induction assumption is $2^{d-1} - 1$.

The statement about the last step follows by symmetry, since every minimal solution from 0^n to 2^n is a minimal solution from 2^n to 0^n taken in inverse move order.

The statement about the disc moved in step k in Theorem 2.1 then follows a posteriori from the characterization of the Gros sequence given in (1.6).

Exercise 2.3   Hint: Use the recursive solution of the problem and induction.

Exercise 2.4   Hint: Start with h.

Exercise 2.5   Let us assume that disc n and all discs of the same parity are made of gold, the others from silver. Moreover, the bases of pegs 0 and 2 should be silver, that of peg 1 golden. Then, apart from the divine rule, of course, follow just two commandments:

(o) Never undo the move just made.

(i) Never place a disc onto a disc or base of the same metal.

Rule (o) is obvious for a shortest solution path and was the only rule to be followed to get from $\alpha^{(n)}$ to $\omega^{(n)}$ in the CR. Rule (i) will guarantee by Proposition 2.3 that you stay on the optimal path.

Exercise 2.6   Let b_n, $n \in \mathbb{N}_0$, be the number of legal arrangements on the source peg i in the optimal solution with n discs. Note that b_n is at the same time the number of legal arrangements of discs on the goal peg j. Then $b_0 = 1 = F_2$ and $b_1 = 2 = F_3$. For $n \geq 2$, we first move discs 1 to $n-1$ onto the intermediate peg $3 - i - j$, yielding b_{n-1} arrangements on the source peg. After disc n is moved to the goal peg, discs 1 to $n-2$ are moved from peg $3-i-j$ to peg i, yielding b_{n-2} arrangements on the source peg. Now prove that these two sets of arrangements are disjoint and observe that after that no new arrangement on the source peg is obtained. It follows that $b_n = b_{n-1} + b_{n-2}$. The argument is completed by the fact that $a_n = b_{n-1}$.

A second proof can be based on Proposition 2.3. By that result, b_n is also the number of admissible arrangements on peg i. Now let $\nu \in \mathbb{N}_0$. If $n = 2\nu$, then peg i is either empty (we only count movable discs) or the bottom disc is $2k + 2$, $k \in [\nu]_0$, leading to b_{2k+1} admissible arrangements on i, i.e.

$$b_{2\nu} = 1 + \sum_{k=0}^{\nu-1} b_{2k+1}. \tag{9.1}$$

If $n = 2\nu+1$, then again peg i is either empty or the bottom disc is $2k+1$, $k \in [\nu+1]_0$, leading to b_{2k} admissible arrangements on peg i, i.e.

$$b_{2\nu+1} = 1 + \sum_{k=0}^{\nu} b_{2k}. \tag{9.2}$$

From (9.1) it follows that $b_{2\nu} + b_{2\nu+1} = b_{2\nu+2}$, and from (9.2) we get $b_{2\nu+1} + b_{2\nu+2} = b_{2\nu+3}$, i.e. the recurrence relation of the Fibonacci sequence is satisfied by $(b_n)_{n \in \mathbb{N}_0}$. (In fact, (9.1) and (9.2) correspond to an equivalent characterization of that sequence; cf. [288, p. 5].)

Exercise 2.7 The proof is by induction on n. The case $n = 1$ is clear. Let $n \in \mathbb{N}$. Then disc $n + 1$ moves by $(j - i) \bmod 3$ only once in move number 2^n. Before that move, disc $d \in [n]$ is relocated in moves $(2k + 1) \, 2^{d-1}$, $k \in [2^{n-d}]_0$, by

$$(((3 - i - j) - i)\,((n - d) \bmod 2 + 1)) \bmod 3$$
$$= ((i - j)\,((n - d) \bmod 2 + 1)) \bmod 3$$
$$= ((j - i)\,(((n + 1) - d) \bmod 2 + 1)) \bmod 3$$

from induction assumption. Similarly for the moves after the largest disc's move, when $k \in [2^{n+1-d}]_0 \setminus [2^{n-d}]_0$.

Exercise 2.8 We assume the initial needle to be 0, the goal to be 2. The age of the universe (in *moves*) can be determined if we know the state $s \in T^{64}$ of the Tower of Brahma *after* the current move of disc δ. If $\delta \neq 1$, this move takes place between the two pegs which are not occupied by disc 1, and by Proposition 2.4 the parity of δ decides on the direction of the move. If $\delta = 1$, we know that $\mathrm{d}(s, 0^n)$ has to be odd. Although we do not know the value of s_1, we can run the recursion in (2.5) down to $d = 2$ and put $\ell_0 = 1$.

Disc 51, being odd, necessarily moves from 0 via 1 to 2. Therefore the goal of the current move is needle 2 because disc 1 lies on needle 0. Therefore, $s = 0^5 1^2 0^5 2^2 0^{50}$ and application of (2.5) yields that the move number is $2^{58} + 2^{57} + 2^{51} + 2^{50}$ (one may countercheck with equations (2.3) and (2.4)), such that the age of the universe is about 13.8 Gyr (billion years). (So it seems that the Brahmins have not yet touched disc 60 and that the universe will survive for a while.)

Exercise 2.9 The task is to get from state 01210021 to perfect state 1^8. We apply Algorithm 10, starting with $k = 1$ corresponding to the goal peg. Each of the discs, except disc 5, leads to a change of state of the P1-automaton of Figure 2.6, such that the algorithm returns $\delta = 1$ and $\mu = 239$; it ends in state 2. Hence we know that the minimal solution has length 239 and starts with the move of disc 1 from peg 1 to peg 0. Thereafter, one can use the statement of Proposition 2.9 to obtain the rest of the optimal path.

Exercise 2.10 a) Apply Algorithm 10 for i and j, i.e. on two copies if the P1-automaton, starting one in state i, the other in state j. Then, as long as $s_d = k$, the states of the automata will just be switched and 2^{d-1} be added to μ. As soon as $s_d \neq k$, however, only one of the automata will change its state and add 2^{d-1} to μ. This can not be compensated during the rest of the performance.

b) The initial state is $s = 0^n 1^n$, and the result will depend on the parity of n. If either n is odd and the goal peg is 1 or n is even and the goal is 2, then the P1-automaton changes its state with every input of an s_d, such that all in all $4^n - 1$ moves are necessary. Otherwise, the automaton stays in the same state after the inputs from the n largest discs, such that only $4^n - 2^n$ moves are necessary. Quite obviously, the distance of s to the perfect state 0^{2n} is just $2^n - 1$.

Exercise 2.11 Moving one step from s to t, say, s and t differ in precisely one tit: $s_d \neq t_d$. But then $\mathrm{d}(s, k^n)$ and $\mathrm{d}(t, k^n)$ can differ by at most 1, and by Lemma 2.8

they cannot all be pairwise equal. The rest follows from Proposition 2.13 as applied to s and t, respectively.

Exercise 2.12   We may assume that $j = 1$. Then, by Proposition 2.13, $d(s, 0^n) + d(s, 2^n) = 2^n - 1$. Hence s lies on the shortest path from perfect state 0^n to perfect state 2^n. There are 2^n such states.

Exercise 2.13   The number of moves can be obtained from Proposition 2.20 as the value of $d(s, 2^8)$, which is 182. This number can also be found by applying the P1-automaton to the task to get from $(02)^4$ to 1^8. We cannot use the information about the idle peg for this task directly, because we are dealing with the inverse task. However, since disc 8 will only move once from peg 1 to peg 0, the optimal solution will start with the transfer of a 7-tower from 1 to 2, which means that disc 1 is moved to peg 2 at the beginning. The rest of the solution should also be clear.

Exercise 2.14   We may assume that $i = 0$ and $j = 2$. According to Algorithm 10 we start in state 2 of the P1-automaton and enter the inputs s_d from $d = n$ down to $d = 1$. If during execution state 0 of the automaton is never reached, then $d(s, 2^n) = \Psi\left(s^{-1}(\{0\})\right)$ by (2.8), because the same powers of 2 have been summed up. Otherwise, on the first transit to state 0 of the automaton caused by disc δ, say, $s_\delta \neq 0$ and so $2^{\delta-1}$ is only counted for the distance between s and 2^n, such that $d(s, 2^n) > \Psi\left(s^{-1}(\{0\})\right)$.

Let $\nu \in [n+1]_0$ and $d_0 := n + 1 > d_1 > \cdots > d_\nu > 0 =: d_{\nu+1}$, $E = \{d_k \mid k \in [\nu]\}$. Then equality in (2.42) can hold for an s only if $s_d = 0$ for $d \in E$ and otherwise $s_d = 1 + (d_{2\ell+1} < d < d_{2\ell})$ for $\ell \in \left[\lfloor \frac{\nu}{2}\rfloor + 1\right]_0$. Hence, $s = 2^{n_0}01^{n_1}0\ldots0i^{n_\nu}$ with $n_k = d_k - d_{k+1} - 1$, $k \in [\nu+1]_0$, and $i = 2 - (\nu \bmod 2)$ is the idle peg in the best first move from s to 2^n in which disc d_ν moves from peg 0 to peg $3 - i$.

Remark. Note that the mapping $E \mapsto s$ is injective, so that equality holds in (2.42) for exactly 2^n vertices in H_3^n. The right-hand side of (2.42) is 0 if and only if $s^{-1}(\{i\}) = \varnothing$, i.e. for 2^n vertices, but we actually have

$$\forall s \in T^n : \ d(s, j^n) \geq \max\left\{\Psi\left(s^{-1}(\{i\})\right) \mid i \in T \smallsetminus \{j\}\right\}, \tag{9.3}$$

the right-hand side of which is 0 if and only if $s = j^n$, in which case the left-hand side is 0 too. Except for this trivial case, the maximum is attained for one i only; consequently, equality in (9.3) holds for exactly $2^{n+1} - 1$ vertices ($s = j^n$ occuring twice).

The estimate in (2.42) is [404, Theorem 1] and [62, (5.1)]; our (2.43) is [404, Theorem 2]. Stockmeyer gives an ad hoc induction proof for his Theorem 1, thus obtaining an independent proof of $d(0^n, 1^n) \geq 2^n - 1$ [404, Corollary] (cf. Theorem 0.26).

Formula [62, (5.2)] says that

$$\forall s, t \in T^n : \ d(s, t) \geq \Psi([n]) - \Psi\left(s^{-1}(\{j, 3-i-j\})\right) - \Psi\left(t^{-1}(\{i, 3-i-j\})\right). \tag{9.4}$$

Obviously, (2.42) follows from (9.4) by putting $t = j^n$. But (9.4) also follows from (2.42). To see this, we may assume that $i = 0$ and $j = 2$, i.e. we show that

$$\forall\, s, t \in T^n : \; \mathrm{d}(s,t) \ge \Psi([n]) - \Psi\big(s^{-1}(\{2,1\})\big) - \Psi\big(t^{-1}(\{0,1\})\big). \tag{9.5}$$

The case $n = 0$ is trivial (with $=$), so we may assume $n \in \mathbb{N}$. Then, with $s = s_n \bar{s}$ and $t = t_n \bar{t}$,

$$\Psi([n]) - \Psi\big(s^{-1}(\{2,1\})\big) - \Psi\big(t^{-1}(\{0,1\})\big)$$

$= ((s_n = 0) + (t_n = 2) - 1)\cdot 2^{n-1} + \Psi([n-1]) - \Psi(\bar{s}^{-1}(\{2,1\})) - \Psi(\bar{t}^{-1}(\{0,1\}))$

$\big(\le \Psi([n-1]) - \Psi(\bar{s}^{-1}(\{2,1\})) - \Psi(\bar{t}^{-1}(\{0,1\}))$, if $s_n = t_n$, with $<$, if and only if $s_n = 1 = t_n$; so we may assume $s_n \ne t_n\big)$

$= ((s_n = 0) + (t_n = 2) - 2)\cdot 2^{n-1} + 1 + 2\Psi([n-1]) - \Psi(\bar{s}^{-1}(\{2,1\})) - \Psi(\bar{t}^{-1}(\{0,1\}))$

$= ((s_n = 0) + (t_n = 2) - 2)\cdot 2^{n-1} + 1 + \Psi(\bar{s}^{-1}(\{0\})) + \Psi(\bar{t}^{-1}(\{2\})). \tag{$*$}$

From (2.43) we have

$$\Psi(\bar{s}^{-1}(\{0\})) \le \mathrm{d}(\bar{s}, 2^{n-1}), \; \Psi(\bar{s}^{-1}(\{0\})) \le \mathrm{d}(\bar{s}, 1^{n-1});$$

$$\Psi(\bar{t}^{-1}(\{2\})) \le \mathrm{d}(\bar{t}, 0^{n-1}), \; \Psi(\bar{t}^{-1}(\{2\})) \le \mathrm{d}(\bar{t}, 1^{n-1}).$$

Moreover (see (2.8)),

$$\mathrm{d}(\bar{s}, 1^{n-1}) < \mathrm{d}(\bar{s}, 0^{n-1}) + 2^{n-1}, \; \mathrm{d}(\bar{t}, 1^{n-1}) < \mathrm{d}(\bar{t}, 2^{n-1}) + 2^{n-1},$$

whence from ($*$) we obtain for $k, \ell \in T$:

$$\Psi([n]) - \Psi\big(s^{-1}(\{2,1\})\big) - \Psi\big(t^{-1}(\{0,1\})\big)$$
$$\le ((s_n = 0) + (t_n = 2) - 2 + (k = 0) + (\ell = 2))\cdot 2^{n-1} + 1 + \mathrm{d}(\bar{s}, k^{n-1}) + \mathrm{d}(\bar{t}, \ell^{n-1}),$$

where the inequality is strict, if $k = 0$ or $\ell = 2$.

Now $\mathrm{d}(s,t)$ is either $\mathrm{d}(\bar{s}, (3 - s_n - t_n)^{n-1}) + 1 + \mathrm{d}(\bar{t}, (3 - s_n - t_n)^{n-1})$ or $\mathrm{d}(\bar{s}, t_n^{n-1}) + 1 + 2^{n-1} + \mathrm{d}(\bar{t}, s_n^{n-1}) > \mathrm{d}(\bar{s}, t_n^{n-1}) + 1 + \mathrm{d}(\bar{t}, s_n^{n-1})$. An inspection of all 6 combinations of s_n and t_n proves (9.5), with equality implying $s_n = 0$ and $t_n = 2$. (Recall that n has replaced the largest d for which $s_d \ne t_d$; (9.4) is also true for $s = k^n = t$, $k \in T$, with equality if and only if $k \ne 1$.)

This discussion also shows that equality in (9.5) can, apart from the case $s = 2^n = t$, hold only if $s_n = 0$ and $t_n = 2$, when the right-hand side is $\Psi(\bar{s}^{-1}(\{0\})) + 1 + \Psi(\bar{t}^{-1}(\{2\}))$ and $\mathrm{d}(s,t) = \mathrm{d}(\bar{s}, 1^{n-1}) + 1 + \mathrm{d}(\bar{t}, 1^{n-1})$, so that $\Psi(\bar{s}^{-1}(\{0\})) = \mathrm{d}(\bar{s}, 1^{n-1})$ and $\Psi(\bar{t}^{-1}(\{2\})) = \mathrm{d}(\bar{t}, 1^{n-1})$, and we know from above that this is the case for precisely 4^n pairs (s,t).

The right-hand side of (9.5) is 0, if and only if $\forall\, d \in [n]: s_d = 0 \Leftrightarrow t_d \ne 2$, and otherwise for $D := \max\{d \in [n] \mid (s_d = 0) \ne (t_d \ne 2)\}$ the right-hand side is < 0, if and only if $s_D \ne 0$ and $t_D \ne 2$.

As before, (9.4) actually gives rise to several estimates, namely depending on which of the 6 pairs (i,j) has been chosen, i.e.

$$\mathrm{d}(s,t) \ge \max\Big\{\Psi([n]) - \Psi\big(s^{-1}(\{j, 3 - i - j\})\big) - \Psi\big(t^{-1}(\{i, 3 - i - j\})\big) \mid (i,j) \in T^2\Big\}.$$

A further combinatorial analysis can decide whether this is a good estimate or not.

Exercise 2.15 There are exactly $2 \cdot 3^{n-k}$ states $s \in T^n \setminus \{0^n\}$ for which the algorithm stops after $k \in [n]$ inputs. We therefore have to calculate

$$\sum_{k=1}^{n} k \cdot 3^{n-k} = \sum_{k=0}^{n-1} (n-k) \cdot 3^k = n \sum_{k=0}^{n-1} 3^k - \sum_{k=1}^{n-1} k \cdot 3^k \, .$$

The first sum on the right-hand side is equal to $\dfrac{3^n - 1}{2}$ by Lemma 2.18, and for the second sum we get $\dfrac{3 + (2n-3) \cdot 3^n}{4}$ from formula [162, (2.26)]. So the number of tits entered for all s adds up to $\frac{3}{2}(3^n - 1) - n$. This means that on the average $\frac{3}{2} - n(3^n - 1)^{-1}$ tits are checked, i.e. asymptotically, for large n, just $\frac{3}{2}$.

Exercise 2.16 By the definition of TH it follows easily that the perfect states are of degree 2 and all the other vertices are of degree 3.

Exercise 2.17 We may, of course, assume that $n \in \mathbb{N}$. The vertices can be colored with colors 0, 1, and 2 according to Proposition 2.24. Each of the 3^{n-1} subgraphs induced by the three vertices $s0$, $s1$, and $s2$ with $s \in H_3^{n-1}$ (they are isomorphic to the complete graph K_3) can obviously be totally colored with colors 0, 1, and 2, respecting the vertex colors already assigned. For $n > 1$, the remaining edges, representing moves of discs 2 to n, can then be colored using color 3.

Exercise 2.18 One way to prove this is to combine the Handshaking Lemma with Exercise 2.16. Other approaches are to solve the recurrence

$$\|H_3^0\| = 0, \ \|H_3^{1+n}\| = 3(\|H_3^n\| + 1) \, ,$$

stemming from (2.12), or to use (2.11) directly.

Exercise 2.19 **a)** The proof is by induction on n, the case $n = 1$ being obvious. For the induction step note first that any hamiltonian path from i^{1+n} to j^{1+n} cannot use the edge between ik^n and jk^n. But then the hamiltonian path is obtained uniquely from induction assumption, putting together the unique hamiltonian paths from i^{1+n} to ij^n on iH_3^n, from kj^n to ki^n on kH_3^n, and from ji^n to j^{1+n} on jH_3^n, respectively, using the edges $\{ij^n, kj^n\}$ (with idle peg j) and $\{ki^n, ji^n\}$ (with idle peg i). Also by induction assumption none of the other edges involved use k as the idle peg. As an illustration, the hamiltonian paths between i^4 and j^4, can be seen in Figure 2.10 by deleting the edges colored like vertex k^4.
b) Hint: Use part (a). In Figure 9.5, follow the 27(!) letters of the English alphabet on the graph H_3^3.
c) Clearly, H_3^0 and H_3^1 are eulerian. All the others are not even semi-eulerian, because there are $3^n - 3 > 2$ vertices of odd degree 3.

Exercise 2.20 Hint: Every move changes the parity of the number of discs on peg 0.

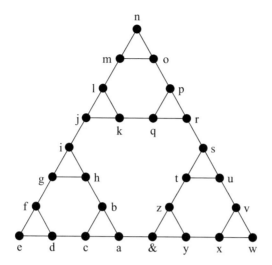

Figure 9.5: Hamiltonian labeling of H_3^3

Exercise 2.21 For fixed $d \in \mathbb{N}$ use induction on $n - d \in \mathbb{N}_0$. The number of moves of disc d is $2 \cdot 3^{n-d}$.

Exercise 2.22 a) Let us write the right-hand side in (2.13) as $\widetilde{d}_r$. We will prove $d_r = \widetilde{d}_r$ by induction on n. The case $n = 0$ is trivial. Let $s = s_{n+1}\overline{s}$ with $\overline{s} \in T^n$ for some $n \in \mathbb{N}_0$. Since $H_{3,\mathrm{lin}}^{1+n}$ is a path graph, we have the following cases depending on the value of s_{n+1}:

$$
\begin{aligned}
d(0\overline{s}) &= d(\overline{s}, 2^n) + 3^n = 3^n + (3^n - 1 - d(\overline{s})), \\
d(1\overline{s}) &= d(\overline{s}, 1^n) = d(\overline{s}), \\
d(2\overline{s}) &= d(\overline{s}, 1^n) + 2 \cdot 3^n = 2 \cdot 3^n + d(\overline{s}).
\end{aligned}
$$

For the calculation of $\widetilde{d}$ we note that $\widetilde{\overline{s}}_{n+1} = 0$, whence $\widetilde{d}(s)_{n+1} = (1 - s_{n+1}) \bmod 3$, in accordance with $d(s)_{n+1}$. For $r \in [n]$ we observe

$$
\widetilde{\overline{s}}_r = (s_{n+1} = 0) + \widetilde{\overline{s}}_r,
$$

which is $\widetilde{\overline{s}}_r$, if $s_{n+1} \neq 0$ and $1 - \widetilde{\overline{s}}_r$ otherwise. We therefore get $\widetilde{d}(s)_r = (1 + (2\widetilde{\overline{s}}_r - 1)s_r) \bmod 3$, if $s_{n+1} \neq 0$, and $\widetilde{d}(s)_r = (1 - (2\widetilde{\overline{s}}_r - 1)s_r) \bmod 3$ otherwise. So the cases $s_{n+1} \neq 0$ follow immediately from the induction assumption, and in the case $s_{n+1} = 0$ we remark that the r-th component of $3^n - 1 - d(\overline{s})$ is $2 - d(\overline{s})_r$.

b) We set $s := s(d)$ and $t := s(3^n - 1 - d)$. For the term on the left side of the equality to be proved we have from Proposition 2.26:

$$
d(t)_r = (3^n - 1 - d)_r = 2 - d_r = 2 - d(s)_r = (1 - (2\widetilde{\overline{s}}_r - 1)s_r) \bmod 3.
$$

Let $\sigma \in T^n$ be defined by $\forall r \in [n] : \sigma_r = s_r \vartriangle 0$. Then $\sigma_r = (-s_r) \bmod 3$ and since $\sigma_r = 0 \Leftrightarrow s_r = 0$, we get $\widetilde{\sigma}_r = \widetilde{s}_r$. It follows that

$$\mathrm{d}(\sigma)_r = (1 + (2\widetilde{\sigma}_r - 1)\sigma_r) \bmod 3 = (1 - (2\widetilde{s}_r - 1)s_r) \bmod 3 = \mathrm{d}(t)_r .$$

Therefore $t = \sigma$.

Exercise 2.23 $\gamma(G) = |G|$ means that $V(G)$ is the only dominating set. If $\Delta(G) = 0$, then no vertex can be dominated by any other vertex. If $\Delta(G) \in \mathbb{N}$, then delete some v with $\deg(v) \neq 0$ such that other vertices are dominated by themselves and v by its neighbor(s), whence $V(G)$ is not minimal.

Exercise 2.24 For $\Delta(G) = 0$ we have $G = K_0$ or $G = K_1$ with $\gamma(K_0) = 0$ and $\gamma(K_1) = 1$, respectively. For $\Delta(G) = 1$ we get $G = K_2$ with $\gamma(K_2) = 1$. Now let $\Delta(G) = 2$. Then, by Proposition 0.11, G is either a path P_k or a cycle C_k, $k \in \mathbb{N}_3$, and $\gamma(P_k) = \left\lceil \frac{k}{3} \right\rceil = \gamma(C_k)$. For the cycles this follows from the fact that at least every third vertex has to be in a dominating set, and for the paths perfect codes can be constructed as in Exercise 1.4.

Exercise 2.25 Let $a_n := |A_n|$ and $c_n := |C_n|$. By construction, $a_0 = 0$ and $c_0 = 1$. Moreover,

$$a_{n+1} = a_n + 2\left((n \text{ even})a_n + (n \text{ odd})c_n\right),$$

$$c_{n+1} = c_n + 2\left((n \text{ even})a_n + (n \text{ odd})c_n\right),$$

such that in particular $c_{n+1} - a_{n+1} = c_n - a_n$, whence $c_n = a_n + 1$. Therefore,

$$\begin{aligned}
a_{n+1} &= a_n + 2((n \text{ even})a_n \\
&\quad + (n \text{ odd})(a_n + 1)) \\
&= 3a_n + 2(n \text{ odd}).
\end{aligned}$$

It follows that

$$\begin{aligned}
a_n &= 2 \sum_{k=0}^{n-1} 3^k (n - 1 - k \text{ odd}) \\
&= \sum_{k=0}^{n-1} 3^k \left(1 - (-1)^{n-1-k}\right)
\end{aligned}$$

and consequently

$$\begin{aligned}
a_n &= \frac{3^n - 1}{2} - \frac{3^n - (-1)^n}{4} \\
&= \tfrac{1}{4}\left(3^n - 2 + (-1)^n\right)
\end{aligned}$$

and $c_n = \tfrac{1}{4}\left(3^n + 2 + (-1)^n\right)$.

A more direct proof can be based on the fact that the number of perfect states which are codewords is $2 + (-1)^n$. Each of them covers 3 vertices, while the non-perfect codewords cover 4. So we only have to solve the equation

$$(2 + (-1)^n) \cdot 3 + (c_n - (2 + (-1)^n)) \cdot 4 = 3^n.$$

Exercise 2.26 **a)** Any matching of H_3^{1+n} is the union of matchings of H_3^n or graphs obtained from H_3^n by deleting perfect states. In Figure 9.6 one can find all possible configurations depending on the number of connecting edges included in the matching, the two middle ones occurring in three symmetric versions each; the little triangles represent copies of matchings of H_3^n with perfect states deleted shown as unfilled dots.

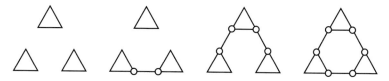

Figure 9.6: Composing matchings for H_3^{1+n}

We get the following recurrence relation for the numbers m_n of matchings of H_3^n, depending on the number ℓ_n of matchings of H_3^n with one, k_n with two, and j_n with three perfect states deleted:

$$m_{1+n} = m_n^3 + 3m_n\ell_n^2 + 3\ell_n^2 k_n + k_n^3.$$

In turn, a matching of an H_3^{1+n} with two perfect states deleted will also always be composed as in Figure 9.6, which now has to be read with the two lower perfect states deleted. The remaining unsymmetric cases are shown in Figure 9.7.

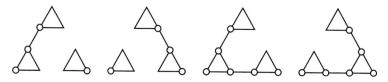

Figure 9.7: Composing matchings for a reduced H_3^{1+n}

This leads to

$$k_{1+n} = m_n\ell_n^2 + m_nk_n^2 + 2k_n\ell_n^2$$
$$+ k_n^3 + 2j_nk_n\ell_n + j_n^2k_n;$$

similar arguments yield two more equations obtained by interchanging m with j and k with ℓ simultaneously. The seeds of this recurrence are

$$m_0 = 1 = \ell_0, \quad k_0 = 0 = j_0.$$

b) The recurrence for the number of respective perfect matchings is the same as for matchings in general, except for the seeds, which are $m_0 = k_0 = j_0 = 0$, $\ell_0 = 1$. Obviously, $m_n = 0 = k_n$ for all n, because the corresponding graphs are of odd order. Therefore the recurrence reduces to

$$\ell_{1+n} = j_n \ell_n^2 + \ell_n^3 \,,$$
$$j_{1+n} = j_n^3 + \ell_n^3 \,.$$

Since $\ell_1 = 1 = j_1$, we see that $\ell_n = j_n$ for all $n \in \mathbb{N}$. So we are left with solving the recurrence

$$\ell_1 = 1, \ \forall\, n \in \mathbb{N} : \ \ell_{1+n} = 2\ell_n^3 \,.$$

It is obvious that $\ell_n = 2^{\lambda_n}$ with the sequence $(\lambda_n)_{n \in \mathbb{N}}$ fulfilling $\lambda_1 = 0$ and $\lambda_{1+n} = 3\lambda_n + 1$. By Lemma 2.18, we get $\lambda_n = \frac{1}{2}\left(3^{n-1} - 1\right)$, whence $\ell_n = \frac{1}{\sqrt{2}}\sqrt[6]{2}^{\,3^n}$. (Note that $\lambda_n = \frac{1}{3}\|H_3^{n-1}\|$; cf. Exercise 2.18.)

Exercise 2.27 By definition, for any two vertices $\bar{s}$ and $\bar{t}$ of H_3^∞, there is an $n \in \mathbb{N}_0$, such that $\bar{s} = 0^\infty s$, $\bar{t} = 0^\infty t$, $s, t \in H_3^n$. Hence there is a path between s and t in H_3^n and consequently there is a (finite) path between $\bar{s}$ and $\bar{t}$ in H_3^∞.

Exercise 2.28 From the formula in Proposition 2.33 we see that $\varepsilon(is)$ becomes minimal, if $\mathrm{d}(s, i^n) - \mathrm{d}(is)$ gets maximal. Remark 2.12 tells us that $\mathrm{d}(is) = 0 \Leftrightarrow s = i^n$, in which case $\varepsilon(i^{1+n}) = 2^{n+1} - 1$. Otherwise, $\varepsilon(is) \geq 2^{n+1} - 1 - \frac{1}{2}(2^n - 1 - 1) = 3 \cdot 2^{n-1}$. The latter value is strictly less than $2^{n+1} - 1$ for $n > 1$ and then attained only for vertices is with $\mathrm{d}(s, i^n) = 2^n - 1$ and $\mathrm{d}(is) = 1$. Since $s \in T^n$ is uniquely determined by the two values $\mathrm{d}(s, i^n)$ and $\mathrm{d}(s; k, j)$, there are precisely 6 vertices fulfilling this, namely iji^{n-1}, $i, j \in T$, $i \neq j$.

The **periphery** of H_3^{1+n} consists of those vertices is for which, without loss of generality, $\mathrm{d}(s, i^n) = \mathrm{d}(s; k, j)$, or $\mathrm{d}(is, j^{1+n}) = 2^{n+1} - 1$, i.e. all vertices which have maximal distance to one of the perfect states. For $n \leq 1$, center and periphery coincide.

Exercise 2.29 We perform the induction proof for the last statement; the others are easy. For $n = 0$ we have $1 = z_0(\mu) = (\mu = 0) \Leftrightarrow \mu = 0$, i.e. $k = 0$. For the induction step we may assume that $\mu \in [2^{n+1}]_0$. From (2.23) we get:

$$z_{n+1}(\mu) = 1 \Leftrightarrow 1 = z_n(\mu - 2^n) + z_n(\mu) + z_n(\mu + 2^n)\,.$$

The equality on the right holds if and only if exactly one of the three summands is 1, the others 0, where $z_n(\mu + 2^n) = 0$ anyway. We have $z_n(\mu) = 0$ iff $\mu = \nu + 2^n$ for some $\nu \in [2^n]_0$ and by induction assumption $z_n(\nu) = 1 \Leftrightarrow \exists\, k \in [n+1]_0 : \nu = 2^n - 2^k$, i.e. $\mu = 2^{n+1} - 2^k$. The other case, $z_n(\mu - 2^n) = 0$ happens iff $\mu = 0$, i.e. for $k = n+1$.

Exercise 2.30 By definition of $z_n(\mu)$ for a fixed n, every $s \in T^n$ is counted for exactly one $\mu \in \mathbb{Z}$, which proves the first equation in Lemma 2.35.

The second equation follows immediately from this and (2.26).

Let $a_n := \sum\limits_{\mu \in \mathbb{N}} \mu z_n(\mu)$. Then, by (2.23), $a_0 = 0$ and

$$
\begin{aligned}
a_{n+1} &= \sum_{\mu \in \mathbb{N}} \mu z_{n+1}(\mu) \\
&= \sum_{\mu \in \mathbb{N}} \mu \left(z_n(\mu - 2^n) + z_n(\mu) + z_n(\mu + 2^n) \right) \\
&= \sum_{\mu \in \mathbb{N}} (\mu < 2^n) \mu \left(z_n(\mu - 2^n) + z_n(\mu) \right) \\
&\quad + 2^n z_n(0) + \sum_{\mu \in \mathbb{N}} (2^n < \mu) \mu z_n(\mu - 2^n) \\
&= \sum_{\mu \in \mathbb{N}} \mu z_n(\mu) + 2^n \sum_{\mu \in \mathbb{Z}} z_n(\mu) = a_n + 6^n ,
\end{aligned}
$$

where use has been made of (2.26). Lemma 2.18 yields the third identity in Lemma 2.35.

Exercise 2.31 The vertices of a diametrical pair have to lie in different subgraphs iH_3^n and jH_3^n, $\{i,j\} \in \binom{T}{2}$, by virtue of Theorem 2.31. We may therefore assume that $i = 0$ and $j = 1$, and we consider the sets

$$
\begin{aligned}
A &:= \{\{0s, 1t\} \mid s, t \in T^n, \, d_1(0s, 1t) = M_{n+1}\} , \\
B &:= \{\{0s, 1t\} \in A \mid d_2(0s, 1t) = M_{n+1}\} , \\
C &:= \{\{0s, 1t\} \in A \mid d_2(0s, 1t) < M_{n+1}\} ,
\end{aligned}
$$

where

$$
\begin{aligned}
d_1(0s, 1t) &= d(0s, 02^n) + 1 + d(12^n, 1t) \\
&= d(s, 2^n) + d(t, 2^n) + 1
\end{aligned}
$$

and

$$
\begin{aligned}
d_2(0s, 1t) &= d(0s, 01^n) + 1 + d(21^n, 20^n) + 1 + d(10^n, 1t) \\
&= d(s, 1^n) + d(t, 0^n) + M_n + 2 .
\end{aligned}
$$

Thus, making use of (0.2) and Theorem 2.31 again:

$$
\begin{aligned}
\{0s, 1t\} \in A \quad &\Leftrightarrow \quad d(s, 2^n) + d(t, 2^n) = 2M_n \\
&\Leftrightarrow \quad d(s, 2^n) = M_n = d(t, 2^n) .
\end{aligned} \tag{9.6}
$$

From Algorithm 10 it follows that $|A| = 4^n$. Moreover, again by (0.2), the equality in B amounts to

$$
d(s, 1^n) + d(t, 0^n) + M_n + 2 = M_{n+1} \Leftrightarrow d(s, 1^n) + d(t, 0^n) = M_n - 1 .
$$

There are M_n choices for $d(s, 1^n)$ to fulfil the latter equation. For these individual choices the value of $d(t, 0^n)$ is then fixed and (9.6), together with Remark 2.14, guarantees that these are the only s and t, whence $|B| = M_n$. Similarly, for C we have

$$\mathrm{d}(s,1^n) + \mathrm{d}(t,0^n) + M_n + 2 < M_{n+1} \Leftrightarrow \mathrm{d}(s,1^n) + \mathrm{d}(t,0^n) < M_n - 1.$$

Here for every $\mu \in [M_n - 1]_0$ the distance $\mathrm{d}(s,1^n)$ can have values in $[\mu + 1]_0$ resulting in

$$|C| = \sum_{\mu=0}^{M_n-2} (\mu + 1) = M_n - 1 + \Delta_{M_n-2} = \Delta_{M_n-1}.$$

So, together with symmetry, we get

$$\mathrm{Diam}(H_3^{1+n}) = 3(|A| - |C|) = 3(4^n - \Delta_{M_n-1}) = \tfrac{3}{2}(4^n + 3 \cdot 2^n - 2).$$

Note that $|B| + |C| = (\Delta \circ M)_n$; cf. p. 20.

Exercise 2.32 We first remark that by an easy induction argument it follows from either (2.30) or (2.31) that $\forall\, n \in \mathbb{N}_0 : b(2^n) = 1$.

Now assume that (2.30) holds. Then $b(2) = b(1)$ by the previous statement and from case $n = 1$ of (2.30) we have $b(3) = b(1) + b(1) = b(1) + b(2)$. Let (2.31) be true up to $\nu - 1 \in \mathbb{N}$. Since we may assume that $2^n < 2\nu < 2^{n+1}$ for some $n \in \mathbb{N}$, we have $2^{n-1} < \nu < 2^n$. Then

$$\begin{aligned}
b(2\nu) &= b(2^{n+1} - 2\nu) + b(2\nu - 2^n) \\
&= b\big(2(2^n - \nu)\big) + b\big(2(\nu - 2^{n-1})\big) \\
&= b(2^n - \nu) + b(\nu - 2^{n-1}) \\
&= b(\nu).
\end{aligned}$$

Similarly,

$$\begin{aligned}
b(2\nu + 1) &= b\big(2^{n+1} - (2\nu + 1)\big) \\
&\quad + b\big((2\nu + 1) - 2^n\big) \\
&= b\big(2(2^n - (\nu + 1)) + 1\big) \\
&\quad + b\big(2(\nu - 2^{n-1}) + 1\big) \\
&= b\big(2^n - (\nu + 1)\big) + b(2^n - \nu) \\
&\quad + b(\nu - 2^{n-1}) + b\big((\nu + 1) - 2^{n-1}\big) \\
&= b(\nu) + b(\nu + 1),
\end{aligned}$$

where we note that the equations in (2.31) are trivially valid for $\nu = 0$.

Now assume that (2.31) holds. The case $n = 0$ is clear. Let (2.30) be valid up to $n - 1 \in \mathbb{N}_0$ and let $2^n < \nu < 2^{n+1}$. If $v = 2\mu$, then $2^{n-1} < \mu < 2^n$ and we get

$$\begin{aligned}
b(\nu) = b(2\mu) &= b(\mu) \\
&= b(2^n - \mu) + b(\mu - 2^{n-1}) \\
&= b(2^{n+1} - \nu) + b(\nu - 2^n).
\end{aligned}$$

For $\nu = 2\mu + 1$, where $2^{n-1} \le \mu < 2^n$, we have

$$
\begin{aligned}
b(\nu) = b(2\mu + 1) &= b(\mu) + b(\mu + 1) \\
&= b(2^n - \mu) + b(\mu - 2^{n-1}) \\
&\quad + b\left(2^n - (\mu + 1)\right) + b\left((\mu + 1) - 2^{n-1}\right) \\
&= b\left(2(2^n - (\mu + 1)) + 1\right) \\
&\quad + b\left(2(\mu - 2^{n-1}) + 1\right) \\
&= b(2^{n+1} - \nu) + b(\nu - 2^n).
\end{aligned}
$$

Exercise 2.33 a) Since ij^n is a perfect state in iT^n, we have $X(ij^n) = X(ij^n) \setminus iT^n$, whence from Proposition 2.41 we obtain

$$
|X(ij^n)| = b\left(\mathrm{d}(ij^n)\right) = b(2^n - 1) = z_n(1) = n,
$$

the latter following easily from (2.23).

b) Induction on n. The case $n = 1$ follows from (a), as does the case $s = j^{n-1}$, $j \in T \setminus \{i\}$ in the step from $n - 1$ to n, $n \ge 2$. If $s \in T^{n-1}$ is not perfect, the induction assumption guarantees that $|X(is) \cap iT^{n-1}| \ge n - 1$ and another element of $X(is)$ is given by Proposition 2.40.

c) For $n \le 2$ we have $|X(is)| \le n$. For $n = 1$ this follows from (a), and for $n = 2$ we have for $s \notin \{0^2, 1^2, 2^2\}$ that $\mathrm{d}(s) = 1 \le \mathrm{d}(is) \le 2$, whence $b(\mathrm{d}(s)) = 1 = b(\mathrm{d}(is))$ in Corollary 2.43. An example in H_3^{1+3} is $X(0102) = \{0012, 0120, 1002, 1021\}$.

d) We already know that $X(01^n) = X(01^n) \cap jT^n$ with $j \in T$ fulfilling $\mathrm{d}(1^n, (0 \vartriangle j)^n) > \mathrm{d}(1^n, j^n)$, i.e. $j = 1$. Moreover, the condition on $1t \in X(01^n)$ in Theorem 2.42 reads $\mathrm{d}(t, 2^n) = \mathrm{d}(t, 0^n) + 1$. Looking at (2.8), this means that there is a $d \in [n]$ for which the coefficient of 2^{d-1} in $\mathrm{d}(t, 2^n)$ is 1, followed by 0s and the one in $\mathrm{d}(t, 0^n)$ is 0 followed by 1s; the coefficients for higher powers of 2 must all be 1 in both representations. In terms of Stockmeyer's P1-automaton of Figure 2.6 we enter t, from t_n to t_1, starting in state 2 (automaton A) or starting in state 0 (automaton B) for the target pegs 2 and 0, respectively. At the beginning we enter $n - d$ tits 1. Depending on whether $n - d$ is even or odd we will end in state 2 or 0 in automaton A and in 0 or 2 in automaton B, respectively. Entering t_d must lead to a change of state of automaton A and no change for automaton B. If $n - d$ is even, this means that $t_d = 0$, leaving automaton A in state 1 and automaton B in state 0. If $n - d$ is odd, then $t_d = 2$ leaving automaton A in state 1 and automaton B in state 2. The remaining entries have to be 1s again. Writing $\nu := n - d + 1 \in [n]$, we arrive at the desired representation of it.

Exercise 2.34 We show by induction on $n \in \mathbb{N}_0$ that the inequality is true if x and y lie in the union σ^n of all constituting subtriangles of side-lengths greater than or equal to 2^{-n}; it then follows for the remaining pairs by continuity. If $n = 0$, the points x and y lie on the triangle with side-lengths 1. Among the corners of that triangle choose z such that the triangle xzy has a $60°$ angle at

z. Let $a := \|x - y\|$, $b := \|x - z\|$, and $c := \|z - y\|$. Then, by the cosine theorem, $a^2 = b^2 + c^2 - 2bc\cos(60°) = b^2 + c^2 - bc$. It follows that $4a^2 = (b+c)^2 + 3(b-c)^2 \geq (b+c)^2$, whence $2a \geq b + c$. Now $\mathrm{d}(x, y) \leq \mathrm{d}(x, z) + \mathrm{d}(z, y) = \|x - z\| + \|z - y\| \leq 2\|x - y\|$.

For the induction step we look at the straight line section between the two points x and y on σ^{n+1}. Starting in x let $\tilde{x}$ be the first point on σ^n (if any) and $\tilde{y}$ the last point on σ^n before we reach y (if any). Then for the (at most) one segment between $\tilde{x}$ and $\tilde{y}$ we can use the induction assumption, and the remaining segments can be treated like the case $n = 0$, just in a smaller triangle.

If we choose x and y to be the center points of two sides of the middle triangle, i.e. the one to be added in the step from $n = 0$ to $n = 1$, then $\mathrm{d}(x, y) = \frac{1}{2} = 2\|x - y\|$.

Exercise 2.35　By isomorphy, the number we are looking for is equal to the number of states which are at a distance of μ from a perfect state of the TH. So the statement follows immediately from Proposition 2.16.

An alternative proof can be based on (0.17); this was Glaisher's argument in [160, p. 156].

Exercise 2.36　As $b(\nu)$ is the number of odd entries in the νth subdiagonal of AT mod 2, all entries of which add up to the Fibonacci number F_ν, $b(\nu)$ is even if and only if F_ν is (cf. Figure 2.27). From the recurrence for the Fibonacci numbers it is easy to see that this is the case if and only if 3 divides ν.

Chapter 3

Exercise 3.1 As the start vertex take the state $\sigma = 1 \ldots (n - 2)\, n\, (n - 1)\, |\,|$ from Remark 3.4. Then we have seen in this remark that it takes 2^{n-2} moves to get disc n away from peg 0. After this first move of the largest disc the state is either $n - 1\,|\, n\,|\, 1 \ldots n - 2 = 102^{n-2}$ or $n - 1\,|\, 1 \ldots n - 2\,|\, n = 201^{n-2}$. If we choose $t = 0^n$ as the target state, then it is clear that disc n has to be moved (at least twice) if we want to get from σ to t, and from formula (2.8) we deduce that the distance from the intermediate state to t is $2^n - 1$.

Exercise 3.2 Hint: Assume that $n \geq 3$ and distinguish three cases for the initial state $\sigma \in V(\overrightarrow{H}_3^n)$:

- Disc n lies on bottom of t_n.

- $s_n \neq t_n$ and either $s_{n-1} = 3 - s_n - t_n$ or n does not lie on bottom of s_n.

- $s_n \neq t_n$, $s_{n-1} \neq 3 - s_n - t_n$, and n lies on bottom of s_n.

Make use of Lemma 3.2 and Theorem 3.5.

Exercise 3.3 The optimal solutions are, respectively,

$$53\,|\,|\,421 \xrightarrow{4} 3\,|\,124\,|\,5 = 21011 \xrightarrow{7} 21^4,$$

$$53\,|\,|\,421 \xrightarrow{1} 3\,|\,5\,|\,421 \xrightarrow{6} 1234\,|\,|\,5 = 20^4 \xrightarrow{15} 21^4,$$

$$53\,|\,|\,421 \xrightarrow{1} 3\,|\,5\,|\,421 \xrightarrow{2} 5\,|\,|\,3421 \xrightarrow{6} 21^4.$$

The figures on top of the arrows indicate the move numbers of the intermediate steps. So the lengths of the respective paths are 11, 22, and 9. It turns out that the path where disc n moves from peg 0 to 1, back to 0, and finally to its goal peg 2 represents the unique optimal solution. This shows that they *do* come back (cf. Lemma 2.32)!

Exercise 3.4 We are asked to get from $\sigma = 6\,3\,8 \mid 4\,1\,7\,9\,2 \mid 10\,5 \, \in \mathfrak{T}^{10}$ to $2^{10} \in T^{10}$ in $\vec{H}_3^{10}$, which is a special case task. Therefore, in the first call of Algorithm 14, n' is identified to be 9, such that $\nu = 0$ and consequently $i = \mathrm{sp}(\sigma) = 0$, $k = 1$. The best buffer disc b is

$$\mathrm{bb}\,(1; D(0), 10{\uparrow}) = \mathrm{bb}\,(1; \{3,6,8\}, \varnothing) = 9$$

such that a recursive call of the algorithm is made for the standard case task $6\,3\,8 \mid 4\,1\,7 \mid \;\rightarrow\; \mid 1\,3\,4\,6\,7\,8 \mid$, where $j = 1$, $i = 0$, and $k = 2$. Here the best buffer is

$$\mathrm{bb}\,(2; 8{\uparrow}, D(1)) = \mathrm{bb}\,(2; \{3,6\}, \{1,4,7\})\,,$$

i.e. the bottom of peg 2 and a recursive call is made for the task $6\,3 \mid 4\,1\,7 \mid \;\rightarrow\; \mid\mid 1\,3\,4\,6\,7$. This, in turn, is a standard case task which the algorithm analogously solves in 12 moves. Then disc 8 moves from peg 0 to 1 and another recursive call is made for $\mid\mid 1\,3\,4\,6\,7 \;\rightarrow\; \mid 1\,3\,4\,6\,7 \mid$, which is equivalent to a regular P0-type task for 5 discs, needing 31 moves.

Back in the main loop after 44 moves, disc 10 is transferred from peg 2 to 0 and another best buffer disc

$$b' = \mathrm{bb}\,(1; \varnothing, D(2)) = \mathrm{bb}\,(1; \varnothing, \{5\}) = 6$$

is chosen, leading to a recursive call for $\mid 1\,3\,4 \mid 5 \;\rightarrow\; \mid 1\,3\,4\,5 \mid$, a regular task equivalent to $2111 \rightarrow 1^4$, taking 15 moves. Move 61 is now made by disc 10 from 0 to 2. Finally, the algorithm is called again for $\mid 1\,3\,4\,5\,6\,7\,8\,9\,2 \mid \;\rightarrow\; 2^9$, a standard case P3 task which is solved similarly by the algorithm taking another 382 moves, the relatively large number stemming from the fact that intermediate states are much more regular by now. Altogether 443 moves have been made. (A path using peg 1 instead of the special peg for the intermediate move of disc 10 takes at least 710 moves.)

Exercise 3.5 The upper bound follows immediately from Theorem 3.5. For the lower bound, note that for 1/12 of all states we have $s_{n-1} = j = s_n$ with disc n above disc $n - 1$. (Disregarding discs 1 to $n - 2$, there are $4!/2 = 12$ arrangements of the two bars and the two largest discs in a state $\sigma \in \mathfrak{T}^n$ (the order of the bars being irrelevant). Only one of these, namely $\mid\mid n(n-1)$, has disc $n-1$ underneath disc n on goal peg $j = 2$. Cf. also $\vec{H}_3^2$ in Figure 3.2.) In these (special) cases, a perfect $(n-1)$-tower must be moved after the second move of disc n, which takes precisely $2^{n-1} - 1$ moves of discs from $[n-1]$.

Numerical experiments by S. Finsterwalder (cf. [147, Tabelle 4]) indicated that $\overline{d}_n \approx a \cdot 2^n$ for large n with an a between 0.3 and 0.4. M. A. Schwarz was able to prove in [376, Proposition 4.1] that the sequence given by $a_n = 2^{-n}\overline{d}_n$ converges to $a > 0$ and to improve the estimate of this exercise to $\frac{1}{12} \le a_n \le \frac{5}{8}$ in [376, Corollary 4.6]. The asymptotic behavior of a_n was further analyzed by simulations summarized in [376, Table 2].

Chapter 4

Exercise 4.1 The vertices in question are shown in Figure 9.8. With the aid of the H_3-to-S_3-automaton one finds that the start vertex corresponds to 0110 in Sierpiński labeling. Similarly, the goal vertices in (a) to (c) transform into 2100, 2212, and 2211, respectively. For (d) it suffices to enter the first two tits, leading to 20 * *; for (e), the first three tits become 211*.

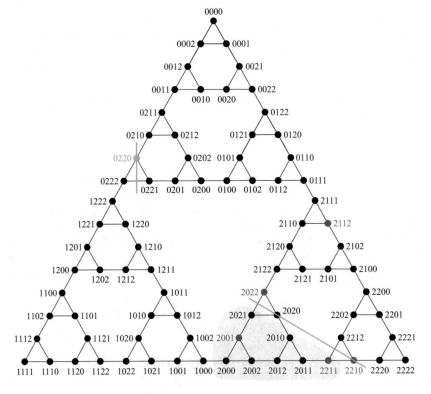

Figure 9.8: Graph H_3^4 for the tasks in Exercise 4.1 with the starting vertex in green, the goal vertices in red

The first pair of input $(0, 2)$ defines the P2 decision automaton, namely with $i = 0$, $j = 2$, and $k = 1$. Entering the other pairs leads to the states with the corresponding capital letter, such that for (a) and (d), disc 4 moves once only, for (c) and (e) it has to move twice and for (b) there are two optimal solutions including one or two moves of the largest disc, respectively.

For tasks involving a relatively small number of discs, one might, of course, find the type of solution easily by an inspection of the corresponding graph. In Figure 9.8 the set of those goal states which need two moves of disc 4 to be reached from the green vertex 0220 is highlighted in the lower right subgraph; on the green line next to that area one can find the only two goal vertices with two shortest paths from 0220.

Exercise 4.2 If $t_{n+1} = i$, then disc $n + 1$ will not move at all. If $t_{n+1} = k$, then we have to use Romik's P2 decision automaton in Figure 2.28 with letters j and k switched and we will always end up in state D. If $t_{n+1} = j$, then we employ the automaton as it is and realize that, although we might reach state B eventually if the pair (k, k) is to be entered, the j in the first position of all subsequent input pairs will never lead to either C or E.

By isomorphy, non-perfect initial states exist for the TH as well such that all tasks can be solved optimally with at most one move of the largest disc.

Exercise 4.3 Hint: Induction on n.

Exercise 4.4 Apart from the trivial cases S_p^0, every S_p^n has p (extreme) vertices of degree $p - 1$ and $p^n - p$ vertices of degree p. Hence the graphs S_p^n are eulerian, whereas S_2^n are non-eulerian, but semi-eulerian. For $p \geq 3$, $S_p^1 \cong K_p$ is (semi-)eulerian if and only if p is odd, but S_p^n is not semi-eulerian for $n \geq 2$ because then both $p > 2$ and $p^n - p > 2$ hold.

Exercise 4.5 For $n = 1$ hamiltonicity is trivial since S_p^1 is a complete graph. Let $n \geq 2$ and consider the sequence of paths $P_0, P_1, \ldots, P_{p-1}$, where P_0 is a path between the vertices $0(p-1)^{n-1}$ and 01^{n-1}, P_{p-1} between the vertices $(p-1)(p-2)^{n-1}$ and $(p-1)0^{n-1}$, and for $i = 1, 2, \ldots, p-2$, P_i is a path between the vertices $i(i-1)^{n-1}$ and $i(i+1)^{n-1}$. We claim that the paths P_i can be constructed in such a way that they include all the vertices from iS_p^n, $i \in [p]_0$.

To prove the claim it is enough to see that for any i, j and g, $j \neq g$, there is a path between ij^{n-1} and ig^{n-1} which goes through all vertices from iS_p^n. Obviously that reduces the inductive argument to the statement that j^n and g^n, $j \neq g$, may be connected in S_p^n by a path going through all vertices (for all n). Without loss of generality assume that $j = 0$ and $g = p-1$. By the inductive hypothesis we may find a path from 0^n to 01^{n-1} through all vertices of $0S_p^n$. Add the edge between 01^{n-1} and 10^{n-1} to the path. By the same argument we may find a path from 10^{n-1} to 12^{n-1} through all vertices of $1S_p^n$. Continue this procedure until $(p-2)(p-1)^{n-1}$ is joined to $(p-1)(p-2)^{n-1}$ and a path from $(p-1)(p-2)^{n-1}$ to $(p-1)^n$ through all vertices of $(p-1)S_p^n$ is added at the end.

Exercise 4.6 Hint: Determine the size of a corresponding perfect code using the fact that every vertex of it, except possibly some extreme vertices, dominates $p+1$ vertices.

Exercise 4.7 A planar drawing of S_4^2 is given in Figure 9.9.

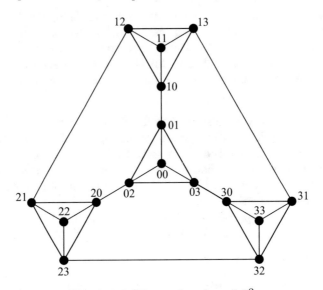

Figure 9.9: Planar drawing of S_4^2

Every subgraph ijS_4^1 of S_4^3 with $i, j \in [4]_0$, $i \neq j$, is isomorphic to K_4. Adding a vertex ikj with $i \neq k \neq j$ it can easily be seen from Figure 4.5 that these five vertices induce a K_5-subdivision in S_4^3. Hence all S_4^n with $n \geq 3$ are non-planar and so are S_p^n with $p > 4$ and $n \geq 1$ since they contain a K_5.

Exercise 4.8 Hint: Mimic the proof of Proposition 2.25.

Exercise 4.9 First check the cases $p = 1, 2$. For $p \geq 3$, the p-cliques $\underline{s}S_p^1$ (cf. Theorem 4.3) cover all vertices of S_p^n. As there are p^{n-1} such cliques and each can contain at most one vertex of an independent set, $\alpha\left(S_p^n\right) \leq p^{n-1}$. On the other hand, it is well-known (and not difficult to see) that $\chi(G) \geq |G|/\alpha(G)$; cf. [432, Proposition 5.1.7]. Since $\chi(S_p^n) = p$ it follows that $\alpha\left(S_p^n\right) \geq p^{n-1}$. (A direct argument for the latter inequality would be to observe that vertices of the form $s_n s_{n-1} \ldots s_2 0$ are not adjacent; cf. the canonical vertex-coloring.)

Chapter 5

Exercise 5.1 Let $x = 0$ (red case in Figure 5.3). Then $m_n = \Delta_{\nu-1}$. For $m = m_n$, we have $\mu + (y \neq 0) = \nu - 1 < \nu = n - m$. If $m < m_n$, then $\mu + (y \neq 0) < \nu - 1 + 1 = \nu < n - m$.

For $m > m_n$, necessarily $\mu = \nu - 1$ and $y \in [\nu - 1]$, such that $\mu + (y \neq 0) = \nu - 1 + 1 = \nu > \nu - y = n - m$.

Turning to the case $x \neq 0$ (green in Figure 5.3), m_n has to be written as Δ_ν, if $x = \nu$, and $\Delta_{\nu-1} + x$ otherwise. In the former case we have for $m = m_n$: $\mu + (y \neq 0) = \nu = n - m$, in the latter $\mu + (y \neq 0) = \nu - 1 + 1 = \nu = n - m$. If $m < m_n$, then $\mu + (y \neq 0) \leq \mu + 1 \leq \nu < n - m$. So let finally $m > m_n$. Then either $\mu = \nu - 1$ and $y > x$ or $\mu = \nu$ and $y \in [x]_0$. In the former case, $\mu + (y \neq 0) = \nu - 1 + 1 = \nu > n - m$ and in the latter case $\mu + (y \neq 0) = \nu + (y \neq 0) \geq \nu > n - m$.

Exercise 5.2 The case $\nu = 0$ is trivial; for $\nu = 1$ it is clear that each disc moves at least once. For $\nu = 2$ we have $n = 3 + x$ discs, $x \in [3]_0$, which can be transferred to two pegs in $2(2 + x)$ moves. But this number is also needed, because if all discs go to one and the same peg, then $2 + x$ of them have to move at least twice, resulting in at least $2(2 + x) + 1$ moves. With $3 + x$ discs on two pegs, $1 + x$ move at least twice, such that $2(1 + x) + 2 = 2(2 + x)$ moves are necessary.

Exercise 5.3 a) For every $w \in V(G)$ we have $d(u, w) \leq d(u, v) + d(v, w) = 1 + d(v, w) \leq 1 + \varepsilon(v)$, whence also $\varepsilon(u) \leq 1 + \varepsilon(v)$, and the statement of (a) follows by symmetry.

b) Since $C(G) \neq \varnothing \neq P(G)$ and G is connected, there is a path $v_0, \ldots, v_K$ in G with $v_0 \in C(G)$ and $v_K \in P(G)$. On that path there must be a minimal $k \in [K]$ with $\varepsilon(v_k) > \varepsilon$. But then $\varepsilon(v_{k-1}) \leq \varepsilon$ and since $d(v_{k-1}, v_k) = 1$ we get $\varepsilon(v_{k-1}) = \varepsilon$ from (a).

Exercise 5.4 First select k non-empty pegs in $\binom{p}{k}$ ways. Top discs for these pegs can be selected in $\binom{n-1}{k-1}$ ways because disc 1 is always one of them. After the top discs are chosen, there are $k!$ ways to assign them to the non-empty pegs.

Exercise 5.5 This is proved by induction on ν. Quite surprisingly, the case $\nu = 0$ is not trivial as usual but requires another straightforward induction on q to get

$$\sum_{k=0}^{q} (-2)^k \binom{q}{k} = (-1)^q.$$

The induction step is then by showing that both polynomials fulfill the recurrence relation

$$P(\nu + 1) = \frac{1}{2}\left(P(\nu) + \binom{\nu + q}{q}\right),$$

which in turn follows from the standard properties of combinatorial numbers.

Exercise 5.6 Consider an optimal path. By Theorem 5.19, disc 8 moves precisely once (from peg 0 to peg 1). Before that move, the 7-tower on 0 has to be distributed over pegs 2 and 3, which takes at least $\overline{FS}_4^7 = 16$ moves; cf. Table 5.1. We now show that $d(t, 1^3 3^4) > 7$ for $t \in \{2, 3\}^7 \setminus \{\tau\}$, $\tau := 2^3 3^4 \in Q^7$; this means that 1τ is unavoidable on an optimal $0^8, 1^4 3^4$-path. (It is clear that the indicated solution passes 1τ and takes $16 + 1 + 7 = 24$ moves.)

Look at the task $t \to 1^3 3^4$ in H_4^7. It needs at least 4 moves of discs from $\{5, 6, 7\}$. If one of these discs lies on peg 3 in t, then all discs from $\{1, 2, 3, 4\}$ have to move once at least (namely to peg 3 after the move of that disc), such that $\mathrm{d}(t, 1^3 3^4) \geq 8$. We may therefore assume that all discs from $\{5, 6, 7\}$ are on peg 2 in t and consequently need at least $FS_4^3 = 5$ moves altogether. If a disc from $\{1, 2, 3, 4\}$ were on 2 in t, then it moves at least once and all smaller discs at least twice, such that discs 2, 3, and 4 may be considered to lie on peg 3. If for $t = 2^3 3^3 2$ disc 1 would move only once, the 3-tower made up from $\{5, 6, 7\}$ would have to make at least 7 moves. The possibility of 2 moves of disc 1 can be excluded by analyzing the tree rooted in t and containing only the respective paths.

So any optimal path goes $0^8 \to 0\tau \to 1\tau \to 1^4 3^4$.

Exercise 5.7 The formula for FS follows from Proposition 5.20 and the paragraph preceding it. From this we get the the first formula by Corollary 5.19.

Exercise 5.8 Let s be an arbitrary regular state and consider a sequence $\mathcal{M}$ of moves in which each disc is moved at least once. Just before disc n is moved, discs $1, 2, \ldots, n - 1$ are on the same peg. Without loss of generality assume that disc $n - 1$ is moved after the move of disc n. It follows that after the move of n, we need to solve a P0 problem for $n - 2$ discs using at least $2^{n-2} - 1$ moves and to move disc $n - 1$. Together with the move of disc n, the sequence $\mathcal{M}$ is of length at least $2^{n-2} + 1$. Hence $g(3, n) \geq 2^{n-2} + 1$. To see that equality holds, consider the regular state 001^{n-2}.

Exercise 5.9 Let $k \in [p]$ and consider the regular states in which exactly k pegs are non-empty. We can select k pegs in $\binom{p}{k}$ ways. The n discs can be partitioned into k non-empty parts in $\left\{ {n \atop k} \right\}$ ways, these parts can be distributed onto k pegs in $k!$ ways. Hence there are exactly $\binom{p}{k} \left\{ {n \atop k} \right\} k! = \left\{ {n \atop k} \right\} p^{\underline{k}}$ regular states with exactly k non-empty pegs.

Consider next an arbitrary regular state with k non-empty pegs. There are $k(p - k)$ legal moves between non-empty and empty pegs. Moreover, between two non-empty pegs, exactly one move is legal. We conclude that the degree of an arbitrary regular state with k non-empty pegs is $k(p - k) + \binom{k}{2} = \frac{1}{2} k(2p - k - 1)$, the extra factor $\frac{1}{2}$ in the desired formula stemming from double-counting of edges.

Exercise 5.10 Hint: First use induction on n to show that there is a hamiltonian path starting and ending in distinct perfect states. (Say, move disc $n + 1$ stepwise from peg 0 to 1, from peg 1 to peg 2, and so on to $p - 1$. Before each such move, a hamiltonian path in the corresponding n discs subgraph can be performed to transfer the tower consisting of the n smallest discs to a peg allowing disc $n + 1$ to move. Note that this sequence visits all the states in which disc $n + 1$ is on a fixed peg. Act similarly after the last move of disc $n + 1$.) Now start in the state 01^n of H_p^{1+n} and transfer the tower of n smallest discs on hamiltonian paths according to the above in a cyclic fashion from peg to peg, each complete transfer being followed by a single move of disc $n + 1$ to the next peg in the same direction.

Exercise 5.11 Assume that one of the $p(p-1)$ non-perfect states, say 01, is a codeword. Then $2H_p^1$ contains no codeword, because otherwise 21 would be covered twice. But this means that 22 is not covered.

Being surrounded by non-codewords, all perfect states must be codewords themselves. This results in a (unique) perfect code.

Exercise 5.12 Hint: Look at the 24 flat states, i.e. those where the three discs are distributed on three different pegs.

The subgraph of H_4^3 shown in Figure 9.10 is a subdivision of $K_{3,3}$. Therefore, H_4^3 is not planar.

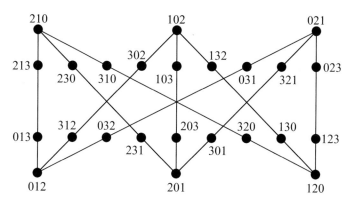

Figure 9.10: $K_{3,3}$-subdivision in H_4^3

Exercise 5.13 The proof of Proposition 5.60 is by induction on n. The cases $n = 0$ and $n = 1$ are clear. Now let $s, t \in [p]_0^{n+1}$, $n \in [p-2]$, with $\forall\, d \in [n+1] : t_d = d - 1$. Starting in s, move as many discs as possible directly to their goal pegs. Then, by virtue of Lemma 5.59, some peg $k \in \mathbb{N}_{n+1}$ is empty in the resulting state. Let d be any top disc which is not already in its end position. (If there is no such disc, then t is reached and each disc moved at most once.) Move it to peg k. Now peg $d - 1$, the goal of disc d, is not empty and all discs on it are not in their end position. We may therefore consider the task which remains after ignoring pegs k and $d - 1$ and all discs lying there. This is a task in H_{p-2}^ν with $\nu \in [n]_0$ of the same type as the original one. (The cases $p = 3$ and 4 can easily be sorted out.) By induction assumption, this can be solved with at most $\lfloor \frac{\nu}{2} \rfloor \le \lfloor \frac{n-1}{2} \rfloor$ discs moving precisely twice and all other discs moving at most once, those in end position being fixed. All discs on peg $d - 1$ can now be moved directly to their goals and finally disc d too in its second move. The former ones therefore all moved precisely once, the latter one twice; no disc in end position was moved. Since $\lfloor \frac{n-1}{2} \rfloor + 1 = \lfloor \frac{n+1}{2} \rfloor$, Proposition 5.60 is proved.

For (a) we show that

$$\forall\, s \in [p]_0^n \ \exists\, t \in [p]_0^n : \mathrm{d}(s,t) \ge n + \lfloor \tfrac{n}{2} \rfloor.$$

For $s \in [p]_0^n$ let $\forall d \in \left[\left\lfloor \frac{n}{2} \right\rfloor\right]$:

$$t_d = (s_d + (s_d = s_{n-d+1})) \bmod p = t_{n-d+1}.$$

Then all discs from $[n] \setminus \left[\left\lfloor \frac{n}{2} \right\rfloor\right]$ move at least once, the others at least twice, summing up to at least $n + \left\lfloor \frac{n}{2} \right\rfloor$ moves.

Now $\varepsilon(s) \geq n + \left\lfloor \frac{n}{2} \right\rfloor$ for every s, such that $\operatorname{rad}(H_p^n) \geq n + \left\lfloor \frac{n}{2} \right\rfloor$.

For (b) we deduce from Proposition 5.60 that

$$\exists t \in [p]_0^n \; \forall s \in [p]_0^n : \; d(s,t) \leq n + \left\lfloor \frac{n}{2} \right\rfloor,$$

such that there is a t with $\varepsilon(t) \leq n + \left\lfloor \frac{n}{2} \right\rfloor$, whence $\operatorname{rad}(H_p^n) \leq n + \left\lfloor \frac{n}{2} \right\rfloor$.

Note that t and in fact all flat states belong to the center of H_p^n if $n < p$. The sequence $x_n = n + \left\lfloor \frac{n}{2} \right\rfloor$ is A032766; it fulfills the recurrence

$$x_0 = 0, \; x_1 = 1, \; \forall n \in \mathbb{N} : \; x_{n+2} = x_n + 3.$$

Exercise 5.14 Hint: Observe that the states in question are precisely the states in which disc n is alone on peg i and j is the only empty peg. (For the sequence, cf. A180119.)

Exercise 5.15 Assume, w.l.o.g., that peg 1 is empty in an optimal half-way state $s \in [p-2]^n$ and consider an optimal $0^n, s$-path. Then either peg 1 has been empty throughout the path or there has been a (last) move onto the bottom of peg 1. In the first case replace the first move of some disc that moves twice on the path considered (there must be one since otherwise only up to $p-2$ discs have been moved) by a move to peg 1 and leave the disc there, leading to a strictly shorter path. In the second case leave the last disc moved to the bottom of peg 1 in that position; again this engenders a shorter path.

Exercise 5.16 The n-tower 0^n can be spread to a flat distribution of discs on the pegs from $[p-1]$, where we make sure that some disc $d \in [p-3]$ goes to peg $p-1$. Then in the final of optimal $n+1$ moves we transfer disc d onto any of the discs from $[p-1] \setminus [d+1]$ to be misplaced. So we can construct $\sum_{d=1}^{p-3}(p-d-2) = \binom{p-2}{2} = \Delta_{p-3}$ non-isomorphic non-subtower half-way states. Lemma 5.63 tells us that these are all.

Exercise 5.17 See Figure 9.11.

Exercise 5.18 Sufficiency follows from the solutions given in Figure 9.12. In fact, as found out by computer search, the 3-LDMs solution on the right is the unique optimal solution for the task. That means that no solution can be shorter than 10 and that no solution using less than 3 LDMs can be shorter than 11 moves. Therefore, it suffices to prove that no solution with precisely 1 LDM exists making 11 moves only. Let us assume in the contrary that there *is* such a solution. Then,

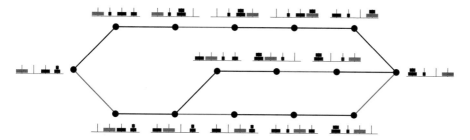

Figure 9.11: Optimal solutions for 0233 → 3001

since disc 6 necessarily moves once only from peg 0 to peg 3, there must be an $s \in [2]^5$ such that, with $r := 22333$ and $t := 00101$, we have $d(r,s) + d(s,t) = 10$ in H_4^5. As any solution for the task $r \to s$ would distribute the 3-tower on peg 3 in r to the pegs 1 and 2 in s and similarly for the tower consisting of 3 discs on peg 0 in t, we know that $x := d(r,s) \in \{4,5,6\}$, because $\overline{FS}(H_4^3) = 4$.

We now show by combinatorial arguments that each of the three values for x leads to a contradiction. First observe that $s_5 = 2$, since otherwise to get from r to s discs from $\{4,5\}$ would move at least 3 times altogether and discs from $\{1,2,3\}$ at least 4 times, such that $x > 6$.

If $x = 4$, then necessarily $s_4 = 2$ and the three smaller discs are not all on the same peg in s. So there are three cases each for the options $s_3 = 2$ and $s_3 = 1$; they are $s = 22211, 22212, 22221$ and $s = 22112, 22121, 22122$, respectively. In the first group of cases, discs 1 and 4 have to move at least twice in a solution for $s \to t$ and all others at least once, which would lead to 7 moves. In the second group of cases, disc 3 does not move; otherwise we would be in the same kind of situation as before. But then the discs 1 and 2 move at least 4 times altogether.

If $x = 6$, then the same argument as before shows that $s_3 = 1 = s_1$. If disc 1 does not move, then it would take 5 moves to transfer the 3 discs on peg 0 in state t to peg 2 in state s. If disc 1 does move, then these 3 discs must make at least another 4 moves. In any case the task $s \to t$ would take at least 5 moves.

So we are left with $x = d(r,s) = 5 = d(s,t)$. Let us again look at the task $t \to s$ with $s_5 = 2$. If disc 5 were to move to peg 1 first, it would make at least 2 moves and all the other discs at least 1, altogether 6. If it were to go to peg 3 first, this could happen at the earliest in move number 4, in which case disc 5 can not go to peg 2 immediately. So, together with the boxer rule, disc 5 moves precisely once, namely from peg 0 to peg 2. But then it is easy to see that it is not possible for the other discs to get out of its way and back to pegs 1 and 2 in just 4 moves.

An alternative proof for the non-existence of an admissible state s can be based on numerical computations. Figure 9.13 shows all paths of length 10 from r to t. It turns out that the intermediate states s with $x \in \{4,5,6\}$ on these paths are 01301, 01310, 02330, 02332, 22101, 01323, 02302, 02320, 23101, 00311, 00313, 00323, 02102, 02120, and 23131; none of them has all discs on pegs 1 or 2.

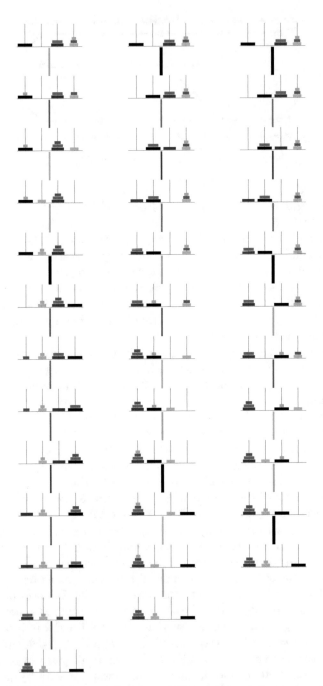

Figure 9.12: Optimal solutions using, from left to right, 1, 2 or 3 LDMs

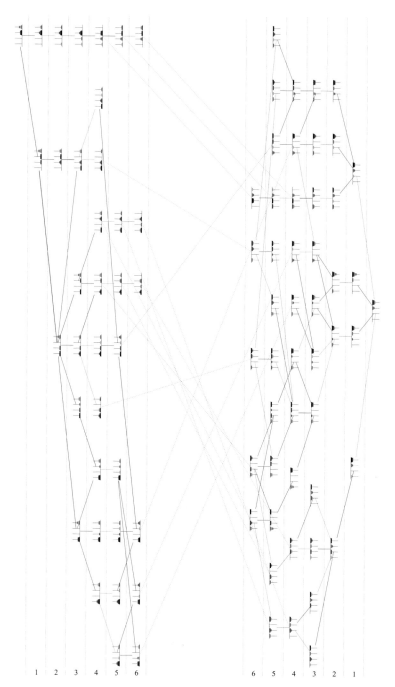

Figure 9.13: Distance layered graph detecting intermediate states

Chapter 6

Exercise 6.1 Hint: Observe that in any optimal sequence of moves, after a move of a disc that is not one of the two smallest, the next move must be either a move of the smallest white disc or a move of the smallest black disc. Also, any solution must start and end with a move of a smallest disc. Hence the number of moves of the two smallest discs is at least one more than the number of moves of all the other discs. Then use induction to reach the required conclusion.

Exercise 6.2 A solution making 11 moves is presented in Figure 9.14. By (6.2), the solution is optimal. To distinguish the colors of the three stacks we denote their discs by d, d', and d'', $d \in \{1,2\}$, respectively.

```
 1       1'      1''
 2  _ 2'    _ 2''     _        →

 1               1''
 2  _ 2'   1' 2''     _        →

 1
 2  _ 2'   1' 2''  1''         →

          1
 2  _ 2'   1' 2''  1''         →

          1
 _  2 2'   1' 2''  1''         →

          1
2'' 2 2'   1'     1''          →

 1
2'' 2 2'   1'     1''          →

 1
2'' 2      1' 2'  1''          →

 1
2''   _ 2  1' 2'  1''          →

          1
2''   _ 2  1' 2'  1''          →

          1   1'
2''   _ 2  _ 2'  1''           →

          1   1'
1''   1    1'
2''   _ 2  _ 2'   _
```

Figure 9.14: A solution making 11 moves

Exercise 6.3 By Theorem 6.3 it suffices to find two different solutions with 22 moves. This can be achieved in the following way (cf. Figure 6.6; the pegs are labelled 0, 1, 2 from left to right, as usual):

1. Move two small discs to peg 1.

2. Move large discs in the order $0 \to 2$, $1 \to 0$, $2 \to 1$, $2 \to 1$, $0 \to 2$, $1 \to 0$.

3. To make these moves of large discs possible, transfer the 3-tower of all small discs in the order $1 \to 2 \to 0 \to 1 \to 2$ making 3 moves, respectively. As it is transported an even number of times, the order of the discs inside the tower has not changed.

4. Move two small discs from peg 2.

To complete the argument note that in step 1 one has two alternative choices which disc to move first.

Exercise 6.4 As in the case of an even number of discs, the optimal solution for the task $1(01)^n \to 0(10)^n$ can be decomposed into the transfer of $2n$ smaller discs to peg 2, the single move of the extra disc to peg 0 and the sorting of the $2n$-tower on 2 into odd and even on pegs 0 and 1, respectively. This time, the first three input pairs are $(1,0)$, $(0,1)$, and $(1,0)$ for the Hanoi labeling, leading to $(1,0),(2,2),(0,1)$ in the Sierpiński notation, such that in the P2 decision automaton $i = 1$, $j = 0$, and $k = 2$ and again the two input pairs are (k,k) and (j,i). Therefore, the total number of moves is $2y_{2n} + 1 = \left\lfloor \frac{5}{7}2^{2n+1} \right\rfloor + (n \bmod 3 = 2)$. Putting the results together, the swapping of the positions of $N \in \mathbb{N}$ discs, initially sorted according to parity, needs $2y_{N-1} + 1 = \left\lfloor \frac{5}{7}2^N \right\rfloor + (N \bmod 3 = 2)$ moves.

Exercise 6.5 Translated into the language of the TH, the task is $(01)^4 \to (1001)^2$. Applying the H_3-to-S_3-automaton of Figure 4.4, this becomes $0(210)^2 2 \to (1221)^2$, and Romik's automaton in Figure 2.28 tells us that the largest disc 8, i.e. the largest golden disc, moves twice in the shortest path, i.e. we have a class II task. Before the first move of disc 8, disc 7 is not moved, because it is not in the way. Between the two moves of disc 8, the 7-tower of smaller discs is transferred with one move of disc 7. This is followed by the last move of disc 8. After that the 7-tower is decomposed into its final distribution with no move of disc 7 because it is already in its terminal position and the starting configuration is perfect. Thus, the largest silver disc moves only once, although it is the "lighter" one. The shortest path needs 208 moves.

Exercise 6.6 With the aid of the H_3-to-S_3-automaton of Figure 4.4 the task translates into the task $s := 02020202 \to 12121212 =: t$ in S_3^8. Romik's automaton in Figure 2.28, with $i = 0$, $j = 1$, and $k = 2$ moves to state B on the input (k,k), remains there after the next input (i,j) and then, on (k,k) again, changes to state E, such that disc 8 necessarily moves twice. It follows that

$$d(s,t) = d(s,02^7) + 1 + (2^7 - 1)$$
$$+ 1 + d(12^7, t) = 213.$$

The first sub-task is $201201 \to 1^6$ in H_3^6. The P1-automaton in Figure 2.6 yields 1 as the idle peg of the best first move, such that disc 2 moves from peg 0 to peg 2. (With only one move of the largest disc, the full 255 moves are needed; in that case, disc 1 moves to peg 2 first, such that this would destroy optimality already.)

Exercise 6.7 Since $t = 2 = r$ and $N = 3$, Theorem 6.4 asserts that the optimal number of moves is 29, which is indeed the length of the given path. The sequence of moves differs from the one produced by Algorithm 22 starting from the state $217|465|3$. After this state disc 2 is moved while Algorithm 22 moves disc 3.

Chapter 7

Exercise 7.1 As G is hamiltonian, $G - X$ has at most $|X|$ connected components consisting of the sections of an arbitrary hamiltonian cycle C between the removed vertices of X. Hence if two vertices of X would be consecutive on C, $G - X$ would consist of at most $|X| - 1$ connected components.

For $G = L$, consider X as the set of the 12 vertices of degree 4.

Exercise 7.2 a) The correspondence between s and σ is given by $s_{n+l} = l + \sum_{k=1}^{l} \sigma_k$ for all $l \in [p+1]_0$. Figure 9.15 refers to this situation: on the left one can see the positions s_{n+k} of peg bottoms dividing the set of n (uncolored) balls, and on the right the corresponding distribution of the balls on the pegs is depicted.

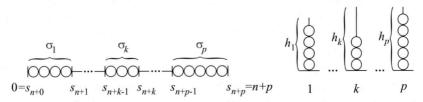

Figure 9.15: Correspondence between positions of peg bottoms and the distribution of balls

b) We have, in fact for $n \in \mathbb{Z}$,

$$l_h^n = \left| \left\{ \sigma \in \mathbb{N}_0^p \mid \sigma \le h, \sum_{k=1}^{p} \sigma_k = n \right\} \right|,$$

where $\sigma \le h$ is meant componentwise, i.e. $\sigma_k \le h_k$ for every $k \in [p]$. (Of course, $l_h^n = 0$ for $n < 0$.) Then for $p = 0$ we obviously get

$$l_{\varnothing}^n = (n = 0). \tag{9.7}$$

Moreover, if $g \in \mathbb{N}_0^p$ and $\alpha \in \mathbb{N}_0$, then

$$l_{g\alpha}^n = \sum_{\nu=n-\alpha}^{n} l_g^\nu; \tag{9.8}$$

this is because necessarily $\sigma_{p+1} = n - \sum\limits_{k=1}^{p} \sigma_k$ and therefore $\sum\limits_{k=1}^{p} \sigma_k$ runs through all (non-negative) values from $n - \alpha$ to n.

With equation (9.8), we can break down l_h^n recursively to a sum consisting only of terms of the form $l_\emptyset^\mu$, which can then be evaluated using equation (9.7). Explicit formulas for l_h^n can only be given in specific cases; cf. Exercise 7.3. As an example we take the classical London graph L, where $p = 3 = n$ and $h = 123$, such that

$$l_{123}^3 = \sum_{\nu=0}^{3} l_{12}^\nu = \sum_{\nu=0}^{3} \sum_{\mu=\nu-2}^{\nu} \sum_{\lambda=\mu-1}^{\mu} (\lambda = 0)$$

$$= \sum_{\nu=0}^{3} \sum_{\mu=\nu-2}^{\nu} (\mu \le 1) = 1 + 2 + 2 + 1 = 6.$$

c) Each σ corresponds to a distribution of n uncolored balls on the pegs. Then there are $n!$ different colorings of the balls.

Exercise 7.3 a) Induction on p. For $p = 1$, we have

$$l_n^n = \sum_{\nu=0}^{n} l_\emptyset^\nu = 1 = \binom{1 + n - 1}{n}.$$

Furthermore,

$$l_{n^{p+1}}^n = \sum_{\nu=0}^{n} l_{n^p}^n = \sum_{\nu=0}^{n} \binom{p + \nu - 1}{\nu}$$

$$= \binom{p + 1 + n - 1}{n},$$

the latter by induction on n. It follows that

$$|O_p^n| = n!\binom{p + n - 1}{n} = \frac{(p + n - 1)!}{(p - 1)!}.$$

b) Of all permutations $s \in \mathrm{Sym}_{n+p}$ only those with $s_{p+n} = p + n$ are admissible, i.e. at most $|\mathrm{Sym}_{n+p-1}| = (p + n - 1)!$. Only $1/(p - 1)!$ of these fulfil the necessary condition of strict monotonicity $s_{n+1} < s_{n+2} < \cdots < s_{n+p-1} < s_{n+p}$. But the latter also implies $k \le s_{n+k} \le n + k$ for all $k \in [p + 1]_0$, whence $1 \le s_{n+k} - s_{n+k-1} \le n + 1$ for all $k \in [p]$. Therefore the condition in Definition 7.2 is automatically fulfilled for all remaining permutations.

Exercise 7.4 Choose $q \in [p]$ pegs. The number of states with all balls distributed among *all* these pegs is, for $n \ge q$,

$$\frac{n!}{(n - q)!}|O_q^{n-q}| = n!\binom{n - 1}{q - 1}.$$

This is so because there are $\frac{n!}{(n-q)!}$ choices to fill the bottom line of balls on all q pegs, and the remaining $n - q$ balls can be distributed arbitrarily, such that the number of these distributions can be taken from the previous exercise.

Each of these states has degree $q(p-1)$, whence by the Handshaking Lemma

$$\|O_p^n\| = \frac{1}{2}(p-1)\,n!\sum_{q=1}^{p} q\binom{p}{q}\binom{n-1}{q-1}$$

$$= \Delta_{p-1}\,n!\sum_{q=1}^{p}\binom{p-1}{q-1}\binom{n-1}{q-1}$$

$$= \Delta_{p-1}\,n!\sum_{\lambda=0}^{n-1}\binom{p-1}{\lambda}\binom{n-1}{n-1-\lambda}$$

$$= \Delta_{p-1}\,n!\binom{p-2+n}{n-1}$$

$$\left(= \Delta_{p-1}\,n!\binom{p-2+n}{p-1}\right)$$

$$= \frac{np}{2}\frac{(p-2+n)!}{(p-2)!}\,,$$

where in the fourth step use has been made of formula (0.14).

Exercise 7.5 a) The steps from the proof of Theorem 7.3 are illustrated in Figure 9.16.

b) Since no ball is in right position at the start, it is clear that each one has to move at least once and ball 3 at least twice. Moreover, an easy combinatorial analysis shows that if ball 2 moves once only, then 4 has to move three times and if ball 4 moves only once, then 2 moves thrice. Therefore, an optimal solution needs at least 7 steps. Here is one:

$$1|2|43 \to 21||43 \to 21|4|3 \to 1|24|3 \to 1|324| \ \to |324|1 \to |24|31 \to |4|231\,.$$

Exercise 7.6 Any three vertices with maximal degree (= 4) and mutual distances 2 form an O_3^2-dominating set, whence $\gamma(O_3^2) \le 3$. On the other hand, any three vertices with minimal degree (= 2) and mutual distance 3 have non-overlapping neighborhoods, such that $\gamma(O_3^2) \ge 3$ by Proposition 2.27. There is no perfect code for O_3^2 though, because a dominating set with 3 vertices will have to contain at least one vertex of degree 4, but then their neighborhoods will overlap.

Exercise 7.7 The lower bound follows from $1 = \mathrm{g}\left(L_{222}^3\right) \le \mathrm{cr}\left(L_{222}^3\right)$. For the upper bound, see Figure 9.17. Better bounds are not known so far.

Exercise 7.8 This can be done by adding all missing vertices and their adjacent edges to the drawing of L_{222}^3 in Figure 7.9 without introducing any crossings. In Figure 9.18 the mixed graph of Lucas's second problem for 3 discs $\overrightarrow{H}_3^3$ is drawn on the torus; the underlying graph, completed by the edges in red, is O_3^3.

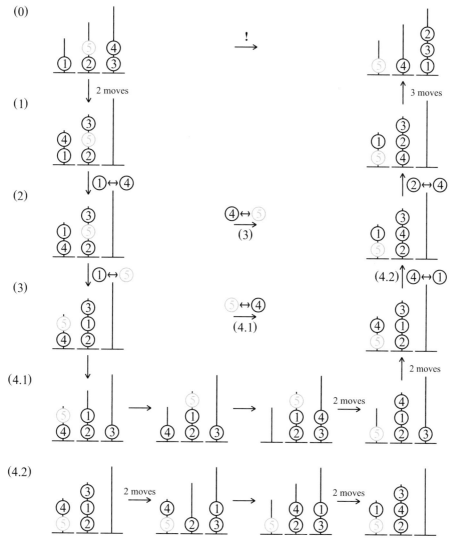

Figure 9.16: Solution for $1|2|43 \to |4|231$ in L^4_{234}

In fact, we know from Exercise 7.4 that $\|O^n_3\| = \frac{3}{2}n(n+1)! = 3\Delta_n n!$ (cf. [147, Satz 3.7]). Finsterwalder noted in [147, Satz 3.8] that $3\Delta_{n-1}n!$ out of the $2\|O^n_3\|$ arcs of the *directed* Oxford graph are missing in $\overrightarrow{H}^n_3$ because these moves do not obey the divine rule; hence it has only $\frac{3}{2}n(n+3)n!$ arcs. Finally, $\Delta_{n-2}n!$ edges of O^3_n, $n \in \mathbb{N}$, i.e. 6 for $n = 3$, are missing in the underlying graph of $\overrightarrow{H}^n_3$, because the corresponding moves do not conform with the divine rule in both directions;

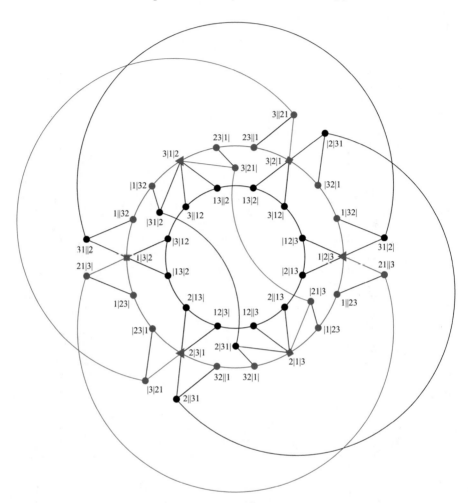

Figure 9.17: The graph L_{222}^3 with 8 crossings in the plane

hence, this graph has $(n^2 + 3n - 1)n!$ edges (cf. [147, Satz 3.9]).

Exercise 7.9 Let us first determine the fixed points for the action of $(\Gamma_{33}, \cdot, 1_{33})$ on $V(L_{222}^3)$. Apart from the identity 1_{33}, only "flat" distributions of the three balls can lead to a fixed point, and they can only be a fixed point of (χ, π) if both χ and π are rotations (derangements) or both are reflections. The distribution of fixed points to the group elements is shown in Table 9.1.

We can read the following from that table. The sum of the sizes of the fixed point sets is

$$\sum_{g \in \Gamma_{33}} |V(L_{222}^3)^g| = 42 + 4 \times 3 + 9 \times 2 = 72,$$

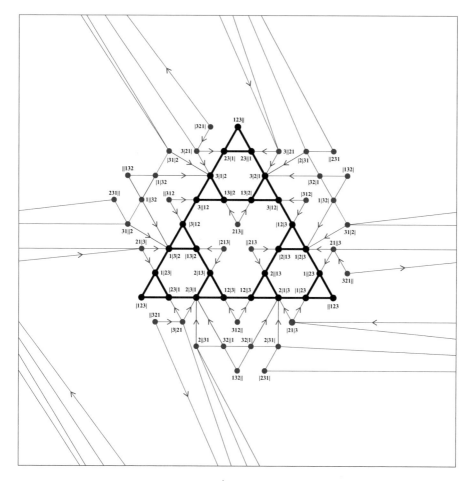

Figure 9.18: The mixed graph $\overrightarrow{H}_3^3$ and extra edges of O_3^3 on the torus

such that Burnside's lemma Corollary 0.24 yields (with $|\Gamma_{33}| = 3! \times 3! = 36$) that the number of equisets of vertices is just 2. From Theorem 0.23 we learn that one of the classes consists of the 36 non-flat states, which stay fixed only under the identity, and the other one of the 6 flat distributions of balls, each of which is fixed under the action of 6 group elements. We therefore only have to consider two representatives, $12|3|$ and $1|2|3$, say. They have degrees 3 and 6, respectively, such that the statements relating to degrees are obvious.

We now turn to the conclusions about metric properties of L_{222}^3. The only tedious step is to determine all distances to other vertices from the two prototype states, i.e. to span shortest-path trees rooted in $12|3|$ and $1|2|3$. This can be done, e.g., by recourse to Figure 9.17.

χ \ π	123	231	312	132	321	213
123	V	∅	∅	∅	∅	∅
231	∅	1|2|3 2|3|1 3|1|2	1|3|2 2|1|3 3|2|1	∅	∅	∅
312	∅	1|3|2 2|1|3 3|2|1	1|2|3 2|3|1 3|1|2	∅	∅	∅
132	∅	∅	∅	1|2|3 1|3|2	2|1|3 3|1|2	2|3|1 3|2|1
321	∅	∅	∅	2|1|3 2|3|1	1|2|3 3|2|1	1|3|2 3|1|2
213	∅	∅	∅	3|1|2 3|2|1	1|3|2 2|3|1	1|2|3 2|1|3

Table 9.1: Fixed point sets for symmetries on $V(L_{222}^3)$

In Figure 9.19 all distances from the respective start vertices are indicated. It turns out that the height of both trees is 5, which proves the statements about eccentricities and diameter, the latter being attained, e.g. for the task $1\,|\,2\,|\,3 \rightarrow 3\,|\,|\,12$. (From the Handshaking Lemma it also follows that $\|L_{222}^3\| = 72$.) Moreover, the sums of lengths of all shortest paths in the trees are 140 and 133, respectively. This leads to the total sum of $36 \cdot 140 + 6 \cdot 133 = 5838$ and the average distance of $\frac{139}{41}$.

Finally, in order to arrange tasks into equisets, we have to let the group act on the set $V(L_{222}^3) \dot{\times} V(L_{222}^3)$. A task (s,t) can only be invariant under the action of a group element g, if both s and t are in the fixed point set of g. This means that there are just 2 fixed points for each combination of reflections and 6 for the combinations of non-trivial rotations. That makes a total of $1722 + 4 \cdot 6 + 9 \cdot 2 = 1764$, such that Corollary 0.24 tells us that there are 49 equisets of tasks. Moreover, Table 9.1 shows that tasks being fixed by a non-trivial rotation are fixed by another one, whereas those fixed by a reflection are not invariant under any other non-trivial group element. Therefore, the former constitute the only equiset of size 12, the latter the only one of size 18. Representatives of these two classes are, e.g., $1\,|\,2\,|\,3 \rightarrow 2\,|\,3\,|\,1$ and $1\,|\,2\,|\,3 \rightarrow 2\,|\,1\,|\,3$, respectively. All other task are invariant only under the identity, such that Theorem 0.23 puts them into equisets of size 36.

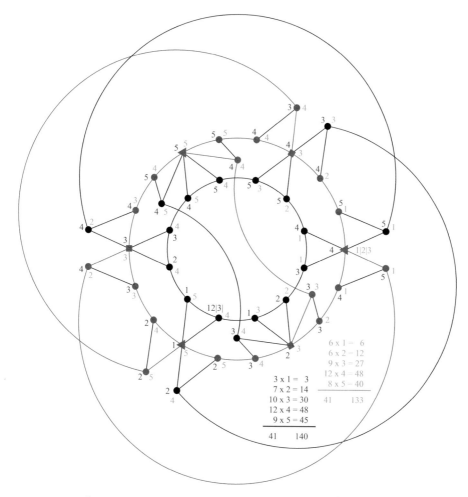

Figure 9.19: L_{222}^3 with distances from vertex $1\,|\,2\,|\,3$ (green) and from vertex $12\,|\,3\,|$ (blue)

Chapter 8

Exercise 8.1 It suffices to prove that the tower of n discs can be moved from peg 0 to peg 1. Design an algorithm for this task and prove its correctness by induction.

Exercise 8.2 First show by induction that $|0 \xrightarrow{[n]} 1| = |2 \xrightarrow{[n]} 0|$ and $|0 \xrightarrow{[n]} 2| = |1 \xrightarrow{[n]} 0|$. Setting $a_n = |1 \xrightarrow{[n]} 2|$, $b_n = |1 \xrightarrow{[n]} 0|$, $c_n = |2 \xrightarrow{[n]} 1|$, and $d_n = |2 \xrightarrow{[n]} 0|$, we read from Algorithm 23:

$$a_{n+1} = 2b_n + 1,$$
$$b_{n+1} = 2b_n + d_n + 2,$$
$$c_{n+1} = 2d_n + 1,$$
$$d_{n+1} = b_n + c_n + 1.$$

The initial conditions are left to the reader.

Exercise 8.3 This is done as in Exercise 8.2. Now, with $a_n = |2 \xrightarrow{[n]} 1|$, $|0 \xrightarrow{[n]} 1| = b_n = |2 \xrightarrow{[n]} 0|$, $c_n = |1 \xrightarrow{[n]} 2|$, and $|0 \xrightarrow{[n]} 2| = d_n = |1 \xrightarrow{[n]} 0|$, we get:

$$a_{n+1} = 2a_n + c_n + 2,$$
$$b_{n+1} = a_n + d_n + 1,$$
$$c_{n+1} = 2d_n + 1,$$
$$d_{n+1} = b_n + c_n + 1.$$

The base of the asymptotic exponential behavior of these sequences is then the greatest (with respect to absolute value) solution of $\lambda^3 - \lambda^2 - 4\lambda + 2 = 0$, which is approximately 2.342923.

Exercise 8.4 The first assertion follows from the definitions of legal moves in TH and in $\mathrm{TH}(\vec{C}_3)$, for the second use Proposition 8.6 and Theorem 8.7.

Exercise 8.5 To solve the shift Cyclic Tower of Antwerpen, apply first the Transpose Cyclic Tower of Antwerpen for the yellow tower and the red tower, and then apply another instance of the Transpose Cyclic Tower of Antwerpen for the black tower and the (new) red tower. Proceed similarly in the case of the Double Shift Cyclic Tower of Antwerpen.

Exercise 8.6 That these graphs are planar is obvious: St_p^0 is trivial, St_p^1 are star graphs, St_3^n are path graphs, and the planarity of St_4^2 can be seen in Figure 8.4. It has also been seen that St_4^3 is not planar and hence all supergraphs St_4^n, $n \in \mathbb{N}_3$. Moreover, the Star Hanoi graph St_5^2 (see Figure 9.20) contains a subdivision of $K_{3,3}$ engendered by the two sets of vertices $\{00, 20, 40\}$ (red in the figure) and $\{01, 02, 03\}$ (green in the figure).
Remark. So the "smallest" non-planar cases are St_4^3 and St_5^2. The drawings of St_4^3 in Figure 8.5 and St_5^2 in Figure 9.20 make them accessible for investigation. We conjecture that $\mathrm{cr}(St_4^3) = 3$, $\mathrm{cr}(St_5^2) = 2$, and $\mathrm{g}(St_4^3) = 1 = \mathrm{g}(St_5^2)$.

Exercise 8.7 Double induction on p and n. For $p = 3$ we have for all $n \in \mathbb{N}_0$: $st_3^0(n) = \Sigma T_n \le 2\Sigma T_n = st_3^1(n)$. So let $p \ge 4$. If $n = 0$, then $st_p^0(0) = 0 = st_p^1(0)$. For $n \ge 1$ and any $m \in [n]_0$ the induction assumptions on n and p yield

$$st_p^0(m) + st_p^1(m) + st_{p-1}^0(n-m) \le 2st_p^1(m) + st_{p-1}^1(n-m),$$

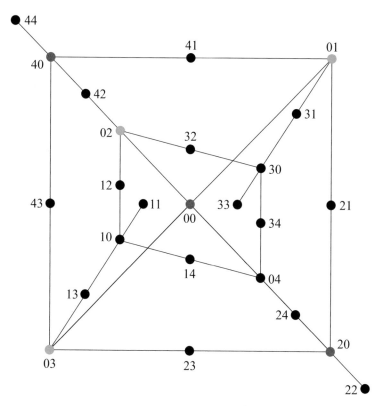

Figure 9.20: Drawing of the graph St_5^2 with 2 crossings

whence $st_p^0(n) \leq 2st_p^1(m) + st_{p-1}^1(n-m)$ and consequently $st_p^0(n) \leq st_p^1(n)$.

Exercise 8.8 We proceed as in the proof of Theorem 8.18, i.e. by (strong) induction on $n \in \mathbb{N}$. For $n = 1$ only $m = 0 = n - 1 - \lfloor \text{lb}(A_n) \rfloor$ is admissible and the only disc makes $1 = A_n$ move.

For $n \geq 2$ let $\ell := n - m$ ($\in [n]$). Discs $n - \mu$, $\mu \in [\ell]_0$, are transferred according to the classical TH, i.e. as in H_3^ℓ, whence by Proposition 2.4 they move $2^\mu = B_{\mu+1}$ times, respectively. On the other hand, by induction assumption the smallest m discs make $3A_{\nu+1}$ moves, where ν runs through $[m]_0$.

Note that the sequences given by $a_\mu = B_{\mu+1}$ and $b_\nu = 3A_{\nu+1}$ are disjoint and that they partition the sequence $A \upharpoonright \mathbb{N}$. Indeed, let $A_i = 2^j 3^k$, $j, k \in \mathbb{N}_0$. If $k = 0$ then $A_i = 2^j$ is a term of a and if $k \geq 1$ then $A_i = 3 \cdot 2^j 3^{k-1}$ is a term of b. The sum of the first ℓ terms of a and the first m terms of b will be minimized if their union consists of the terms of $A \upharpoonright [n]$. This will happen if and only if for the terms of a we have $a_{\ell-1} \leq A_n < a_\ell$. For the minimum to be attained we thus get $2^{\ell-1} \leq A_n < 2^\ell$, i.e. $(\ell - 1) \ln(2) \leq \ln(A_n) < \ell \ln(2)$, which in turn implies that $\ell = 1 + \lfloor \text{lb}(A_n) \rfloor$.

We have thus proved that $m_n = n - 1 - \lfloor \text{lb}(A_n) \rfloor$ is the unique value that minimizes the Matsuura strategy for the task to transfer an n-tower from peg 1 to peg 2, say, in $\mathrm{TH}(K_4 - e)$. Moreover, the argument also shows that discs move A_i, $i \in [n]$, times, respectively, hence the algorithm makes $\sum_{i=1}^{n} A_i = \Sigma_n$ moves..

Exercise 8.9 Starting from 1^5 in the Diamond TH move discs 1 and 2 to peg 3 making 3 moves. Then discs 3, 4 and 5 are transferred to peg 2, avoiding peg 3, i.e. in the classical way taking 7 moves. Finally, the 2-tower on peg 3 can be brought to peg 2 in 4 moves (disc 1 to peg 1, disc 2 to peg 0, then to peg 2, and disc 1 to peg 2). Altogether, these are 14 moves, which is strictly less than $\widetilde{\Sigma}_5 = \Sigma_5 = 16$. (In Algorithm 25 the last transfer of a 2-tower takes 6 moves.)

Exercise 8.10 Hint: Arrange the pegs in the natural order 0–1–$\cdots$–$(p-1)$ and consider the states of the puzzle as the points of the lattice $[p]_0 \times [p]_0$. Optimal solutions of the puzzle correspond to paths in the grid between $(0,0)$ and $(p-1, p-1)$ that meet a diagonal point (i,i) for precisely one $i \in [p-2]$. Then take a closer look at the shape of these paths. (For details, see [240, p. 375f].)

Exercise 8.11 We first prove the identity by induction on n. The case $n = 0$ is trivial. Starting from 0^{1+n} it is unavoidable to reach the state 20^n before any of the pegs $p \geq 3$ have been involved, because discs can never return to the pegs in T if they have left this set of pegs. This happens on a classical undirected 3-pegs Linear TH, so it needs $2 \cdot 3^n$ moves, namely 2 moves of disc $n + 1$ and $2(3^n - 1)$ moves of the other discs. The rest of a shortest path to $(p-1)^{1+n}$ will include $p - 3$ moves of disc $n + 1$, because it has no other choice, and a transfer of the n-tower from peg 0 to peg $p - 1$, which takes another $3^n - 1 + n(p - 3)$ moves by induction assumption. Altogether $2 \cdot 3^n + p - 3 + 3^n - 1 + n(p - 3) = 3^{n+1} - 1 + (n + 1)(p - 3)$ moves are necessary and sufficient.

For the inequality we note that if $j \in T$ it is clear from the classical Linear TH and otherwise if the starting peg i is 1 or 2, then the intermediate states to be passed are 20^n again or 21^n, which are reached in $\frac{1}{2}(3^n + 1)$ or $\frac{1}{2}(3^n - 1)$ moves, respectively, in any case in less than $2 \cdot 3^n$ moves. If the goal is $3 \leq j < p - 1$, then we may just ignore the pegs with labels larger than j, i.e. we do not move there.

Exercise 8.12 Let a_n be the number of moves for $n \in \mathbb{N}_0$ discs. It is easy to see that $a_0 = 0$ and $a_n = 3a_{n-1} + 2$ for $n \geq 1$. This leads to $a_n = 3^n - 1$. To get a solution using only 18 moves (instead of the 26 needed by the algorithm) use the fact that when disc 3 is moved from peg 0 to peg 1, discs 1 and 2 need not be both on peg 2; and similarly in the later stages.

Exercise 8.13 Hint: It is easy to see that for any vertex v of a moderately enhanced cycle D, the digraph $D - v$ cannot contain a strong subdigraph on at least three vertices. For the converse suppose that D has no vertex w such that the digraph $D - w$ contains a strong subdigraph on at least three vertices and consider two cases. The first is that there are vertices $u, v \in V(D)$ such that $(u, v) \in A(D)$ and $(v, u) \notin A(D)$, the second being that this is not so.

Chapter 10

The End of the World

With the theory developed in Chapter 2 it was not very difficult to determine the age of the universe, see Exercise 2.8. Similarly, the time when the end of the universe is to be expected can be computed. The situation with the end of time is more delicate though, especially now that The Reve's Puzzle has been solved and the solution is readily available in the present edition of our book. It cannot be excluded that the Brahmins will be exposed to the book and consequently come to the idea to save some time and use one more peg—the Devil's one. Luckily for us, the Brahmins are men of great honor and will therefore not deviate from "the fixed and immutable laws" imposed by the *Trimurti* of creation, maintenance, and destruction of the world represented as the three "diamond needles" corresponding to the triad of dieties Brahma, Vishnu, and Shiva. So we are safe for some time until we might be faced with dramatic events in the history of the universe.

These events have already been anticipated in 1966 at precisely 21 minutes and 33 seconds into the fourth part, called *The Final Test*, of the episode *The Celestial Toymaker* from the British science fiction television series *Doctor Who*. This is the moment when the First Doctor in his 1023rd move (optimally) finishes the classical three-peg TH with 10 discs, named the *Trilogic game* in this story. Luckily for us, with the last move the Doctor only destroys the world of Toymaker, whereas he himself and his two companions manage to escape the vanishing world with the time machine TARDIS.

So we may continue, in our world, to play the *Polylogic game* and to advance the mathematical theory of the Tower of Hanoi. We hope that the readers are now highly motivated to join us in this endeavor. As an incentive, we list here some challenging problems encountered throughout the book that are waiting to be solved. We repeat the open questions from the first edition, even when they have been answered meanwhile. (*Emphasized text* shows updates with respect to the first edition.) Even if you can not be the *first*, you might still come forward with the *best* solutions.

© Springer International Publishing AG, part of Springer Nature 2018
A.M. Hinz et al., *The Tower of Hanoi – Myths and Maths*,
https://doi.org/10.1007/978-3-319-73779-9_11

Guy's Conjecture (p. 47) on crossing numbers of complete graphs (cf. [125]) in connection with

Köhler's Conjecture (p. 194) on crossing numbers of reduced complete graphs.

Kauffman's Ring Conjecture (p. 87) Mechanical and topological exchange numbers coincide.

   Confirmed by Andreas Höck [216]; cf. our Theorem 1.12.

Double Task Conjecture (p. 160) The optimal number of moves for the double P2 problem $0^n \to 2^n \parallel 2^n \to 0^n$ is $\frac{1}{3}\left(2^{n+2} - (-1)^n\right)$.

Frame-Stewart Conjecture (p. 209) $\forall\, n \in \mathbb{N}_0 : \mathrm{d}(0^n, 2^n) = FS_4^n$.

   Solved by Tierry Bousch [62]; cf. our Section 5.5.

Korf-Felner Conjecture (p. 224) For any $n \geq 20$, $ex(n) > 0$.

Monotonicity Conjecture (p. 225) The function $EX(n) = \mathrm{diam}(H_4^n) - \mathrm{d}(0^n, 3^n)$ is (eventually strictly) monotone increasing.

Strong Frame-Stewart Conjecture (p. 231)

$$\forall\, p \geq 3, \ \forall\, n \in \mathbb{N}_0 : \mathrm{d}\left(0^n, (p-1)^n\right) = FS_p^n.$$

Non-subtower Conjecture Non-subtower solutions between two perfect states in H_p^{n+1} exist if and only if $p \leq n+1 < \binom{p}{2}$.

   "Only" the absence of such solutions for $n + 1 \geq \binom{p}{2}$ is still open; see Conjectures 5.65 and 5.66. The rest has been confirmed by D. Parisse; see Theorem 5.64. A particular question is whether the number of non-isomorphic non-subtower solutions is $\frac{1}{2}p(p-3)$ if $n + 2 = \binom{p}{2}$; see page 278.

LDM Pattern Conjecture (p. 281) If an LDM pattern $b \in [2^{p-1}]_0$ occurs in H_p^n, then it occurs in every H_p^m for $m \geq n$.

Rainbow Conjecture (p. 295) Procedure Rainbow is optimal for any odd $m \geq 5$ when the discs of the initial configuration are alternately colored with m colors.

Stockmeyer's Automata Conjecture (p. 319) The solutions in Theorem 8.8 can all be realized by automata.

Allouche-Sapir Conjecture (p. 323) The two solvable TH with oriented disc moves on three pegs that are not the classical, the cyclic, and the linear TH, are not k-automatic for any k.

   The case of C_3^+ has been confirmed by Caroline Holz auf der Heide [217]; cf. above, p. 323.

Stockmeyer's Conjecture (p. 336) Algorithm 24 makes the smallest number of moves among all procedures that solve the Star puzzle.

Solved by Tierry Bousch [63]; cf. our Section 8.3.3. Has to be replaced by:

Generalized Stockmeyer Conjecture (p. 333) The Stewart-Stockmeyer strategy makes the smallest number of moves among all procedures that solve the Star puzzle for any number of pegs and any combination of start and destination pegs. An even stronger claim is **Conjecture 8.30**.

Ferme's Conjecture (p. 342) The best lower bound for n in Conjecture 8.17.

Some other open problems include (cf. [203]):

- Solve the recurrence from the solution to Exercise 2.26 a.

- Analyze average distances in $\overrightarrow{H}_3^n$ further. In particular, determine the value of a from the solution of Exercise 3.5.

- Find an optimal algorithm for type P4 tasks in Lucas's second problem.

- Design an automaton analogous to Romik's automaton for a "P2 task" in S_p^n, $p \geq 4$.

  Done; see page 190.

- (Ambiguous perfect goal) In H_p^n, $p \in \mathbb{N}_4$, $n \in \mathbb{N}$, and for $\overline{s} \in [p]_0^{n-1}$ find (all) $j \in [p-1]$ for which $d(0\overline{s}, j^n)$ is minimal. (For the case $p = 3$, see Proposition 2.22.)

- Find a formula for the average distance on Hanoi graphs for $p \geq 4$.

- Find the diameter of these graphs.

- Find the metric dimension of Hanoi graphs. (For the Sierpiński case, see p. 191.)

- Find genera and crossing numbers for non-planar Hanoi and Sierpiński graphs.

- Find an optimal algorithm for Black and white TH Variant C.

- Find optimal algorithms for the Shift and Double Shift Cyclic Tower of Antwerpen; see p. 325.

- Find the domination numbers for general Hanoi (p. 262) and London (p. 308) graphs. Is it true that $\gamma(H_p^n) \leq \gamma(S_p^n)$ for $n \in \mathbb{N}_3$ and (even) $p \in \mathbb{N}_4$?

- Investigate (computationally) the numerical relations between the lenghts of solutions for different tasks in Stewart-type algorithms; cf. Section 8.3.4.

- Find minimal solutions for the Linear TH (Section 8.3.5) and for the (directed) Cyclic TH (Section 8.4).

So it turns out that Zhuge Liang not only entertained his wife by asking her to solve the Chinese rings and Édouard Lucas the Brahmins working on the Tower of Benares, but generations of mathematicians are occupied with trying to find sound statements about the mathematical theories these innocent looking puzzles engender.

"Most mathematicians would argue that as the scale of our mathematical investigations increases so does the depth and beauty of our discoveries, illuminating patterns within patterns ad infinitum; that our collective mathematical vision is not an artifact of the scale of our view but instead it is a glimpse of a world beyond."

W. Hugh Woodin, 1998, [445, p. 350]

"Un dernier mot. Si ces pages plaisent à quelques savants, si elles intéressent quelques gens du monde, si elles inspirent à quelques jeunes intelligences le goût du raisonnement et le désir des jouissances abstraites, je serais satisfait."

Édouard Lucas, 1881, [283, p. viii]

(A last word. If these pages please some scholars, if they are of interest for some men of the world, if they inspire in some young minds the taste of reasoning and the desire for abstract enjoyments, I will be satisfied.)

Glossary

adjacency matrix of a graph The matrix associated with graph G given by

$$V(G) \times V(G) \to B, \ (u,v) \mapsto (\{u,v\} \in E(G)) .$$

binary tree A rooted tree where every vertex has at most two adjacent vertices which are further away from the root.

bit Short for **binary dig**it.

bridge An edge whose deletion increases the number of connected components of a graph. Also called *cut-edge*. In particular, edge whose deletion leaves a connected graph disconnected.

center The set of those vertices of a graph whose eccentricities are equal to its radius.

clique in a graph A complete subgraph.

complete bipartite graph A graph with vertex set $V_1 \dot\cup V_2$ and edge set $\{\{v_1, v_2\} \mid v_1 \in V_1, \ v_2 \in V_2\}$; the class is denoted by K_{p_1,p_2}, if $|V_1| = p_1$ and $|V_2| = p_2$.

complete graph A graph with vertex set V and edge set $\binom{V}{2}$; the class is denoted by K_p, if $|V| = p$.

connectivity The connectivity of a graph G is the smallest size $|X|$ among those sets $X \subseteq V(G)$ for which $G - X$ is disconnected or has only one vertex.

cyclic permutation A permutation which moves the elements of a subset cyclically: $c_1 \mapsto c_2 \mapsto \cdots \mapsto c_k \mapsto c_1$ and leaves the others fixed; its length is k, and it is denoted by $(c_1 c_2 \ldots c_k)$.

derangement A fixed-point-free permutation.

diameter of a graph The maximum over all distances in a connected graph G, denoted by $\mathrm{diam}(G)$.

digraph Ordered pair $D = (V(D), A(D))$ consisting of the set of vertices $V(D)$ and the set of arcs $A(D)$ which are ordered pairs of vertices. (Cf. [35].)

© Springer International Publishing AG, part of Springer Nature 2018
A.M. Hinz et al., *The Tower of Hanoi – Myths and Maths*,
https://doi.org/10.1007/978-3-319-73779-9

eccentricity of a vertex The maximum distance between a given vertex and all the other vertices.

edge coloring A function from the edge set of a graph to a set of colors such that any pair of adjacent edges receives different colors.

eulerian graph A graph which contains an eulerian circuit.

forest A graph F without cycles. It has $|F| - \|F\|$ components [422, Theorem I.35].

geodesic A shortest path between two vertices of a graph G is called a geodesic in G.

graph Ordered pair $G = (V(G), E(G))$ consisting of the set of vertices $V(G)$ and the set of edges $E(G)$, the latter consisting of unordered pairs of vertices.

graph distance The distance function on the vertex set of a connected graph given by the length of a shortest path between any two vertices.

hamiltonian cycle A cycle containing all the vertices of a given graph.

hamiltonian path A path containing all the vertices of a given graph.

Handshaking Lemma Asserts that the sum of the degrees in a graph is twice the number of its edges.

height of a rooted tree The largest distance from the root.

inclusion-exclusion principle For two finite sets X and Y, $|X \cup Y| = |X| + |Y| - |X \cap Y|$. The principle has an extension to any finite family of finite sets.

independence number A subset of vertices of a graph G which are pairwise non-adjacent is called *independent*. The maximum size among all such sets is the independence number of the graph, denoted by $\alpha(G)$.

matching A set of pairwise non-adjacent edges of a graph.

mixed graph Triple $G = (V(G), E(G), A(G))$ consisting of the set of vertices $V(G)$, the set of edges $E(G)$ which are unordered pairs of vertices, and the set of arcs $A(G)$ which are ordered pairs of vertices.

natural number A non-negative integer. The set of natural numbers is denoted by $\mathbb{N}_0$, while $\mathbb{N}$ stands for the set of positive integers.

octit Short for oct**ernary dig**it.

partition of a set $P \subseteq 2^V$ is a partition of the set V, if $\varnothing \notin P$, $V = \bigcup P$, and $\forall \{X, Y\} \in \binom{P}{2} : X \cap Y = \varnothing$.

path in a digraph A subdigraph with vertex set $\{x_i \mid i \in [\ell + 1]_0\}$ of size $\ell + 1$ and arc set $\{(x_{i-1}, x_i) \mid i \in [\ell]\}$, where $\ell \in \mathbb{N}_0$ is its length, x_0 its start vertex and x_ℓ its end vertex.

perfect matching A matching M of a graph G such that every vertex of G is *saturated*, i.e. is incident with (precisely) one element of M.

periphery The set of those vertices of a graph whose eccentricities are equal to its diameter.

permutation A bijective mapping from a finite set to itself.

Pigeonhole principle There is no injective mapping from $[n]$ to $[m]$ if $\mathbb{N}_0 \ni m < n \in \mathbb{N}$ (as can be proved by induction on m).

pit Short for **p**-ary d**i**gi**t**, the entities of the number system with base p; cf. bit, octit, qit, and tit.

qit Short for **q**uaternary dig**it**.

radius of a graph The minimum over all eccentricities of vertices in a connected graph.

regular graph Graph all of whose vertices have the same degree. Called k-regular if their degree is k.

resolving set A subset $\{r_1, \ldots, r_m\}$, $m \in \mathbb{N}_0$, of the vertex set of a connected graph such that the *distance vector* $(\mathrm{d}(v, r_1), \ldots, \mathrm{d}(v, r_m))$ determines every vertex v uniquely.

root, rooted tree A rooted tree is a tree one of whose vertices has been specified as its root.

semi-eulerian graph A graph which contains an eulerian trail.

1-skeleton of a polyhedron The graph whose vertex set is the set of vertices of the polyhedron and where two vertices form an edge if the corresponding vertices on the polyhedron are the endpoints of an edge.

spanning subgraph A subgraph with the same vertex set as its host graph.

Stirling numbers of the second kind The number of partitions of an n-set into k non-empty parts; denoted by $\left\{{n \atop k}\right\}$, cf. [246].

subdivision A subdivision of a graph is a graph obtained from it by successively finitely often replacing an edge $\{x, y\} \in E$ by two edges $\{x, z\}$ and $\{z, y\}$ with a new vertex z.

subgraph A graph $G' = (V', E')$ is a subgraph of graph $G = (V, E)$, denoted by $G' \subseteq G$; if $V' \subseteq V$ and $E' \subseteq E$. If $E' = \{\{x, y\} \in E \mid x, y \in V'\}$, then G' is called *induced* (by V').

suffix of a sequence, finite or infinite, $(x_k)_{k \in [\ell]_K}$ or $(x_k)_{k \in \mathbb{N}_K}$, respectively, is a subsequence $(x_k)_{k \in [\ell - L + K]_L}$ or $(x_k)_{k \in \mathbb{N}_L}$ for some $L \geq K$ with $L \leq K + \ell$ if the original sequence has length $\ell \in \mathbb{N}_0$.

tit Short for **t**ernary dig**it**.

total coloring A vertex coloring together with an edge coloring such that each edge receives a color different from the colors of its end vertices.

total eccentricity The sum of eccentricities $\varepsilon(v)$ over all vertices v of a (connected) graph.

transposition A permutation leaving all but two elements fixed, i.e. a cyclic permutation of length 2.

tree A connected graph without cycles, i.e. a connected forest.

vertex coloring A function from the vertex set of a graph to a set of colors such that adjacent vertices receive different colors.

Bibliography

[1] Afriat, S. N., The Ring of Linked Rings, Duckworth, London, 1982.

[2] Ahrens, W., Mathematische Spiele, B. G. Teubner, Leipzig, 1907.

[3] Alekseyev, M. A., Berger, T., Solving the Tower of Hanoi with Random Moves, in: [42], Chapter 5.

[4] Allardice, R. E., Fraser, A. Y., La Tour d'Hanoï, Proceedings of the Edinburgh Mathematical Society 2 (1883/4) 50–53.

[5] Allouche, J.-P., Note on the cyclic towers of Hanoi, Theoretical Computer Science 123 (1994) 3–7.

[6] Allouche, J.-P., Astoorian, D., Randall, J., Shallit, J., Morphisms, Square-free Strings, and the Tower of Hanoi Puzzle, The American Mathematical Monthly 101 (1994) 651–658.

[7] Allouche, J.-P., Bacher, R., Toeplitz sequences, paperfolding, Towers of Hanoi and progression-free sequences of integers, L'Enseignement Mathématique (2) 38 (1992) 315–327.

[8] Allouche, J.-P., Bétréma, J., Shallit, J. O., Sur des points fixes de morphismes d'un monoïde libre, RAIRO Informatique Théorique et Applications 23 (1989) 235–249.

[9] Allouche, J.-P., Dress, F., Tours de Hanoï et automates, RAIRO Informatique Théorique et Applications 24 (1990) 1–15.

[10] Allouche, J.-P., Sapir, A., Restricted towers of Hanoi and morphisms, Lecture Notes in Computer Science 3572 (2005) 1–10.

[11] Allouche, J.-P., Shallit, J., The ubiquitous Prouhet-Thue-Morse sequence, in: C. Ding, T. Helleseth, H. Niederreiter (eds.), Sequences and their Applications, Springer, London, 1999, 1–16.

[12] Allouche, J.-P., Shallit, J., Automatic Sequences, Cambridge University Press, Cambridge, 2003.

[13] Alonso Ruiz, P., Dirichlet forms on non self-similar sets: Hanoi attractors and the Sierpiński gasket, Ph.D. thesis, Universität Siegen, Siegen, 2013.

© Springer International Publishing AG, part of Springer Nature 2018
A.M. Hinz et al., *The Tower of Hanoi – Myths and Maths*,
https://doi.org/10.1007/978-3-319-73779-9

[14] Alonso-Ruiz, P., Kelleher, D. J., Teplyaev, A., Energy and Laplacian on Hanoi-type fractal quantum graphs, Journal of Physics A: Mathematical and Theoretical 49 (2016) Art. 165206.

[15] Amarel, S., On representations of problems of reasoning about actions, in: D. Michie (ed.), Machine Intelligence 3, Elsevier/North-Holland, Amsterdam, 1968, 131–171.

[16] Anderson, I., Combinatorial set theory, in: [437], Chapter 13.

[17] Anderson, J. R., Albert, M. V., Fincham, J. M., Tracing Problem Solving in Real Time: fMRI Analysis of the Subject-paced Tower of Hanoi, Journal of Cognitive Neuroscience 17 (2005) 1261–1274.

[18] André, J., Girou, D., Father Truchet, the typographic point, the *Romain du roi*, and tilings, TUGboat 20(1) (1999) 8–14.

[19] Arett, D., Coding Theory on the Generalized Towers of Hanoi, Proceedings of the REU Program in Mathematics, Oregon State University, Corvallis OR, 1999, 1–12.

[20] Arett, D., Dorée, S., Coloring and Counting on the Tower of Hanoi Graphs, Mathematics Magazine 83 (2010) 200–209.

[21] Armstrong, M. A., Groups and Symmetry, Springer, New York NY, 1988.

[22] Atkinson, M. D., The cyclic Towers of Hanoi, Information Processing Letters 13 (1981) 118–119.

[23] Aumann, S., Über die Anzahl der Stege auf Geodäten von Hanoi-Graphen, Diploma thesis, Ludwig-Maximilians-Universität, München, 2009.

[24] Aumann, S., Götz, K. A. M., Hinz, A. M., Petr, C., The number of moves of the largest disc in shortest paths on Hanoi graphs, The Electronic Journal of Combinatorics 21 (2014) P4.38.

[25] Autebert, J.-M., Décaillot, A.-M., Schwer, S. R., Henri-Auguste Delannoy et la publication des œuvres posthumes d'Édouard Lucas, Gazette des Mathématiciens 95 (2003) 51–62.

[26] Azriel, D., Berend, D., On a question of Leiss regarding the Hanoi Tower problem, Theoretical Computer Science 369 (2006) 377–383.

[27] Azriel, D., Solomon, N., Solomon, S., On an Infinite Family of Solvable Hanoi Graphs, ACM Transactions on Algorithms 5 (2009) Art. 13.

[28] Backhouse, R., Fokkinga, M., The associativity of equivalence and the Towers of Hanoi problem, Information Processing Letters 77 (2001) 71–76.

[29] Badger, L:, Williams, M., The Tower of Stanford, Problems and solutions: Solutions: 10956, The American Mathematical Monthly 111 (2004) 364f.

[30] Bailly, B., Les jeux scientifiques, Cosmos, Revue des sciences, Nouvelle série (15) 39 (1890) 156–159.

[31] Ball, W. W. R., Mathematical recreations and problems of past and present times, Second edition, Macmillan, London, 1892.

[32] Ball, W. W. R., Mathematical Recreations & Essays, Macmillan, London, 1956.

[33] Ballot, C., Anderson, P. G., Komatsu, T. (eds.), Proceedings of the 17th International Conference on Fibonacci Numbers and Their Applications (= The Fibonacci Quarterly 55(5) (2017)).

[34] Bandt, C., Kuschel, T., Self-similar sets 8. Average interior distance in some fractals, Supplemento ai Rendiconti del Circolo Matematico di Palermo, serie II, 28 (1992) 307–317.

[35] Bang-Jensen, J., Gutin, G. Z., Digraphs, Theory, Algorithms and Applications, 2nd Edition, Springer, London, 2009.

[36] Barlow, M. T., Perkins, E. A., Brownian Motion on the Sierpinski Gasket, Probability Theory and Related Fields 79 (1988) 543–623.

[37] Beauquier, J., Burman, J., Clavière, S., Sohier, D., Space-Optimal Counting in Population Protocols, Lecture Notes in Computer Science 9363 (2015) 631–646.

[38] Beck, M., Geoghegan, R., The Art of Proof, Springer, New York NY, 2010.

[39] Becquaert, C., Les tours de Hanoï, Pentamino 2 (1976) 73–84.

[40] Becquaert, C., Les tours de Hanoï, Pentamino 3 (1977) 55–71.

[41] Beineke, J., Beineke, L., Some ABCs of graphs and games, in: [42], Chapter 4.

[42] Beineke, J., Rosenhouse, J. (eds.), The Mathematics of Various Entertaining Subjects, Research in Recreational Math, Princeton University Press, Princeton NJ, 2016.

[43] Beineke, J., Rosenhouse, J. (eds.), The Mathematics of Various Entertaining Subjects, Volume 2, Research in Games, Graphs, Counting, and Complexity, Princeton University Press, Princeton NJ, 2017.

[44] Bell, E. T., Mathematics Queen and Servant of Science, Mathematical Association of America, Washington DC, 1987.

[45] Benjamin, A. T., Ornstein, J., A bijective proof of a derangement recurrence, in: [33], 28–29.

[46] Bennish, J., The Tower of Hanoi problem and mathematical thinking, Missouri Journal of Mathematical Sciences 11 (1999) 164–166.

[47] Berend, D., Sapir, A., The Cyclic Multi-Peg Tower of Hanoi, ACM Transactions on Algorithms 2 (2006) 297–317.

[48] Berend, D., Sapir, A., The diameter of Hanoi graphs, Information Processing Letters 98 (2006) 79–85.

[49] Berend, D., Sapir, A., Exponential vs. Subexponential Tower of Hanoi Variants, Journal of Graph Algorithms and Applications 20 (2016) 461–479.

[50] Berend, D., Sapir, A., Solomon, S., The Tower of Hanoi problem on $Path_h$ graphs, Discrete Applied Mathematics 160 (2012) 1465–1483.

[51] Berg, W. K., Byrd, D. L., The Tower of London spatial problem-solving task: Enhancing clinical and research implementation, Journal of Clinical and Experimental Neuropsychology 24 (2002) 586–604.

[52] Berlekamp, E. R., Conway, J. H., Guy, R. K., Winning Ways for Your Mathematical Plays, Vol. 4, Second edition, A K Peters, Wellesley MA, 2004.

[53] Biggs, N., The icosian calculus of today, Proceedings of the Royal Irish Academy 95A Supplement (1995) 23–34.

[54] Biggs, N. L., Discrete Mathematics, Oxford University Press, Oxford, 2002.

[55] Biggs, N. L., Lloyd, E. K., Wilson, R. J., Graph Theory 1736–1936, Clarendon Press, Oxford, 1986.

[56] Bode, J.-P., Hinz, A. M., Results and open problems on the Tower of Hanoi, Congressus Numerantium 139 (1999) 113–122.

[57] Boettcher, S., Gonçalves, B., Anomalous diffusion on the Hanoi networks, Europhysics Letters 84 (2008) Art. 30002.

[58] Boettcher, S., Gonçalves B., Guclu H., Hierarchical regular small-world networks, Journal of Physics A: Mathematical and Theoretical 41 (2008) Art. 252001.

[59] Boettcher, S., Li, S., Real-space renormalization group for spectral properties of hierarchical networks, Journal of Physics A: Mathematical and Theoretical 48 (2015) Art. 415001.

[60] Bondy, J. A., Murty, U. S. R., Graph Theory, Springer, New York NY, 2008.

[61] Bosma, W., Wiskundige Verpoozingen, Nieuw Archief for Wiskunde (5) 10 (2009) 42–47.

[62] Bousch, T., La quatrième tour de Hanoï, Bulletin of the Belgian Mathematical Society Simon Stevin 21 (2014) 895–912.

[63] Bousch, T., La Tour de Stockmeyer, Séminaire Lotharingien de Combinatoire 77 (2017) Article B77d.

[64] Brendel, J., Spanning Trees of Hanoi Graphs, B.Sc. thesis, Ludwig-Maximilians-Universität, München, 2012.

[65] Brocot, A., Calcul des rouages par approximation, Revue Chronométrique 3 (1859-61) 186–194. (Reprinted in: R. Chavigny, Les Brocot, une dynastie d'horlogers, Antoine Simonin, Neuchâtel, 1991, 187–197.)

[66] Brousseau, B. A., Tower of Hanoi with more pegs, Journal of Recreational Mathematics 8 (1975/6) 169–176.

[67] Buckley, F., Harary, F., Distance in graphs, Addison-Wesley, Redwood City CA, 1990.

[68] Calkin, N., Wilf, H. S., Recounting the Rationals, The American Mathematical Monthly 107 (2000) 360–363.

[69] Cantor, G., Ueber eine Eigenschaft des Inbegriffs aller reellen algebraischen Zahlen, Journal für die reine und angewandte Mathematik 77 (1874) 258–262.

[70] Cantor, G., Ueber eine elementare Frage der Mannigfaltigkeitslehre, Jahresberichte der Deutschen Mathematikervereinigung 1 (1890–91) 75–78.

[71] Cardanus, H., De Subtilitate, Guillaume Rouillé, Lyon, 1551.

[72] Carlitz, L., Single variable Bell polynomials, Collectanea Mathematica 14 (1962) 13–25.

[73] Carr, G. S., Question 3575, in: Miller, W. J. C. (ed.), Mathematical Questions with their Solutions. From the "Educational Times.", Vol. XVIII, C. F. Hodgson & Sons, London, 1873, 31–32.

[74] Chan, T.-H., A statistical analysis of the Towers of Hanoi problem, International Journal of Computer Mathematics 28 (1989) 57–65.

[75] Chang, S.-C., Chen, L.-C., Yang, W.-S., Spanning Trees on the Sierpinski Gasket, Journal of Statistical Physics 126 (2007) 649–667.

[76] Chappelon, J., Matsuura, A., On generalized Frame-Stewart numbers, Discrete Mathematics 312 (2012) 830–836.

[77] Chen, G.-H., Duh, D.-R., Topological Properties, Communication, and Computation on WK-Recursive Networks, Networks 24 (1994) 303–317.

[78] Chen, H., Wu, R., Huang, G., Deng, H., Dimer-monomer model on the Towers of Hanoi graphs, International Journal of Modern Physics B 29 (2015) Art. 1550173.

[79] Chen, H., Wu, R., Huang, G., Deng, H., Independent sets on the Towers of Hanoi graphs, Ars Mathematica Contemporanea 12 (2017) 247–260.

[80] Chen, X., Shen, J., On the Frame-Stewart conjecture about the Towers of Hanoi, SIAM Journal on Computing 33 (2004) 584–589.

[81] Chen, X., Tian, B., Wang, L., Santa Claus' Towers of Hanoi, Graphs and Combinatorics 23 (Supplement 1) (2007) 153–167.

[82] Claus (= É. Lucas), N., La Tour d'Hanoï, Véritable casse-tête annamite, P. Bousrez, Tours, 1883.

[83] Claus (= É. Lucas), N., La Tour d'Hanoï, Jeu de calcul, Science et Nature 1 (1884) 127–128.

[84] Cook, S., The P Versus NP Problem, in: J. Carlson, A. Jaffe, A. Wiles (eds.), The Millennium Prize Problems, Clay Mathematics Institute and American Mathematical Society, Cambridge MA and Providence RI, 2006, 87–104.

[85] Crowe, D. W., The n-dimensional cube and the tower of Hanoi, The American Mathematical Monthly 63 (1956) 29–30.

[86] Culin, S., Korean games with notes on the corresponding games in China and Japan, University of Pennsylvania, Philadelphia PA, 1895.

[87] Cull, P., Ecklund, E. F., On the Towers of Hanoi and Generalized Towers of Hanoi Problems, Congressus Numerantium 35 (1982) 229–238.

[88] Cull, P., Gerety, C., Is Towers of Hanoi Really Hard?, Congressus Numerantium 47 (1985) 237–242.

[89] Cull, P., Nelson, I., Error-Correcting Codes on the Towers of Hanoi Graphs, Discrete Mathematics 208/209 (1999) 157–175.

[90] Cyvin, S. J., Gutman, I., Kekulé Structures in Benzenoid Hydrocarbons, Springer, Berlin, 1988.

[91] D'Angeli, D., Donno, A., Self-similar groups and finite Gelfand pairs, Algebra and Discrete Mathematics 2007 (2) 54–69.

[92] D'Angeli, D., Donno, A., Weighted spanning trees on some self-similar graphs, The Electronic Journal of Combinatorics 18(1) (2011) Art. 16.

[93] Décaillot, A.-M., L'arithméticien Édouard Lucas (1842–1891): théorie et instrumentation, Revue d'Histoire des Mathématiques 4 (1998) 191–236.

[94] Décaillot, A.-M., Les *Récréations Mathématiques* d'Édouard Lucas: quelques éclairages, Historia Mathematica 41 (2014) 506–517.

[95] Décaillot-Laulanglet, A.-M., Edouard Lucas (1842–1891) : le parcours original d'un scientifique français dans la deuxième moitié du XIXe siècle, Thèse de doctorat, Université René Descartes, Paris, 1999.

[96] Delahaye, J.-P., Dérivées et intégrales dans le monde des 0 et des 1, in: Le calcul intégral (Tangente HS 50), POLE, Paris, 2014, 106–111.

[97] Delahaye, J.-P., Les tours de Hanoï, plus qu'un jeu d'enfants, Pour la Science 457 (2015) 108–114. (Also [98, Chapitre 7].)

[98] Delahaye, J.-P., Les mathématiciens se plient au jeu, Belin, Paris, 2017.

[99] Della Vecchia, G., Sanges, C., A Recursively Scalable Network VLSI Implementation, Future Generation Computer Systems 4 (1988) 235–243.

[100] De Moivre, A., The Doctrine of Chances: or A Method of Calculating the Probability of Events in Play, W. Pearson, London, 1718.

[101] Devlin, K., The Man of Numbers, Bloomsbury, London, 2012.

[102] Deza, M. M., Deza, E., Encyclopedia of Distances, Second Edition, Springer, Heidelberg, 2013.

[103] Diaconis, P., Graham, R., Magical Mathematics, Princeton University Press, Princeton NJ, 2012.

[104] Diestel, R., Graph Theory, Springer, New York NY, 2010.

[105] Dinitz, Y., Solomon, S., On Optimal Solutions for the Bottleneck Tower of Hanoi Problem, Lecture Notes in Computer Science 4362 (2007) 248–259.

[106] Dinitz, Y., Solomon, S., Optimality of an algorithm solving the Bottleneck Tower of Hanoi problem, ACM Transactions on Algorithms 4 (2008) Art. 25.

[107] Döll, S., Hinz, A. M., *Kyū-renkan*—the Arima sequence, Advanced Studies in Pure Mathematics, to appear.

[108] Domoryad, A. P., Mathematical Games and Pastimes, Pergamon Press, Oxford, 1964.

[109] Dubrovsky, V., Nesting puzzles Part I: Moving oriental towers, Quantum 6(1) (1996) 53–59, 49–51.

[110] Dubrovsky, V., Nesting puzzles Part II: Chinese rings produce a Chinese monster, Quantum 6(2) (1996) 61–64, 58–59.

[111] Dudeney, H. E. ("Sphinx"), Puzzles and prizes, No. 68. The Tower of Bramah, The Weekly Dispatch (1896–08–09) ??; (1896–08–23) ??.

[112] Dudeney, H. E. ("Sphinx"), Puzzles and prizes, No. 124. Sphinx Junior's Latest, The Weekly Dispatch (1896–11–15) ??; (1896–11–29) ??.

[113] Dudeney, H. E., The Canterbury puzzles, The London Magazine 8(46)(May 1902) 367–371; 8(47) (June 1902) 480–482.

[114] Dudeney, H. E., Puzzles and prizes, No. 447. The Reeve's puzzle, The Weekly Dispatch (1902–05–25) 13; (1902–06–15) 13.

[115] Dudeney, H. E., Puzzles and prizes, No. 494. Five stools and cheeses, The Weekly Dispatch (1903–03–15) 13; (1903–03–29) 13; (1903–04–05) 13.

[116] Dudeney, H. E., The Canterbury puzzles and other curious problems,[96] William Heinemann, London, 1907.

---

[96]Usually cited with "E. P. Dutton and Company, New York NY, 1908". P. K. Stockmeyer was able to locate a copy of the London edition (private communication, 2016).

[117] Dudeney, H. E., The world's best puzzles, The Strand Magazine 36(1908) 779–787.

[118] Dudeney, H. E., Perplexities, No. 27. The ten counters, The Strand Magazine 41 (1911) 116, 235.

[119] Dudeney, H. E., Amusements in Mathematics, T. Nelson & Sons, London, 1917.

[120] Dudeney, H. E., Perplexities, No. 725. Transferring the counters, The Strand Magazine 68 (1924) 530, 670.

[121] Dunkel, O., Editorial note, The American Mathematical Monthly 48 (1941) 219.

[122] Eccles, P. J., An Introduction to Mathematical Reasoning, Cambridge University Press, Cambridge, 1997.

[123] Edwards, A. W. F., Pascal's Arithmetical Triangle, Charles Griffin and Oxford University Press, London and New York NY, 1987.

[124] Eisenstein, G., Eine neue Gattung zahlentheoretischer Funktionen, welche von zwei Elementen abhängen und durch gewisse lineare Funktional-Gleichungen definirt werden, Bericht über die zur Bekanntmachung geeigneten Verhandlungen der Königlich Preussischen Akademie der Wissenschaften zu Berlin, 1850, 36–42.

[125] Elkies, N. D., Crossing numbers of complete graphs, in: [43], Chapter 13.

[126] Emert, J. W., Nelson, R. B., Owens, F. W., Multiple Towers of Hanoi with a Path Transition Graph, Congressus Numerantium 188 (2007) 59–64.

[127] Emert, J. W., Nelson, R. B., Owens, F. W., Multiple Towers of Hanoi with a Directed Cycle Transition Graph, Congressus Numerantium 193 (2008) 89–95.

[128] Er, M. C., An iterative solution to the generalized Towers of Hanoi problem, BIT 23 (1983) 295–302.

[129] Er, M. C., An analysis of the generalized Towers of Hanoi problem, BIT 23 (1983) 429–435.

[130] Er, M. C., The Cyclic Towers of Hanoi: A Representation Approach, The Computer Journal 27 (1984) 171–175.

[131] Er, M. C., The Generalized Colour Towers of Hanoi: An Iterative Algorithm, The Computer Journal 27 (1984) 278–282.

[132] Er, M. C., The Complexity of the Generalised Cyclic Towers of Hanoi Problem, Journal of Algorithms 6 (1985) 351–358.

[133] Er, M. C., The Towers of Hanoi and Binary Numerals, Journal of Information & Optimization Sciences 6 (1985) 147–152.

[134] Er, M. C., The Cyclic Towers of Hanoi and Pseudoternary Codes, Journal of Information & Optimization Sciences 7 (1986) 271–277.

[135] Er, M. C., A Time and Space Efficient Algorithm for the Cyclic Towers of Hanoi Problem, Journal of Information Processing 9 (1986) 163–165.

[136] Er, M. C., A General Algorithm for Finding a Shortest Path between two n-Configurations, Information Sciences 42 (1987) 137–141.

[137] Er, M. C., A Loopless and Optimal Algorithm for the Cyclic Towers of Hanoi Problem, Information Sciences 42 (1987) 283–287.

[138] Er, M. C., Counter Examples to Adjudicating a Towers of Hanoi Contest, International Journal of Computer Mathematics 21 (1987) 123–131.

[139] Erickson, M., Pearls of discrete mathematics, CRC Press, Boca Raton FL, 2010.

[140] Euler, L., Calcul de la probabilité dans le jeu de rencontre, Histoire de l'Académie Royale des Sciences et Belles Lettres 7 (1753) 255–270.

[141] Euler, L., Solutio quaestionis curiosae ex doctrina combinationum, Mémoires de l'Académie Impériale des Sciences de St. Pétersbourg 3 (1811) 57–64.

[142] Эйлер, Л., Письма к Ученым, В. И. Смирнов (ред.), Издательство Академии Наук СССР, Москва, 1963.

[143] Eves, H., Starke, E. P. (eds.), The Otto Dunkel memorial problem book, supplement to the American Mathematical Monthly 64 (1957).

[144] Faber, A. H., Kühnpast, N., Sürer, F., Hinz, A. M., Danek, A., The iso-effect: Is there specific learning of Tower of London iso-problems?, Thinking & Reasoning 15 (2009) 237–249.

[145] Fang, J.-F., Liang, W.-Y., Chen, H.-R., Ng, K.-L., Novel Broadcasting Algorithm of the Complete Recursive Network, Informatica 31 (2007) 131–136.

[146] Fauvel, J., Gray, J. (eds.), The History of Mathematics: A Reader, Macmillan, Houndsmills, 1987.

[147] Finsterwalder, S., Der Turm von Hanoi mit irregulären Anfangszuständen, Diploma thesis, Ludwig-Maximilians-Universität, München, 2008.

[148] Fischer, H., Zastrow, A., Word calculus in the fundamental group of the Menger curve, Fundamenta Mathematicae 235 (2016) 199–226.

[149] Fischer, I. S., The Star Tower of Hanoi, BSc thesis, Ludwig-Maximilians-Universität, München, 2015.

[150] Frame, J. S., Problems and solutions: Advanced problems: Solutions: 3918, The American Mathematical Monthly 48 (1941) 216–217.

[151] Franci, R., Jealous Husbands Crossing the River: A Problem from Alcuin to Tartaglia, in: Y. Dold-Samplonius, J. W. Dauben, M. Folkerts, B. van Dalen (eds.), From China to Paris: 2000 years transmission of mathematical ideas, Franz Steiner, Stuttgart, 2002, 289–306.

[152] Fréchet, M., Sur quelques points du calcul fonctionnel, Rendiconti del Circolo Matematico di Palermo 22 (1906) 1–74.

[153] Fréchet, M., Les espaces abstraits, Gauthiers-Villars, Paris, 1928.

[154] Fritsch, R., Fritsch, G., The Four-Color Theorem, Springer, New York NY, 1998.

[155] Fu, J.-S., Hamiltonian connectivity of the WK-recursive network with faulty nodes, Information Sciences 178 (2008) 2573–2584.

[156] Garnier, R., Taylor, J., 100% Mathematical Proof, John Wiley & Sons, Chichester, 1996.

[157] Gauss, C. F., Mathematisches Tagebuch 1796–1814, 2. Auflage, Akademische Verlagsgesellschaft Geest & Portig, Leipzig, 1979.

[158] Gedeon, T. D., The Cyclic Towers of Hanoi: An Iterative Solution Produced by Transformation, The Computer Journal 39 (1996) 353–356.

[159] Gillings, R. J., Mathematics in the Time of the Pharaohs, Dover, New York NY, 1982.

[160] Glaisher, J. W. L., On the residue of a binomial-theorem coefficient with respect to a prime modulus, The Quarterly Journal of Pure and Applied Mathematics 30 (1899) 150–156.

[161] Götz, K. A. M., Äquivalenzklassen von Turm-Aufgaben, Diploma thesis, Ludwig-Maximilians-Universität, München, 2008.

[162] Graham, R. L., Knuth, D. E., Patashnik, O., Concrete mathematics, Addison-Wesley, Reading MA, 1989.

[163] Graves, R. P., Life of Sir William Rowan Hamilton, Vol. III, Dublin University Press, Dublin, 1889.

[164] Gravier, S., Klavžar, S., Mollard, M., Codes and $L(2,1)$-labelings in Sierpiński graphs, Taiwanese Journal of Mathematics 9 (2005) 671–681.

[165] Gray, F., Pulse code communication, US-Patent Nr. 2,632,058, 1953.

[166] Grigorchuk, R., Šunić, Z., Schreier spectrum of the Hanoi Towers group on three pegs, in: P. Exner, J. P. Keating, P. Kuchment, T. Sunada, A. Teplyaev (eds.), Analysis on Graphs and Its Applications, American Mathematical Society, Providence RI, 2008, 183–198.

[167] Grigorchuk, R., Šuniḱ, Z., Asymptotic aspects of Schreier graphs and Hanoi Towers groups, Comptes Rendus de l'Académie des Sciences Paris, Ser. I, 342 (2006) 545–550.

[168] Gros, L., Théorie du Baguenodier, Aimé Vingtrinier, Lyon, 1872.

[169] Grosu, C., A New Lower Bound for the Towers of Hanoi Problem, The Electronic Journal of Combinatorics 23 (2016) P1.22.

[170] Guan, D.-J., Generalized Gray Codes with Applications, Proceedings of the National Science Council, Republic of China - Part A 22 (1998) 841–848.

[171] Guibert, O., Stack words, standard Young tableaux, permutations with forbidden subsequences and planar maps, Discrete Mathematics 210 (2000) 71–85.

[172] Guy, R. K., Crossing numbers of graphs, Lecture Notes in Mathematics 303 (1972) 111–124.

[173] de Guzmán, M., The role of games and puzzles in the popularization of mathematics, L'Enseignement Mathématique (2) 36 (1990) 359–368.

[174] Halphén, G., Sur des suites de fractions analogues à la suite de Farey, Bulletin de la Société Mathématique de France 5 (1877) 170–175.

[175] Hamilton, W. R., Memorandum respecting a new System of Roots of Unity, The London, Edinburgh and Dublin Philosophical Magazine and Journal of Science, 4th series, 12 (1856) 446.

[176] Hamilton, W. R., Account of the Icosian Calculus, Proceedings of the Royal Irish Academy 6 (1858) 415–416.

[177] Harary, F., Graph Theory, Addison-Wesley, Reading MA, 1969.

[178] Harary, F., Palmer, E. M., Graphical enumeration, Academic Press, New York NY, 1973.

[179] Harkin, D., On the mathematical work of François-Édouard-Anatole Lucas, L'Enseignement Mathématique (2) 3 (1957) 276–288.

[180] Hartsfield, N., Ringel, G., Pearls in Graph Theory, Academic Press, Boston MA, 1990.

[181] Hausdorff, F., Grundzüge der Mengenlehre, Veit, Leipzig, 1914.

[182] Hayes, B., Introducing a department concerned with the pleasures of computation, Scientific American 249 (1983) 22–36.

[183] Hayes, B., Gauss's Day of Reckoning, American Scientist 94 (2006) 200–205.

[184] Hayes, B., Foolproof, and Other Mathematical Meditations, The MIT Press, Cambridge MA, 2017.

[185] Hayes, P. J., A note on the Towers of Hanoi problem, The Computer Journal 20 (1977) 282–285.

[186] Heard, R. J., The Tower of Hanoi as a Trivial Problem, The Computer Journal 27 (1984) 90.

[187] Heeffer, A., Hinz, A. M., "A difficult case": Pacioli and Cardano on the Chinese Rings, Recreational Mathematics Magazine 4(8) (2017) 5–23.

[188] Hering, H., Zur Mathematisierung des »Turm von Hanoi«-Spiels, Der mathematisch-naturwissenschaftliche Unterricht 26 (1973) 408–411.

[189] Hering, H., Minimalstrategien in endlichen Systemen, Der Mathematikunterricht 23 (1977) 25–35.

[190] Hering, H., Genetische Mathematisierung am Beispiel einer Variante des Turm-von-Hanoi-Spiels, Didaktik der Mathematik 4 (1978) 307–317.

[191] Hering, H., Varianten zum "Turm von Hanoi-Spiel" – ein Weg zu einfachen Funktionalgleichungen, Der Mathematikunterricht 25 (1979) 82–92.

[192] Heroor, V. D., The history of mathematics and mathematicians of India, Vidya Bharathi Karnataka, Bangalore, 2006.

[193] Hinz, A. M., An Iterative Algorithm for the Tower of Hanoi with Four Pegs, Computing 42 (1989) 133–140.

[194] Hinz, A. M., The Tower of Hanoi, L'Enseignement Mathématique (2) 35 (1989) 289–321.

[195] Hinz, A. M., Pascal's Triangle and the Tower of Hanoi, The American Mathematical Monthly 99 (1992) 538–544.

[196] Hinz, A. M., Shortest Paths Between Regular States of the Tower of Hanoi, Information Sciences 63 (1992) 173–181.

[197] Hinz, A. M., Square-free Tower of Hanoi sequences, L'Enseignement Mathématique (2) 42 (1996) 257–264.

[198] Hinz, A. M., The Tower of Hanoi, in: K.-P. Shum, E. J. Taft, Z.-X. Wan (eds.), Algebras and Combinatorics, Springer, Singapore, 1999, 277–289.

[199] Hinz, A. M., A straightened proof for the uncountability of $\mathbb{R}$, Elemente der Mathematik 65 (2010) 26–28.

[200] Hinz, A. M., Tower of Hanoi, Nieuw Archief for Wiskunde (5) 12 (2011) 239.

[201] Hinz, A. M., Graph theory of tower tasks, Behavioural Neurology 25 (2012) 13–22.

[202] Hinz, A. M., The Lichtenberg sequence, The Fibonacci Quarterly 55 (2017) 2–12.

[203] Hinz, A. M., Open Problems for Hanoi and Sierpiński Graphs, Electronic Notes in Discrete Mathematics 63 (2017) 23–31.

[204] Hinz, A. M., Holz auf der Heide, C., An efficient algorithm to determine all shortest paths in Sierpiński graphs, Discrete Applied Mathematics 177 (2014) 111–120.

[205] Hinz, A. M., Klavžar, S., Milutinović, U., Parisse, D., Petr, C., Metric properties of the Tower of Hanoi graphs and Stern's diatomic sequence, European Journal of Combinatorics 26 (2005) 693–708.

[206] Hinz, A. M., Klavžar, S., Zemljič, S. S., Sierpiński graphs as spanning subgraphs of Hanoi graphs, Central European Journal of Mathematics 11 (2013) 1153–1157.

[207] Hinz, A. M., Klavžar, S., Zemljič, S. S., A survey and classification of Sierpiński-type graphs, Discrete Applied Mathematics 217 (2017) 565–600.

[208] Hinz, A. M., Kostov, A., Kneißl, F., Sürer, F., Danek, A., A mathematical model and a computer tool for the Tower of Hanoi and Tower of London puzzles, Information Sciences 179 (2009) 2934–2947.

[209] Hinz, A. M., Movarraei, N., The Hanoi graph H_4^3, submitted (2018).

[210] Hinz, A. M., Parisse, D., On the Planarity of Hanoi Graphs, Expositiones Mathematicae 20 (2002) 263–268.

[211] Hinz, A. M., Parisse, D., Coloring Hanoi and Sierpiński graphs, Discrete Mathematics 312 (2012) 1521–1535.

[212] Hinz, A. M., Parisse, D., The Average Eccentricity of Sierpiński Graphs, Graphs and Combinatorics 28 (2012) 671–686.

[213] Hinz, A. M., Petr, C., Computational Solution of an Old Tower of Hanoi Problem, Electronic Notes in Discrete Mathematics 53 (2016) 445–458.

[214] Hinz, A. M., Schief, A., The average distance on the Sierpiński gasket, Probability Theory and Related Fields 87 (1990) 129–138.

[215] Hinze, R., Functional Pearl: La Tour D'Hanoï, ACM SIGPLAN Notices-ICFP '09, 44(9) (2009) 3–10.

[216] Höck, A. C., The Chinese String—Its solution and analogy to the Chinese Rings, Ph.D. thesis, Ludwig-Maximilians-Universität, München, 2017.

[217] Holz auf der Heide, C., Distances and automatic sequences in distinguished variants of Hanoi graphs, Ph.D. thesis, Ludwig-Maximilians-Universität, München, 2016.

[218] Hoseiny Farahabady, M., Imani, N., Sarbazi-Azad, H., Some topological and combinatorial properties of WK-recursive mesh and WK-pyramid interconnection networks, Journal of Systems Architecture 54 (2008) 967–976.

[219] Huang, J., Mickey, M., Xu, J., The Nonassociativity of the Double Minus Operation, Journal of Integer Sequences 20 (2017) Art. 17.10.3.

[220] Hunter, W. S., Human behavior, The University of Chicago Press, Chicago IL, 1928.

[221] Imrich, W., Klavžar, S., Rall, D. F., Topics in Graph Theory, A K Peters, Wellesley MA, 2008.

[222] Jakovac, M., Klavžar, S., Vertex-, edge-, and total-colorings of Sierpiński-like graphs, Discrete Mathematics 309 (2009) 1548–1556.

[223] Joyner, D., Adventures in Group Theory, Second edition, The Johns Hopkins University Press, Baltimore MD, 2008.

[224] Jullien, P., Trois jeux, in: E. Galion, La mathématique et ses applications, CEDIC, Paris, 1972, 23–43.

[225] Kak, S., The Golden Mean and the Physics of Aesthetics, in: B. S. Yadav, M. Mohan (eds.), Ancient Indian Leaps into Mathematics, Birkhäuser, New York NY, 2011, 111–119.

[226] Kaplan, R., The nothing that is, Penguin Books, London, 2000.

[227] Katz, V. J., Jewish combinatorics, in: [437], Chapter 4.

[228] Kauffman, L. H., Tangle complexity and the topology of the Chinese rings, in: J. P. Mesirov, K. Schulten, D. W. Sumners (eds.), Mathematical Approaches to Biomolecular Structure and Dynamics, Springer, New York NY, 1996, 1–10.

[229] Keister, W., Locking disc puzzle, US-Patent Nr. 3,637,215, 1972.

[230] Keister, W., Pattern-matching puzzle, US-Patent Nr. 3,637,216, 1972.

[231] Kimberling, C., Problem proposals, in: [33], 213–221.

[232] Klahr, D., Goal formation, planning, and learning by pre-school problem solvers or: "My socks are in the dryer", in: R. S. Siegler (ed.), Children's Thinking: What Develops?, Lawrence Erlbaum, Hillsdale NJ, 1978, 181–212.

[233] Klahr, D., Robinson, M., Formal assessment of problem-solving and planning processes in preschool children, Cognitive Psychology 13 (1981) 113–148.

[234] Klavžar, S., Coloring Sierpiński graphs and Sierpiński gasket graphs, Taiwanese Journal of Mathematics 12 (2008) 513–522.

[235] Klavžar, S., Milutinović, U., Graphs $S(n,k)$ and a variant of the Tower of Hanoi problem, Czechoslovak Mathematical Journal 47(122) (1997) 95–104.

[236] Klavžar, S., Milutinović, U., Simple Explicit Formulas for the Frame-Stewart Numbers, Annals of Combinatorics 6 (2002) 157–167.

[237] Klavžar, S., Milutinović, U., Petr, C., Combinatorics of topmost discs of multi-peg Tower of Hanoi problem, Ars Combinatoria 59 (2001) 55–64.

[238] Klavžar, S., Milutinović, U., Petr, C., On the Frame-Stewart algorithm for the multi-peg Tower of Hanoi problem, Discrete Applied Mathematics 120 (2002) 141–157.

[239] Klavžar, S., Milutinović, U., Petr, C., 1-perfect codes in Sierpiński graphs, Bulletin of the Australian Mathematical Society 66 (2002) 369–384.

[240] Klavžar, S., Milutinović, U., Petr, C., Hanoi graphs and some classical numbers, Expositiones Mathematicae 23 (2005) 371–378.

[241] Klavžar, S., Mohar, B., Crossing Numbers of Sierpiński-Like Graphs, Journal of Graph Theory 50 (2005) 186–198.

[242] Klavžar, S., Zemljič, S. S., On distances in Sierpiński graphs: almost-extreme vertices and metric dimension, Applicable Analysis and Discrete Mathematics 7 (2013) 72–82.

[243] Klein, C. S., Minsker, S., The super towers of Hanoi problem: large rings on small rings, Discrete Mathematics 114 (1993) 283–295.

[244] Knuth, D., All Questions Answered, Notices of the American Mathematical Society 49 (2002) 318–324.

[245] Knuth, D. E., Computer Science and Mathematics, American Scientist 61 (1973) 707–713.

[246] Knuth, D. E., Two Notes on Notation, The American Mathematical Monthly 99 (1992) 403–422.

[247] Knuth, D. E., The Art of Computer Programming. Vol. 1, Third edition, Addison-Wesley, Reading MA, 1997.

[248] Knuth, D. E., The Art of Computer Programming. Vol. 4 A (Combinatorial Algorithms, Part 1), Addison-Wesley, Upper Saddle River NJ, 2011.

[249] Köhler, T., Überschneidungszahlen spezieller Hanoi- und Sierpiński-Graphen, Diploma thesis, Ludwig-Maximilians-Universität, München, 2011.

[250] König, D., Theorie der endlichen und unendlichen Graphen, B. G. Teubner, Leipzig, 1986.

[251] Korf, R. E., Best-First Frontier Search with Delayed Duplicate Detection, in: Nineteenth National Conference on Artificial Intelligence (AAAI-2004), The MIT Press, Cambridge MA, 2004, 650–657.

[252] Korf, R. E., Linear-Time Disk-Based Implicit Graph Search, Journal of the ACM 55 (2008) Art. 26.

[253] Korf, R. E., Felner, A., Recent Progress in Heuristic Search: A Case Study of the Four-Peg Towers of Hanoi Problem, in: M. M. Veloso (ed.), Proceedings of the Twentieth International Joint Conference on Artificial Intelligence (IJCAI-07), AAAI Press, Menlo Park CA, 2007, 2324–2329.

[254] Kotovsky, K., Hayes, J. R., Simon, H. A., Why Are Some Problems Hard? Evidence from Tower of Hanoi, Cognitive Psychology 17 (1985) 248–294.

[255] Kotovsky, K., Simon, H. A., What Makes Some Problems Really Hard: Explorations in the Problem Space of Difficulty, Cognitive Psychology 22 (1990) 143–183.

[256] Kummer, E. E., Über die Ergänzungssätze zu den allgemeinen Reciprocitätsgesetzen, Journal für die reine und angewandte Mathematik 44 (1852) 93–146.

[257] Kusuba, T., Plofker, K., Indian combinatorics, in: [437], Chapter 1.

[258] Legendre, A. M., Essai sur la théorie des nombres, Seconde édition, Courcier, Paris, 1808.

[259] Leibnitz, G. W., Explication de l'arithmétique binaire, Histoire de l'Académie Royale des Sciences 1703, 85–89.

[260] Leiss, E. L., Solving the "Towers of Hanoi" on Graphs, Journal of Combinatorics, Information & System Sciences 8 (1983) 81–89.

[261] Leiss, E. L., Finite Hanoi problems: how many discs can be handled?, Congressus Numerantium 44 (1984) 221–229.

[262] Leiss, E. L., The worst Hanoi graphs, Theoretical Computer Science 498 (2013) 100–106.

[263] Levy, H., Lessman, F., Finite Difference Equations, Macmillan, New York NY, 1961.

[264] Li, C.-K., Nelson, I., Perfect codes on the Towers of Hanoi graph, Bulletin of the Australian Mathematical Society 57 (1998) 367–376.

[265] Liao, Y., Hou, Y., Shen, X., Tutte polynomial of the Apollonian network, Journal of Statistical Mechanics: Theory and Experiment (2014) P10043.

[266] G. C. L. (= Georg Christoph Lichtenberg), Ueber das Spiel mit den künstlich verflochtenen Ringen, welches gewöhnlich Nürnberger Tand genannt wird, Göttingische Anzeigen von gemeinnützigen Sachen 1 (1769) 637–640.

[267] Lin, C.-H., Liu, J.-J., Wang, Y.-L., Yen, W. C.-K., The Hub Number of Sierpiński-Like Graphs, Theory of Computing Systems 49 (2011) 588–600.

[268] Lindley, E. H., A study of puzzles with special reference to the psychology of mental adaption, American Journal of Psychology 8 (1897) 431–493.

[269] Lindquist, N. F., Solution to Problem 1350, Mathematics Magazine 64 (1991) 200.

[270] van Lint, J. H., Wilson, R. M., A Course in Combinatorics, Second edition, Cambridge University Press, Cambridge, 2001.

[271] Lipscomb, S. L., A universal one-dimensional metric space, Lecture Notes in Mathematics 378 (1974) 248–257.

[272] Lipscomb, S. L., On imbedding finite-dimensional metric spaces, Transactions of the American Mathematical Society 211 (1975) 143–160.

[273] Lipscomb, S. L., Fractals and Universal Spaces in Dimension Theory, Springer, New York NY, 2009.

[274] Lipscomb, S. L., The Quest for Universal Spaces in Dimension Theory, Notices of the American Mathematical Society 56 (2009) 1418–1424.

[275] de Longchamps, G., Variétés, Journal de Mathématiques Spéciales (2) 2 (1883) 286–287.

[276] Lovász, L., Random Walks on Graphs: A Survey, in: D. Miklós, V. T. Sós and T. Szőnyi (eds.), Combinatorics, Paul Erdős is eighty, Vol. 2, János Bolyai Mathematical Society, Budapest, 1996, 353–397.

[277] Lu, X., Towers of Hanoi Graphs, International Journal of Computer Mathematics 19 (1986) 23–38.

[278] Lu, X.-M., Towers of Hanoi Problem with Arbitrary $k \geq 3$ Pegs, International Journal of Computer Mathematics 24 (1988) 39–54.

[279] Lucas, É., Sur la théorie des nombres premiers, Atti della Reale Accademia delle Scienze di Torino 11 (1875-1876) 928–937.

[280] Lucas, É., Note sur l'application des séries récurrentes à la recherche de la loi de distribution des nombres premiers, Comptes Rendus des Séances de l'Académie des Sciences 82 (1876) 165–167.

[281] Lucas, É., Théorie des Fonctions Numérique Simplement Périodiques, American Journal of Mathematics 1 (1878) 184–240, 289–321.

[282] Lucas, É., Récréations scientifiques sur l'arithmétique et sur la géométrie de situation (1), Revue Scientifique (2) 19 (1880) 36–42.

[283] Lucas, É., Récréations Mathématiques, Gauthier-Villars et fils, Paris, 1882, 1891^2.

[284] Lucas, É., Récréations Mathématiques II, Gauthier-Villars et fils, Paris, 1883, 1896^2.

[285] Lucas, É., Jeux Scientifiques, Première Série, N° 3, La Tour d'Hanoï, Chambon & Baye/Édouard Lucas, Paris, 1889.

[286] Lucas, É., Nouveaux Jeux Scientifiques de M. Édouard Lucas, La Nature 17 (1889) 301–303.

[287] Lucas, É., Les appareils de calcul et les jeux de combinaisons, Revue Scientifique de la France et de l'Étranger 45 (1890) 1–13.

[288] Lucas, É., Théorie des nombres, Tome premier, Gauthiers-Villars et fils, Paris, 1891.

[289] Lucas, É., Récréations Mathématiques III, Gauthier-Villars et fils, Paris, 1893.

[290] Lucas, É., Récréations Mathématiques IV, Gauthier-Villars et fils, Paris, 1894.

[291] Lucas, É., L'arithmétique amusante, Gauthiers-Villars et fils, Paris, 1895.

[292] Lucas, É., Notes sur M. Edouard Lucas Arithméticien, Yvert et Tellier, Amiens, 1907.

[293] Lunnon, W. F., The Reve's Puzzle, The Computer Journal 29 (1986) 478.

[294] Luo, C., Zuo, L., Metric properties of Sierpiński-like graphs, Applied Mathematics and Computation 296 (2017) 124–136.

[295] Majumdar, A. A. K., A Note on the Iterative Algorithm for the Reve's Puzzle, The Computer Journal 37 (1994) 463–464.

[296] Majumdar, A. A. K., The generalized four-peg Tower of Hanoi problem, Optimization 29 (1994) 349–360.

[297] Majumdar, A. A. K., The divide-and-conquer approach to the generalized p-peg Tower of Hanoi problem, Optimization 34 (1995) 373–378.

[298] Majumdar, A. A. K., Halder, A., A Recursive Algorithm for the Bottleneck Reve's Puzzle, Sūrikaisekikenkyūsho Kōkyūroku 947 (1996) 150–161.

[299] Mallion, R. B., The Six (or Seven) Bridges of Kaliningrad: a Personal Eulerian Walk, 2006, MATCH Communications in Mathematical and in Computer Chemistry 58 (2007) 529–556.

[300] Mandelbrot, B., Les objets fractals : forme, hasard et dimension, Flammarion, Paris, 1975.

[301] Mandelbrot, B. B., Fractals: form, chance, and dimension, W. H. Freeman, San Francisco CA, 1977.

[302] Mandelbrot, B. B., The fractal geometry of nature, W. H. Freeman, New York NY, 1982.

[303] Martin, G. E., Counting: The Art of Enumerative Combinatorics, Springer, New York NY, 2001.

[304] Martínez, A. A., The cult of Pythagoras—Math and Myths, University of Pittsburgh Press, Pittsburgh PA, 2012.

[305] Matsuura, A., Exact Analysis of the Recurrence Relations Generalized from the Tower of Hanoi, SIAM Proceedings in Applied Mathematics 129 (2008) 228–233.

[306] Maurer, S. B., Ralston, A., Discrete Algorithmic Mathematics, Third edition, A K Peters, Wellesley MA, 2004.

[307] McCarthy, J., The Tower of Stanford, Problems and solutions: Problem 10956, The American Mathematical Monthly 109 (2002) 664.

[308] Meier, M., Exzentrizitäten in Hanoi-Graphen, Diploma thesis, Ludwig-Maximilians-Universität, München, 2009.

[309] Meinhardt, H., The Algorithmic Beauty of Sea Shells, Springer, Berlin, 1995.

[310] Milutinović, U., Completeness of the Lipscomb universal space, Glasnik Matematički, Serija III 27(47) (1992) 343–364.

[311] Minsker, S., The Towers of Hanoi Rainbow Problem: Coloring the Rings, Journal of Algorithms 10 (1989) 1–19.

[312] Minsker, S., The Towers of Antwerpen problem, Information Processing Letters 38 (1991) 107–111.

[313] Minsker, S., The Little Towers of Antwerpen problem, Information Processing Letters 94 (2005) 197–201.

[314] Minsker, S., The Linear Twin Towers of Hanoi Problem, ACM SIGCSE Bulletin 39 (2007) 37–40.

[315] Minsker, S., Another Brief Recursion Excursion to Hanoi, ACM SIGCSE Bulletin 40 (2008) 35–37.

[316] Minsker, S., The Classical/Linear Hanoi Hybrid Problem: Regular Configurations, ACM SIGCSE Bulletin 41 (2009) 57–61.

[317] Minsker, S., The Cyclic Towers of Antwerpen problem—A challenging Hanoi variant, Discrete Applied Mathematics 179 (2014) 44–53.

[318] Mintz, D. J., 2,3 sequence as binary mixture, Fibonacci Quarterly 19 (1981) 351–360.

[319] Mohar, B., Thomassen, C., Graphs on Surfaces, The Johns Hopkins University Press, Baltimore MD, 2001.

[320] Montucla, J. F., Histoire des Mathématiques, Nouvelle édition, Tome premier, Henri Agasse, Paris, 1758.

[321] Neale, R., The temple of Hanoi—An old puzzler with a new twist, Games 6 (1982) 70.

[322] Newell, A., Simon, H. A., Human problem solving, Prentice-Hall, Englewood Cliffs NJ, 1972.

[323] Noland, H., Problem 1350, Mathematics Magazine 63 (1990) 189.

[324] Нурлыбаев, А. Н., К задаче Люка о башнях и ее обобщении, Известия Национальной Академии Наук Республики Казахстан, Серия физико-математическая 182 (1995) 24–30.

[325] Obara, S., Hirayama, Y., The "Red and White Towers of Hanoi" Puzzle, Journal of Hokkaido University of Education, Section II A 46 (1995) 61–69.

[326] Obara, S., Hirayama, Y., The "Distribution of the Tower of Hanoi" Puzzle, Journal of Hokkaido University of Education, Section II A 47 (1997) 175–179.

[327] Ozanam, J., Récréations mathématiques et physiques, Nouvelle édition, Claude Jombert, Paris, 1735.

[328] Pacioli, L., De Viribus Quantitatis, Raccolta Vinciana, Milano, 1997.

[329] Pan, S., Richter, R. B., The Crossing Number of K_{11} Is 100, Journal of Graph Theory 56 (2007) 128–134.

[330] Papin, P., Histoire de Hanoi, Fayard, Paris, 2001.

[331] Parisse, D., The Tower of Hanoi and the Stern-Brocot Array, Ph.D. thesis, Ludwig-Maximilians-Universität, München, 1997.

[332] Parisse, D., On Some Metric Properties of the Sierpiński Graphs $S(n, k)$, Ars Combinatoria 90 (2009) 145–160.

[333] Park, S. E., The Group of Symmetries of the Tower of Hanoi Graph, The American Mathematical Monthly 117 (2010) 353–360.

[334] Parreau, A., Problèmes d'identification dans les graphes, Ph.D. thesis, Université de Grenoble, Grenoble, 2012.

[335] de Parville, H., La tour d'Hanoï et la question du Tonkin, La Nature 12 (1884) 285–286.

[336] Petersen, T. K., Eulerian Numbers, Springer, New York, 2015.

[337] Peterson, J., Lanier, L. H., Studies in the comparative abilities of whites and negroes, Williams and Wilkins, Baltimore MD, 1929.

[338] Petr, C., Kombinatorika posplošenih Hanojskih stolpov, Ph.D. thesis, Univerza v Mariboru, Maribor, 2004.

[339] Pettorossi, A., Derivation of efficient programs for computing sequences of actions, Theoretical Computer Science 53 (1987) 151–167.

[340] Piaget, J., La prise de conscience, Presses Universitaires de France, Paris, 1974.

[341] Poole, D., The Bottleneck Towers of Hanoi problem, Journal of Recreational Mathematics 24 (1992) 203–207.

[342] Poole, D. G., Solution to Problem 1350, Mathematics Magazine 64 (1991) 200–201.

[343] Poole, D. G., The Towers and Triangles of Professor Claus (or, Pascal Knows Hanoi), Mathematics Magazine 67 (1994) 323–344.

[344] Posamentier, A. S., Lehmann, I., The (Fabulous) Fibonacci Numbers, Prometheus, Amherst NY, 2007.

[345] Pressman, I., Singmaster, D., The jealous husbands and The missionaries and cannibals, The Mathematical Gazette 73 (1989) 73–81.

[346] Przytycki, J. H., Sikora, A. S., Topological insights from the Chinese Rings, Proceedings of the American Mathematical Society 130 (2002) 893–902.

[347] Purkiss, H. J., Question 1632, in: Miller, W. J. (ed.), Mathematical Questions with their Solutions. From the "Educational Times.", Vol. III, C. F. Hodgson & Son, London, 1865, 66–67.

[348] Quint, J.-F., Harmonic analysis on the Pascal graph, Journal of Functional Analysis 256 (2009) 3409–3460.

[349] Rankin, R. A., The first hundred years (1883–1983), Proceedings of the Edinburgh Mathematical Society 26 (1983) 135–150.

[350] Reid, C. R., Sumpter, D. J. T., Beekman, M., Optimisation in a natural system: Argentine ants solve the Towers of Hanoi, The Journal of Experimantal Biology 214 (2011) 50–58.

[351] Richeson, D. S., Euler's Gem, Princeton University Press, Princeton NJ, 2008.

[352] Ringel, G., Map color theorem, Springer, New York NY, 1974.

[353] Romik, D., Shortest paths in the Tower of Hanoi graph and finite automata, SIAM Journal on Discrete Mathematics 20 (2006) 610–622.

[354] Rosenhouse, J., Taalman, L., Taking Sudoku Seriously, Oxford University Press, Oxford, 2011.

[355] Roth, T., The Tower of Brahma revisited, Journal of Recreational Mathematics 7 (1974) 116–119.

[356] Rowe, D. E., An *Intelligencer* Quiz on Gauss and Gaussian Legends, The Mathematical Intelligencer 37(4) (2015) 45–47.

[357] Rowe, D. E., Looking Back on Gauss and Gaussian Legends: Answers to the Quiz from 37(4), The Mathematical Intelligencer 38(4) (2016) 39–45.

[358] Ruger, H. A., The psychology of efficiency, The Science Press, New York NY, 1910.

[359] Rukhin, A., On the Generalized Tower of Hanoi Problem: An Introduction to Cluster Spaces, Master thesis, University of Maryland, College Park MD, 2004.

[360] Sachs, H., Stiebitz, M., Wilson, R. J., An Historical Note: Euler's Königsberg Letters, Journal of Graph Theory 12 (1988) 133–139.

[361] Sagan, H., Space-Filling Curves, Springer, New York NY, 1994.

[362] Saint-Cyr, J. A., Taylor, A. E., Lang, A. E., Procedural learning and neo-striatal dysfunction in man, Brain 111 (1988) 941–959.

[363] Sainte-Laguë, A., Géométrie de situation et jeux, Gauthier-Villars, Paris, 1929.

[364] Sapir, A., The Tower of Hanoi with Forbidden Moves, The Computer Journal 47 (2004) 20–24.

[365] Sartorius v. Waltershausen, W., Gauss zum Gedächtnis, S. Hirzel, Leipzig, 1856.

[366] Sartorius von Waltershausen, W., Carl Friedrich Gauss—a memorial, H. Worthington Gauss (transl.), Colorado Springs CO, 1966.

[367] Scarioni, F., Speranza, H. G., A probabilistic analysis of an error-correcting algorithm for the Towers of Hanoi puzzle, Information Processing Letters 18 (1984) 99–103.

[368] Scarioni, F., Speranza, M. G., The density function of the number of moves to complete the Towers of Hanoi puzzle, Annals of Operations Research 1 (1984) 291–303.

[369] Schaal, S., Das Verhalten des Abstands perfekter Ecken und des Durchmessers von Hanoi-Graphen H_p^n für große n, Diploma thesis, Ludwig-Maximilians-Universität, München, 2009.

[370] Scheinerman, E. R., Mathematics: A Discrete Introduction, Second edition, Thomson Brooks/Cole, Belmont CA, 2006.

[371] Schmid, R. S., Überschneidungszahlen und Geschlecht bei Hanoi- und Sierpiński-Graphen, Diploma thesis, Ludwig-Maximilians-Universität, München, 2010.

[372] Schoute, P. H., De ringen van Brahma, Eigen Haard 10 (1884) 274–276, 286–287.

[373] Schröder, E., Ueber Algorithmen und Calculn, Archiv der Mathematik und Physik, Zweite Reihe 5 (1887) 225–278.

[374] Schuh, F., Wonderlijke Problemen, Tweede druk, Thieme, Zutphen, 1943.

[375] Schützenberger, M., Une sotie sur les nombres parfaits, in: Les Nombres, Pole, 2008, 58–61.

[376] Schwarz, M. A., The average distance in Lucas's Second Problem, B.Sc. thesis, Ludwig-Maximilians-Universität, München, 2012.

[377] Schwenk, A. J., Trees, trees, so many trees, in: [43], Chapter 12.

[378] Scorer, R. S., Grundy, P. M., Smith, C. A. B., Some binary games, Mathematics Magazine 280 (1944) 96–103.

[379] Shallice, T., Specific impairments of planning, Philosophical Transactions of the Royal Society of London Series B 298 (1982) 199–209.

[380] Sierpinski, W., Sur une courbe dont tout point est un point de ramification, Comptes Rendus Hebdomadaires des Séances de l'Académie des Sciences 160 (1915) 302–305.

[381] Sierpiński, W., Oeuvres choisies, PWN–Éditions Scientifiques de Pologne, Warszawa, 1975.

[382] Simon, H. A., The Functional Equivalence of Problem Solving Skills, Cognitive Psychology 7 (1975) 268–288.

[383] Singmaster, D., The Utility of Recreational Mathematics, in: R. K. Guy, R. E. Woodrow (eds.), The Lighter Side of Mathematics, The Mathematical Association of America, Washington DC, 1994, 340–345.

[384] Soifer, A., The Mathematical Coloring Book, Springer, New York NY, 2009.

[385] Spencer, D. D., Game Playing with BASIC, Hayden Book, Rochelle Park NJ, 1977.

[386] Stadel, M., Another nonrecursive algorithm for the Towers of Hanoi, ACM SIGPLAN Notices 19 (1984) 34–36.

[387] Stanley, R. P., Catalan Numbers, Cambridge University Press, New York NY, 2015.

[388] Staples, E., The Tower of Hanoi problem with arbitrary start and end positions, SIGACT News 18 (1987) 61–64.

[389] Stern, M. A., Ueber eine zahlentheoretische Funktion, Journal für die reine und angewandte Mathematik 55 (1858) 193–220.

[390] Stewart, B. M., Problems and solutions: Advanced problem 3918, The American Mathematical Monthly 46 (1939) 363.

[391] Stewart, B. M., Problems and solutions: Advanced problems: Solutions: 3918, The American Mathematical Monthly 48 (1941) 217–219.

[392] Stewart, I., Le lion, le lama et la laitue, Pour la Science 142 (1989) 102–107.

[393] Stewart, I., Another Fine Math You've Got Me Into..., W. H. Freeman and Company, New York NY, 1992.

[394] Stewart, I., Four Encounters With Sierpiński's Gasket, The Mathematical Intelligencer 17 (1995) 52–64.

[395] Stewart, I., How to Cut a Cake, Oxford University Press, New York NY, 2006.

[396] Stewart, I., Sierpiński's Pathological Curve and Its Modern Incarnations, Wiadomości Matematyczne 48(2) (2012) 239–246.

[397] Stewart, I., Professor Stewart's incredible numbers, Profile Books, London, 2015.

[398] Stewart, I., Tall, D., The foundations of mathematics, Second Edition, Oxford University Press, Oxford, 2015.

[399] Stierstorfer, Q., Perfekte Codes auf Hanoi- und Sierpiński-Graphen, Diploma thesis, Ludwig-Maximilians-Universität, München, 2007.

[400] Stirzaker, D., Stochastic Processes & Models, Oxford University Press, Oxford, 2005.

[401] Stockmeyer, P. K., Variations on the four-post Tower of Hanoi puzzle, Congressus Numerantium 102 (1994) 3–12.

[402] Stockmeyer, P. K., The Average Distance between Nodes in the Cyclic Tower of Hanoi Digraph, in: Y. Alavi, D. R. Lick, A. Schwenk (eds.), Combinatorics, Graph Theory, and Algorithms, New Issues Press, Kalamazoo MI, 1999, 799–808.

[403] Stockmeyer, P. K., The Tower of Hanoi: A Bibliography, manuscript, 2005.

[404] Stockmeyer, P. K., A Cute Observation Concerning the Tower of Hanoi Puzzle, manuscript, 2016.

[405] Stockmeyer, P. K., Solving the Stewart Recursion for the Multi-Peg Tower of Hanoi, manuscript, 2016.

[406] Stockmeyer, P. K., An exploration of sequence A000975, in: [33], 174–185.

[407] Stockmeyer, P. K., The Tower of Hanoi for humans, in: [43], Chapter 4.

[408] Stockmeyer, P. K., Bateman, C. D., Clark, J. W., Eyster, C. R., Harrison, M. T., Loehr, N. A., Rodriguez, P. J., Simmons, J. R. III, Exchanging disks in the Tower of Hanoi, International Journal of Computer Mathematics 59 (1995) 37–47.

[409] Stockmeyer, P. K., Lunnon, F., New variations on the Tower of Hanoi, Congressus Numerantium 201 (2010) 277–287.

[410] Strichartz, R. S., Differential Equations on Fractals, Princeton University Press, Princeton NJ, 2006.

[411] Strohhäcker, S., A program to find distances in Tower of Hanoi and related graphs, Technical report, Technische Universität München, Garching, 2008.

[412] Šunić, Z., Twin Towers of Hanoi, European Journal of Combinatorics 33 (2012) 1691–1707.

[413] Swetz, F. J., Leibniz, the *Yijing*, and the Religious Conversion of the Chinese, Mathematics Magazine 76 (2003) 276–291.

[414] Szegedy, M., In How Many Steps the k Peg Version of the Towers of Hanoi Game Can Be Solved?, Lecture Notes in Computer Science 1563 (1999) 356–361.

[415] Takes, F. W., Kosters, W. A., Determining the Diameter of Small World Networks, in: B. Berendt et al. (eds.), Proceedings of the 20th ACM International Conference on Information and Knowledge Management CIKM '11, Association for Computing Machinery, New York NY, 2011, 1191–1196.

[416] Teufl, E., Wagner, S., Enumeration of matchings in families of self-similar graphs, Discrete Applied Mathematics 158 (2010) 1524–1535.

[417] Teufl, E., Wagner, S., Resistance Scaling and the Number of Spanning Trees in Self-Similar Lattices, Journal of Statistical Physics 142 (2011) 879–897.

[418] Toyota, B. (丰田文景), Shūki sanpō (拾玑算法), 5 fascicles, Kyoto, 1769.

[419] Truchet, S., Mémoire sur les combinaisons, Mémoires de Mathématique et de Physique, tirés des registres de l'Acádeemie Royale des Sciences, 1704, 363–373.

[420] Tunstall, J. R., Improving the utility of the Tower of London, a neuropsycholgical test of planning, M.Phil. thesis, Griffith University, Nathan QLD, 1999.

[421] Tutte, W. T., Graph Theory As I Have Known It, Clarendon Press, Oxford, 1998.

[422] Tutte, W. T., Graph Theory, Cambridge University Press, Cambridge, 2001.

[423] Urbiha, I., Some properties of a function studied by De Rham, Carlitz and Dijkstra and its relation to the (Eisenstein-)Stern's diatomic sequence, Mathematical Communications 6 (2001) 181–198.

[424] Vajda, S., Fibonacci & Lucas numbers, and the Golden Section, Ellis Horwood, Chichester, 1989.

[425] Varghese, S., Vijayakumar, A., Hinz, A. M., Power domination in Knödel graphs and Hanoi graphs, Discussiones Mathematicae Graph Theory 38 (2018) 63–74.

[426] Wallis, J., De Algebra Tractatus, Theatrum Sheldonianum, Oxford, 1693.

[427] Walsh, T. R., The Towers of Hanoi revisited: moving the rings by counting the moves, Information Processing Letters 15 (1982) 64–67.

[428] Walsh, T. R., A case for iteration, Congressus Numerantium 40 (1983) 409–417.

[429] Walsh, T. R., Iteration strikes back—at the Cyclic Towers of Hanoi, Information Processing Letters 16 (1983) 91–93.

[430] Ward, G., Allport, A., Planning and problem-solving using the Five-disc Tower of London task, The Quarterly Journal of Experimental Psychology 50A (1997) 49–78.

[431] Weil, A., Souvenirs d'apprentissage, Birkhäuser, Basel, 1991.

[432] West, D. B., Introduction to Graph Theory, Second edition, Prentice Hall, Upper Saddle River NJ, 2001.

[433] Wexelblat, R. L., Editorial, ACM SIGPLAN Notices 20 (1985) 1.

[434] Wiesenberger, H. L., Stochastische Eigenschaften von Hanoi- und Sierpiński-Graphen, Diploma thesis, Ludwig-Maximilians-Universität, München, 2010.

[435] Williams, H. C., Édouard Lucas and Primality Testing, John Wiley & Sons, New York NY, 1998.

[436] Wilson, R., Four Colors Suffice, Revised Color Edition, Princeton University Press, Princeton NJ, 2014.

[437] Wilson, R., Watkins, J. J. (eds.), Combinatorics: ancient and modern, Oxford University Press, Oxford, 2013.

[438] Wilson, R. J., An Eulerian Trail Through Königsberg, Journal of Graph Theory 10 (1986) 265–275.

[439] Wilson, R. J., Graph Theory, in: I. M. James (ed.), History of Topology, Elsevier, Amsterdam, 1999, 503–529.

[440] Wilson, R. J., Euler, lecture held at Gresham College, London, 2002.

[441] Wolfram, S., Geometry of binomial coefficients, The American Mathematical Monthly 91 (1984) 566–571.

[442] Wolfram, S., A new kind of science, Wolfram Media, Champaign IL, 2002.

[443] Wood, D., The towers of Brahma and Hanoi revisited, Journal of Recreational Mathematics 14 (1981/82) 17–24.

[444] Wood, D., Adjudicating a Towers of Hanoi Contest, International Journal of Computer Mathematics 14 (1983) 199–207.

[445] Woodin, W. H., The tower of Hanoi, in: H. G. Dales, G. Oliveri (eds.), Truth in Mathematics, Oxford University Press, New York NY, 1998, Chapter 18.

[446] Yap, H. P., Total Colourings of Graphs, Lecture Notes in Mathematics 1623, Springer, Berlin, 1996.

[447] van Zanten, A. J., The complexity of an optimal algorithm for the Generalized Tower of Hanoi problem, International Journal of Computer Mathematics 36 (1990) 1–8.

[448] van Zanten, A. J., An Optimal Algorithm for the Twin-Tower Problem, Delft Progress Report 15 (1991) 33–50.

[449] Zhang, J., Sun, W., Xu, G., Enumeration of spanning trees on Apollonian networks, Journal of Statistical Mechanics: Theory and Experiment (2013) P09015.

[450] Zhang, Z., Wu, S., Li, M., Comellas, F., The number and degree distribution of spanning trees in the Tower of Hanoi graph, Theoretical Computer Science 609 (2016) 443–455.

[451] Zhou Y., I Ching (Book of Changes), translated by James Legge, Bantam, New York NY, 1969.

Name Index

Alcuin of York, 48
Alekseyev, M. A., 118
Allardice, R. E., 60
Allouche, J.-P., 98, 100, 322–324, 402
Allport, A., 301
Alonso-Ruiz, P., 202
Apollonios of Perge, 133
Archimedes, 108
Arett, D., 259, 262
Arima, Y., 58, 74
Astoorian, D., 98
Atkinson, M. D., 324
Aumann, S., 232, 278, 279
Azriel, D., 353

Badger, L., 360
Ball, W. W. R., 1
Bandt, C., 197
Barlow, M. T., 152
Becquaert, C., 61
Behzad, M., 48
Bell, E. T., 14, 15
Bennish, J., 160
Berend, D., 324, 349–353
Berg, W. K., 304
Berger, T., 118, 191
Berlekamp, E. R., 86
Binet, J. P. M., 355
Bode, J.-P., 221
Boettcher, S., 88, 133
Bousch, T., 235, 243, 247, 251, 252,
    327, 336, 343–347, 402, 403
Brendel, J., 130
Brocot, L.-A., 27, 28, 139, 155

Brousseau, Brother A., 231, 238
Byrd, D. L., 304

Calkin, N., 155
Cantor, G., 21, 36, 51, 155, 199, 338,
    341, 343
Cardano, G., 4, 58, 74, 75
Cattin, A., 65
Chan, T.-H., 65, 150
Chang, S.-C., 130
Chappelon, J., 214
Chaucer, G., 63, 239
Chen, G.-H., 196
Chen, H., 130
Chen, H.-R., 196
Chen, L.-C., 130
Chen, Xiao, 234
Chen, Xiaomin, 298, 299
Claus (de Siam), N., 1–3, 7, 10, 13,
    15, 60
Comellas, F., 130
Conway, J. H., 86
Coxeter, H. S. M., 2
Crowe, D. W., 104
Culin, S., 4
Cull, P., 62, 127, 197, 232

D'Angeli, D., 160
De Moivre, A., 25
Décaillot, A.-M., 13, 15
Della Vecchia, G., 196
Deng, H., 130
Dinitz, Y., 298–300
Domoryad, A. P., 161

© Springer International Publishing AG, part of Springer Nature 2018
A.M. Hinz et al., *The Tower of Hanoi – Myths and Maths*,
https://doi.org/10.1007/978-3-319-73779-9

Subject Index

abelian square, 91
absorbing Markov chain, 181
absorbing state, 181
accelerated Chinese rings, 74
adjacency matrix, 130, 405
admissible, 351
admissible arrangement, 99
admissible state, 99
algorithm
    Fleury's, 39
    Frame's, 238
    greedy, 48, 83
    human, 144, 160
    Olive's, 96
    Stewart's, 240
Allouche-Sapir conjecture, 323, 402
alphabet, 83
alternating group, 42
ape, 119
Apollonian network, 133
archimedian cylinder, 108
Argentine ants, 67
Arima sequence, 74
Arithmetical triangle, 22, 25
arrangement, 68
    admissible, 99
association, 51
atom, 26
attractor
    Hanoi, 203
automaton
    cellular, 28
    Gros code, 76

P1-, 105
    Romik's, 157, 180
    Romik's H_3-to-S_3, 178
automorphism group, 56, 128
average distance, 78
average eccentricity, 78
    normalized, 141

ba gua 八卦, 5
baguenaudier, 5
base, 175
basic double task, 160
benzenoid hydrocarbon, 147
best buffer (disc), 171
binary carry sequence, 80
binary exclusive or, 20
binary inclusive or, 20
binary logarithm, 154
binary tree, 120, 405
Binet's formula, 355
binomial theorem, 23
bipartite graph, 47
bit, 5, 405
Black and white Tower of Hanoi, 287
black sheep formula, 24
Bottleneck Tower of Hanoi, 296
bounded, 199
Bousch sequence, 252
boxer rule, 135, 188, 232, 329
breadth-first search (BFS), 41, 221, 358
bridge, 39, 405
Brocot sequence, 28, 139
Burnside's lemma, 56, 129

© Springer International Publishing AG, part of Springer Nature 2018
A.M. Hinz et al., *The Tower of Hanoi – Myths and Maths*,
https://doi.org/10.1007/978-3-319-73779-9

Symbol Index

Printed in the United States
By Bookmasters